Georgios Maltidis

Seismic soil structure interaction of navigation locks

Karlsruher Reihe

Massivbau
Baustofftechnologie
Materialprüfung

Heft 82

Institut für Massivbau und Baustofftechnologie
Materialprüfungs- und Forschungsanstalt, MPA Karlsruhe

Univ.-Prof. Dr.-Ing. Harald S. Müller
Univ.-Prof. Dr.-Ing. Lothar Stempniewski

Seismic soil structure interaction of navigation locks

by
Georgios Maltidis

KIT Scientific Publishing

Karlsruher Institut für Technologie
Institut für Massivbau und Baustofftechnologie

Seismic soil structure interaction of navigation locks

Zur Erlangung des akademischen Grades eines Doktor-Ingenieurs
von der KIT-Fakultät für Bauingenieur-, Geo- und Umweltwissenschaften
des Karlsruher Instituts für Technologie (KIT) genehmigte Dissertation

von Georgios Maltidis M.Sc. M.Sc. M.Sc. aus Athen

Tag der mündlichen Prüfung: 11. Mai 2017
Referenten: Prof. Dr.-Ing. Lothar Stempniewski
Prof. Dr.-Ing. habil. Theodoros Triantafyllidis

Impressum

KIT Scientific Publishing

Karlsruher Institut für Technologie (KIT)
KIT Scientific Publishing
Straße am Forum 2
D-76131 Karlsruhe

KIT Scientific Publishing is a registered trademark
of Karlsruhe Institute of Technology.
Reprint using the book cover is not allowed.

www.ksp.kit.edu

Print on Demand 2017 – Gedruckt auf FSC-zertifiziertem Papier

ISSN 1869-912X
ISBN 978-3-7315-0718-5
DOI 10.5445/KSP/1000073630

Seismic soil- and water-structure interaction of navigation locks

Zur Erlangung des akademischen Grades eines

DOKTOR-INGENIEURS

Von der Fakultät für
Bauingenieur-, Geo- und Umweltwissenschaften

des Karlsruher Instituts für Technologie

genehmigte

DISSERTATION

von

Georgios Maltidis M.Sc. M.Sc. M.Sc.

aus Athen

Tag der mündlichen
Prüfung:
11.05.2017

Referent: Prof. Dr.-Ing. Lothar Stempniewski
Korreferent: Prof. Dr.-Ing. habil. Theodoros Triantafyllidis

Karlsruhe 2017

Abstract

The subject of this work is the investigation of the seismic behaviour of concrete hydraulic structures considering the soil-structure and fluid-structure interaction with an emphasis on navigation locks. The scope of this thesis is to validate via finite element analyses and extend further the existing theories referring to the dynamic soil and water pressures and to give further useful information about the proper design of embedded structures with respect to earthquake safety.

The first chapter describes the problem statement, introduces the field of investigation to the reader and sets the aims of this work.

The second and third chapters consist of the literature research on this subject; they gather information about the dynamic behaviour of concrete hydraulic structures, emphasizing the

a. hydrodynamic pressures on structures
b. dynamic soil pressures on structures
c. dynamic behaviour of structures with soil-structure interaction.

The fourth and fifth chapters describe the finite element models used for the aforementioned investigation. Here, modelling concepts and assumptions are discussed and parametric studies are conducted. These chapters aim to validate the correctness of the theoretical background and to extend it for cases where an analytical solution is difficult to be conducted. The fourth chapter deals with the hydrodynamic pressures of soil-water-structure systems and the fifth chapter with dynamic soil pressures of soil-structure systems.

The sixth chapter differs from the concept of the two former chapters and investigates another topic of numerical analysis for soil structure interaction, i.e. the appropriate boundaries of the numerical model. For embedded structures or for general soil-structure interaction problems, where the numerical model has to include the soil, the boundaries given at the model can influence the wave propagation dramatically and thus an incorrect acceleration time history is applied at the structure. This chapter compares the responses of models using the most common boundaries with the predicted response of 1-dimensional wave propagation.

The seventh chapter presents two case studies of a seismic analysis of two navigation locks. The first navigation lock has gravity chamber walls and the second one has a U-frame section. This chapter aims to validate the applicability of the findings of the third and fourth chapters.

The eighth and last chapter summarizes the conclusions drawn by this numerical investigation and indicates some subjects for further investigation.

Kurzfassung

Das Thema dieser Arbeit ist die Untersuchung des seismischen Verhaltens von massiven Wasserbauwerken mit Berücksichtigung der Boden-Bauwerk- und Flüssigkeits-Bauwerk-Wechselwirkung mit dem Schwerpunkt Schiffsschleusen. Das Ziel der Arbeit ist die Validierung der vorhandenen Theorien und ihrer Weiterentwicklung mittels Finite-Elemente-Analysen. Die Ergebnisse geben nützliche Informationen über die Bemessung von eingebetteten Bauwerken im Hinblick auf die Sicherheit gegen Erdbebeneinwirkungen.

Das erste Kapitel beschreibt die Problematik, führt den Leser ins Untersuchungsfeld ein und setzt die Ziele dieser Arbeit.

Das zweite und das dritte Kapitel erfasst die Literaturrecherche dieser Thematik; es erfasst Informationen über das dynamische Verhalten von massiven Wasserbauwerken mit folgenden Schwerpunkten:

a. hydrodynamische Drücke auf Bauwerken
b. dynamische Erddrücke auf Bauwerken
c. dynamisches Verhalten von Bauwerken mit Boden-Bauwerk-Wechselwirkung

Das vierte und fünfte Kapitel beschreiben die Finite-Elemente-Modelle der Untersuchung. Hier werden Modellierungstechniken und Annahmen diskutiert und parametrische Studien durchgeführt. Beide Kapitel zielen auf die Validierung der Richtigkeit des theoretischen Unterbaus und seiner Erweiterung für Fälle, in denen die Ableitung einer analytischen Lösung schwierig ist. Das vierte Kapitel behandelt die hydrodynamischen Drücke von Boden-Flüssigkeit-Bauwerk-Systemen und das fünfte Kapitel die dynamischen Erddrücke von Boden-Bauwerk-Systemen.

Das sechste Kapitel weicht von dem Konzept der vorherigen zwei Kapitel ab und untersucht eine andere Problematik der numerischen Analyse von Boden-Bauwerk-Wechselwirkung: die Problematik der geeigneten Randbedingungen des numerischen Modells. Für eingebettete Baukonstruktionen oder allgemein für Probleme der Boden-Bauwerk-Wechselwirkung, bei denen der Boden mitmodelliert werden muss, können die eingegebenen Randbedingungen das Verhalten der Wellenausbreitung sehr stark beeinflussen, sodass eine falsche Bodenbeschleunigung am Bauwerk angebracht wird. Das Kapitel vergleicht die Antworten der verschiedenen Modelle mit den am häufigsten angewendeten Randbedingungen mit der erwarteten Antwort der eindimensionalen Wellenausbreitung.

Last but not least I would like to express my gratitude to my wife Maria for her support over the past ten years in every decision I made regarding my further scientific and professional development.

Finally, I would like to refer to three Greek dictums and one German dictum:

"Εις μεν τον πατέρα μου οφείλω το ζην, εις δε τον διδάσκαλό μου το ευ ζην"

(To my father I own the fact that I live, to my teacher I own the fact that I live better – Alexander the great referring to Aristoteles)

"ἓν οἶδα, ὅτι οὐδὲν οἶδα" (I know only one thing, that I know nothing - Socrates)

"Γηράσκω δ' αεί πολλά διδασκόμενος" (As I grow older, I constantly learn more - Solon)

"Ich habe viel Mühe, ich bereite meinen nächsten Irrtum vor." (B. Brecht)

With these four dictums I would like to indicate that a scientific work still leaves many questions open, cannot deepen or cannot describe the state of knowledge of all the fields it refers to in the specific time given and there is much room for improvement. However, a scientific thesis aims to improve this state of knowledge, to contribute to a better understanding of a scientific field, to give some engineering solutions and to improve the engineering standards used for the design of structures. I also believe a scientific work is work in progress and always the start for further development.

Contents

Figures

Tables

List of abbreviations

AF	Amplification Factor
BAW	Bundesanstalt für Wasserbau (German Federal Waterways Engineering and Research Institute)
BEM	Boundary Element Method
FEM	Finite Element Method
FSI	Fluid-Structure-Interaction
IFMB	Institut für Massivbau und Baustofftechnologie
KIT	Karlsruher Institut für Technologie (Karlsruhe Institute of Technology)
M-O	Mononobe-Okabe method
NEHRP	National Earthquake Hazards Reduction Program
PGA	Peak Ground Acceleration
SASSI	System for Analysis of Soil Structure Interaction
SDOF	Single-Degree-Of-Freedom
SF	Safety Factor
SSI	Soil-Structure-Interaction
V-Y	Veletsos-Younan (referring to their analytical solution)

Chapter 1
Preface

1.1 Problem statement

Navigation locks are very important for the waterways, as the suspension of navigation through them hinders the further traffic of ships and causes enormous financial loss for the transportation section of a country. The navigation locks can be seen as bridges in unique roads. If the bridge collapses, then traffic on the road is totally suspended. Only Germany has about 326 navigation locks, which serve 7,290 km of river waterways. The following table shows the importance of the inland waterways for the transport economy of a country.

Table 1 Waterways networks of several countries

Country/River	USA	China	Brazil	Rhine	Danube	Nile	Volga
Length of waterways (Tkm)	20	110	60	1.32	2.86	6.671	3.7
Traffic (billion t/year)	0.6	1200	0.045	0.330	0.04	0.003	0.02

Despite the great importance of navigation locks, most of the countries have no design standards for them. Germany and China have standards that refer to the design of hydraulic structures, whereas in the USA several guidelines are provided for each responsible authority (i.e. USACE etc.). However, the Chinese standards refer only to the design of dams but not to other hydraulic structures. Apart from the American guidelines, the other standards do not refer to seismic loading and earthquake-resistant design of navigation locks. Civil engineers based on their judgement often refer to similar constructions with similar functions, such as retaining walls and fluid tanks. Apart from this, the majority of navigation locks were built without taking seismic loading into consideration or using oversimplified and obsolete theories, which sometimes are only partially appropriate for the problem considered. Only to mention that the formulas provided in the aseismic design of retaining walls in EN 1998-5 originate from the theories of Mononobe-Okabe and Westergaard, which are about 80 years old.

Moreover, with the introduction of Eurocodes in the European countries, the already existing design codes have to be updated and conform with the Eurocodes, where applicable. This attempt for common design principles among the European countries raised further problems, as for example the seismic zones of the European countries have to be integrated into a European seismic map where the seismic zones conform to neighbouring countries (SHARE project (SHARE)). This fact can lead to an increase of the seismic loads for some countries. For example, the hydraulic structures along the Rhine River would be designed for different seismic loads, if they were designed in Germany or in France. Given that according to the seismic codes Germany has the lowest seismicity compared with the neighbouring countries, an increase in seismic loads may have to be expected.

Another reason for investigating the seismic loading of such structures is that many of the structures built at non seismic areas may experience dynamic or seismic loading not due to tectonic earthquakes, but due to other man-made causes, such as hydraulic fracturing of the underlying bedrock in order to obtain natural gas for energy purposes.

1.2 Aim of this work

The aim of this thesis is to investigate the seismic loading of hydraulic structures with an emphasis on the dynamic water and soil pressures and to provide further information for the design and analysis of navigation locks taking into account the soil-structure and fluid-structure interaction.

The outline of this thesis is as follows: first a quick review of the existing standards and guidelines is given. A literature research about the hydrodynamic pressures and the dynamic earth pressures on navigation locks' chamber walls follows. The existing theories are further developed and validated through finite element analyses. The investigated theories are applied to two sections of navigation locks, and parametric analyses are carried out. At the end an outlook of the present investigation is given.

1.3 State of the standards

In Germany the DIN 19702 (19702) standard specifies the provisions for the design of hydraulic structures. In this design code the lifetime for navigation locks is specified to be 100 years. Regarding seismic loads, DIN 19702 refers to DIN EN 1998-1 (EN 1998-1:2004 Eurocode 8) and DIN EN 1998-5 (EN 1998-5:2004 Eurocode 8). A distinction is made between navigation locks with a height of more than 15.00 m from the foundation level and the smaller ones, assigning to the higher navigation locks an importance factor of 1.2. Nevertheless, the paradox of assigning seismic loads which refer to buildings with a lifetime of 50 years to navigation locks that have the double lifetime is noticeable. A reason for assigning smaller seismic loads may be that a possible damage to a navigation lock can hardly lead to human losses and leads primarily to financial losses. In the author's opinion this has to be taken into consideration when drawing up the next version of the code if the seismic safety needs to be updated.

In USA the best known guidelines or engineering manuals are those of the U.S. Corps of Engineers (U.S. Army Corps of Engineers 2003, 1999), who are responsible for the biggest part of the navigation locks of the east states. Design and analysis provisions are given for navigation locks among other hydraulic structures such as spillways, water intake towers, dams, etc.

In Europe, the PIANC guidelines generally refer to the seismic design of port quay walls without referring especially to the seismic design of navigation locks. In China the code (DL 5073-2000) for the seismic design of hydraulic structures refers only to different types of dams. Moreover, engineers can use the Eurocodes (DIN EN 1990:2010-12; DIN EN 1992-1; EN 1998-1:2004 Eurocode 8; EN 1998-4:2006 Eurocode 8; EN 1998-5:2004 Eurocode 8) and combine them in order to design or assess navigation locks in interaction with earthquakes.

1.4 Observed seismic failures of hydraulic structures

The term "hydraulic structures" refers to a big family of structures (dams, levees, navigation locks, quay walls, intake towers, etc.). When focusing on navigation locks, it is remarkable how few engineering and scientific reports are available compared with other hydraulic structures. The number of scientific contributions on the analysis of navigation locks increases dramatically if they are treated as retaining structures and/or as fluid tanks or water reservoirs. On the other hand, despite the large number of scientific reports related to the analysis and design of retaining structures, there are a small number of such contributions related to the seismic analysis and behaviour of embedded structures, which is what navigations locks are most times.

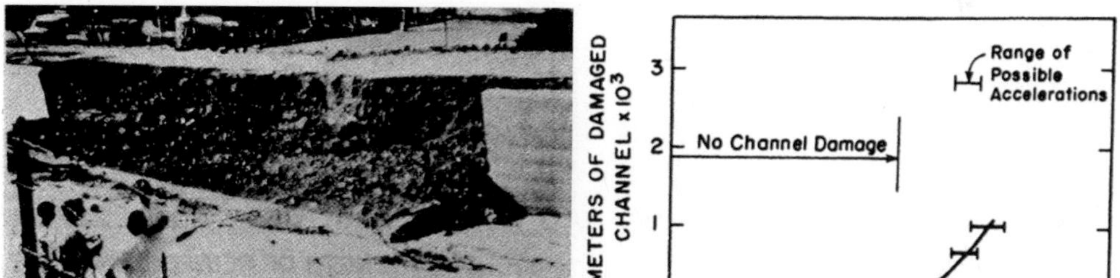

Fig. 1-1 Photograph of toppled open channel wall and estimation of the peak ground acceleration up to which no damage was observed (Clough, G. W., Fragaszy, R. F. 1977).

It is generally observed that concrete dams have performed very well during earthquake events (Committee on Seismic Aspects of Dam Design 2012; USSD 2014; Wieland 2007) with no failure until now, and it is also reported that retaining structures have also coped well, even if they were not designed for seismic loading, especially for small ground acceleration up to 0.2-0.3 g (Clough, G. W., Fragaszy, R. F. 1977; Seed H. B., Whitman R. V. 1970; Gazetas, G., Klonaris, G., Psarropoulos, P. N. 2005).

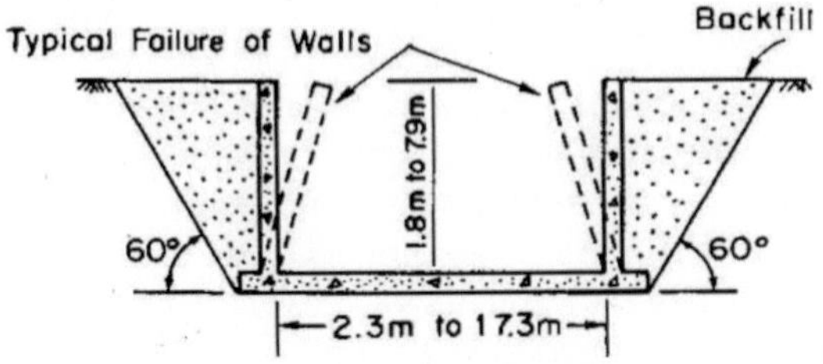

Fig. 1-2 Typical mode of failure of chamber walls due to earthquakes (Clough, G. W., Fragaszy, R. F. 1977).

On the other hand some failures, especially in water channels, have been reported (Wood 1973). These failures have been related to inadequate design or state of knowledge and were the reason for many scientific studies.

Fig. 1-3 Failures at retaining channels' walls (Wood 1973).

As it can be seen later, there are two sides, as it is always the case in scientific studies. One side supports the state of practice with further knowledge and the other side tries to prove the inadequacy of the existing methods. As it will be more obvious later, the different sides also rely on different assumptions and often represent the two limit cases for this problem. Another reason is that the two limit cases are easier to solve analytically with purely mathematical or mechanics-based relations than the coupled problem, which in most cases can hardly be solved analytically. The aim of this thesis is not to support one of these limit cases but to try to bridge their differences in a scientific way and to explain the observed response of this type of structures under dynamic load.

Fig. 1-4 Rotated retaining wall during the 2014 earthquake in Iquique, Chile, – Courtesy of G. Candia (Sitar, N., Wagner, N. 2015).

1.5 Description of the engineering problem to be investigated

There are several types of navigations locks, concerning the shape and the load-bearing function. The most common cases are (i) single gravity retaining walls and (ii) U- or W-frame chambers. The W-frame sections can be seen as wide U-frame sections for the boundary problem with the soil. Typical chamber lock heights are between 5 to 30 m and chamber widths between 10 and 60 m. The L/H ratio fluctuates between 1/3 and 3/1. Generally, the chamber walls are stiff enough (the wall's cross section can vary from 1 m to 10 m) and can be massive or have openings for water supply. The same holds for the chamber base. Moreover, the chamber can be partly submerged in water or embedded in soil.

During dynamic events like earthquakes the concrete structure experiences, apart from its own mass forces, additional dynamic forces from the soil and the water. Depending on the direction of action the dynamic water pressures can have a favourable effect on the concrete structure, as they can reduce the hydrostatic pressures, or an unfavourable action, as they can be added to them. When adding the hydrodynamic water pressures to the hydrostatic pressures the engineer has to consider if they have a favourable effect on the dynamic soil pressures acting on the other side of the chamber wall. The development of hydrodynamic pressures is nevertheless investigated here and it is left to the analyst to decide whether they must be considered or not. Moreover, the following investigation of the hydrodynamic pressures on structures with intense soil-structure interaction can be used for other structures apart from navigation locks, for example at quay walls.

The chamber walls of navigation locks act at the same time as retaining walls. Their static design already takes into account the rigidity of the structure and it is recommended to design such walls not for active conditions of the earth pressures but for a reduced at-rest earth pressure (DIN 4085:2011-05). The seismic codes (for example (EN 1998-5:2004 Eurocode 8), however, do not take into account this state and give provisions only for the extreme cases of dynamic active earth pressures and dynamic earth pressures for immovable walls. An effort to bridge these two extreme cases is made in (EN 1998-5:2004 Eurocode 8) by adapting (reducing) the design acceleration for the calculation of the dynamic earth pressures. If the retaining wall can move, the design acceleration can be reduced. Even though the dynamic at-rest pressure taken by (Wood 1973) refers to the "static" force of 1 g (which equals the gravitational force of a segment of the soil with dimensions H×H acting towards the wall). As it is shown later, this force is either too conservative if a neighbouring wall is at a close distance and the silo effect applies, or not conservative enough as a possible resonance of the soil stratum is not taken into consideration. The investigation done here tries to illuminate some further fields, which have not be taken into account at former investigations, and to provide information for a better knowledge of the problem.

At the conclusion it is explicitly expressed which further significant phenomena need to be taken into account for further investigations. These additional information given here may under certain circumstances support the state of the design of such structures as it is done nowadays, but this is a coincidental result that takes into account different phenomena (for example an excitation at the resonance frequency of the soil increases the dynamic soil pressures but at the same time the increased soil strains reduce its shear modulus and its damping which can lead to the same dynamic earth pressures as the static 1 g pressures).

Important for the following text is to understand the terms "statically" excited systems and "static" case. These terms do not refer to gravitational forces or static forces at all, but to the static component of the steady state response of dynamically excited systems ($\omega \rightarrow 0$) when the dynamic effects are ignored. Further information can be found in Annex A.

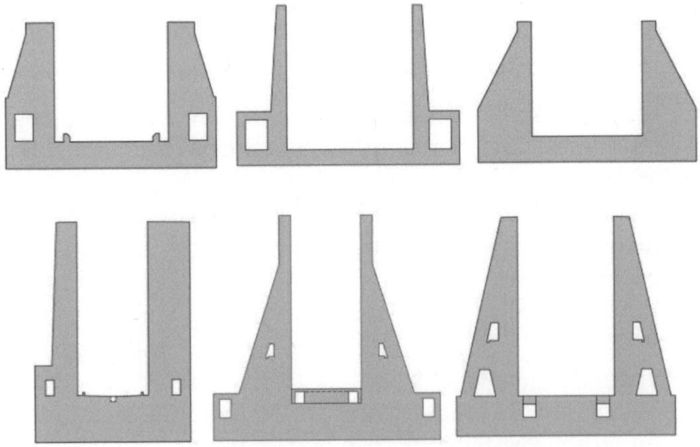

Fig. 1-5 Typical U-sections of navigation locks.

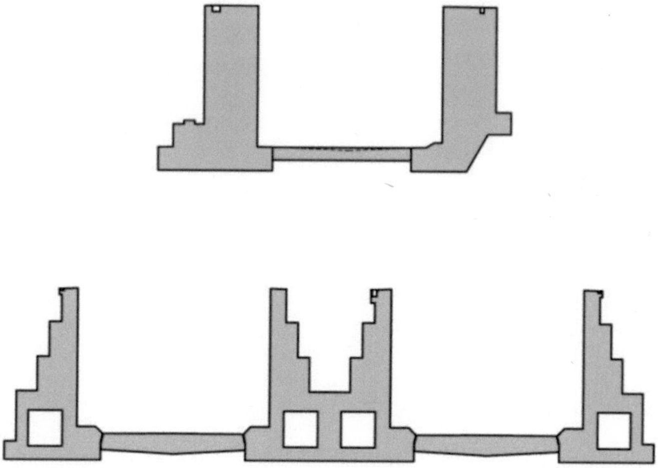

Fig. 1-6 Typical navigation lock sections with gravity retaining walls.

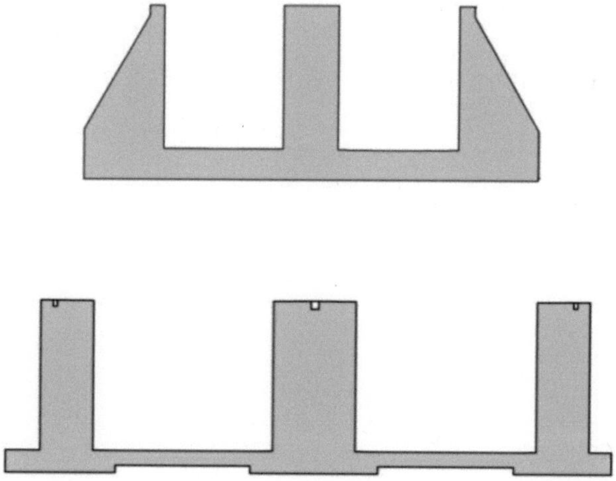

Fig. 1-7 Typical W-sections of navigation locks.

Chapter 2
Hydrodynamic pressures on structures

Summary

The second chapter presents the theories referring to the calculation of hydrodynamic pressures and their evolution in time. The chapter is divided into three parts; one referring to dams, one referring to tanks and one referring briefly to submerged strucutres.

The first part is handled more in detail than the second part because of the plane strain conditions, which characterize the system and the similarity of the problem with the navigation locks. The historical review begins with the theory of Westergaard and its assumptions. Sequently the influence of a finite reservoir is presented as described by Brahtz and Heilbron and Werner and Sundquist for incrompressible water. Following, the influence of an inclined wall on the hydrodynamic pressures as investigated by Zangar is presented. After that, the change of the hydrodynamic pressures due to compressible water is shown (Chopra et al., Bustamante and Flores) for both infinite and finite reservoirs. Sequently, the depedance of the dynamic water pressures on the dam or wall flexibility is presented through several analytical and numerical studies. The first part of the second chapter ends with the influence of a flexible wall support on the hydrodynamic pressures and a very short reference to the influence of the surface waves.

The second part refers to the hydrodynamic pressures applied on tanks. The historical review begins with the experiments of Jacobsen et al. for submerged cylinders and continues with the theories of Housner, Haroun and Veletsos et al.. A short reference to a few numerical studies is also given. The second part ends with the comparison of the hydrodynamic pressures on tanks as prescribed by the american standard standard ACI 350:3 and the european standard EN 1998-4. The comparison intends to show the similarity of the pressures resulting from formulas with much different niveau of difficulty. The chapter ends with a short reference of the hydrodynamic pressures on submerged structures.

2.1 Dams and channel walls

The hydrodynamic pressures on navigation locks seem to have many similarities with the hydrodynamic pressures acting on other structures. And that is because of the many different layouts that a navigation lock can have. When the chamber wall stands alone between two equal reservoir levels we can speak about a submerged structure, when the standalone chamber wall retains water only at one side, we can speak about a dam-like behaviour, and when the chamber has a U-shape containing the water, we can speak about a fluid tank. Distinguishing between these three main categories, a historical review is given here.

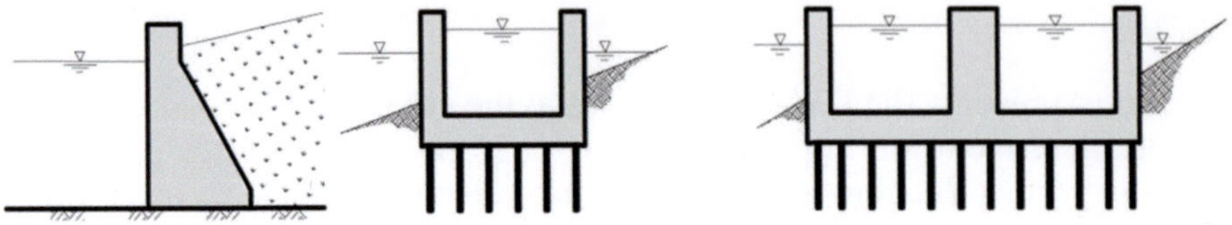

Fig. 2-1 Three types of chamber wall behaviour concerning hydrodynamic pressures: dam-like (left), tank-like (middle), submerged wall (right) (US Army Corps of Engineers 2003).

This work investigates these configurations and tries to discretize their fields of validity. A literature research about the hydrodynamic pressures on dams is presented initially, followed by a literature research about the hydrodynamic pressures in tanks. After shortly summarizing the milestones of research in each field, the theories of interest for navigation locks are chosen and investigated further. The results of the different theories are compared to the results of a numerical analysis performed with the finite element program Abaqus, for a contemporaneous validation of the theory and the numerical analysis.

The fundamental solutions are based on the following simplified assumptions (Newmark, N. M., Rosenblueth, E. 1971):

- the water's viscosity is neglected (Euler formulation of equations)
- small displacements are considered
- air trapping is omitted

2.1.1 Westergaard's theory

The most commonly used approach for the calculation of the hydrodynamic structures goes back to 1933 and the theory of the added mass approach developed by Westergaard (Westergaard 1933). Although at most times the conditions for which the theory was developed are not met, engineers still use the added mass calculated by Westergaard's formula. These conditions are:

- rigid dam
- infinite reservoir
- vertical surface of the dam
- water is incompressible.

The formula of Westergaard for the hydrodynamic pressures is:

$$p(y) = \frac{8\rho a H}{\pi^2} \sum_{1,3,5,\dots}^{n} \frac{1}{n^2 c_n} \sin\left(\frac{n\pi y}{2H}\right) \tag{2-1}$$

where α is the ground acceleration and ρ the density of water, H the height of the reservoir and y the depth of the reservoir counted from the free surface.

The factor:

$$c_n = \sqrt{n^2 - \frac{\omega^2 H^2}{c^2}} \tag{2-2}$$

has to be a real number (Chopra 1967). In order for c_n to be a real number of the period T, the earthquake must be bigger than a value which depends also on the height of the reservoir H. The first period of the reservoir can be estimated as follows:

$$T = \frac{4H}{c} = \frac{4H}{1497\ m/s} = 2.67 \times 10^{-3} H \tag{2-3}$$

Table 2 Computed first periods of an infinite reservoir for different depths

Depth of reservoir H (m)	First period T (sec)	Depth of reservoir H (m)	First period T (sec)
5	0.013	50	0.134
10	0.027	100	0.267
15	0.040	150	0.401
20	0.053	200	0.534

The computed added mass is obtained from the formula:

$$m = \frac{7}{8}\sqrt{Hy}\frac{\gamma_w}{g}A \tag{2-4}$$

where m stands for the added mass, H and y for the height and the depth of the reservoir respectively, γ_w for the density of the water, g for the gravitational acceleration and A for the contributing area of the surface. An approximate formula for the hydrodynamic pressures has also been given by Westergaard:

$$p(y) = \frac{7}{8}\rho a\sqrt{Hy} \tag{2-5}$$

The maximum hydrodynamic pressure takes the value at the bottom of the dam and equals:

$$P_o = \frac{8\rho aH}{\pi^2}\sum_{1,3,5,\ldots}^{n}\frac{(-1)^{\frac{n-1}{2}}}{n^2 c_n} \approx 0.743\rho aH \tag{2-6}$$

In his discussion of Prof. Westergaard's paper, Theodor von Karman (Karman 1933b) has provided another approximate formula where the hydrodynamic pressure distribution has the shape of a quadrant of an ellipse:

$$P_o = 0.707\rho aH \tag{2-7}$$

2.1.2 Influence of reservoir finite boundaries

The companion papers of Westergaard's paper indicated the influence of some other characteristics, such as the compressibility of the water and the finite length of the reservoir, on the hydrodynamic pressures. Among the papers, the work of Brahtz and Heilbron (Brahtz, H. A., Heilbron, C. H. 1933) is of great interest, as they first proposed correction (reduction) factors for the hydrodynamic pressures due to a finite reservoir's length.

Table 3 Correction factor for Q_o for walls moving in 0° phase (Brahtz, H. A., Heilbron, C. H. 1933)

Ratio L/H	Correction factor	Ratio L/H	Correction factor
0.5	0.397	2.0	0.921
1.0	0.670	3.0	0.983
1.5	0.835	4.0	0.996
		∞	1.000

Table 4 Correction factor for Q_o for one moving wall and one immovable wall (Brahtz, H. A., Heilbron, C. H. 1933)

Ratio L/H	Correction factor	Ratio L/H	Correction factor
0	∞	1.0	1.095
0.4	1.80	1.5	1.020
0.6	1.37	2.0	1.005
0.8	1.18	∞	1.000

where

$$Q_c = \frac{16}{\pi^3} waH^2 q \tag{2-8}$$

$$q = \sum_{1,3,5,\ldots}^{n} \frac{1}{n^3 c_n} \tag{2-9}$$

Werner and Sundquist (Sundquist, K. J., Werner, P. W. 1949) also investigated the influence of the reservoir's length on the hydrodynamic pressures and concluded that there is no change in the hydrodynamic pressures assuming a semi-infinite reservoir if the ratio *L/H* is bigger than 2.7, where *H* stands for the height and *L* for the length of the reservoir. They further investigated the influence of an immovable wall upstream of the dam and the influence of a wall moving with a phase angle of 180°. The reduction factor for two rigid walls moving without phase is given by (Halabian 2015):

$$C_n = \begin{cases} \frac{4}{3}\frac{L/H}{(1+L/H)}, & when\ L/H < 2.7 \\ 1, & when\ L/H \geq 2.7 \end{cases} \tag{2-10}$$

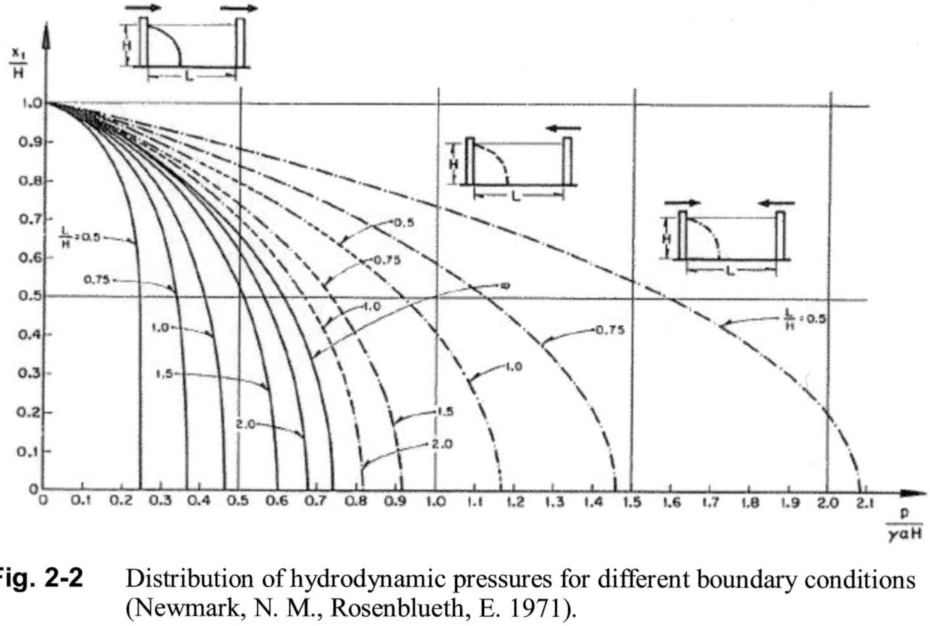

Fig. 2-2 Distribution of hydrodynamic pressures for different boundary conditions (Newmark, N. M., Rosenblueth, E. 1971).

Of great interest are the papers of (Bustamante, J. I., Flores, A., E. Herrera, Rosenblueth I. 1963; Bustamante, J. I., Flores, A. 1966), who conducted research on the error when ignoring the water compressibility in terms of the height and length of the reservoir. They showed that reservoirs with low ratios of L/H have natural periods that are well separated, and that the error is negligible (less than 5%) when neglecting the water compressibility for small values of *H/T* (up to a value of 100), where *T* is the predominant period of the earthquake excitation. Considering a predominant period between 0.3-0.5 sec of a strong earthquake motion in the magnitude range of engineering interest (Bray, J. D., Faraj, F., Rathje, E. M., Russell, S. 2004), the error is less than 5% for reservoirs with up to 30 m depth. The natural periods of the finite reservoir can be calculated by (Bustamante, J. I., Flores, A., E. Herrera, Rosenblueth I. 1963):

$$T = \frac{4H}{c}\frac{L}{\sqrt{(2mH)^2 + L^2(2n-1)^2}} = T_\infty \frac{L}{\sqrt{(2mH)^2 + L^2(2n-1)^2}} \qquad \textbf{(2-11)}$$

$$n = 1,2,3,\ldots \quad m = 0,1,2,\ldots.$$

The former equation also delivers:

$$\omega_{m,n} = \frac{2\pi}{T} = \frac{\pi}{2}\frac{c}{H}\frac{\sqrt{(2mH)^2 + L^2(2n-1)^2}}{L} = \omega_\infty \frac{\sqrt{(2mH)^2 + L^2(2n-1)^2}}{L}$$

(2-12)

As it can be seen the limit of the equation where L tends to infinity gives the natural periods of an infinite reservoir.

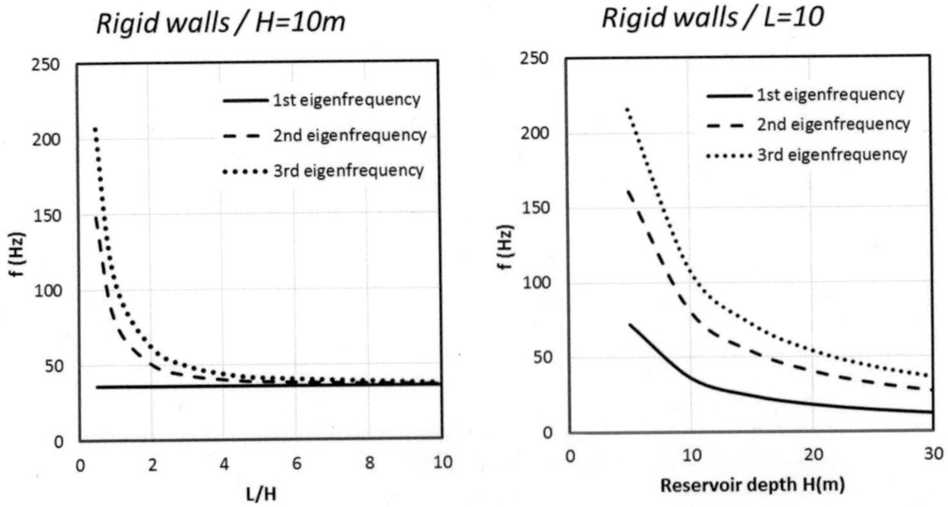

Fig. 2-3 Natural frequencies of finite reservoirs as functions of their length (*L*) and depth (*H*).

Rashed (Rashed 1982) researched on the hydrodynamic pressures for narrow reservoirs taking the effect of the existent transverse boundaries into account. He provided results for different ratios *B/D*, where *B* is the depth and *D* the width of the reservoir.

Kotsubo (Kotsubo 1959, 1961, 1965a, 1965b) indicated the differences in the hydrodynamic pressures due to irregular earthquakes (not sinusoidal excitation) and investigated the influence of the shape of the reservoir. He also inserted the Bessel functions in the solution of the hydrodynamic pressures, which was first indicated by (Brahtz, H. A., Heilbron, C. H. 1933).

2.1.3 Influence of the wall's inclination

Some years later Zangar (Zangar, C. N., Haefelri, J. 1952; Zangar 1952) investigated the influence of the inclination of the dam's surface and proposed a correction factor by using an electrical analogue. His results were validated two years later by Housner (Housner 1954) and 25 years later by Chwang, and Chwang and Housner (Chwang 1977; Chwang, A. T., Housner, G. W. 1978). The correction factor C is to be taken from a given diagram, but varies almost linearly between 0.735 for θ=75° and 0.165 for θ=0°. *C* is the correction factor calculated for the pressure at the dam's base and C_m is its maximum value at a height above the dam's base.

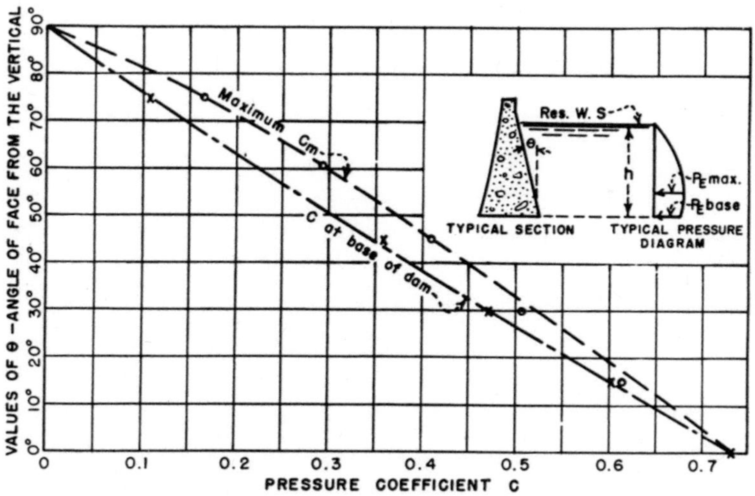

Fig. 2-4 Pressure coefficients for constant sloping faces (Zangar 1952).

$$m = 0.5HC_m\left[\frac{y}{H}\left(2-\frac{y}{H}\right)+\sqrt{\frac{y}{H}\left(2-\frac{y}{H}\right)}\right]\frac{\gamma_w}{g}A \quad \textbf{(2-13)}$$

The parabolic distribution is given by (Zangar 1952):

$$C(\eta) = 0.5C_m[\eta(2-\eta) + \sqrt{\eta(2-\eta)}] \quad \textbf{(2-14)}$$

$$\eta = \frac{y}{H} \quad \textbf{(2-15)}$$

And the water pressure distribution by (Zangar 1952):

$$P = \alpha w H C \quad \textbf{(2-16)}$$

Where the correction factor C_m for the inclination can be estimated by the relation (Halabian 2015):

$$C_m \approx 0.012\,\theta(degrees) \approx \left[\frac{\theta(degrees)}{90}\right]^{0.85} \approx 2.0\frac{\alpha}{\pi} \quad \textbf{(2-17)}$$

2.1.4 Influence of water's compressibility

Brahtz and Heilbron (Brahtz, H. A., Heilbron, C. H. 1933) first indicated the necessity of using Bessel functions for the calculation of water pressures with compressible water. Kotsubo (Kotsubo 1959) used Bessel functions to calculate the hydrodynamic pressures on a dam caused by an earthquake excitation and not by sinusoidal excitation.

When neglecting the water's compressibility, the hydrodynamic problem becomes much simpler, as it leads to a solution independent of the vibration's frequency. In the case of incompressible water the solution, i.e. the water pressures depend only on the instantaneous values of the ground acceleration (Chopra 1966).

Since the late 1960s till today Chopra and his research fellows have been engaged with the subject of hydrodynamic pressures on dams taking into account effects such as dam flexibility (Chopra 1966), foundation flexibility, the vertical component of the earthquake (Chopra, A. K., Chakrabarti, P. 1973; Chakrabatarti, P., Chopra, A. K. 1974) and sediment absorption (Chopra, A. K., Fenves, G. 1983, 1984a, 1985a, 1985b, 1985c).

2.1.5 Influence of structure's flexibility

Brahtz and Heilbron (Brahtz, H. A., Heilbron, C. H. 1933) showed first that an increased flexibility of the dam leads to increased water pressures. They assumed that a dam has a mixed shear and bending deflection curve and that this increased deflection in comparison to the rigid dam causes increased hydrodynamic pressures.

Housner (Housner 1957) gave an approximate formula for the calculation of water pressures on a flexible wall. He showed that the water pressures decrease with increased wall flexibility. The water pressure distribution on a flexible tank wall is given by (Housner 1957):

$$p_w(\eta) = \rho\alpha H\omega^2\sqrt{3}\sqrt{\frac{1-1.68\beta+1.18\beta^2}{1+2.44\beta+1.63\beta^2}}\left((1-\beta)(\eta-0.5\eta^2) + \left(\frac{2}{\pi}\right)^2\beta\sin\left(\frac{\pi}{2}\eta\right)\right) \quad \text{(2-18)}$$

Where β stands for the wall's flexibility expressed as:

$$\beta = \frac{P}{\alpha_o}\frac{h^3}{\left(\frac{\pi}{4}\right)^4 EI} \quad \text{(2-19)}$$

Bustamante et al. (Bustamante, J. I., Flores, A., E. Herrera, Rosenblueth I. 1963) also showed that the dynamic water pressures decrease if the wall is assumed to be flexible.

Chopra (Chopra 1967) showed that a flexible dam appears to be subjected to smaller hydrodynamic pressures than a rigid dam for an earthquake response. He indicated, however, that for excitations near the resonant period of the infinite reservoir the hydrodynamic pressures are higher on a flexible than on a rigid dam. He drew no firm conclusions as the hydrodynamic response depends strongly on the excitations frequency.

Lee and Tsai (Lee, G. C., Tsai, C. S. 1991) solved, in the time domain, the problem of a vibrating flexible wall fixed at its base, which interacts with a fluid at its one side. To solve this boundary problem, they used the Laplace transformation to solve some differential equations in the frequency domain and then they used the reverse Laplace transformation in order to obtain the solution in the time domain. Their results indicate a great dependence of the hydrodynamic pressures on the wall's structural rigidity. They showed that the hydrodynamic pressures increase with increasing flexibility of the wall. In a second paper the same authors gave a solution for the same boundary problem using the substructure method (Lee, G. C., Tsai, C. S. 1991).

Bouaanani and Miquel (Bouaanani und Miquel 2015) provided a simplified method for determining the dynamic response of coupled flexible beam-fluid systems via modal analysis. They provided solutions for different beam constraints and show that the flexibility of the beam reduces the water pressures on it.

Today the finite element method allows to validate these theories, some of which are older than 80 years, and to extend the results for more complicated boundary conditions. For example, the analytical solution for a wall (or dam) free to move elastically in phase with the ground or free to tilt becomes difficult because of the implicit equations involved.

A short analysis of the literature shows that many researchers support the view that an increased wall's or dam' flexibility decrease the hydrodynamic pressures whereas other researchers support exactly the opposite.

2.1.6 Influence of foundation's flexibility

Bustamante et al. (Bustamante, J. I., Flores, A., E. Herrera, Rosenblueth I. 1963) showed that a wall based on a flexible base is able to rotate and that slide affects the hydrodynamic pressures on it. The base compliance reduces the hydrodynamic pressures on a wall. Moreover, they showed that a relative displacement of the wall or dam with different movement frequencies causes a much different water pressure distribution than that of Westergaard and tension forces can develop at the same time with compression forces.

Chakrabatarti and Chopra (Chakrabatarti, P., Chopra, A. K. 1974) and Fenves and Chopra (Chopra, A. K., Fenves, G. 1984a) showed that a dam based on a flexible rock experiences smaller hydrodynamic pressures than a dam based on a rigid rock. Their investigation was made in the frequency domain using the substructure method. (Chopra, A. K., Fenves, G. 1983, 1984a, 1984b, 1985a, 1985b, 1985c) showed the influence of the foundation's flexibility and the absorption of the reservoir bottom sediments on the hydrodynamic pressures acting on dams. All these effects lead to a reduction of the hydrodynamic pressures. As the foundation's flexibility also lengthens the natural period of the dam, they approximate the influence of the foundation flexibility and sediment absorption as an additional damping, which not only reduces the hydrodynamic pressures but also lengthens the natural period of the dam.

Papazafeiropoulos et al. (Papazafeiropoulos et al. 2011) performed a steady state finite element analysis of a dam based on a flexible base. Their results show that for steady state conditions a compliant base increases the hydrodynamic pressures on the dam. However, considering the foundation's flexibility by adding a soil layer resting on bedrock, the wave propagation in the soil must be also taken into consideration and the modified, mostly amplified acceleration at the free surface of the model must be considered for the comparison of the results. So the results must be treated carefully if the acceleration is assigned at the base of the finite element model, and care must be taken also to ensure compliant boundaries of the soil domain at the sides of the model. Otherwise, it is recommended to assign the acceleration at the nodes of the wall/dam-soil interface in order to avoid the wave propagation, taking into account, however, the foundation's flexibility. A massless foundation would eliminate the problem of wave propagation, but the effect of radiation damping would also be lost.

2.1.7 Influence of surface waves

Bustamante et al. (Bustamante, J. I., Flores, A., E. Herrera, Rosenblueth I. 1963) showed that the error introduced by neglecting the surface waves for compressible water is:

- e < 5% if (H/T) > $4.2H^{1/2}$
- 20% < e < 5% if 4.2 $H^{1/2}$ < (H/T) < 2.6 $H^{1/2}$
- e > 20% if (H/T) < 2.6 $H^{1/2}$

Chopra (Chopra 1966) indicated that earthquakes have significant harmonics with periods less than 3 sec. So the error is less than 5% when neglecting the surface waves for reservoirs up to 158.5 m, covering in that way the case of navigation locks, which have significantly less reservoir height. In the same work, Chopra showed that the difference between the solutions, when taking into account the surface waves, depends on the quantity $g/\omega C$, where C is the sound velocity in water. The quantity $g/\omega C$ takes its maximum value for $\omega=1$, i.e. 6.82E-3. For bigger values of ω the quantity takes smaller values and the influence of the surface waves on the water pressures of an infinite reservoir decreases.

Table 5 Computed first periods of a reservoir for different reservoir depths and values for the quantity $g/\omega C$.

Depth of reservoir H (m)	First period T (sec)	ω (rad/sec)	$\frac{g}{\omega C}$
5	0.013	470.65	1.43E-05
25	0.067	94.13	7.17E-05
50	0.134	47.07	1.43E-04
100	0.267	23.53	2.87E-04

2.2 Tanks

2.2.1 Analytical solutions

Apart from the hydrodynamic pressures on dams with bounded reservoirs another approach for the hydrodynamic pressures acting on locks is the research conducted on the dynamic behaviour of tanks. However, the research field for liquid tanks deals more with cylindrical tanks and the solutions provided for rectangular tanks are only approximate. Moreover, recent research in this field focuses on the soil-structure interaction of slender tanks or of tanks that can uplift (unanchored tanks). The navigations locks should be assumed as rigid rectangular tanks and only the research made in this field should be taken into consideration. Another important feature of this field of research is that the vertical component of ground acceleration is taken into account and that the distribution of water pressures at the base of the fluid container can be calculated if necessary, something that is missing from the research field of dams. A brief literature research is presented here.

Hoskins and Jacobsen (Hoskins, L. M., Jacobsen, L. S. 1934), Jacobsen (Jacobsen 1949) and Jacobsen and Ayre (Ayre, R. S., Jacobsen, L. S. 1951) investigated experimentally and analytically the water pressures in a tank due to earthquakes. Their results agreed with the findings already provided by Brahtz and Heilbron. They tried to introduce a mechanical analogue for the impulsive and compulsive water pressures, and they came up with the known pendulum concept of added masses connected firmly or via springs to the tank.

Housner (Housner 1954, 1957, 1963) used the momentum method (first used by von Karman (Karman 1933b)) and provided approximate solutions for the hydrodynamic pressures on tanks. He separated the hydrodynamic pressures into two parts: an impulsive pressure caused by the amount of water accelerating with the tank, and a convective pressure caused by the sloshing of water in the tank. His mechanical analogue was a mass that is firmly connected to the tank and represents the impulsive pressures, and another smaller mass which is connected to the tank with two springs at a bigger height and represents the convective pressures. His results are similar to those provided by Westergaard for infinite tanks and to Werner and Sundquist for different ratios of tank length to tank height. According to Housner the water pressures of an accelerated fluid container can be calculated by the formula:

$$p(y) = \rho a H\left(\frac{y}{H} - 0.5\left(\frac{y}{H}\right)^2\right)\sqrt{3}\tanh\left(\sqrt{3}\frac{L}{H}\right) \quad \textbf{(2-20)}$$

where ρ is the density of the fluid, α the ground acceleration, H the height of the fluid container, y the depth of the fluid and L the length of the fluid container.

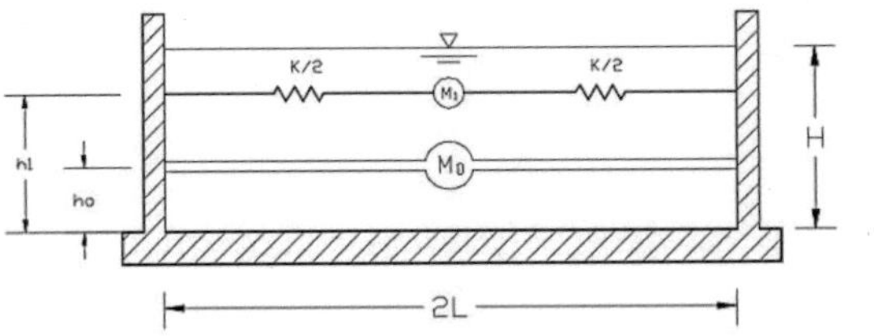

Fig. 2-5 Housner's mathematical model for impulsive and convective hydrodynamic forces (adapted from (US Army Corps of Engineers 2003)).

The equivalent mass M_o of this pressure is given by:

$$M_o = M\frac{\tanh\left(\sqrt{3}\frac{L}{H}\right)}{\sqrt{3}\frac{L}{H}} \approx M\frac{\tanh\left(1.7\frac{L}{H}\right)}{1.7\frac{L}{H}} \quad \textbf{(2-21)}$$

where M is the total mass of the fluid and is acting at an elevation given by:

$$h_o = \frac{3}{8} H \left[1 + \mathrm{a} \left(\frac{\sqrt{3}\frac{L}{H}}{\tanh\left(\sqrt{3}\frac{L}{H}\right)} - 1 \right) \right] \tag{2-22}$$

The impulsive pressures (oscillating fluid) can be replaced by equivalent masses, which are connected to the tank walls with springs. For the first sloshing mode Housner gave the equivalent mass M_1 as:

$$\mathrm{M}_1 = M \frac{1}{3} \sqrt{\frac{5}{2}\frac{L}{H}} \, tanh \left(\sqrt{\frac{5}{2}\frac{H}{L}} \right) \approx M \frac{0.83 \, tanh\left(1.6\frac{H}{L}\right)}{1.6\frac{H}{L}} \tag{2-23}$$

Acting on an elevation:

$$h_1 = H \times \left[1 - \frac{\cos\left(\sqrt{\frac{5}{2}\frac{H}{L}}\right) - 2}{\sqrt{\frac{5}{2}\frac{H}{L}} \sinh\left(\sqrt{\frac{5}{2}\frac{H}{L}}\right)} \right] \approx$$

$$H \left[1 - \frac{1}{3}\frac{M}{M_1}\left(\frac{L}{H}\right)^2 - 0.63b\frac{L}{H}\sqrt{0.28\left(\frac{M}{M_1}\frac{L}{H}\right)^2 - 1} \right] \tag{2-24}$$

Where a=0 and b=1 when the heights h_o and h_1 are to be determined on the basis of the dynamic fluid forces extracted on the walls of the tank only (not on the floor), otherwise a=1.33 and b=2.0. These equations are valid for tanks with ratios of $H/L \leq 1.6$. The first sloshing period is given by the formula:

$$T_1 = 2\pi \sqrt{\frac{M_1}{k_1}} \tag{2-25}$$

And the spring constant for the impulsive mass:

$$k_1 = 3 \frac{M_1^2}{M} \frac{g \times H}{L^2} \tag{2-26}$$

The height of the sloshing wave can be estimated by the formula:

$$d = \frac{0{,}84A_1\left(\frac{k_1}{M_1}\right)}{1 - \frac{A_1}{L}\left(\frac{k_1 L}{M_1 g}\right)^2} \qquad \textbf{(2-27)}$$

Housner's results approximates are in very good accordance with the results provided by Graham and Rodriguez (Graham, E. W., Rodriquez, A. M. 1952), who analysed the system of an oscillating fluid tank in terms of Fourier series.

Haroun (Haroun 1980; Haroun M. A., Housner G. W. 1981, 1982a, 1982b) developed Housner's mechanical model further and provided improved relations for the calculation of the hydrodynamic pressures due to earthquakes. The additional stresses on the tank wall caused by the water pressures can be calculated with the help of the response spectrum method and additional masses.

Veletsos and Veletsos et al. (Shivakumar, P., Veletsos, A. S. 1997; Tang, Y., Veletsos, A. S. 1986; Tang, H. T., Tang, Y., Veletsos, A. S. 1992; Tang, Y., Veletsos, A. S. 1990; Veletsos 1984; Veletsos, A. S., Yang, J. Y. 1977) investigated also the seismic response of rigid and flexible tanks. Their results are in accordance with the results of Haroun and Housner. An important development made by Veletsos and his co-workers was the extension of the seismic response of liquid tanks by taking into account soil-structure interaction. The impulsive and convective periods of the water are affected by the response of the tank resting on a compliant base. The formulas developed by Veletsos and his co-workers have been adopted by EN 1998-4. The formulas, which include Bessel functions, are hard to follow in engineering practice.

2.2.2 Numerical solutions

In recent years, with the development of the finite element method and the dramatic increase of the computational capability of personal computers, there has been a rapid increase in publications in the field of fluid-structure interaction. A comprehensive literature research about the numerical solutions of seismically excited fluid tanks is therefore impossible. Here, only a few publications will be referred to.

Stempniewski (Stempniewski 1990), Eibl and Stempniewski (Eibl, J., Stempniewski, L. 1987a, 1987b, 1987c, 1988, 1989) researched on the damage of reinforced concrete fluid tanks caused by earthquakes, taking into account the nonlinearity of the concrete material using the finite element method.

Doğangün (Dogangün 1995), Doğangün and Livaoğlu (Dogangün, A., Livaoglu, R. 2004, 2007) investigated the hydrodynamic pressures on tanks using the finite element method. Many of their publications account for soil-structure interaction as well.

2.2.3 Codes and Standards

Here only two codes are going to be discussed and compared: Eurocode 1998-4 (EN 1998-4:2006 Eurocode 8) and ACI 350:3 (ACI Committee 350.3-06 2006). In their biggest parts both codes have adopted the research results of Housner and Veletsos. The codes ACI 350:3 and EN 1998-4 have many similarities; however, the American code is in a much simpler form. EN 1998-4 gives the hydrodynamic pressures and the equivalent masses in the form of complex formulas with Bessel functions, which are not appropriate for engineering practice. The same EN 1998-4 2006 (EN 1998-4:2006 Eurocode 8) has some errors in the the graphs, which give the distribution of the convective pressures for the two first eigenfrequencies and the eigenfrequencies dependence on the H/R-ratio (the 1st eigenfrequency should be the second and vice versa). However, the diagrams are presented correctly in a former version of Eurocode (EN 1998-4 1996) indicating a print error.

Eurocode 8, Part 4, in its Annex A provides the procedure for the calculation of the hydrodynamic pressures and resulting shear forces and moments. The biggest part of these equat (Davidson et al. 1986)ions is the same as those used in New Zealand as proposed by Davidson et al. (Davidson et al. 1986) and the research document of Veletsos (Veletsos 1984). For rigid rectangular tanks, as navigation locks could be approximated, the procedure demands the calculation of an impulsive pressure and a convective pressure. The formulas are the same with cylindrical tanks, whereas the radius R of the tank has been replaced by the half-length $0.5L$ ($R=0.5L$).

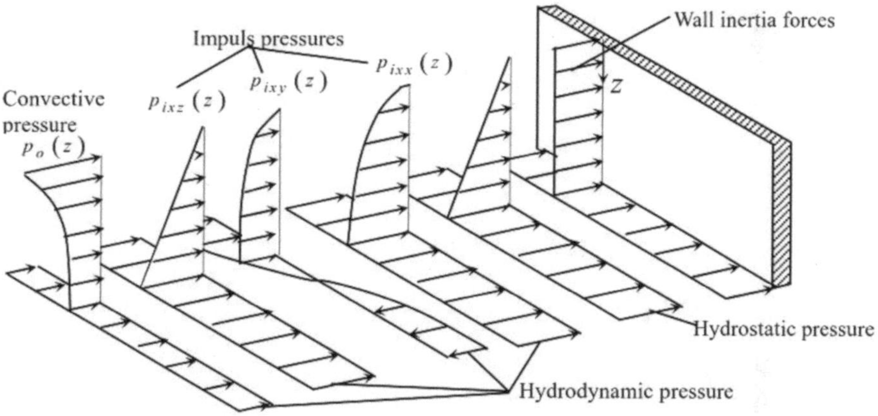

Fig. 2-6 Components of the water pressures acting on a tank's wall (Livaoglu 2008).

For comparison, the impulsive and convective masses and heights according to both standards are presented in the next table, followed by their graphs. It is obvious how much easier the relations of the American standard are for the engineering practice than the European ones, although they deliver the same accuracy.

Table 6 Change of Young and shear modulus with changing Poisson's ratio

	Impulsive mass
ACI-350.3	$$\frac{W_i}{W_L}=\frac{tanh[0.866(L/H_L)]}{0.866(L/H_L)}$$
EC8-4	$$\frac{m_i}{m}=2\gamma\sum_{n=0}^{\infty}\frac{I_1(v_n/\gamma)}{v_n^3 I'_1(v_n/\gamma)}$$
	Convective mass
ACI-350.3	$$\frac{W_c}{W_L}=0.264\,L/H_L\;tanh(3.16\,H_L/L)$$
EC8-4	$$\frac{m_c}{m}=\frac{2tanh(\lambda_n\gamma)}{\gamma\lambda_n(\lambda_n^2-1)}$$
	Height of impulsive masses excluding base pressures
ACI-350.3	$$\frac{L}{H_L}<1.333\rightarrow\frac{h_i}{H_L}=0.5-0.09375\frac{L}{H_L}$$ $$\frac{L}{H_L}\geq 1.333\rightarrow\frac{h_i}{H_L}=0.375$$
EC8-4	$$\frac{h_i}{H}=\frac{\sum_{n=0}^{\infty}\frac{(-1)^n I_1(v_n/\gamma)}{v_n^4 I'_1(v_n/\gamma)}(v_n(-1)^n-1)}{\sum_{n=0}^{\infty}\frac{I_1(v_n/\gamma)}{v_n^3 I'_1(v_n/\gamma)}}$$
	Height of convective masses excluding base pressures
ACI-350.3	$$\frac{\mathrm{h_c}}{\mathrm{H_L}}=1-\frac{\cosh[3.16(\mathrm{H_L}/\mathrm{L})-1]}{3.16(\mathrm{H_L}/\mathrm{L})\sinh[3.16(\mathrm{H_L}/\mathrm{L})]}$$
EC8-4	$$\frac{\mathrm{h_c}}{\mathrm{H}}=1+\frac{1-\cosh(\lambda_n\gamma)}{\lambda_n\gamma\sinh(\lambda_n\gamma)}$$

	Height of impulsive masses including base pressures
ACI-350.3	$\frac{L}{H_L} < 0.75 \rightarrow \frac{h'_i}{H_L} = 0.45$ $\frac{L}{H_L} \geq 0.75 \rightarrow \frac{h'_i}{H_L} = \frac{0.866\, L/H_L}{2\tanh(0.866\, L/H_L)} - \frac{1}{8}$
EC8-4	$\frac{h'_i}{H} = \frac{0.5 + 2\gamma \sum_{n=0}^{\infty} \frac{v_n + 2(-1)^{n+1} I_1(v_n/\gamma)}{v_n^4 I'_1(v_n/\gamma)}}{2\gamma \sum_{n=0}^{\infty} \frac{I_1(v_n/\gamma)}{v_n^3 I'_1(v_n/\gamma)}}$
	Height of convective masses including base pressures
ACI-350.3	$\frac{h'_c}{H_L} = 1 - \frac{cosh[3.16(H_L/L) - 2.01]}{3.16(H_L/L)sinh[3.16(H_L/L)]}$
EC8-4	$\frac{h'_c}{H} = 1 + \frac{2 - cosh(\lambda_n \gamma)}{\lambda_n \gamma\, sinh(\lambda_n \gamma)}$
	$\gamma = H/L; \quad v_n = \frac{2n+1}{2}\pi$ $\lambda_1 = 1.841;\ \lambda_2 = 5.331;\ \lambda_3 = 8.536$ $I'_1(x) = \frac{dI_1(x)}{dx} = I_0(x) - \frac{I_1(x)}{x}$ $I_0(x)$ and $I_1(x)$ the modified Bessel function of 0 and 1st order

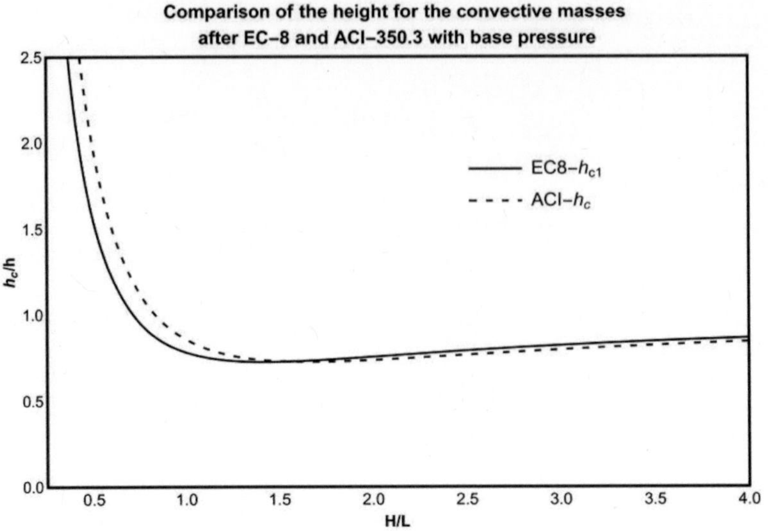

Fig. 2-7 Comparison of the heights for the convective masses considering the base pressures according to EC8-4 and ACI-350.3.

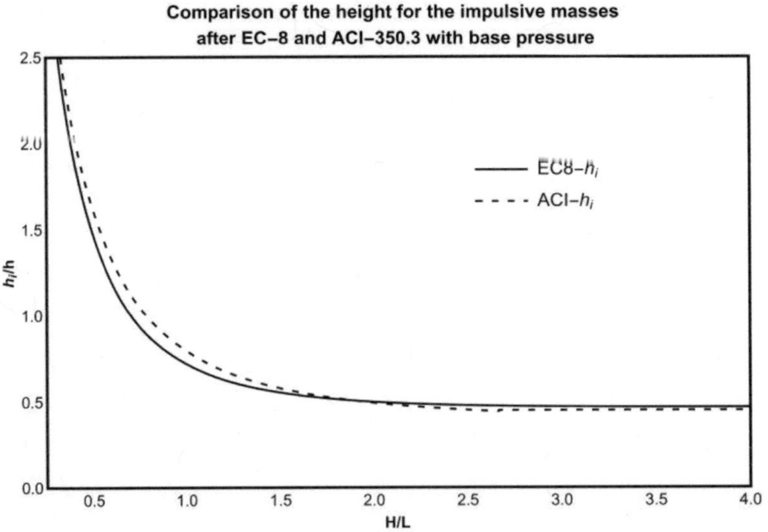

Fig. 2-8 Comparison of the heights for the impulsive masses with base pressures according to EC8-4 and ACI-350.3.

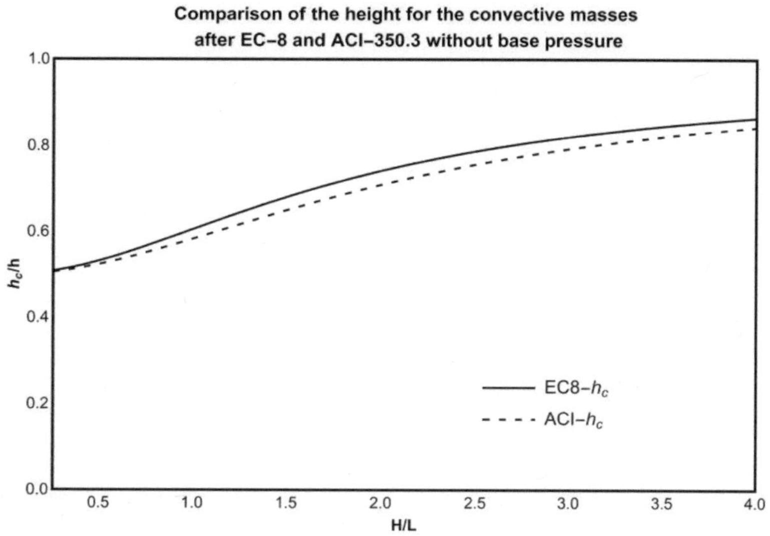

Fig. 2-9 Comparison of the heights for the convective masses without base pressures according to EC8-4 and ACI-350.3.

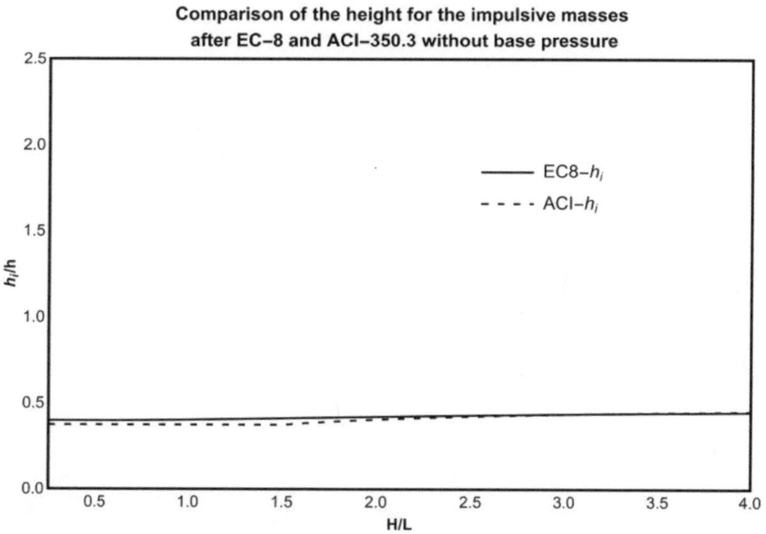

Fig. 2-10 Comparison of the heights for the impulsive masses without base pressures according to EC8-4 and ACI-350.3.

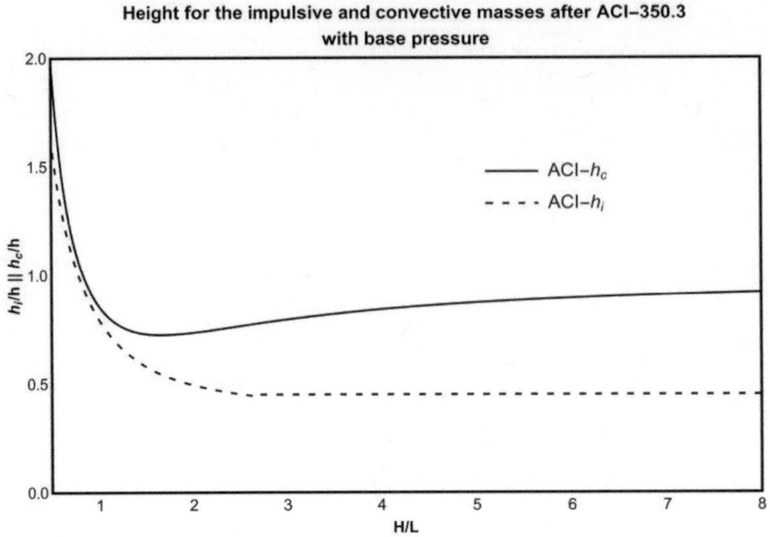

Fig. 2-11 Heights for the impulsive pressures and convective masses with base pressures according to ACI-350.3.

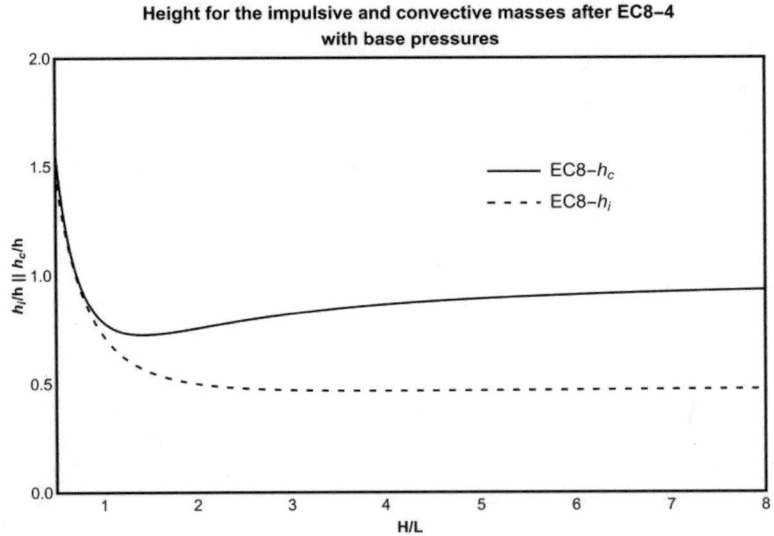

Fig. 2-12 Heights for the impulsive pressures and convective masses with base pressures according to EC8-4.

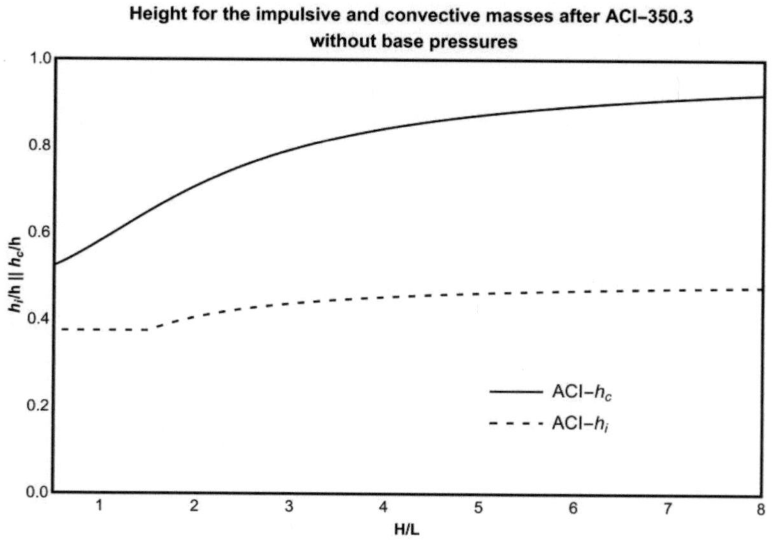

Fig. 2-13 Heights for the impulsive pressures and convective masses without base pressures according to ACI-350.3.

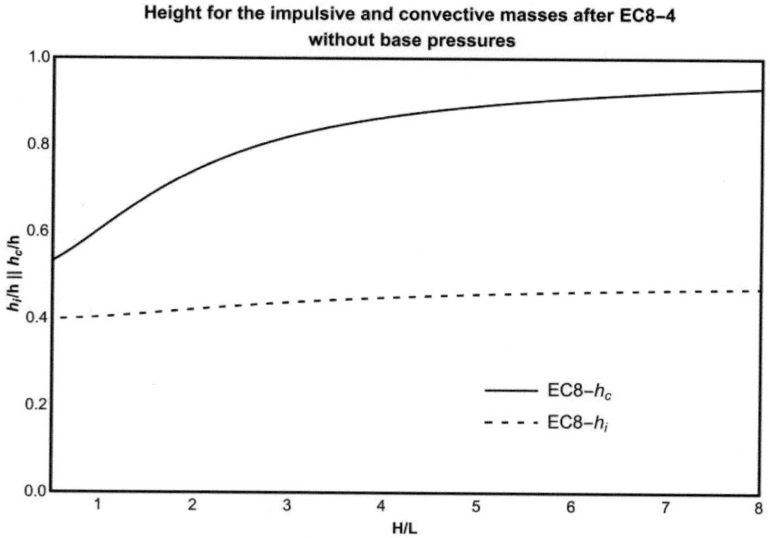

Fig. 2-14 Heights for the impulsive pressures and convective masses without base pressures according to EC8-4.

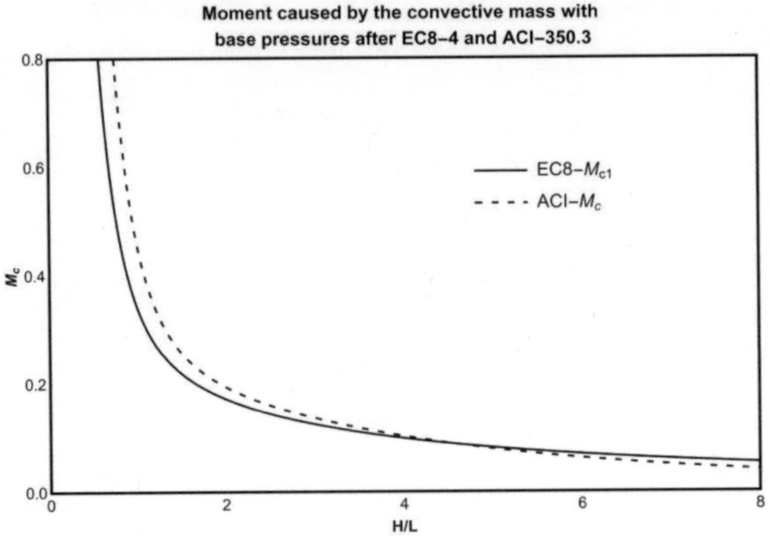

Fig. 2-15 Comparison of moments caused by the convective masses considering the base pressures according to EC8-4 and ACI-350.3.

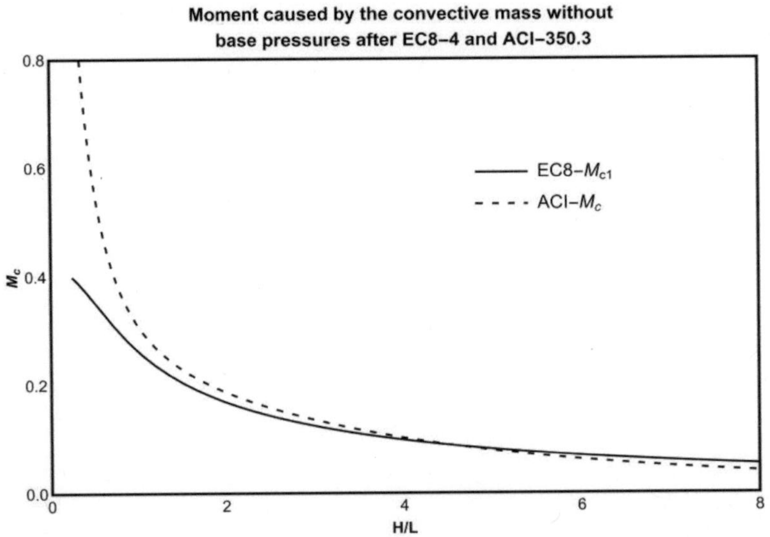

Fig. 2-16 Comparison of moments caused by the convective masses without considering the base pressures according to EC8-4 and ACI-350.3.

2.3 Submerged structures

Most of the research work in this field concerns submerged cylindrical piers such as intake towers and bridge piers. The water pressure distribution of cylindrical structures appears to have many similarities with that of fluid tanks. Although the one-to-one application to middle chamber walls of W-lock sections is not applicable, it must be clear to the analyst that both, pressure and suction, occurs, so the water effect is double.

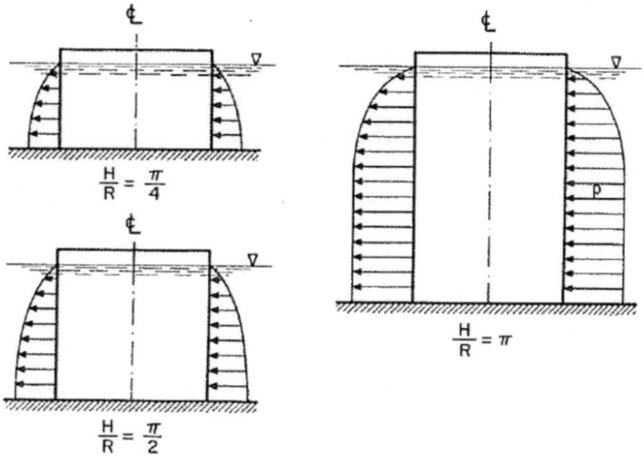

Fig. 2-17 Water pressure distribution on a cylindrical submerged pier according to (Newmark, N. M., Rosenblueth, E. 1971).

The water pressures can be taken as the double value calculated for one chamber after reducing the values due to the reservoir length. The chamber wall's flexibility or water compressibility have a minor contribution to these pressures, as the middle chamber wall is usually very compact and the compressibility of the water has a minor effect on short length reservoirs.

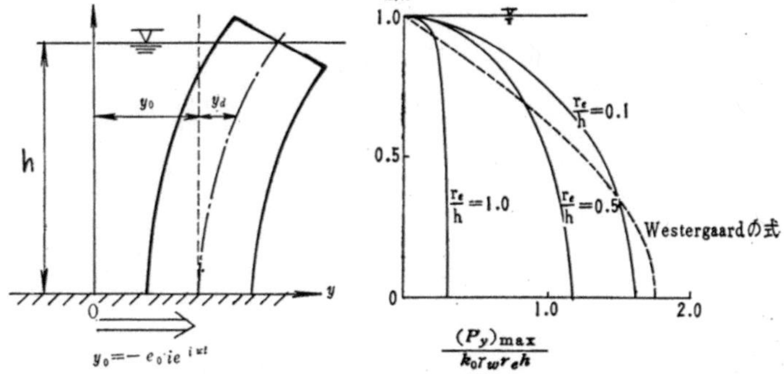

Fig. 2-18 Water pressure distribution on a cylindrical submerged pier according to (Goto und Toki 1963).

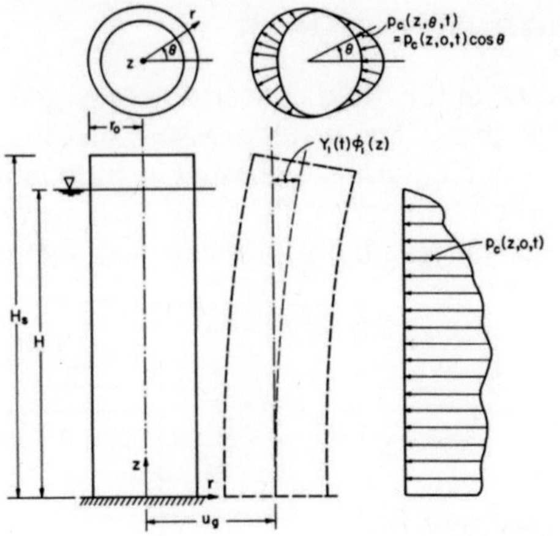

Fig. 2-19 Water pressure distribution on a cylindrical submerged pier according to (Chopra, A. K., Liaw, C.Y. 1973).

Chapter 3
Dynamic soil pressures on structures

Summary

The third chapter contains the literature review of the dynamic soil pressures on structures. At the beginning of the third chapter some solutions for yielding walls are presented. Only the milestones according to the point of view of the author of this thesis are given. Here are referred the solutions of Mononobe and Okabe and the simplification made by Seed and Whitman. The investigation of Steedman and Zeng show the influence of phase effects on the dynamic soil pressures. A detailed analysis of static and dynamic soil pressures under any lateral movement of the retaining wall is given by Zhang et al. The stress plasticity closed form solution of Mylonakis et al. is presented as a recent theory for the calculation of soil dynamic pressures.

The second underchapter handles the elastic solutions or wave propagation analyses for the calculation of the dynamic soil pressures on non yielding walls. As reference analysis the one of Wood is presented. After a short reference of other works (Matsuo and Ohara, Tajimi, Scott, Arias et al.) the analysis of Veletsos and Younan is presented in detail as it constitutes the basis of the numerical investigation of this thesis in chapter 5. Veletsos and Younan take into consideration the ability of the wall to flexure and to rotate at its base. Both of these characteristics lead to reduced dynamic soil pressures. An extension of the theory of Veletsos and Younan is the work of Jung et al., who considered additionally a horizontal and vertical elastic movement of the wall. Another rigorous solution of the problem, the one of Papazafeiropoulos and Psarropoulos is also briefly presented. Kloukinas et al. simplified the theory of Veletsos and Younan for only the first mode shape of the soil stratum. Further, Bradenberg et al. presented a solution based on kinematic equations for the calculation of the dynamic soil pressures. This underchapter ends with the work of Vrettos et al., who based on the methology of Papazafeiropoulos and Psarropoulos gave formulas for the dynamic soil pressures for any soil inhomogeneity.

The third subchapter presents briefly the displacement based design of retaining structures. The works of Elms and Richards, Liao et al., Nadim and Whitman and Elms are referred.

The fourth subchapter deals with recent experimental studies, which aimed to validate or invalidate common methologies predescribed in the standards. Of interest are the results of the experiments of Sitar et al., who showed that the Mononobe-Okabe formula is adequate for the design also for rigid retaining structures up to a PGA of 0,4g.

At the fifth subchapter are given the dynamic soil pressures of saturated soils. Apart from the experimental and analytical study of Matsuo and Ohara, which is also adopted by the EN 1998-5, the elastic solutions of Matsuzawa et al., Chen and Hung and Theodorakopoulos et al. are briefly presented.

The sixth subchapter refers to numerical studies carried out in this field. This underchapter is the fundament of the numerical analysis carried out in this study and presented in chapter 5. The solution of Wood, which valdated by him also numerically, refers only to a homogeneous soil and bounded rigid systems. His solution is adopted by EN 1998-5 for rigid walls. Wu and Finn extended the results of Wood for two classic inhomogeneous profiles, one with parabolic and one with linear distribution of the shear modulus of the soil. They used the finite element method and a frequency domain analysis. Their results are limited to rigid two-wall systems. Psarropoulos used also the finite element method and a time domain analysis to validate numerically the theory of Veletsos and Younan. At his thesis the wall flexibility and the wall flexure are taken into account. He also extended the results for an inhomogeneous soil profile. His results refer only to one wall systems. Jung et al., based on the numerical model of Psarropoulos, added a translational and a vertical spring to account for the elastic horizontal and vertical movement of the wall. They further considered that the wall is able to separate from the soil and they investigated also the influence of the soil-wall friction. Their results are restricted to one-wall systems with homogeneous soil.

A seventh subchapter presents briefly studies referring explicit to navigation locks using several analysis techniques. The next subchapter compares the solutions available in the literature fro the hydrodynamic pressures with the one of the elastic solutions for the dynamic soil pressures. The similarities are obvious and that gives motivation to combine analyses of these two different fields in order to present results missing until now.

Ther last subchapter reviews the literature research and criticizes the results of the presented studies. Additionally, some examples are given in order the reader to understand the practical meaning of the dimensionless parameters used in the analysis of Veletsos and Younan. Finally, the shape and the natural frequency of the first mode is given for flexible walls elastically restrained to rotate and to move horizontally using closed form formulas, contrary to the analysis of Veletsos and Younan and Jung et al., who used the superposition method as the system remains linear.

3.1 Yielding walls – limit equilibrium or failure state methods

3.1.1 Mononobe-Okabe

The most commonly used formulas for retaining walls are those proposed by Mononobe (Matsuo, O., Mononobe, N. 1929) and Okabe (Okabe 1924). These formulas, which have also been adopted by many design standards and the Eurocode (EN 1998-5:2004 Eurocode 8), are quite easy to understand and to apply, as they are a modification of the known Coulomb (Coulomb 1776) formula for the static soil pressures on structures. As point of application of the seismic pressure Mononobe and Okabe considered the same point with the static pressure, hence *H/3* above the base.

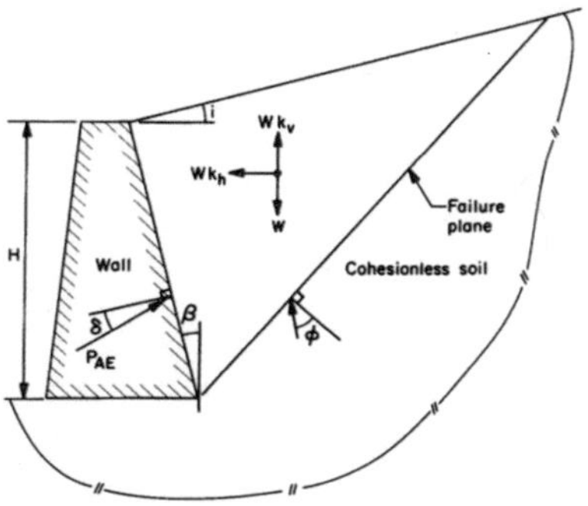

Fig. 3-1 The Mononobe-Okabe proposal for seismic forces on a soil wedge.

The Coulomb formula for the soil active pressures is:

$$P_A = \frac{1}{2}\gamma H^2 K_A \tag{3-1}$$

where

$$K_A = \frac{1}{\cos(\delta+\beta)}\left(\frac{\cos(\varphi-\beta)}{\cos(\beta)\left(1+\sqrt{\dfrac{\sin(\varphi+\delta)\sin(\varphi-i)}{\cos(\delta+\beta)\cos(i-\beta)}}\right)}\right)^2 \tag{3-2}$$

By turning the vertical line after angle θ, where θ is:

$$\theta = \tan^{-1}\left(\frac{k_h}{1 \pm k_v}\right) \tag{3-3}$$

where k_h and k_v are the horizontal and vertical peak ground accelerations in g respectively, and by substituting:

$$i' = i + \theta \tag{3-4}$$

$$\beta' = \beta + \theta \tag{3-5}$$

$$\gamma' = \gamma(1 \pm k_v) \tag{3-6}$$

One gets the Mononobe-Okabe formula for the total (static and seismic) active pressure one a gravity wall:

$$P_{AE} = \frac{1}{2}\gamma' H^2 K_{AE} = \frac{1}{2}\gamma(1 \pm k_v)H^2 K_{AE} \tag{3-7}$$

where

$$K_{AE} = \frac{1}{\cos\theta\cos(\delta + \beta + \theta)} \left(\frac{\cos(\varphi - \beta - \theta)}{\cos(\beta)\left(1 + \sqrt{\frac{\sin(\varphi + \delta)\sin(\varphi - \theta - i)}{\cos(\delta + \beta + \theta)\cos(i - \beta)}}\right)}\right)^2 \tag{3-8}$$

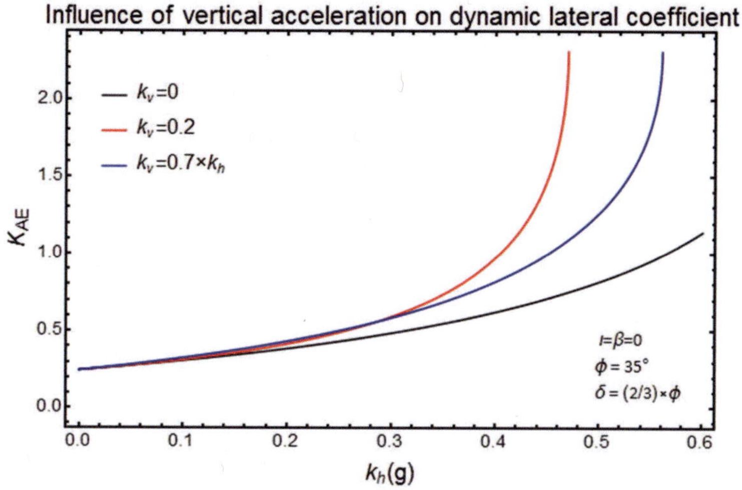

Fig. 3-2 Influence of the vertical acceleration on the dynamic lateral coefficient.

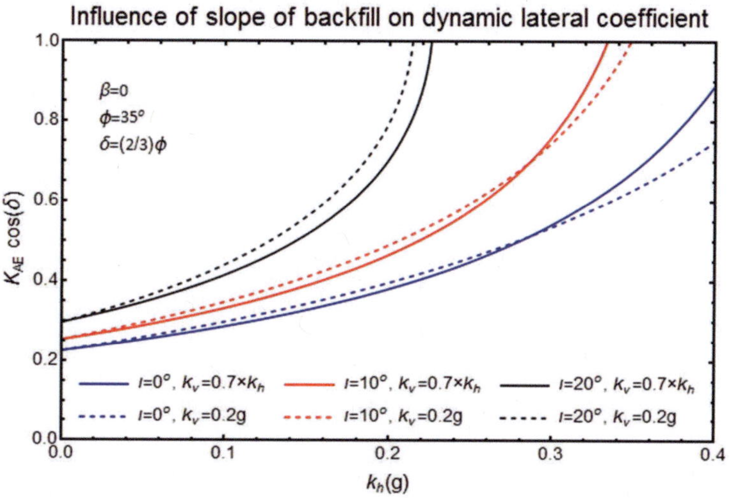

Fig. 3-3 Influence of the slope of bakfill on the dynamic lateral coefficient.

3.1.2 Seed and Whitman

Seed and Whitman (Seed H. B., Whitman R. V. 1970) simplified the M-O formula for the case of horizontal acceleration and concluded that retaining walls adequately designed for static loads can resist earthquakes up to 0.2g. The approximation they made consists of splitting total soil pressure into a static thrust and a dynamic increment, which acts 0.6×H above the base. They further simplified the dynamic increment as 75% of the horizontal acceleration, neglecting the vertical component of the earthquake. However, the simplification they suggested can be safely used in regions with low seismicity, where the seismic loading is not of great importance. This is because the dynamic increment of the soil thrust depends strongly on the vertical component of the earthquake, the angle of friction of the soil, the angle of friction of the wall, the slope and the inclination of the wall. Under the assumption that the vertical component of the earthquake is negligible they suggested that the total pressure consists of a static and dynamic part:

$$P_{AE} = P_{AE} + \Delta P_{AE} \tag{3-9}$$

in which:

$$P_{AE} = \frac{1}{2}\gamma H^2 \left(\frac{3}{4}k_h\right) \tag{3-10}$$

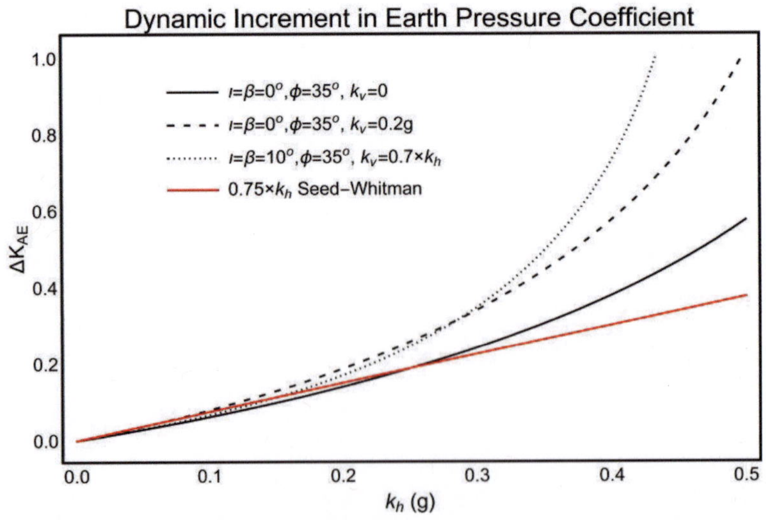

Fig. 3-4 Comparison of the dynamic increment in earth pressure coefficient after Mononobe-Okabe and Seed-Whitman.

3.1.3 Steedman and Zeng

Steedman and Zeng (Steedman, R. S., Zeng, X. 1990a, 1990b) developed relations for the dynamic increment of the soil pressures based on shear wave propagation. They showed that the dynamic increment is a function of the dimensionless quantity $H/(T \times V_s)$ (ratio of time for a wave to travel the whole height of the wall to the period of shaking), that the shear modulus profile of the soil hardly affects the results and that the amplification of the acceleration affects the results significantly.

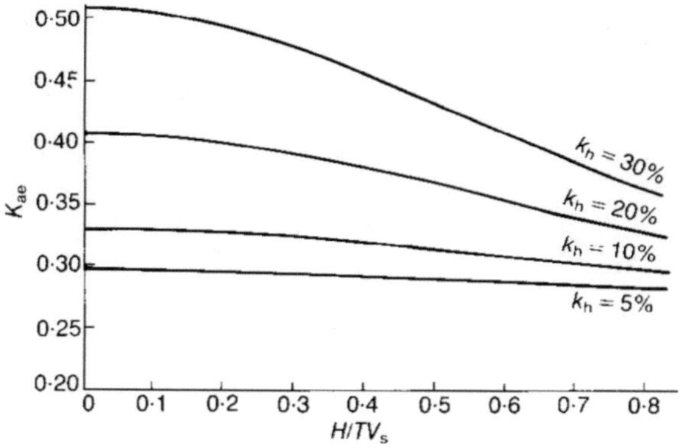

Fig. 3-5 Earth pressure coefficient with different amplification factors according to (Steedman, R. S., Zeng, X. 1990a).

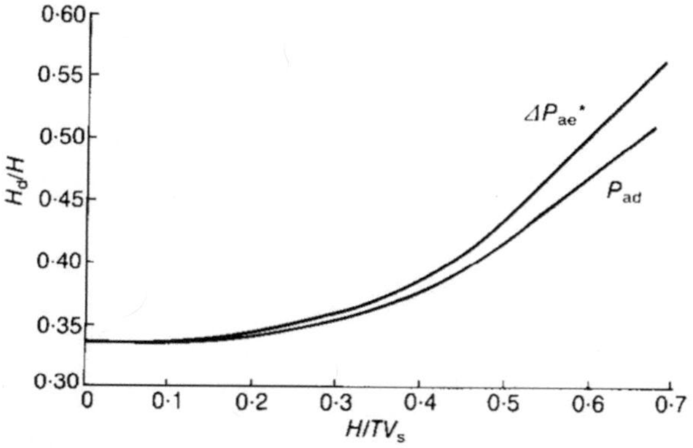

Fig. 3-6 Acting point of dynamic force increment above the base according to (Steedman, R. S., Zeng, X. 1990a).

3.1.4 Zhang et al.

Zhang et al. (Shamoto Y., Tokimatsu, K., Zhang, J. 1998(a), 1998(b); Shamoto, Y., Tokimatsu, K., Zhang, J.-M. 1998(c)) developed new formulas for static and seismic earth pressures on retaining walls under any lateral displacement. Their approach is based on an intermediate soil wedge, the size of which varies depending on the wall displacement and seismic acceleration. They separated seismic earth pressure into four components: i) the effective weight of the soil wedge, ii) seismic inertia force,

iii) surcharge load on the backfill surface and iv) soil vibro-densification effect at or near neutral state. These four components have a different point of application and a different pressure distribution with depth. The authors adopted an equivalent seismic coefficient to account for the non-uniform acceleration with depth in the backfill due to seismic amplification and phase effects. Their newly developed equations can be limited to the M-O formulas for the limiting conditions.

3.1.5 Mylonakis et al.

Mylonakis et al. (Mylonakis G., Kloukinas P., Papantonopoulos C. 2007) presented a closed-form stress plasticity solution for gravitational and earthquake-induced earth pressures on retaining walls, which is much simpler than the M-O method. Mylonakis' approach is essentially an approximate yield-line approach based on the theory of discontinuous stress fields. It takes into account the following parameters: (1) weight and friction angle of the soil material, (2) wall inclination, (3) backfill inclination, (4) wall roughness, (5) surcharge at soil surface and (6) horizontal and vertical seismic acceleration. The investigation of both active and passive conditions is possible by changing the inclination of the stress characteristics in the backfill. Because the solution does not perfectly satisfy equilibrium at certain points in the medium, it cannot be classified in the context of limit analysis theorems. Compared with rigorous numerical results, the method overestimates active pressures and under-predicts the passive ones. Accordingly, it can be viewed as an approximate lower-bound solution, rather than a mere predictor of soil thrust.

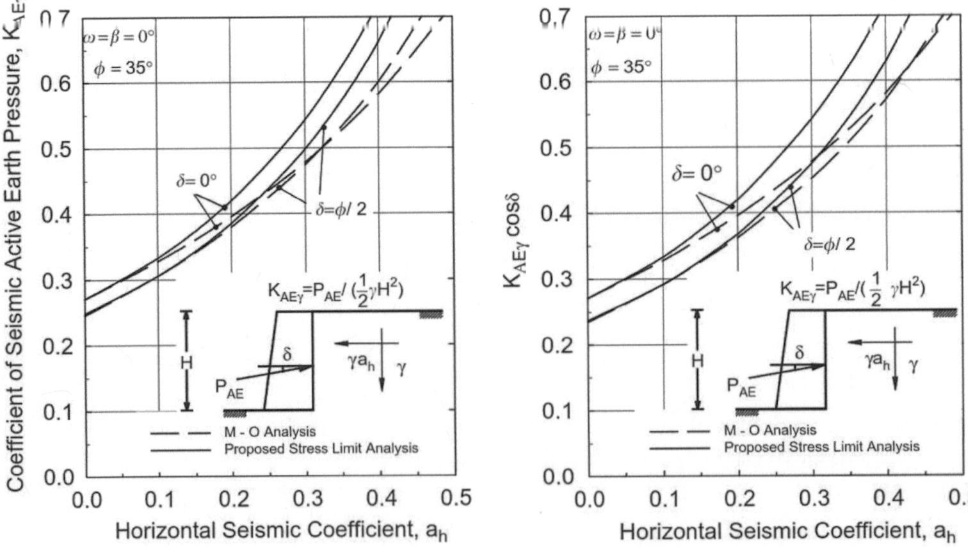

Fig. 3-7 Comparison of the solution provided by Mylonakis et al. and the M-O solution (according to (Mylonakis G., Kloukinas P., Papantonopoulos C. 2007)).

3.2 Non-yielding walls – elastic solutions

Non-yielding retaining walls are generally meant to be the walls of structures that cannot slide or rotate in order to cause a limit or failure state in the soil. Such walls are walls of embedded structures founded on rock, generally constrained massive walls and walls of foundations based on piles. These methods belong to the field of elastodynamics. The soil is supposed to behave elastically and its damping is supposed to be viscous.

3.2.1 Matsuo and Ohara

Matsuo and Ohara (Matsuo, H., Ohara, S. 1960) first proposed a solution for rigid and tilting quay walls based on the wave propagation equation in order to investigate the seismic behaviour of quay walls. They also calculated the dynamic water pressures on the wall when the soil is permeable.

3.2.2 Wood

The reference work about dynamic soil pressures on non-yielding structures is the publication of Wood (Wood 1973). Wood provided an analytical solution and compared it with the results of finite element analyses. According to Wood's solution the dynamic force acts at $0.63H$ from the base. Total dynamic pressure is 2 to 3 times bigger than that proposed by the M-O method. The Wood solution is adopted by many standards, among them by (EN 1998-5:2004 Eurocode 8). Wood investigated also the influence of a rigid rotating wall on dynamic soil pressures as well as the influence of a neighbouring wall on dynamic soil pressures. He showed that for the same value of Poisson's ratio of the elastic contained medium (soil) the dynamic soil pressures on the wall decline with decreasing values of the L/H ratio and he also showed the influence of Poisson's ratio on the dynamic soil pressures.

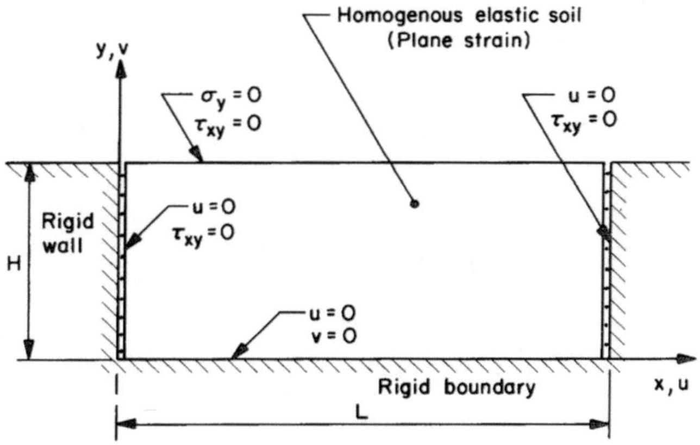

Fig. 3-8 The boundary problem investigated by Wood (Wood 1973).

3.2.3 Tajimi

Tajimi (Tajimi 1973) researched on the dynamic soil pressures acting on embedded structures using the two dimensional wave propagation. He expanded his solution for rocking embedded structures based on elastic rock foundation. He concluded that the forcing moment caused by the soil is considerably larger than the moment due to the mass inertia of the structure itself.

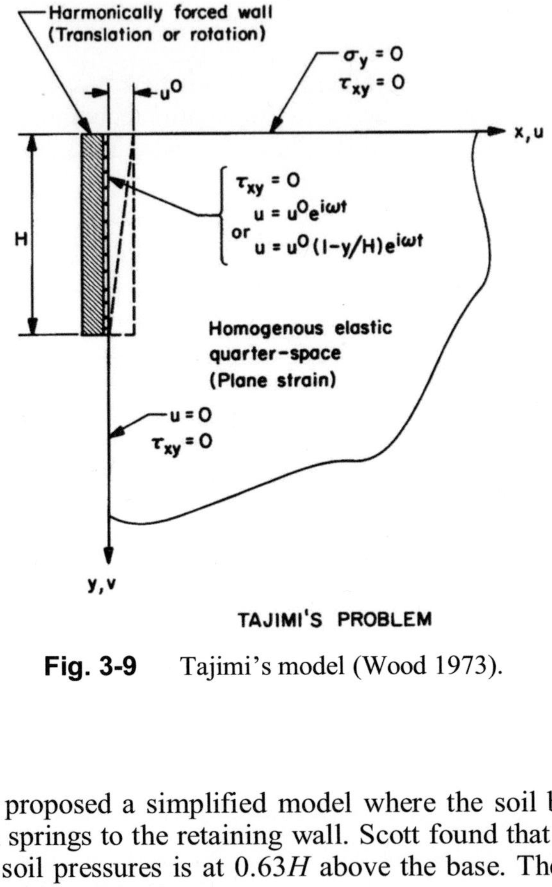

Fig. 3-9 Tajimi's model (Wood 1973).

3.2.4 Scott

Scott (Scott 1974) proposed a simplified model where the soil behaves as a shear beam coupled with springs to the retaining wall. Scott found that the point of application of dynamic soil pressures is at $0.63H$ above the base. The spring's stiffness per unit of length of the wall is given as:

$$K_s = \frac{0.8(1-v)}{1-2v}\frac{G}{H} \tag{3-11}$$

The wall pressure at a given height is expressed by the product of K_s and the relative motions of the shear beam (soil) and the wall at that height. Some drawbacks of this model are that the spring constant is independent from the excitation frequency, the only damping in the model is the damping of the soil and that infinite pressures are predicted when $v \rightarrow 0.5$ (Veletsos, A. S., Younan, A. H. 1992).

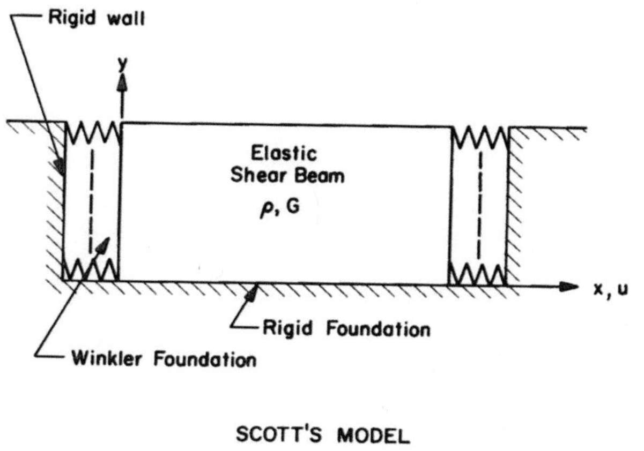

Fig. 3-10 Scott's model (Wood 1973).

3.2.5 Arias et al.

Arias et al. (Arias, A., Sanchez-Sesma, F. J., Ovando-Shelley, E. 1981) proposed a simplified model for dynamic soil pressures within the range of small displacements. The main hypotheses for the development of their model are: a continuous deformable behaviour of the backfill without couple stresses; vertical stresses equal zero; stress and strains in the backfill are related to Hook's law in one dimension.

3.2.6 Veletsos and Younan

Veletsos and Younan (Veletsos, A. S., Younan, A. H. 1993, 1994a, 1994b, 2000a, 1997, 1998b, 1998b, 1997; Parikh, V. H., Veletsos, A. S., Younan, A. H. 1995; Veletsos, A. S., Younan, A. H. 1998a, 2000b, 2000b) researched further on dynamic soil pressures in the field of small displacements, where the soil is supposed to behave elastically. In their approach they considered the flexibility and base rotation of the wall. The results of their investigation showed that smaller dynamic pressures are to be expected if the yielding of the wall in terms of bending flexibility and base rotation increases. They also showed that the simplification they assumed in order to find a solution (i.e. the boundary condition of zero pressure at the free surface is not fulfilled) has no influence on the results. They also provided tables of the calculated shear forces and moments due to the soil pressures as well as the application point of these forces. These forces have to be added to the seismic forces of the wall in order to have the complete section forces of the wall. Their model was also extended for a wall hinged at the top, which can idealize the behaviour of a tie back wall. Their results are a big step forward for a most realistic prediction of dynamic soil pressures and they somehow bridge the results provided by Wood and the M-O method. Their solution is briefly described here:

The value of displacement of a flexible beam able to rotate at its base is given by:

$$w(\eta, t) = \eta H\theta(t) + \sum_{j=1}^{j} \varphi_j(\eta)\, q_j(t) \tag{3-12}$$

where $\eta=y/H$ is the dimensionless height of the wall, H the wall's height, $\theta(t)$ the rotation of the wall at its base as a function of time and the second term with the summation is the response of a flexural beam in form of generalized coordinates q_j and shape functions φ_j. The shape functions can be expressed as:

$$\varphi(\eta) = \sum_{n=1}^{N} c_n \psi_n(\eta) = \sum_{n=1}^{N} c_n \sin\left(\frac{(2n-1)\pi}{2}\eta\right) \tag{3-13}$$

where c_n are dimensionless participation factors defined by appropriate integrals of φ_j and ψ_n, n is the order of shear-beam mode under consideration. The wall's displacements can be rewritten in the form

$$w(\eta, t) = \sum_{j=0}^{j} \varphi_j(\eta)\, q_j(t) \tag{3-14}$$

where the rotational mode of the whole wall at its base is the mode 0. The generalized coordinates can be written as:

$$q_j(t) = Q_j e^{i\omega t} \tag{3-15}$$

with Q_j representing their amplitudes. The equation of motion of the system can be written as (ρ the density of the retained medium/soil, A_g the peak ground acceleration and H the height of the retaining wall):

$$(S - \omega^2 M)Q = -\rho A_g H^2 A \tag{3-16}$$

where M is the mass matrix with the dimensions $(\mathrm{J}+1) \times (\mathrm{J}+1)$ defined by:

$$M = \mu_w H \begin{bmatrix} \frac{1}{3} & \langle \eta, \varphi_1 \rangle & \langle \eta, \varphi_2 \rangle & \cdots & \langle \eta, \varphi_{\xi j} \rangle \\ \langle \eta, \varphi_1 \rangle & 1 & 0 & \cdots & 0 \\ \langle \eta, \varphi_2 \rangle & 0 & 1 & \cdots & 0 \\ \vdots & \vdots & \vdots & \ddots & \vdots \\ \langle \eta, \varphi_j \rangle & 0 & 0 & \cdots & 1 \end{bmatrix} \tag{3-17}$$

where S is the stiffness matrix of the same order defined as:

$$S = S_o + (S_i)_{jk} =$$

$$= \frac{D_w^*}{H^2} \begin{bmatrix} \frac{R_w H}{D_w} & 0 & 0 & \cdots & 0 \\ 0 & \lambda_1^4 & 0 & \cdots & 0 \\ 0 & 0 & \lambda_2^4 & \cdots & 0 \\ \vdots & \vdots & \vdots & \ddots & \vdots \\ 0 & 0 & 0 & \cdots & \lambda_j^4 \end{bmatrix} + H \sum_{n=1}^{N} \frac{\langle \varphi_j, \psi_n \rangle \langle \varphi_k, \psi_n \rangle}{\langle \psi_n, \psi_n \rangle} K_n \tag{3-18}$$

with λ_j for the coefficient in the expression for the jth circular natural frequency of a cantilever beam. The nth circular frequency of the wall (flexural beam) is given by:

$$\omega_{w,j} = \left(\frac{\lambda_j}{H} \right)^2 \sqrt{\frac{D_w}{\mu_w}} \tag{3-19}$$

and A is the matrix with the vectors of the normalized exciting forces:

$$A_j = \langle \varphi_j, 1 \rangle \times \frac{\mu_w}{\rho H} - \frac{1}{\rho A_g H^2} \sum_{n=1}^{N} \langle \varphi_j, \psi_n \rangle K_n U_n \tag{3-20}$$

The quantity K_n expresses the complex-valued impedance or dynamic stiffness of the medium between the wall and the far field when they are both vibrating in the nth shear beam mode (δ is a damping coefficient and equals twice the ratio of critical damping ξ):

$$K_n = \frac{(2n-1)\pi}{2} \sqrt{\frac{2}{1-\nu} \frac{G}{H}} \sqrt{(1+i\delta)[1-(\omega/\omega_n)^2 + i\delta]} \tag{3-21}$$

Where ω_n is the nth circular frequency of the soil (shear beam):

$$\omega_n = \frac{(2n-1)\pi}{2}\frac{v_s}{H} \tag{3-22}$$

With v_s for the shear wave velocity and H for the depth of the soil.

Where μ_w is the wall's distributed mass and D_w represents the flexural rigidity of the wall and equals:

$$D_w = \begin{cases} \dfrac{E_w t_w^3}{12(1-v_w^2)} & \text{if the wall behaves like a plate} \\ \dfrac{E_w t_w^3}{12} & \text{if the wall behaves like a beam} \end{cases} \tag{3-23}$$

The U_n in equation 3.20 represents the amplitude of the displacement of the nth mode of the medium at the far field and equals:

$$U_n = \frac{16}{\pi^3}\frac{\rho A_g H^2}{G}\frac{1}{(2n-1)^3}\frac{1}{(1+i\delta)[1-(\omega/\omega_n)^2+i\delta]} \tag{3-24}$$

Finally, the pressures of the wall can be found by multiplying the differential wall and soil displacement of the nth mode with the complex spring modulus K_n:

$$\sigma(\eta,t) = \sum_{n=1}^{N} K_n \left\{ U_n - \sum_{j=0}^{J} \frac{\langle \varphi_j, \psi_n \rangle}{\langle \psi_n, \psi_n \rangle} Q_j \right\} \psi_n(t) e^{i\omega t} \tag{3-25}$$

By introducing two dimensionless factors d_w and d_θ, which describe the relative flexibility between the wall and the retained soil and the relative flexibility of the rotational base constraint and the retained medium, the soil pressures can be plotted as functions of these two parameters.

$$d_w = \frac{GH^3}{D_w} \tag{3-26}$$

$$d_\theta = \frac{GH^2}{R_\theta} \tag{3-27}$$

A value of d_w=0 and d_θ=0 corresponds to a rigid rotationally constrained wall (corresponds also to the problem researched by Wood). The wall displacements can be calculated from the formula:

$$w_{st}(\eta = 1) = c_2 \frac{A_g H^2}{v_s^2} \tag{3-28}$$

Where c_2 is a value obtained by a table. Veletsos and Younan had also showed that the static displacement (the term static refers to an excitation frequency equal to 0, and not to the gravitational forces) of a flexible wall remains quite small so as not to mobilize passive or active pressures, even if it is multiplied with an amplification dynamic factor of about 2.

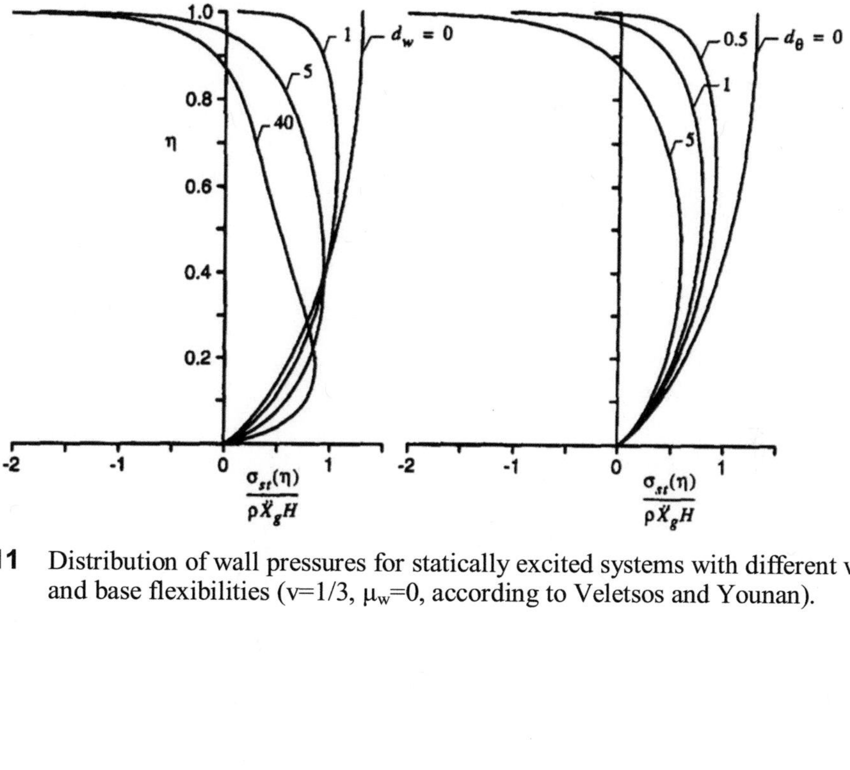

Fig. 3-11 Distribution of wall pressures for statically excited systems with different wall and base flexibilities (v=1/3, μ_w=0, according to Veletsos and Younan).

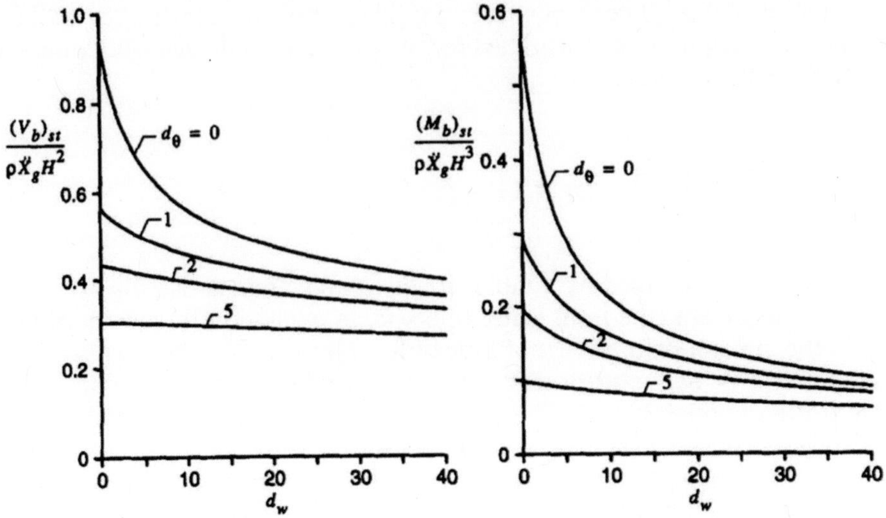

Fig. 3-12 Normalized values of base shear and moments of statically excited systems with different wall and base flexibilities (v=1/3, μ_w=0, according to Veletsos and Younan).

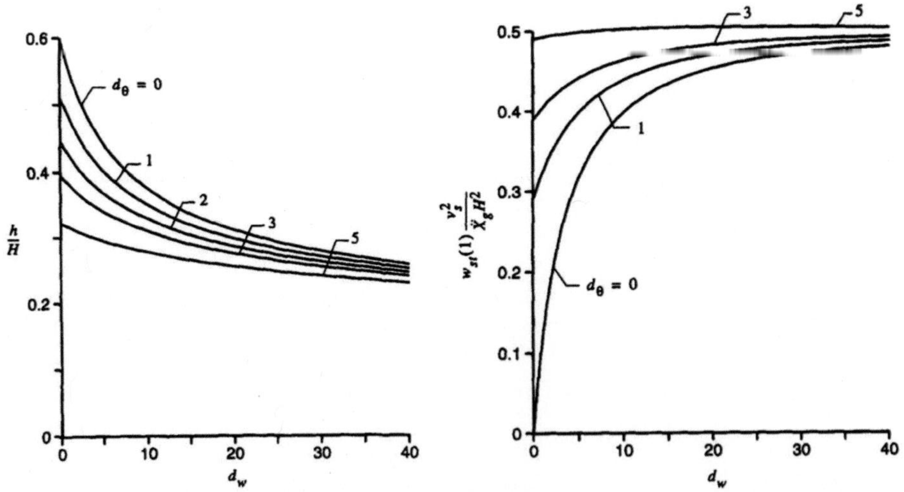

Fig. 3-13 Left: normalized effective heights for statically excited systems with different wall and base flexibilities. Right: normalized top wall displacements relative to base for statically excited systems with different wall and base flexibilities (v=1/3, μ_w=0, according to Veletsos and Younan).

In a same manner, Veletsos et al. (Parikh, V. H., Veletsos, A. S., Younan, A. H. 1995) defined the dynamic soil pressures of a wall-soil system with two walls (bounded system). The pressure distribution on the wall is given as a function of the ratio *L/H* as:

$$\sigma_w(0,h,t) = -\frac{8\psi_o}{\pi^2}\rho A_g H \sum_{n=1,3,....}^{\infty} \frac{1}{n^2}\sqrt{\frac{1+i\delta}{1-\varphi_n^2+i\delta}}\tan\left(\frac{\alpha_n}{2}\frac{L}{H}\right)\sin\left(\frac{n\pi}{2}h\right)e^{i\omega t} \quad \textbf{(3-29)}$$

Where

$$\psi_o = \sqrt{\frac{2}{1-v}}$$

$$\alpha_n = \frac{n\pi}{2\psi_o}\sqrt{1-\frac{\varphi_n^2}{1+i\delta}} \quad \textbf{(3-30)}$$

3.2.7 Ostadan

Ostadan (Ostadan 2005) has proposed an updated approach for the dynamic soil pressures on rigid walls that is also recommended by NEHRP. In his approach he took into account not only the peak ground acceleration but also the frequency content of the excitation. He performed analyses with SASSI (Lysmer et al.) in the frequency domain. As expected, the maximum amplification of the dynamic soil pressures takes place at the frequency corresponding to the soil column frequency, which, for a constant wall height, depends only on the shear wave velocity of the soil. Due to the amplification of the seismic signal through the soil column the maximum soil pressure is observed at the top of the wall. With his simplified method, Ostadan proposes the following computational steps to obtain the soil pressures profile: Firstly, an analysis must be performed in order to obtain the free field acceleration of the soil column with 30% damping at the wall's base. Such an analysis can be performed using SHAKE (Schnabel, P. B., Lysmer, J., Seed, H. B.), and the damping of 30% is explained as the value at which there is the best correlation between the SASSI analyses with the wall and the soil and the spectral value of a soil column. Then the representative SDOF mass is computed with the equation $m = 0.50 \times \rho \times H^2 \times \psi_v$,where ρ is the density of the soil, *H* is the height of the wall and ψ_v is a factor to account for Poisson's ratio. The seismic force can be obtained by multiplication of the representative mass with the acceleration value in the first step. The maximum lateral seismic pressure is obtained by dividing the seismic force by $0.744 \times H$ (area of the normalized seismic soil pressure). The pressure profile is obtained by multiplying the peak pressure with the pressure distribution of the form: $p(y) = -0.0015 + 5.05 \times y - 15.84 \times y^2 + 28.25 \times y^3 - 24.59 \times y^4 + 8.14 \times y^5$. An advantage of this method is that the soil's non-linearity can be accounted for in the form of the equivalent linear method. As is known, soil non-linearity depends on both, the frequency content and the intensity of the excitation affecting the seismic soil pressures.

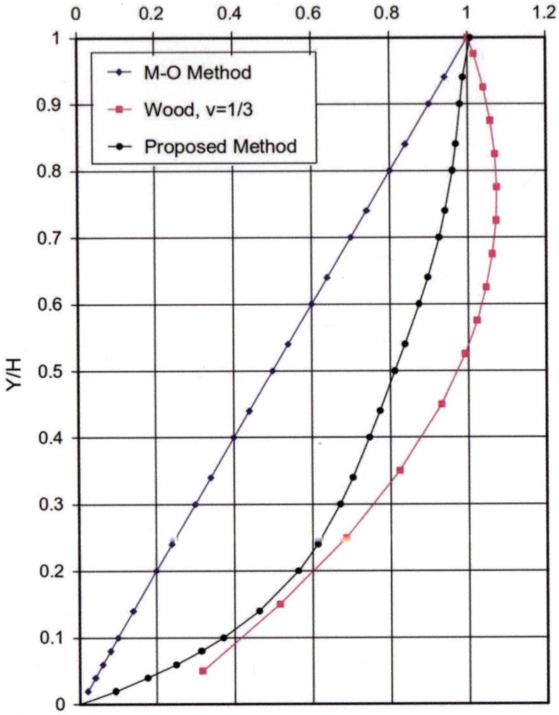

Fig. 3-14 Comparison of normalized pressure profiles according to (Ostadan 2005).

3.2.8 Jung et al.

Jung et al. (Bobet, A., Jung, C. 2008; Bobet, A., Fernández, G., Jung, C. 2010) extended the procedure of Veletsos and Younan taking into account the influence of the elastic soil not only for the rocking of the wall but also for the horizontal and vertical elastic relative movement (they added springs also in the horizontal and vertical direction in order to model the elastic underlying soil more accurately). They showed that the soil pressures are extremely sensitive when the wall is able to move elastically in the horizontal direction. The vertical elastic movement of the wall has been investigated in comparison with the wall's friction and, as expected, it does not affect the soil pressures.

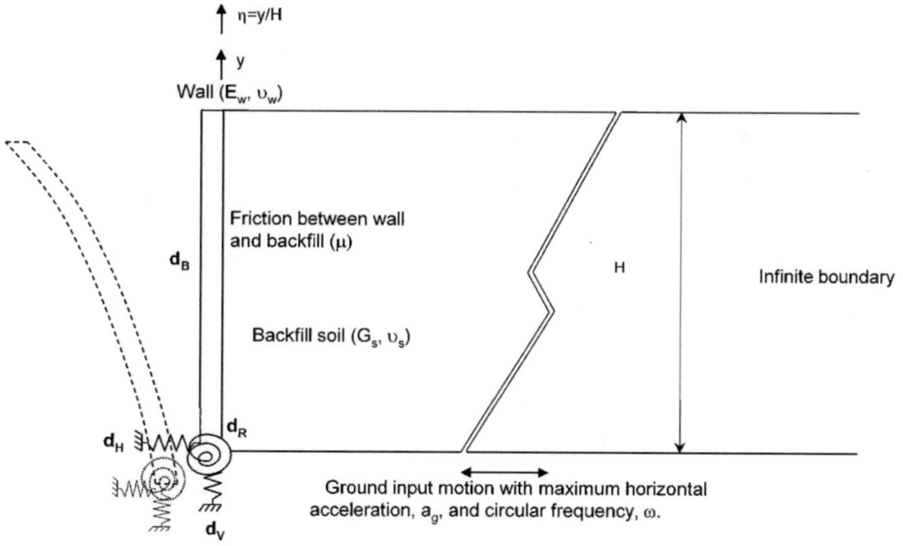

Fig. 3-15 Model of flexible retaining wall with three rigid-body motions at its base (according to (Bobet, A., Fernández, G., Jung, C. 2010) (Bobet, A., Jung, C. 2008)).

Jung et al. added one additional mode of movement apart from the five flexural modes and one rotational mode investigated by Veletsos and Younan. This horizontal rigid-body motion can be approximated according to Jung and Bobet by:

$$\varphi_7(\eta) = 2\sin\left(\frac{\pi}{2}\sqrt{\eta}\right) \tag{3-31}$$

This relation for the rigid mode is false according to the author of this thesis, as it does not describe a rigid body motion (the mode depends on $\eta=y/H$). The rigid body motion should be described by (see also (Bishop, R. E. D., Johnson, D. C. 2011) p. 375):

$$\varphi_7(\eta) = 1 \tag{3-32}$$

The soil pressure distribution for this mode shape (rigid body movement) is shown in the figure below. This pressure distribution is in accordance with the result of the finite element analysis presented in the next chapter.

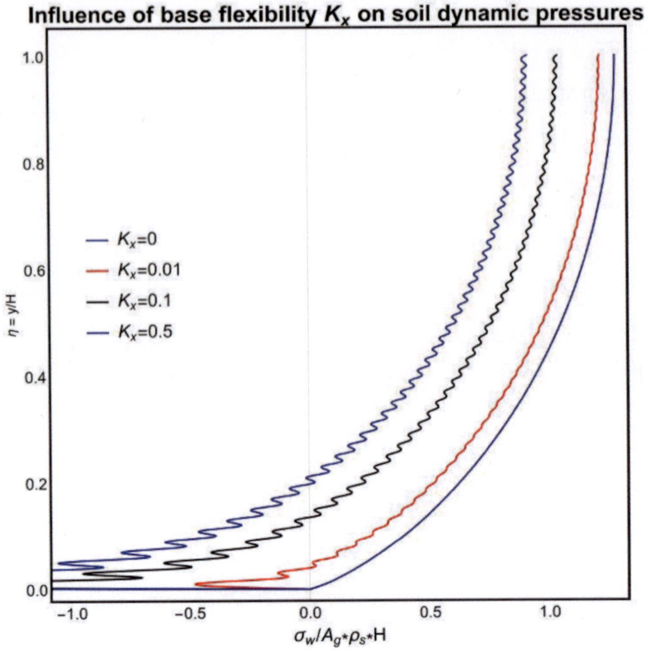

Fig. 3-16 Soil pressure distribution for the mode shape of equation $\varphi_7(\eta) = 1$ for different values of translational spring constants (d_w=0, d_θ=0).

3.2.9 Papazafeiropoulos and Psarropoulos

Papazafeiropoulos and Psarropoulos (Papazafeiropoulos, G., Psarropoulos, P. N. 2010) solved the same problem as Wood (two rigid walls retaining soil) by using a rigorous analytical solution. They used the method of Werner and Sundquist (Sundquist, K. J., Werner, P. W. 1949) for the hydrodynamic pressures and solved the differential equation with separation of variables. The solution for the soil problem is more complicated than that for the hydrodynamic pressures. They provided results and graphs for different length-to-height ratios of the contained soil and for different excitation frequencies. Their analytical solution for higher frequencies makes the problem of wave propagation obvious.

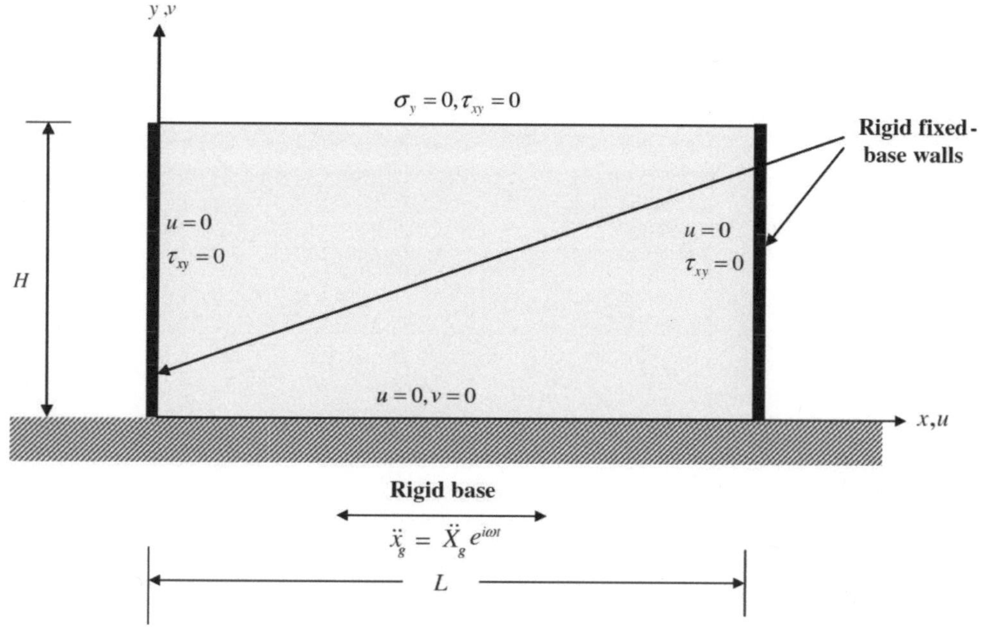

Fig. 3-17 The boundary value problem according to (Papazafeiropoulos, G., Psarropoulos, P. N. 2010).

3.2.10 Kloukinas et al.

Kloukinas et al. (Kloukinas et al. 2012) provided a simple wave solution for the seismic earth pressures on non-yielding walls. They used the technique, which apparently was first addressed by Vlasov and Leontiev (Leontiev, U. N., Vlasov, V. Z.) for the analysis of surface footings to gravity loads, leading to the so-called two-parameter foundation model (Scott 1974). They applied the separation of variables to the two-dimensional wave propagation equation using a shape function for the variable y. After integration over the wall height, they eliminated the variable y and found an ordinary differential equation subjected to given boundary conditions. The solution of this equation describes the dynamic soil pressures. Kloukinas' solution is in good accordance with the results of Veletsos and Younan, however only for small values of the dimensionless rotation stiffness d_θ. Kloukinas' solution is easier than the one derived by Veletsos and Younan because of its closed form.

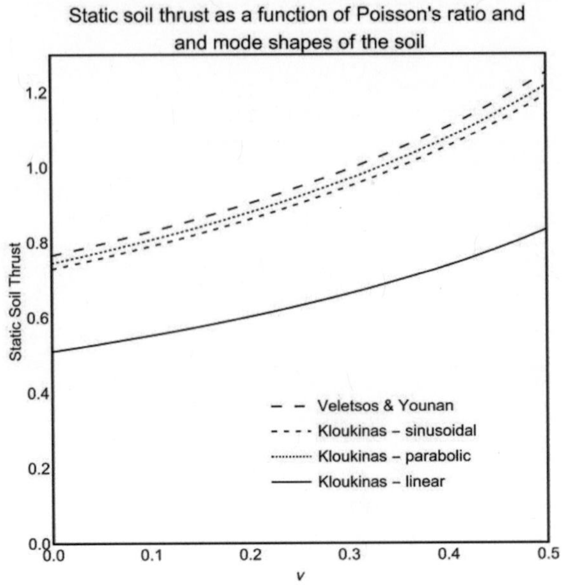

Fig. 3-18 Dynamic soil thrust for different soil mode shapes (according to (Kloukinas et al. 2012)).

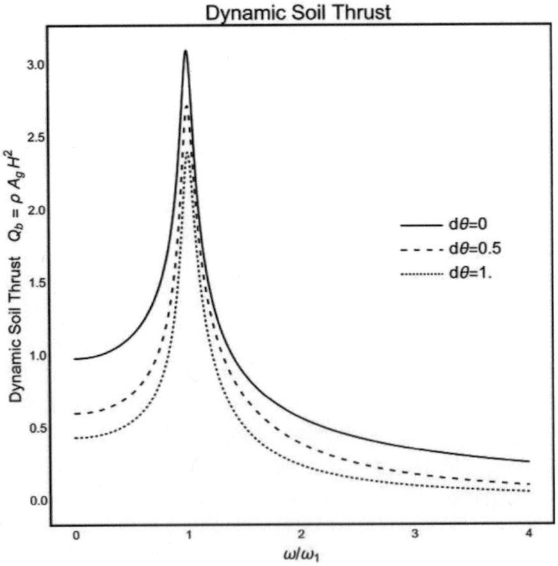

Fig. 3-19 Dynamic soil thrust for different wall base flexibilities (according to (Kloukinas et al. 2012)).

3.2.11 Brandenberg et al.

Brandenberg et al. (Brandenberg, S., Mylonakis, G., Stewart, J. 2015) calculated the resultant pressures acting on the wall of a stiff U-shape embedded structure within a kinematic framework. They used the spring coefficients calculated by Kloukinas et al. for the vertical walls of the U-section, they used spring coefficients from the literature and suggested correction factors in order the total impedance computed with this method to equal the ones found in literature for the whole embedded foundation. Their result is given as a function of the wavelength (inverse analogue of the frequency) of the excitation and show the reduction of the soil pressures acting on the wall as the base becomes more compliant and they take their maximum value when the structure is resting on the bedrock, where the rotation of the structure is restrained. These results are in accordance with other analyses ((Veletsos, A. S., Younan, A. H. 1997), (Ostadan 2005)). They could also partly explain the very small earthquake active pressures found by Al Atik and Sitar and Mikola and Sitar (Mikola, R. G., Sitar, N. 2013). In the experiments carried out by the latter, the U-shape structure rests on sand and the bigger rigid body movements of the structure lead to smaller relative displacements with the soil, resulting in much smaller soil pressures.

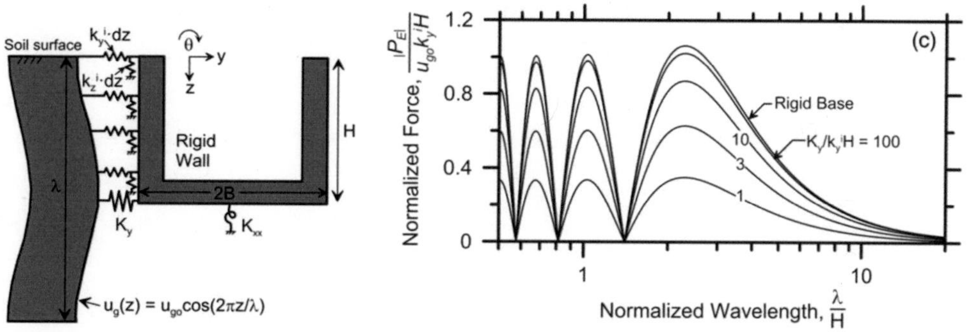

Fig. 3-20 Embedded rigid strip foundation exited by vertically propagated shear wave for the case of no base slab averaging (left) and normalized wall pressure versus normalized wavelength λ/H (right) for various contributions of wall normal stress to translational and rotational stiffness (according to (Brandenberg, S., Mylonakis, G., Stewart, J. 2015)).

Di Laora (Discussion on paper of Brandenberg et al. 2015) discussed the paper of (Brandenberg, S., Mylonakis, G., Stewart, J. 2015) and provided further solutions for non-homogeneous soil and for soil with damping for the special case of no base rotation (structure founded on rock). In a new paper Brandenberg et al. (Brandenberg et al. 2017) extended their method for inhomogeneous soil and presented approximate solutions after making some simplifications regarding the vertical stresses and displacements. Their results are compared with the more rigorous of Vrettos et al. (Vrettos et al. 2016)

3.2.12 Vrettos et al.

Vrettos et al. (Vrettos et al. 2016) solved semi-analytically the Wood problem of two fixed walls with contained soil for the more realistic case of inhomogeneous soil using the method of Papazafeiropoulos and Psarropoulos. The soil has a parabolic shape of shear modulus which is described by parameters such as the gradient of inhomogeneity, α, and the non-homogeneity parameter Ξ_o. His results are in good accordance with the results of other researchers.

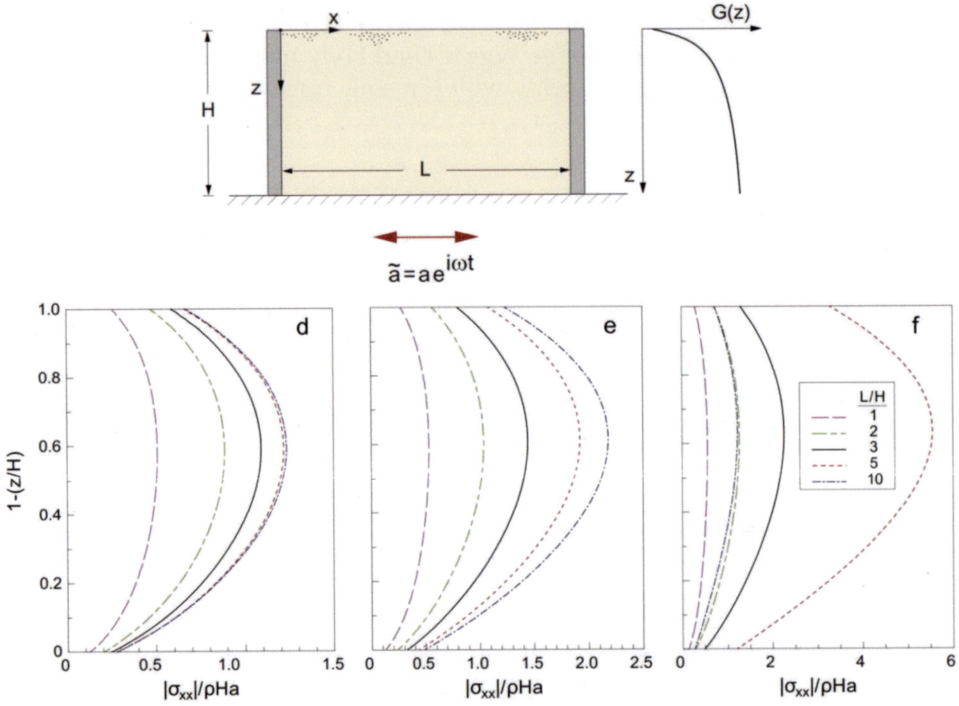

Fig. 3-21 Profile of shear modulus and soil pressure distribution for Ξ_o=0.8 and ω=2.36/5.76/9 rad/s according to (Vrettos et al. 2016).

3.3 Displacements due to earthquakes

3.3.1 Elms and Richards

Elms and Richards (Elms, D. G., Richards R. 1979) gave a formula for the sliding of a gravity retaining wall based on Newmark's sliding block analysis. They took into account the active earth pressures acting on the wall, the inertia forces of the wall itself and the friction force between the wall and the underlying soil and calculated the acceleration at which sliding occurs (yielding acceleration). Furthermore, they gave a formula for checking whether the wall will slide or tilt for the calculated yielding acceleration based on the wall's dimensions. They gave the following formula for the permanent displacements after an earthquake event:

$$d = 0.087 \frac{V^2}{A_g} \left(\frac{N}{A}\right)^{-4} \tag{3-33}$$

Where d is the total relative displacement, V the peak velocity of the earthquake, A_g the peak acceleration, N the coefficient of limiting wall acceleration and A the acceleration seismic coefficient. With this formula not only the displacements can be calculated but reversely the acceleration can be estimated for a desired allowed displacement. The bigger the allowed displacement, the smaller is the earthquake coefficient. EN 1998-5 (EN 1998-5:2004 Eurocode 8) allows a reduction of the seismic coefficient k_h in accordance with the allowed displacement as a ratio of the wall's height.

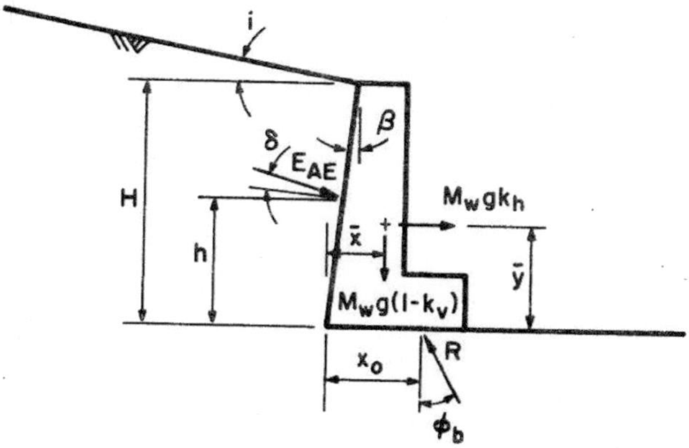

Fig. 3-22 Forces acting on a gravity retaining wall according to (Elms, D. G., Richards R. 1979).

3.3.2 Liao, Whitman, Wong

Liao and Whitman (Liao, S., Whitman, R. V. 1985) and Wong (Wong 1982) proposed another formula for the permanent displacements of gravity walls subjected to an earthquake, which they derived by numerical analyses and statistical processing:

$$d = \frac{37V^2}{A_g} e^{-9.4(N/A)} \tag{3-34}$$

3.3.3 Nadim and Whitman

Nadim and Whitman (Nadim F., Whitman R. V. 1983) researched on the seismic displacements of gravity walls by taking the effect of an amplification of ground motion in the backfill into account. They used a two dimensional plane strain finite element model in order to assess the aforementioned effect. They did not find a new formula predicting the displacement due to earthquakes. They proposed correction factors for A (acceleration coefficient) and V (peak ground velocity) which appear in the formulas of Richard and Elms (Elms, D. G., Richards R. 1979) and Wong (Wong 1982).

Earthquake (1)	A (2)	$f/f1$ (3)	$D2/D1$ (4)	F (Richards-Elms) (5)	F (Wong) (6)
Kern County-Taft	0.2	0.53	4.98	1.38	1.34
		0.81	13.05	1.67	1.65
	0.3	0.53	2.84	1.24	1.30
		0.90	4.79	1.37	1.50
Eureka	0.2	0.50	2.62	1.21	1.19
	0.3	0.50	1.78	1.12	1.17
Puget Sound	0.2	0.86	6.58	1.46	1.45
	0.3	0.86	3.75	1.30	1.37

Fig. 3-23 Correction factors by which A and V should be multiplied in the formulas of Richard and Elms and Wong (according to (Nadim F., Whitman R. V. 1983)).

3.3.4 Elms

Elms (Elms 2000) suggested some refinements to the sliding block model of Newmark. In his model he induced the vertical and lateral components of the earthquake and combined them with several correlations. He concluded that the sliding response can be much more complex and that the lateral and vertical components can lead to significantly bigger displacements.

A state of the art regarding seismic-induced displacements can be found in Wu and Prakash (Prakash S. 2001).

3.4 Experimental studies

The most mentioned experimental study for soil pressures on a retaining wall is the one of Mononobe and Matsuo (Matsuo, O., Mononobe, N. 1929) because it is related to the Mononobe-Okabe (M-O) method. In order to evaluate the analytical results gained by Okabe, Mononobe and Matsuo carried out experiments with relatively loose dry sand and a sinus excitation of 1-g. Their experimental results are in good accordance with the theory of Okabe.

Similar experiments based on 1-g excitation have also been carried out by other researchers such as Matsuo (Matsuo 1941), Matsuo and Ohara (Matsuo, H., Ohara, S. 1960), Sherif et al. (Ishibashi, I., Lee, C. D., Sherif, M. A. 1982), Bolton and Steedman (Bolton M. D., Steedman, R. S. 1982, 1985), Sherif and Fang (Fang,

Y. S., Sherif, M. A. 1984), Steedman (Steedman 1984), Ishibashi and Fang (Fang, Y. S., Ishibashi, I. 1987). An in-depth description of these studies is beyond the purpose of this thesis and can be found elsewhere (Al Atik, L., Sitar, N. 2008; Al Atik, L., Sitar., N. 2007).

Dynamic centrifuge tests have been carried out by many scientists, for instance Ortiz (Lee, J., Ortiz, L. A., Scott, R. F. 1983), Bolton and Steedman (Bolton M. D., Steedman, R. S. 1985), Zeng (Zeng 1990), Steedman and Zeng (Steedman, R. S., Zeng, X. 1991), Stadler (Stadler A. T. 1996). What is interesting is the fact that most of the results of these experimental studies support the Mononobe-Okabe theory. Very important comments have been made, however, by the following scientists: Steedman and Zeng (Steedman, R. S., Zeng, X. 1991), based on the experimental results of the latter (Zeng 1990), indicate the different phases of the propagating waves in the wall and the soil and point out that the amplification or attenuation of these waves are important factors in order to determine the distribution and magnitude of the dynamic soil pressures. Another important observation has been made by Nakamura (Nakamura 2006), who underlines the drawbacks of the M-O method based on his experiments. An important comment of his is that the retaining gravity wall oscillates in phase with the soil so that the active earth pressures remain constant and there is no dynamic increment. More detailed literature about dynamic centrifuge experiments can be found in (Al Atik, L., Sitar, N. 2008; Al Atik, L., Sitar., N. 2007).

3.4.1 Kloukinas et al.

Kloukinas et al. (Kloukinas et al. 2015) performed a series of experiments in a shear beam container of a flexible cantilever retaining wall in order to validate the results of the stress limit analysis (Mylonakis G., Kloukinas P., Papantonopoulos C. 2007). They also investigated the behaviour and stability of walls of this type based on a compliant base. Their results confirm the theoretical predictions as far as yield acceleration and failure mechanisms are concerned. Nevertheless, their interpretation of the experimental results supports the rigid block response of the backfill with constant acceleration.

3.4.2 Al Atik and Sitar

Al Atik and Sitar (Al Atik, L., Sitar, N. 2008; Al Atik, L., Sitar., N. 2007) have recently carried out a series of dynamic centrifuge experiments. The geometry used in their model is of great interest, as it is similar, if not identical with the layout of many twin navigation locks. Moreover, the U-shape retaining walls they used can hardly slide and they are similar to cantilever retaining walls, but the soil-structure interaction is taken into account as well as the influence of a neighbouring chamber wall. Moreover, by assigning different flexibilities to the U-shaped retaining walls, a comparison with the method of Veletsos and Younan can be made. They compared their experimental results also numerically via a nonlinear dynamic finite element model (OpenSees). The experimental and numerical results provided are in good accordance. Based on their results, Al Atik and Sitar commended that the M-O method is inappropriate for the seismic design of retaining walls, because the maximum earth pressure and the inertia force of the walls do not occur simultaneously as the M-O method indicates, and because there is a phase difference between these maximum values (of 90 or 180 degrees). Their results and conclusions are in good accordance with the ones of Nakamura (Nakamura 2006).

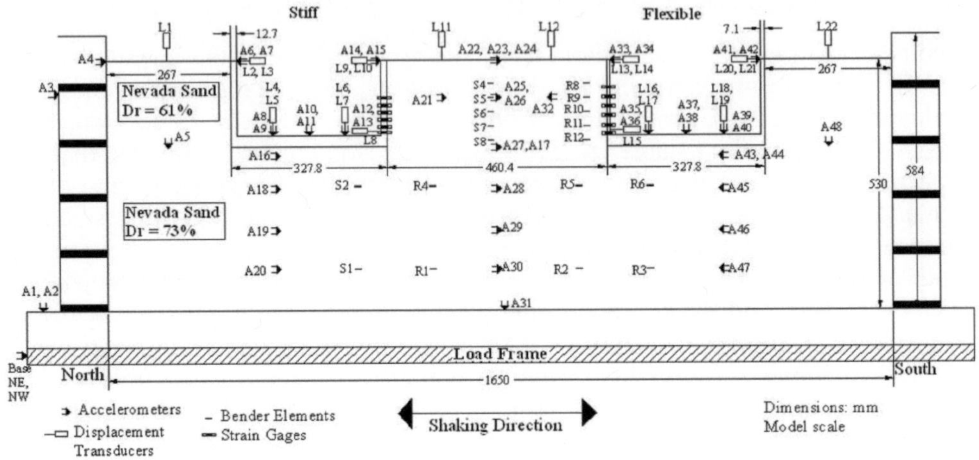

Fig. 3-24 Model configuration: left, stiff U-frame and right, flexible U-frame walls (Al Atik, L., Sitar, N. 2008; Al Atik, L., Sitar., N. 2007).

They also compared the observed maximum bending moments for the stiff and flexible walls with the ones given by the M-O method and the Seed and Whitman approximation. In general, the calculated total moments by these two methods overestimate the bending moments by up to three times for severe earthquakes. The pressure distribution calculated by the aforementioned methods is also very conservative. They calculated reversely the dynamic increment of the active earth pressures by subtracting the inertia moment from the total moment. In this way they showed that the dynamic increment of the soil pressures is zero up to 0.4-*g* accelerations and much less than the one estimated for higher acceleration values. They conclude that the M-O method leads to an over-conservative design of the retaining walls and they suggest adding a much smaller dynamic increment for the soil pressures to the calculated inertia forces of the wall.

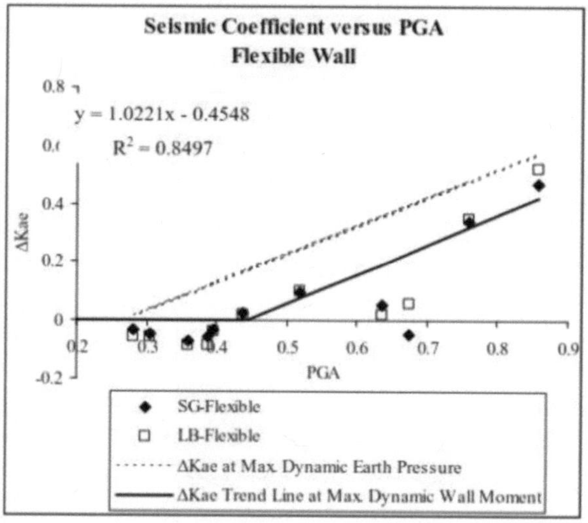

Fig. 3-25 Back-calculated dynamic earth pressure coefficients at time of maximum dynamic wall moments and maximum earth pressures on flexible walls as a function of peak ground acceleration measured in the free field (Al Atik, L., Sitar, N. 2008; Al Atik, L., Sitar., N. 2007).

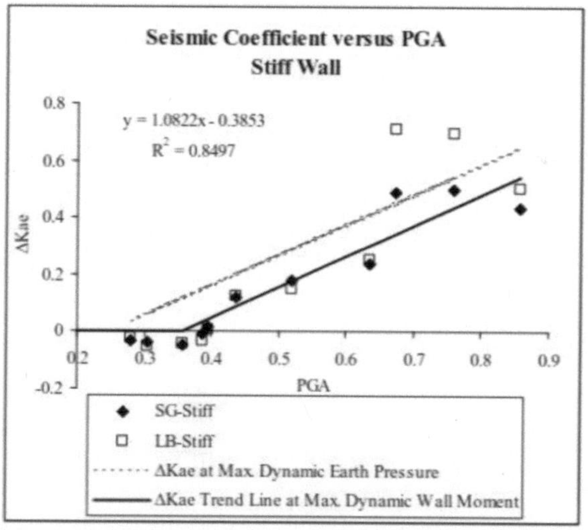

Fig. 3-26 Back-calculated dynamic earth pressure coefficients at time of maximum dynamic wall moments and maximum earth pressures on stiff walls as a function of peak ground acceleration measured in the free field (Al Atik, L., Sitar, N. 2008; Al Atik, L., Sitar., N. 2007).

3.4.3 Mikola and Sitar

Mikola and Sitar (Mikola, R. G., Sitar, N. 2013) performed a series of dynamic centrifuge experiments to measure the magnitude and distribution of seismic earth pressures on both basement and non-displacing and displacing cantilever retaining structures. The U-shape of the centrifuge models, which is similar to the monolithic chambers of navigation locks, is also of great interest here. The authors' experimental results were compared with numerical analyses carried out with FLAC. The results, although not identical, lead to the same conclusions. Mikola and Sitar showed that the M-O method is adequate for the design even of basement, i.e. non-yielding walls. The formula proposed by Wood is far on the safe side. The M-O method delivers no result for high accelerations, however, which makes it also too conservative. The authors also showed that a factor of safety (F.S.) equal to 1.5 is adequate for PGAs up to 0.3*g* and an F.S. of 2.0 is adequate for accelerations up to 0.5 *g*.

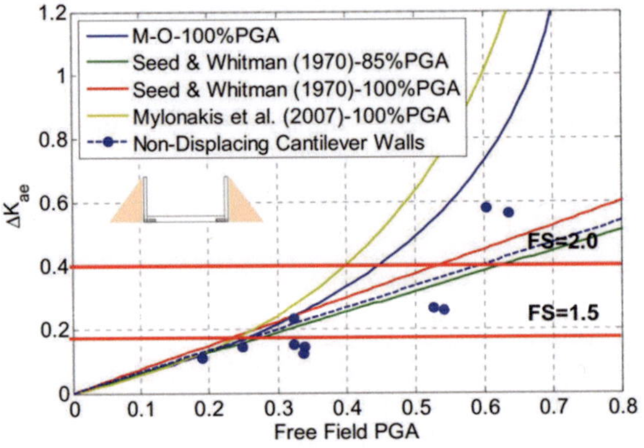

Fig. 3-27 Dynamic earth pressure coefficient as a function of PGA for stiff U-shaped cantilever walls with medium dense backfill (according to (Mikola, R. G., Sitar, N. 2013)).

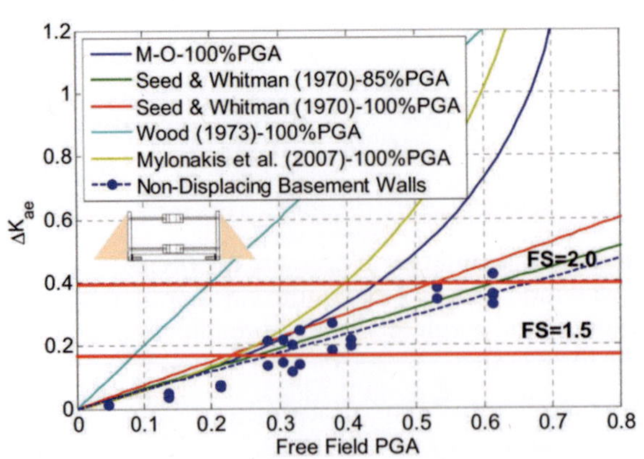

Fig. 3-28 Dynamic earth pressure coefficient as a function of PGA for non-displacing basement walls with medium dense backfill (according to (Mikola, R. G., Sitar, N. 2013)).

3.5 Saturated soils

3.5.1 Matsuo and Ohara

Matsuo and Ohara (Matsuo, H., Ohara, S. 1960) have analytically and experimentally derived the dynamic water pressures on quay walls, when the soil is permeable. They showed that the water pressures have the pressure distribution calculated by Westergaard and their amplitude depends on the soil permeability. They also suggested that an upper bound should be used for the equations derived by Westergaard (factor of 0.7).

3.5.2 Matsuzawa et al.

Matsuzawa et al. (Ishibashi, I., Kawamura, M., Matsuzawa, H. 1985) have conducted experiments in order to calculate the water pressures on walls of submerged soils. They showed that the formula proposed by Matsuo and Ohara for the water pressures of submerged soils on retaining walls is an upper bound and they proposed a reduction factor for the water pressures, which takes the soil permeability into account. Their results have been adopted by Eurocode 8, Part 5, Annex E (EN 1998-5:2004 Eurocode 8) in a simplified form. They also provided a reduction factor, which depends on the permeability of the soil, to calculate the water pressures on a wall (this factor expresses the part of the restricted water able to oscillate):

$$C_e = 0.5 + 0.53 \tanh\left(log \frac{2\,\pi\, n\, \gamma_w\, H_w^2}{7\, K\, k\, T} \right) \tag{3-35}$$

Where n is the porosity of the soil; k is the coefficient of permeability; T is the period of ground motion; K is the compressibility of water (2.25 GPa); and H_w the wall's height.

3.5.3 Chen and Hung

Chen and Hung (Chen und Hung 1993) investigated the dynamic pressures of water and sediment on a rigid dam. What is of interest is the case in which the sediments cover the whole side of the dam so that no water remains to interact freely with the rigid dam. They provided results for different ratios of sediment density to water density and for different values of water compressibility.

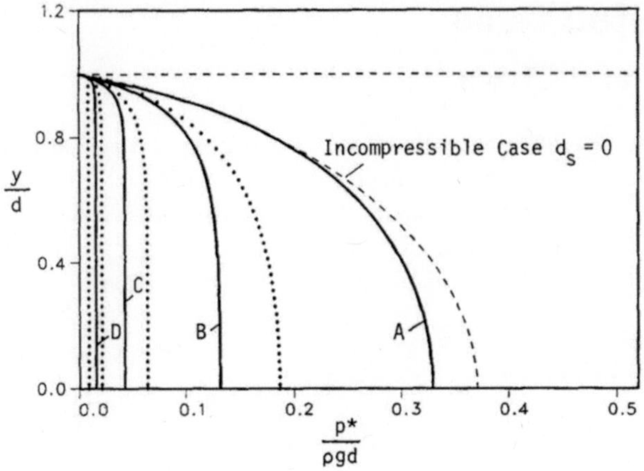

Fig. 3-29 Dynamic pressures on vertical dam face fully covered with sediments for different degrees of sediment permeability (according to (Chen und Hung 1993)).

3.5.4 Theodorakopoulos et al. / Papagiannopoulos et al.

Theodorakopoulos, Theodorakopoulos et al. (Theodorakopoulos und Beskos 2003; Theodorakopoulos et al. 2001a, 2001b; Beskos, D. E., Chassiakos, A. P., Theodorakopoulos, D. D. 2001) and Papagiannopoulos et al. (Beskos, D. E., Papagiannopoulos, G. A., Triantafyllidis, T. 2015) used the method of Veletsos and Younan and extended it for poroelastic soil. They showed the influence of soil permeability, of the amplitudes frequency as well as of the ratio L/H (for a pair of walls) on the dynamic water pressures on rigid retaining walls and on a pair of rigid and rotating retaining walls. For the case of the rigid wall, their results are comparable with the ones provided by Chen and Hung (Chen und Hung 1993).

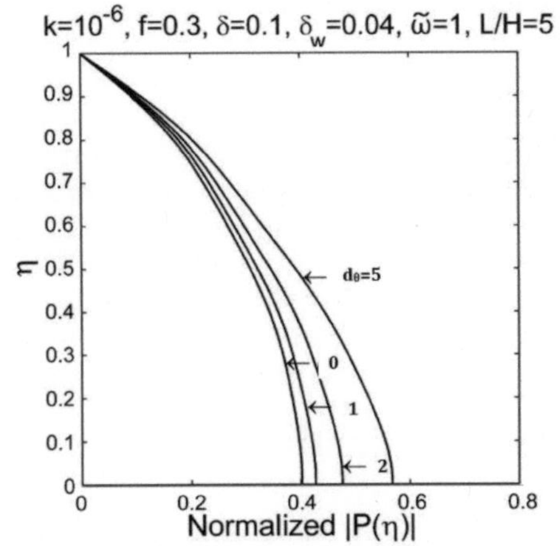

Fig. 3-30 Water pressures on a wall for different rotational base flexibilities according to (Theodorakopoulos et al. 2001a), the direction of y-axis is by mistake mirrored in the original paper.

3.6 Numerical analyses

3.6.1 Wood

Wood (Wood 1973) verified his analytical results via a finite element analysis. Both analyses are in good accordance. The results show the influence of a smooth wall (analytical procedure) against a bonded wall (numerical procedure) on the dynamic soil pressures. The attached wall appears as a singularity at the top layer of the finite elements, where the soil pressures show much higher values.

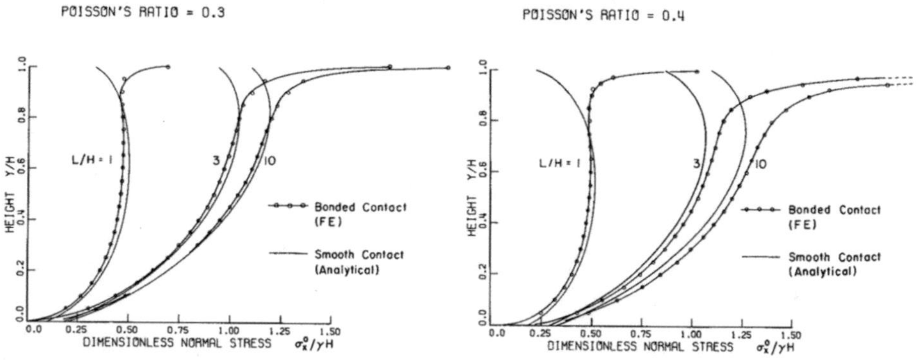

Fig. 3-31 Pressure distributions on rigid wall for different Poisson and L/H ratios according to (Wood 1973).

3.6.2 Wu and Finn

Wu (Wu 1994), Wu and Finn (Wu und Liam Finn 1999) have also investigated numerically the dynamic soil pressures on retaining structures. They verified the results of Wood and with the help of the finite element method they also provided results for other soil profiles with the shear modulus changing in depth. They also showed the amplification factor for the shear forces on the wall in relation to the *L/H* ratio of the contained soil.

3.6.3 Psarropoulos et al.

Psarropoulos et al. (Psarropoulos et al. 2005) verified through numerical analysis the results of Veletsos and Younan. They also investigated how far the assumption of an attached to the wall soil without relative vertical movements is fulfilled in terms of shear and normal stresses on the wall. Moreover, they investigated whether the underlying soil can be replaced by a rotational spring only, as Veletsos and Younan did. Their results show that when modelling the underlying soil (two-layer model) the soil pressures are reduced. As they stated in their work, by modelling the underlying soil an additional degree of freedom is added to the system (horizontal translation). Moreover, their approach to interpreting the different values of the dimensional parameter d_θ by changing the shear modulus of the underlying soil affects the wave propagation and the soil pressures compared arise from different accelerations at the wall's base.

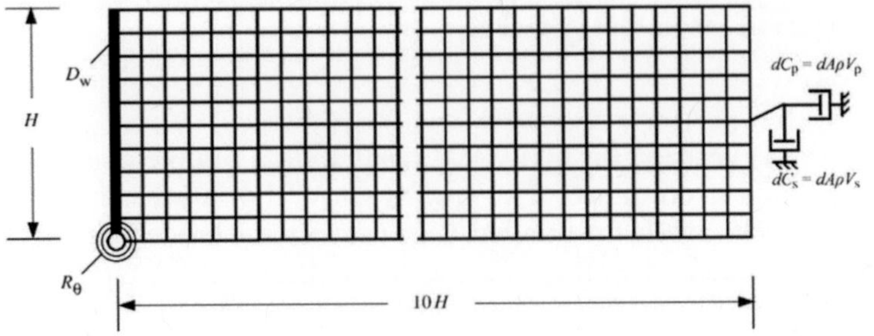

Fig. 3-32 Numerical model used by (Psarropoulos et al. 2005).

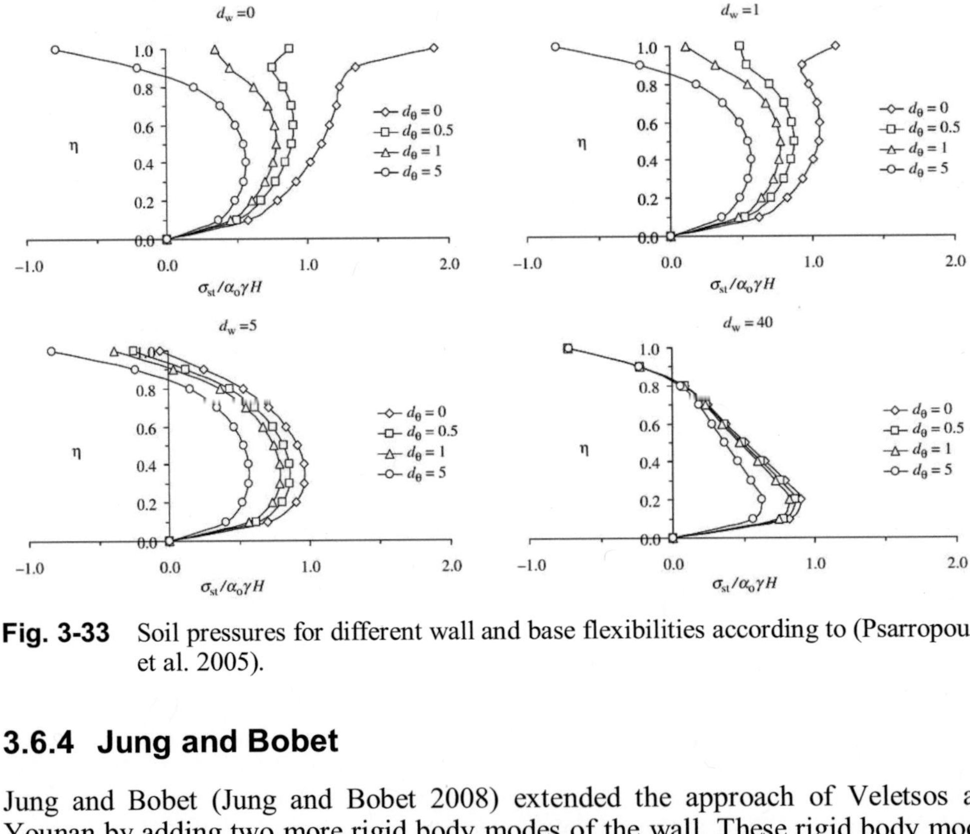

Fig. 3-33 Soil pressures for different wall and base flexibilities according to (Psarropoulos et al. 2005).

3.6.4 Jung and Bobet

Jung and Bobet (Jung and Bobet 2008) extended the approach of Veletsos and Younan by adding two more rigid body modes of the wall. These rigid body modes (one in horizontal and one in vertical direction) refer to the movements in these directions. They also extended the numerical model of Psarropoulos and added two additional translational springs in order to idealize the underlying elastic soil. The results of the numerical study verify their analytical analysis (Jung et al. 2010) and they consider that a tension crack can develop between the soil and the retaining wall (no bonding between the wall and the soil is considered, hence no tension forces can develop against the wall).

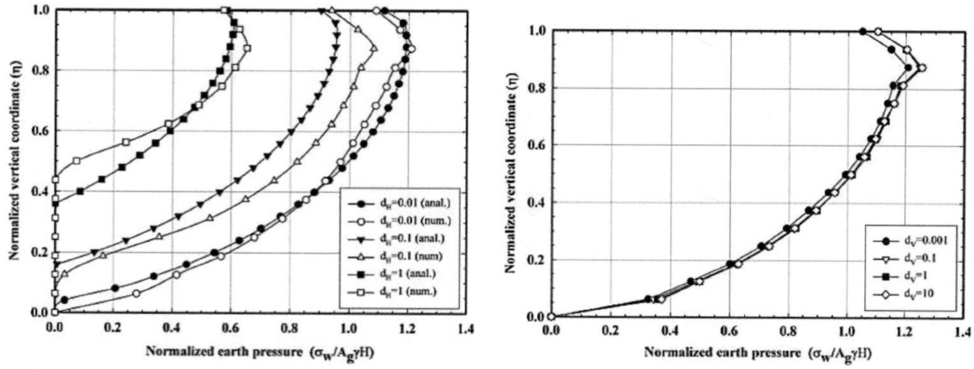

Fig. 3-34 Influence of a horizontal (left) and vertical (right) spring on the soil pressures for a rigid wall (d_w=0, d_θ=0, according to (Bobet, A., Fernández, G., Jung, C. 2010)).

3.7 Literature referring to navigation locks

There is only little literature referring to navigation locks exclusively. Some of the publications are presented here for completeness.

Vallabhan et al. (Rahman,K. R., Sivakumar, J., Vallabhan,C. V. G. 1988) developed the program BEFEC (Boundary Element Finite Element Coupling) in order to analyse the soil structure interaction of U-lock structures embedded in layered soil. The program uses a concept according to which the lock is modelled with finite elements and the soil with boundary elements. The advantage of this procedure is the reduced total number of elements and degrees of freedom making the analysis quite quick.

Truman et al. (Fehl, B., Ferhi, A., Truman, K., Petruska, D. 1991) used the finite element code, ABAQUS, to perform an incremental construction analysis including thermal loads on a pile-founded mass concrete lock and dam structure. Their nonlinear incremental analysis includes the effects of creep, shrinkage, and aging Young's modulus in order to assess the vulnerability of mass concrete structures to thermal stresses and possible cracking during the construction process.

Ebeling et al. (Ebeling, R. M., Mosher, R. L., Peters, J. F. 1997) performed a nonlinear deformation analysis on a U-frame lock embedded in soil in order to investigate the soil-structure interaction of structures embedded in reinforced soils.

Xu and Spyrakos (Xu und Spyrakos 2001) analysed the soil-structure and water-structure interaction using the hybrid BEM-FEM method. In their study the water contained lock is embedded in a layered soil resisting on bedrock. The lock structure is modelled with the FEM whereas the fluid and the soil are modelled with the BEM.

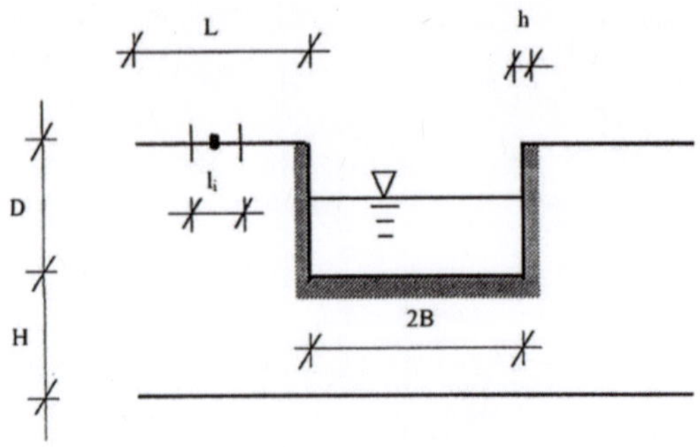

Fig. 3-35 The geometry of the boundary value problem according to (Xu und Spyrakos 2001).

Soares and Mansur (Soares und Mansur 2006) used also the coupled FEM-BEM method in order to investigate the dynamic response of fluid-structure-soil systems such as dams and navigation locks.

Pani and Bhattacharyya (Pani und Bhattacharyya 2007) researched on the fluid-structure interaction of navigation lock gates using the finite element method. They showed the influence that water compressibility and gates flexibility can have on hydrodynamic pressures.

Bouaanani et al. (Bouaanani, N., Goulmot, D., Miquel, B. 2014) gave frequency and time domain solutions for the seismic-fluid-structure interaction of navigation locks and compared their proposed solutions with the results of finite element analyses.

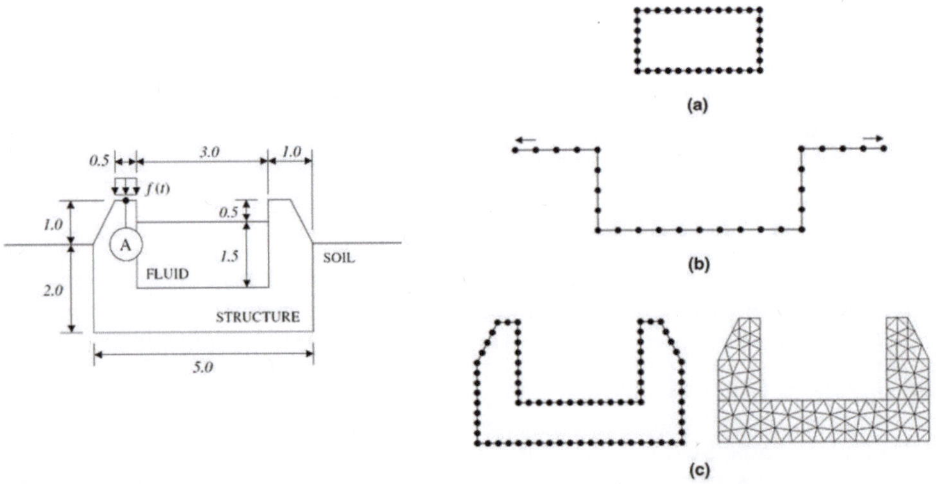

Fig. 3-36 The boundary elements for the soil and water and the finite elements for the lock structure (according to (Mansur, W. J., Soares Jr., D. 2006)).

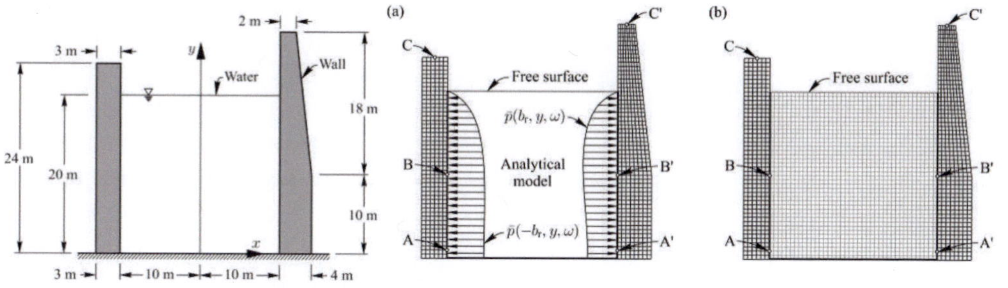

Fig. 3-37 The geometry of the boundary value problem (according to (Bouaanani, N., Goulmot, D., Miquel, B. 2014, 2014)).

Buldgen and Buldgen et al. (Buldgen 2015; Bela, A., Buldgen, L., Philippe, R. 2015) researched on the fluid- structure interaction of flexible navigation lock gates, using both an analytical approach and the FEM. Their results corroborate former researches. They also concluded that the added mass method is non-conservative and should be avoided for flexible structures.

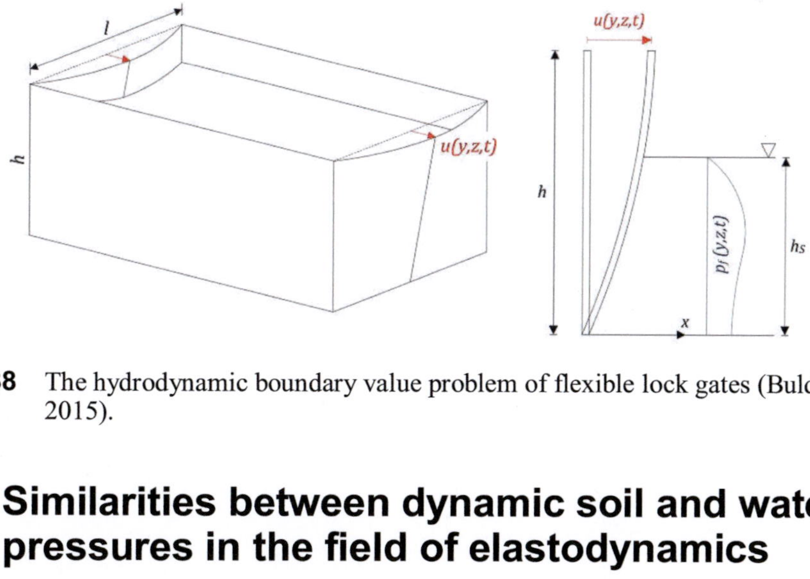

Fig. 3-38 The hydrodynamic boundary value problem of flexible lock gates (Buldgen 2015).

3.8 Similarities between dynamic soil and water pressures in the field of elastodynamics

In both fields of analyses (hydrodynamic pressures on walls or dams and dynamic soil pressures on walls) there are many similarities to observe. The first observation arises for the case of a closed water basin, where the reservoir does not extend to infinity but is bounded by a wall. This is the case investigated by (Brahtz, H. A., Heilbron, C. H. 1933; Sundquist, K. J., Werner, P. W. 1949; Bustamante, J. I., Flores, A., E. Herrera, Rosenblueth I. 1963). The same boundary problem was investigated for the case of two rigid walls containing soil by the (Wood 1973; Prakash S. 2001; Parikh, V. H., Veletsos, A. S., Younan, A. H. 1995; Papazafeiropoulos, G., Psarropoulos, P. N. 2010; Vrettos et al. 2016; Beskos, D. E., Papagiannopoulos, G. A., Triantafyllidis, T. 2015) and many others. In both cases it can be observed that the dynamic pressures on the wall (either water or soil 9pressures) depend on the *L/H* ratio of the boundary problem, where *L* stands for the length of the reservoir or the containing soil and more specifically both pressures decrease with a decreasing *L/H* ratio. This can be better seen for the soil in the case of a constant shear modulus of the soil (homogeneous soil). The maximum responses for the two boundary problems (the first with water, the latter with soil) do not happen for the same excitation frequency because of the different eigenfrequencies of each system.

Another observation arises for the case of the rigid wall able to rotate at his base. For both cases the pressures decrease as the rotational spring stiffness decreases. That is easy to follow because the tilting behaviour of the wall due to the inertia forces causes a smaller relative movement towards the water, something that decreases the water pressures and in the case of soil, the relative movement of the wall away from the retained soil causes the pressures to change from an "at rest" condition to an "active" condition.

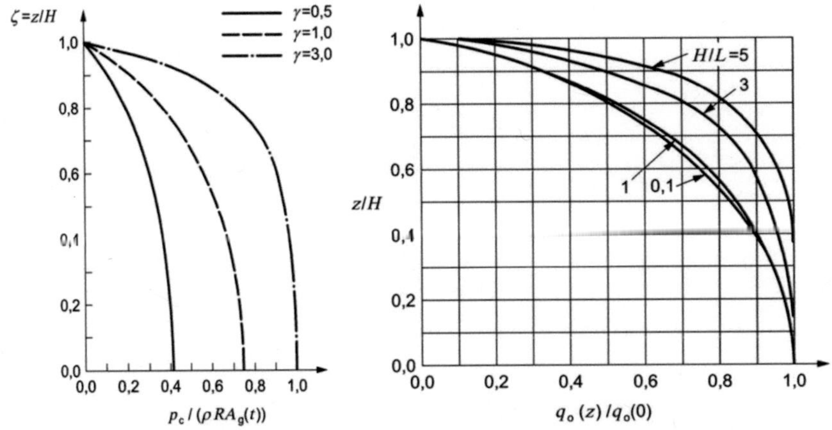

Fig. 3-39 Left: the impulsive water pressure distribution for different H/R ratios. Right: the normalized water pressure distribution for different H/R ratios (according to (EN 1998-4:2006 Eurocode 8)).

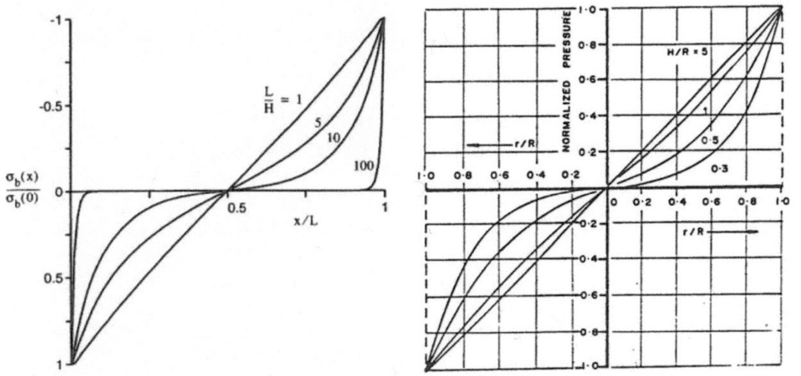

Fig. 3-40 Left: the soil pressure distribution at the base of the boundary problem for different L/H ratios (Parikh, V. H., Veletsos, A. S., Younan, A. H. 1995). Right: the water pressure distribution at the base of the boundary problem for different H/R ratios (Davidson et al. 1986).

Moreover, the formel for the calculation of the natural frequencies of the systems (two-walls-soil or two-walls-water system) are almost identical. For a water domain with depth H and bounded by two walls in distance L, the natural frequencies are calculated with the formula (Bustamante, J. I., Flores, A., E. Herrera, Rosenblueth I. 1963):

$$\omega_{m,n} = \frac{\pi\, c}{2\, H}\sqrt{m^2 + \frac{(2n-1)^2}{4}\left(\frac{L}{H}\right)^2} = \omega_\infty\sqrt{m^2 + \frac{(2n-1)^2}{4}\left(\frac{L}{H}\right)^2} \qquad \textbf{(3-36)}$$

And for a soil domain with depth H bounded by two walls in distance L, the natural frequencies are calculated with the formula (Parikh, V. H., Veletsos, A. S., Younan, A. H. 1995):

$$\omega_{m,n} = \frac{\pi V_s}{2H}\sqrt{(2n-1)^2 + 4m^2\frac{2}{1-v}\left(\frac{H}{L}\right)^2} = \omega_\infty\sqrt{m^2\frac{2}{1-v} + \frac{(2n-1)^2}{4}\left(\frac{L}{H}\right)^2}$$

(3-37)

3.9 Review of the literature research

The literature research on dynamic soil pressures shows that there are two main groups. The first one has shown with experimental results ad observations that the limit state methods or more specifically the M-O method is adequate for the design of retaining structures. The second group has shown experimentally, analytically and with observations that the M-O method has either shortcomings or that it is inadequate for the design of retaining structures. However, both methods, limit state and elastic solutions, do not, or only partially, take into consideration the soil-structure interaction and the ability of the structure to move elastically, slide and/or tilt, which is the decisive factor for the estimations of the seismic loading of it (limit state solutions presume an adequate movement of the retaining wall).

The M-O method has some important shortcomings (Hadjian, A. H., Nazarian, H. N. 1979):

- The amplification of the ground motion, as well as the soil-structure interaction, is not considered
- Wall inertia forces are neglected
- It is based on rigid body motions
- It does not converge for large values of acceleration.

The same authors pointed out that the elastic solutions hold for small amplitudes of lateral motion. In order to hold the elastic solutions, the behaviour of the wall-soil system must remain in an elastic field and the acceleration must be smaller than the yield acceleration, which causes the wall to slip (Fig. 3-41.). Another important factor beyond the amplitude of the acceleration, which concerns the permanent displacements of the wall, is the frequency content of the seismic excitation.

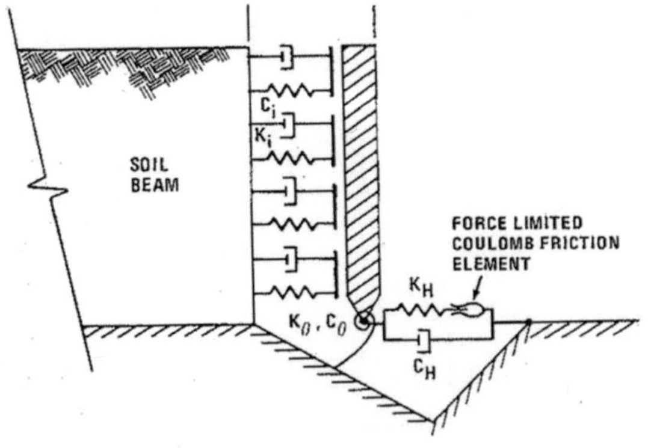

Fig. 3-41 The soil-structure interaction problem regarding the soil pressures (according to (Hadjian, A. H., Nazarian, H. N. 1979)).

It becomes quickly clear that the solution depends not only on the boundary problem but also on the intensity of the seismic forces. If the soil pressures together with the inertia forces of the wall are higher than the sliding resistance of the system, the elastic solutions do not longer hold and limit state solutions are effective. Here it must be mentioned that the Wood elastic solution gives about 2.5-3 times higher soil pressures than the M-O method for the same ground acceleration (Li 1999; Veletsos, A. S., Younan, A. H. 1994a). Another observation is that the M-O equation for e.g. the active pressures does not converge as the ground acceleration increases (the coefficient K_{ae} tends to infinity). From this point of view someone could say that for a non-constrained system, where the sliding of the retaining structure is possible, the Wood elastic solution (or the solutions of Veletsos and Younan or/and Jung and

Bobet) holds for small accelerations and after yield acceleration, when sliding occurs, the limit state solutions become effective. This fact can also be seen from the results of the elastic solutions: when the flexibility of the wall or its base increases and tension forces appear at the soil. Veletsos and Younan pointed out that these tension forces should be firstly superimposed with the pressures due to gravity and if the tension pressures remain, then a wall-soil separation will occur.

3.9.1 Field of application of Veletsos' and Younan's theory

In order to see the field of application and to understand what the different values of the relative flexibilities between the wall and the retained soil mean in practice, some graphs are shown here. The parameters given in the following plots refer to the geometry of the two following systems; a single cantilever wall and a U-shape lock structure.

3.9.1.1 Single wall

A gravity or cantilever retaining chamber wall resting on elastic soil is considered. The wall is able to rotate at its base and to move elastically during a seismic motion. If the motion is severe and the acceleration exceeds the yield acceleration, the wall will slide. The equivalent system at the right of the Figure 3-42 is considered. The wall's flexural rigidity *EI*, the wall height *H*, the width of the wall base *B* (here equal to the section's height *h* but referred to separately in order to distinguish between the contribution of *h* to the flexural rigidity and of *B* to the rotational ability of the wall) and the shear modulus of the retained soil are the parameters which give different values of d_w and d_θ. For simplicity reasons the foundation soil is the same as the retained soil, i.e. $v_s = v_b$.

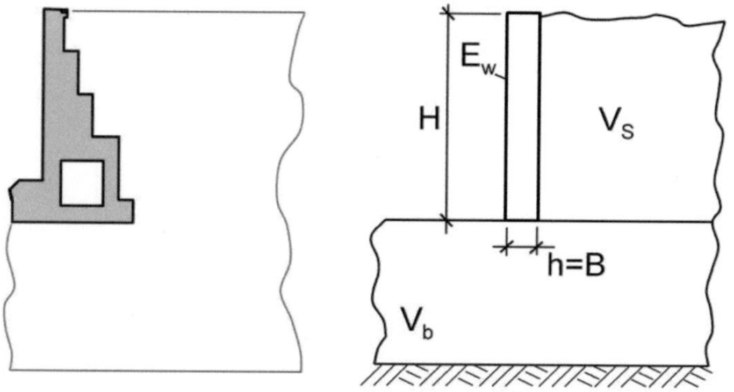

Fig. 3-42 Schematic representation of a single chamber wall.

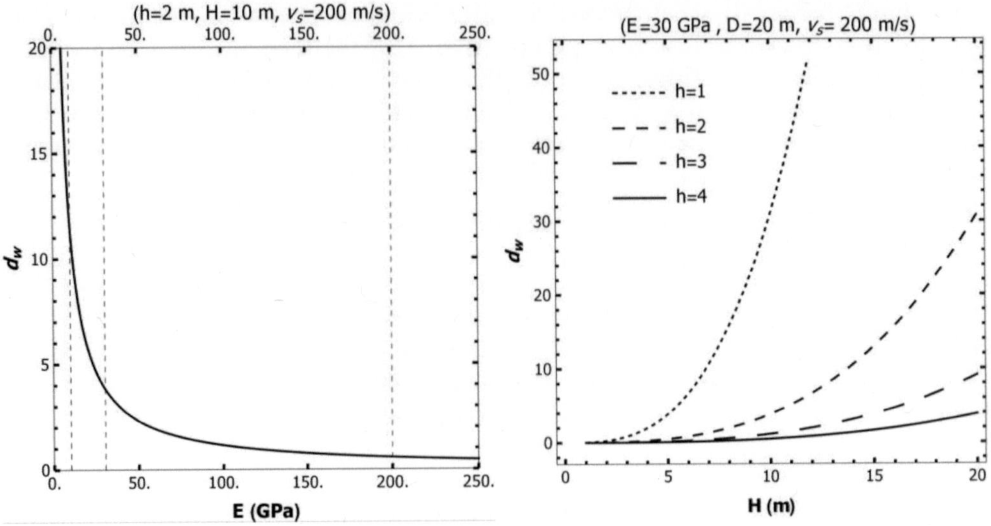

Fig. 3-43 Left: Influence of the wall's modulus of elasticity on the parameter d_w (the vertical lines correspond - from the left to the right - to wood, concrete and steel).
Right: Influence of the wall's height H on the parameter d_w.

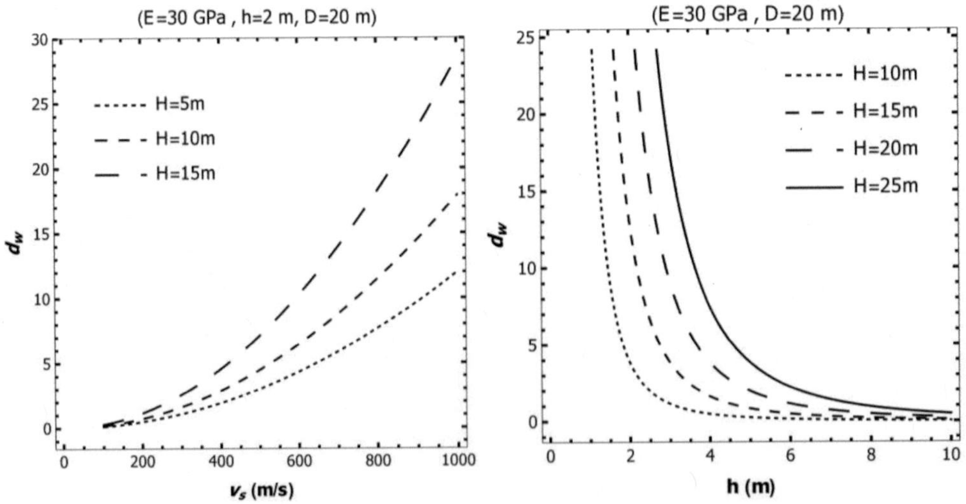

Fig. 3-44 Left: Influence of the shear wave velocity of the backfill on the parameter d_w.
Right: Influence of the section's height h (wall width) on the parameter d_w for different wall heights H.

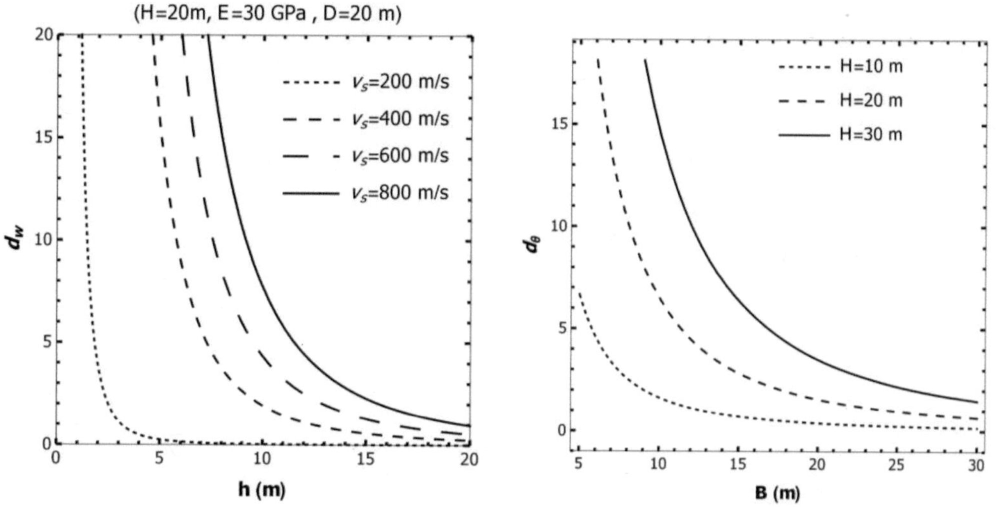

Fig. 3-45 Left: Influence of the section's height h (wall width) on the parameter d_w for different shear wave velocities of the backfill. Right: Influence of the wall width B on the parameter d_θ for different wall heights H.

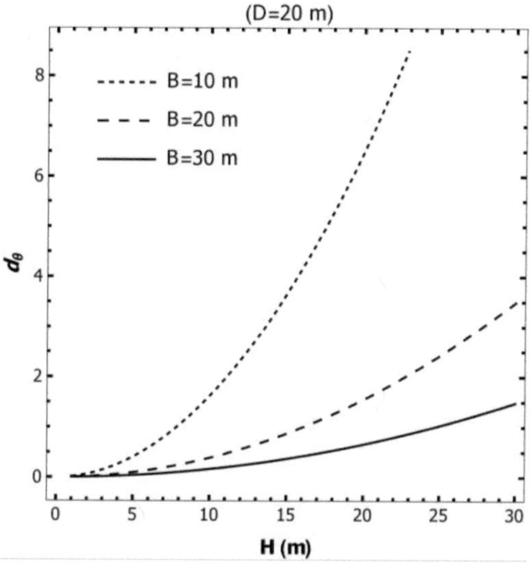

Fig. 3-46 Influence of the wall height H on the parameter d_θ for different wall widths B.

3.9.1.2 U-shape lock structure

A U-section frame chamber embedded in elastic soil is considered. The lock is able to rotate and to move elastically during a seismic motion. The equivalent system at the right of the Figure 3-47 is considered. The lock's height *H* and the width of the lock's base *B* are the parameters which give different values of d_w and d_θ. For simplicity reasons the foundation soil is the same as the retained soil, i.e. v_s=v_b and the distance *D* to the bedrock is taken to be equal to 20 m.

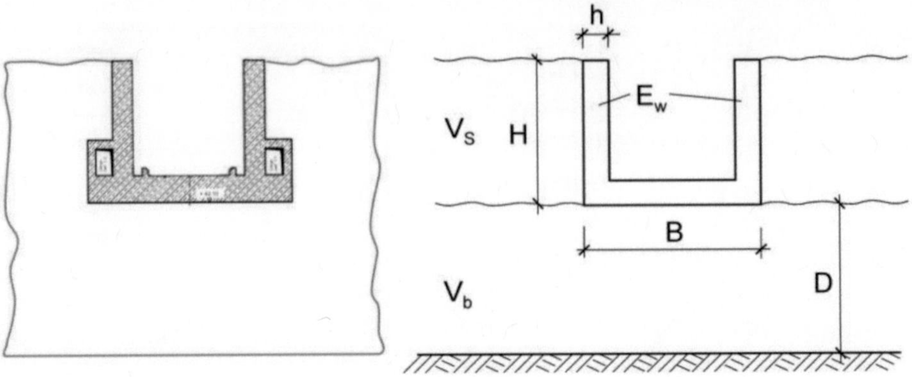

Fig. 3-47 Schematic representation of a U-section embedded lock.

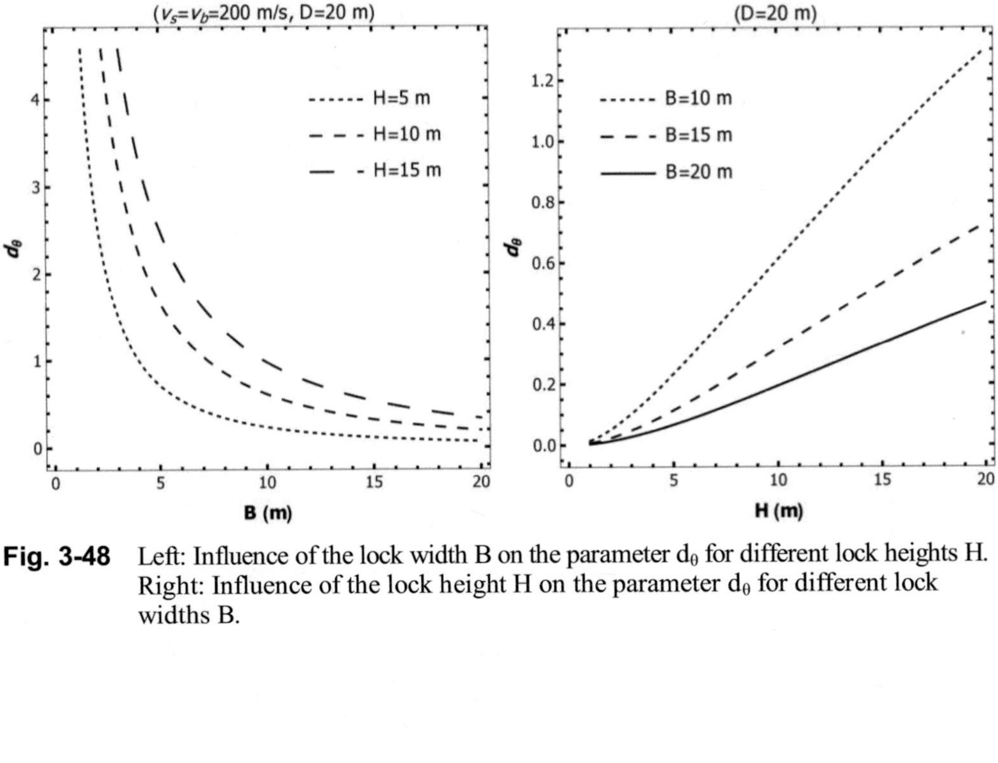

Fig. 3-48 Left: Influence of the lock width B on the parameter d_θ for different lock heights H. Right: Influence of the lock height H on the parameter d_θ for different lock widths B.

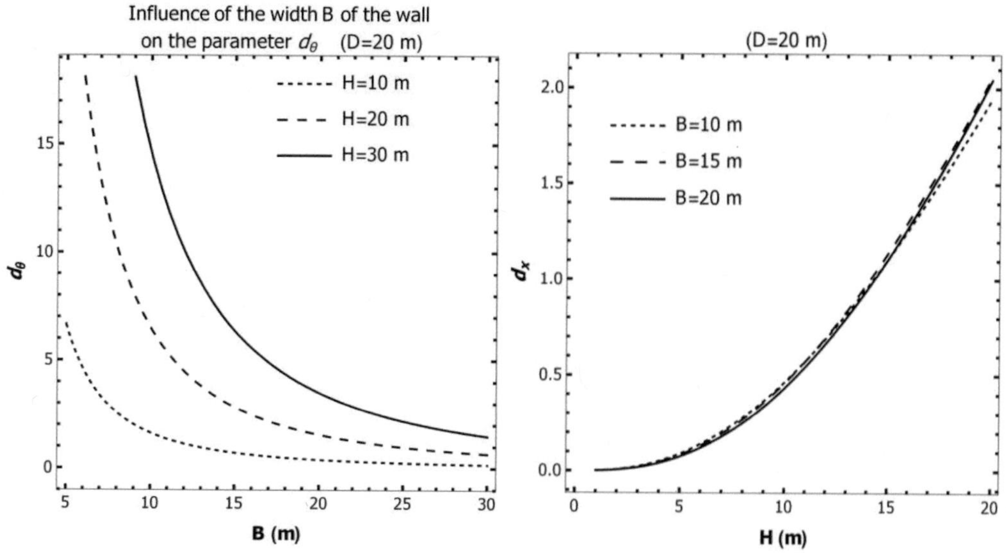

Fig. 3-49 Left: Influence of the lock width B on the parameter d_θ for different lock heights H. Right: Influence of the lock height H on the parameter d_x for different lock widths B.

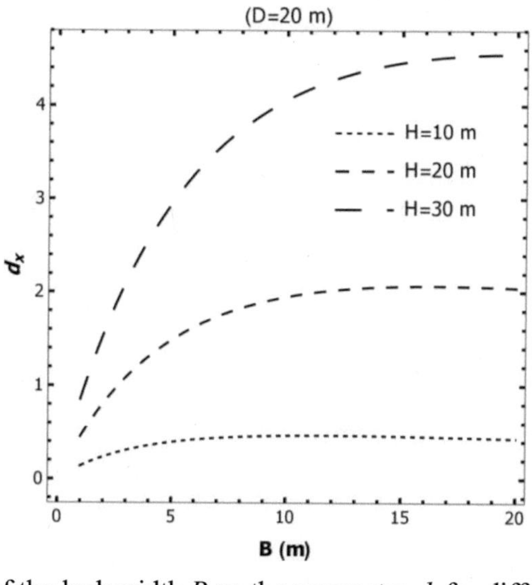

Fig. 3-50 Influence of the lock width B on the parameter d_x for different lock heights H.

3.9.1.3 Natural frequencies of free standing beams based on elastic foundation

In their investigation Veletsos and Younan considered separately the flexural vibration of the retaining wall and the rotation at its base. Due to the linearity of the problem they added the two types of vibration (flexural vibration and rigid body rotation at the base). In order to better understand the results of the following chapter, an estimation of the natural frequencies of a free standing beam based on an elastic foundation able to rotate and move horizontally within the field of elasticity (no sliding occurs) is necessary.

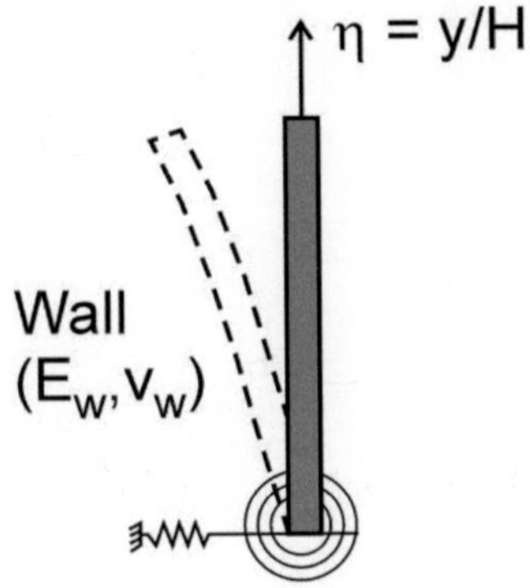

Fig. 3-51 Mathematical model investigated here.

The retaining wall is assumed to be infinitely long and plane strain conditions apply, so the assumption has been made that the wall behaves as a beam with depth equal to unity. The gravitational force of the self-weight of the wall is neglected (there is no normal force due to gravitational force) and the elastic foundation is idealized with a rotational and a translational spring. The vertical spring is neglected, as it is expected that a vibration in the direction of the beam's axis will not affect much the interaction of a free standing beam with a retaining medium (soil or water).

The natural frequency of vibration is given (Karnovskiĭ und Lebed 2001) by:

$$\omega = \frac{\lambda^2}{H^2}\sqrt{\frac{EI}{m}}, \quad \mathrm{m} = \rho\mathrm{A} \tag{3-38}$$

Where λ is a root of the frequency equation. Considering a beam with one end free and the other end constrained elastically by a rotational and translational spring, the frequency is given by (Karnovskiĭ und Lebed 2001):

$$k_x^{*2} + k_x^* \frac{\lambda \frac{k_x^*}{k_r^*}(\sinh\lambda\cos\lambda - \cosh\lambda\sin\lambda) - \lambda^3(\sin\lambda\cosh\lambda + \cos\lambda\sinh\lambda)}{1 + \cosh\lambda\cos\lambda} \tag{3-39}$$

Where the non-dimensional parameters used in the equation are:

$$k_x^* = \frac{k_x H^3}{EI}, \quad k_r^* = \frac{k_r H}{EI} \tag{3-40}$$

For the values of d_w, d_θ and d_x also used in the next chapter the frequencies of an elastically constrained beam are calculated. The values of the spring constants are calculated in the next chapter for the relative stiffness between the retaining wall and the retained soil. Here the non-dimensional parameters are defined in another way. The relations of the non-dimensional parameters defined here with the relative flexibilities as presented by Veletsos et al. are:

$$k_x^* = \frac{\mathrm{G}\, H^3}{EI\, d_x}, \quad k_r^* = \frac{\mathrm{G}\, H^3}{EI\, d_\theta} \tag{3-41}$$

It is expected that for a stiff beam with soft springs the rigid body vibration ruled by the springs will dominate whereas for a flexible beam with hard springs the flexural vibration will dominate. In order to be able to control the solution, the boundary values for the extreme cases of the analytical solution are used. For example, extreme cases are these where the spring constants take either the value 0 (zero) or ∞ (infinite). For these cases the general solution simplifies to another analytical solution for a beam fixed at its one end (k_x=k_r=∞) or for a beam free at its end (k_x=k_r=0) or for intermediate cases.

The next Table shows that the higher the values of E_w, R_θ and K_x, the stiffer are the wall and the springs. The opposite holds for the values of k_r and k_{tr}, where lower values represent stiffer behaviour.

The values of k_r and k_{tr} result from the corresponding values of the EI, R_θ and K_x using the formulas above, assuming a wall section height of 0.2m ($I=6.67\times10^{-4}m^4$) and a wall height of 8 m.

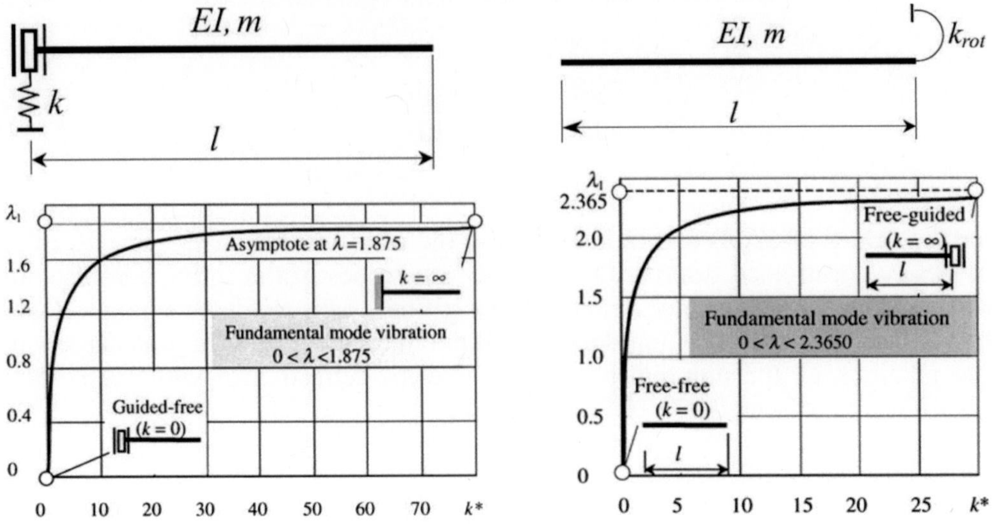

Fig. 3-52 Special cases for the boundary conditions investigated here (Karnovskiĭ und Lebed 2001).

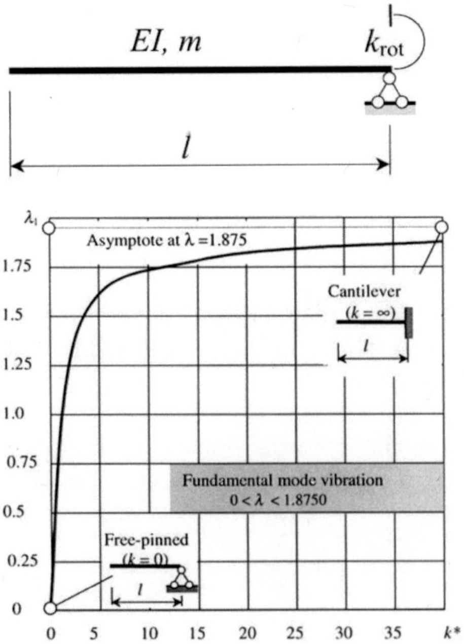

Fig. 3-53 Special cases for the boundary conditions investigated here (Karnovskiĭ und Lebed 2001).

Table 7 Values of R_θ and K_x for different d_w values using the closed form formula

			d_θ	d_θ	d_θ	d_θ	d_x	d_x	d_x	d_x	d_x
			0.01	0.5	1	5	0.001	0.1	0.5	1	5
	H=8m		R_θ	R_θ	R_θ	R_θ	K_x	K_x	K_x	K_x	K_x
	$I=6.66\times10^{-4}\ m^4$		1.15E+11	2.30E+09	2.88E+08	2.30E+08	1.80E+10	1.80E+08	3.60E+07	1.80E+07	3.60E+06
d_w	E_w	EI	k_r*	k_r*	k_r*	k_r*	k_x*	k_x*	k_x*	k_x*	k_x*
0.01	1.33E+15	8.85E+11	1.04	0.02	0.003	0.002	10.42	0.10	0.02	0.01	0.002
0.1	1.3271E+14	8.85E+10	10.42	0.21	0.03	0.02	104.17	1.04	0.21	0.10	0.02
1	1.3271E+13	8.85E+09	104.17	2.08	0.26	0.21	1041.67	10.42	2.08	1.04	0.21
5	2.6542E+12	1.77E+09	520.83	10.42	1.30	1.04	5208.33	52.08	10.42	5.21	1.04
10	1.3271E+12	8.85E+08	1041.67	20.83	2.60	2.08	10416.67	104.17	20.83	10.42	2.08
20	6.6355E+11	4.42E+08	2083.33	41.67	5.21	4.17	20833.33	208.33	41.67	20.83	4.17
30	4.4237E+11	2.95E+08	3125.00	62.50	7.81	6.25	31250.00	312.50	62.50	31.25	6.25
40	3.3178E+11	2.21E+08	4166.67	83.33	10.42	8.33	41666.67	416.67	83.33	41.67	8.33

Table 8 Values of the first natural frequency and mode shape for different d_w, d_θ and d_x values.

	$d_\theta = 0.01$	$d_\theta = 0.5$	$d_\theta = 1$	$d_\theta = 5$	$d_\theta = 0.01$				$d_\theta = 5$
	$d_x = 0.001$				$dx=0.1$	$dx=0.5$	$dx=1$	$dx=5$	$d_x = 5$
	λ_1	λ_1	λ_1	λ_1	λ_1	λ_1	λ_1	λ_1	λ_1
d_w	ω_1	ω_1	ω_1	ω_1	ω_1	ω_1	ω_1	ω_1	ω_1
0.01	1.196	0.494	0.308	0.278	0.558	0.3755	0.316	0.2114	0.1992
	939.8	160.3	62.32	50.89	204.8	92.69	65.64	29.39	26.11
0.1	1.699	0.869	0.546	0.4943	0.981	0.666	0.561	0.376	0.354
	600	157	62.12	50.8	200	92.27	65.49	29.37	26.1
1	1.854	1.431	0.957	0.87	1.569	1.171	0.987	0.667	0.628
	226	134.54	60.25	49.75	161.98	90.14	64.03	29.24	25.99
5	1.8701	1.722	1.314	1.247	1.804	1.576	1.415	0.987	0.932
	102.9	87.2	50.74	45.77	95.71	72.98	58.85	28.66	25.51
10	1.873	1.794	1.481	1.4312	1.839	1.714	1.576	1.17	1.108
	72.92	51.62	45.62	42.6	70.33	61.04	51.65	28.57	25.49
20	1.874	1.8326	1.6234	1.5833	1.8572	1.791	1.714	1.358	1.287
	51.62	4934	38.72	36.83	50.67	47.14	43.18	27.11	24.34
30	1.8744	1.8461	1.6928	1.648	1.8632	1.8185	1.7629	1.4652	1.3897
	42.16	40.91	34.39	32.6	41.66	39.69	37.3	25.76	23.18
40	1.8746	1.8531	1.7227	1.6828	1.8661	1.8302	1.7913	1.5248	1.4536
	36.5	35.67	30.83	29.77	36.17	34.86	33.33	24.15	21.95

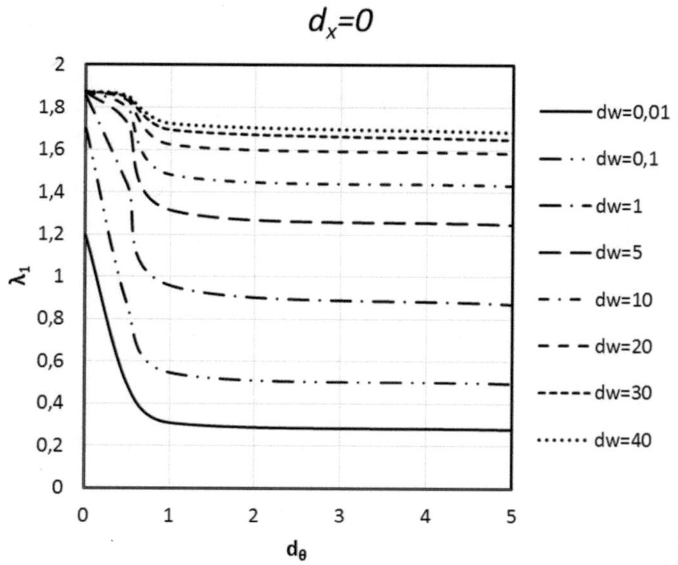

Fig. 3-54 Values of λ_1 for the relative wall flexibilities in Chapter 5 (d_x=0).

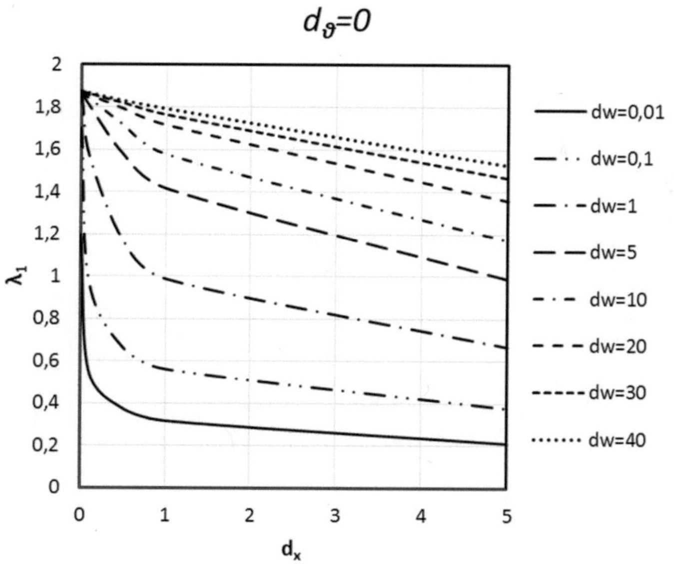

Fig. 3-55 Values of λ_1 for the relative wall flexibilities in Chapter 5 (d_θ=0).

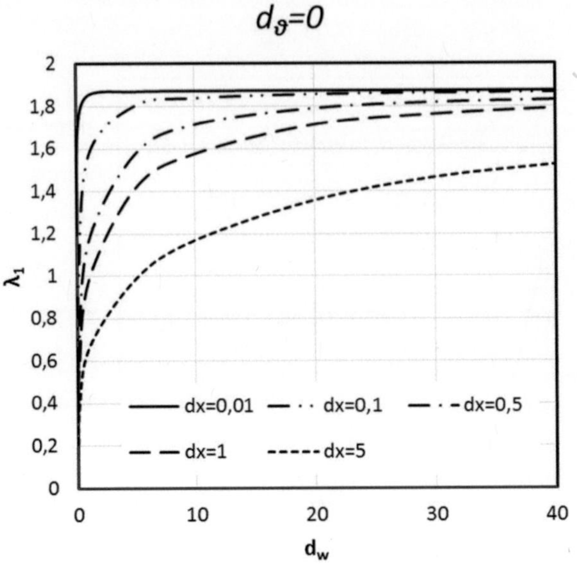

Fig. 3-56 Values of λ_1 for the relative wall flexibilities in Chapter 5 (d_θ=0).

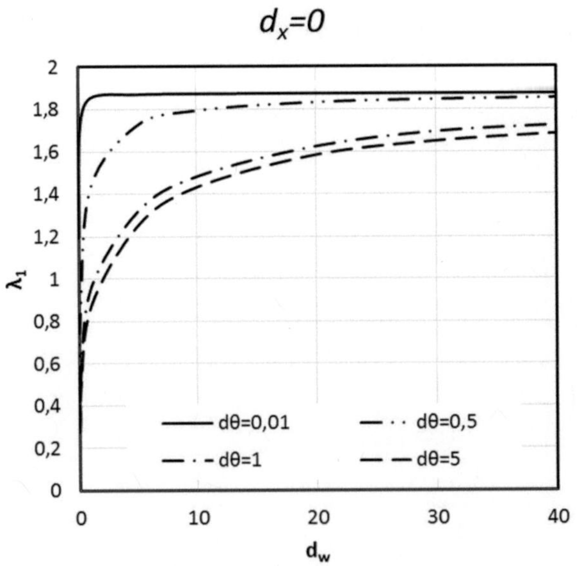

Fig. 3-57 Values of λ_1 for the relative wall flexibilities in Chapter 5 (d_x=0).

Chapter 4
Fluid-structure interaction of navigation locks

Summary

The chapter begins with the presentation of the numerical modelling techniques of water. Followingly, the thories presented in chapter 2 (Westergaard, Zangar, Werner and Sundquist, Bustamante and Flores) are validated numerically and the finite element models used at this chapter are validated in turn. The Finite Element Program used is Simulia Abaqus Standard (ABAQUS 2012). After the verification of the analytical or experimental studies,which most of them handle rigid systems, the influence of the wall and foundation flexibility on the hydrodynamic pressures is investigated. Two main models are used for this parametric study; one with a semi-infinite reservoir (single-wall-system) and one with a finite reservoir (pair-of-walls-system). The water domain is idealized with acoustic elements available in Abaqus, which are the most numerically efficient method for the calculation of the hydrodynamic pressures. The results are presented for both models in form of tables (Annex B) and diagrams.

4.1 Finite element model for Fluid-Structure Interaction (FSI)

4.1.1 Modelling the water with the FEM

There are several ways for modelling the water with the finite element method. These are:

- Added masses
- Continuum elements (Lagrangian method)
- Acoustic elements (Lagrangian method)
- Smooth Particle Hydrodynamics (SPH-Lagrangian method)
- Eulerian method
- Computational Fluid Dynamics

Depending on the field of investigation each method is more or less appropriate. The main criteria for the appropriateness of each method are mainly the computational time, the convergence of the solution and the accuracy of the method. For problems of earthquake engineering, where the conditions mentioned in §1.1 prevail, the first three methods are more appropriate as the water effect is restricted to the pressures acting on the structure and turbulence and other nonlinear effects are neglected or do not take place. Between these three ways of modelling the water the method of added masses may be the easiest to apply with most computer programs, the latter two make the modelling easiest as no masses have to be computed and the water is modelled through finite elements. For these two, the continuum and the acoustic elements methods, it has been shown elsewhere that the latter is more sufficient as its convergence is quicker (Dong J. et al. 2011).

Added mass concept

The added mass method is based on Westergaard's formula for water pressures. The added masses, which cause the same inertia effect with the real hydrodynamic pressures, are computed with Equation 2.4. Although this method is very simple and can be applied with almost every computer program, it has some drawbacks and difficulties; care must be taken that the masses are acting only perpendicularly on the surface as the water causes no tractions, and sometimes the calculation of the added masses as a function of the area and the coordinates of the finite elements nodes is a time-consuming procedure. Moreover, a simplified added mass procedure does not take the influence of the excitation frequency into consideration. Darbre (Darbre 1998) has proposed a sophisticated two-parameter added mass model to account for these effects.

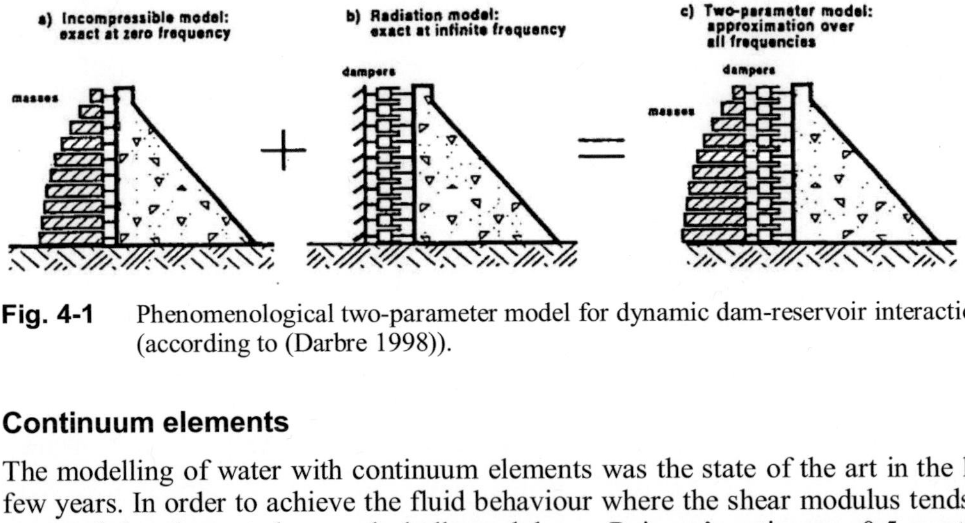

Fig. 4-1 Phenomenological two-parameter model for dynamic dam-reservoir interaction (according to (Darbre 1998)).

Continuum elements

The modelling of water with continuum elements was the state of the art in the last few years. In order to achieve the fluid behaviour where the shear modulus tends to zero and the elements have only bulk modulus, a Poisson's ratio near 0.5 must be given. As the value of 0.5 is not accepted by the FE programs because it leads to numerical instabilities, a value of 0.499 can be given instead (Wilson 2000). The table below shows the change in the bulk modulus of water and the shear modulus as Poisson's ratio tends to 0.5.

Table 9 Change of Young and shear modulus with changing Poisson's ratio

Bulk modulus K (MPa)	Poisson's ratio v	Young modulus E (MPa)	Shear modulus G (MPa)
2.2×10^3	0.4	1320.00	471.43
2.2×10^3	0.49	132.00	44.30
2.2×10^3	0.499	13.20	4.40
2.2×10^3	0.4999	1.32	0.44

With such a shear modulus and Young modulus the continuum elements become very soft and sensible to distortions (hourglassing) (Wilson, E. L., Khalvati, M. 1983). The hourglassing can be treated by adding artificial hourglass stiffness but generally leads to numerical problems and increases the computational time.

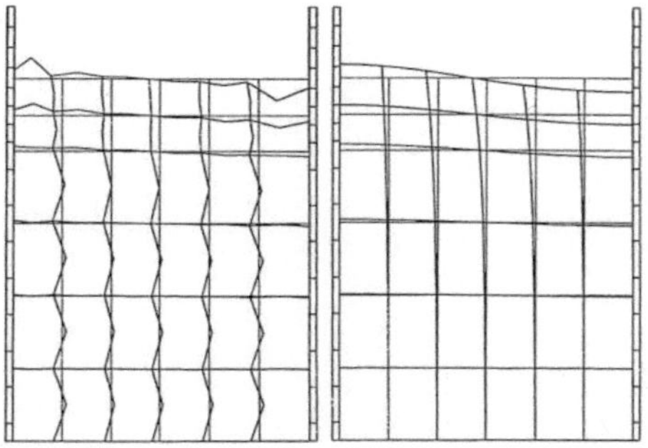

Fig. 4-2 Sloshing of a fluid tank without hourglass control (left) and with hourglass control (right) (according to (Stempniewski 1990)).

Acoustic elements

The acoustic elements are a type of continuum elements that have only pore pressure degrees of freedom. The equilibrium equation for small motions of a compressible, adiabatic fluid with velocity-dependent momentum is given (ABAQUS 2012):

$$\frac{\partial p}{\partial x} + \gamma(x, \theta_i)\dot{u}^f + \rho_f(x, \theta_i)\ddot{u}^f = 0 \quad \textbf{(4-1)}$$

where p is the excess pressure in the fluid (the pressure in excess of any static pressure); x is the spatial position of the fluid particle; $\dot{u}^f$ is the fluid particle velocity; $\ddot{u}^f$ is the fluid particle acceleration; ρ_f is the density of the fluid; γ is the "volumetric drag" (force per unit volume per velocity); θ and i are independent field variables such as temperature, humidity of air, or salinity of water on which ρ_f and γ may depend. The constitutive behaviour of the fluid is assumed to be inviscid, linear, and compressible, so

$$p = -K_f(x, \theta_i)\frac{\partial}{\partial x}u^f \tag{4-2}$$

The equation of motion for the fluid in terms of pressure is also given:

$$\frac{1}{K_f}\ddot{p} + \frac{\gamma}{\rho_f K_f}\dot{p} - \frac{\partial}{\partial x}\left(\frac{1}{\rho_f}\frac{\partial p}{\partial x}\right) = 0 \tag{4-3}$$

For this differential equation and with the appropriate boundary conditions the general solution can be found.

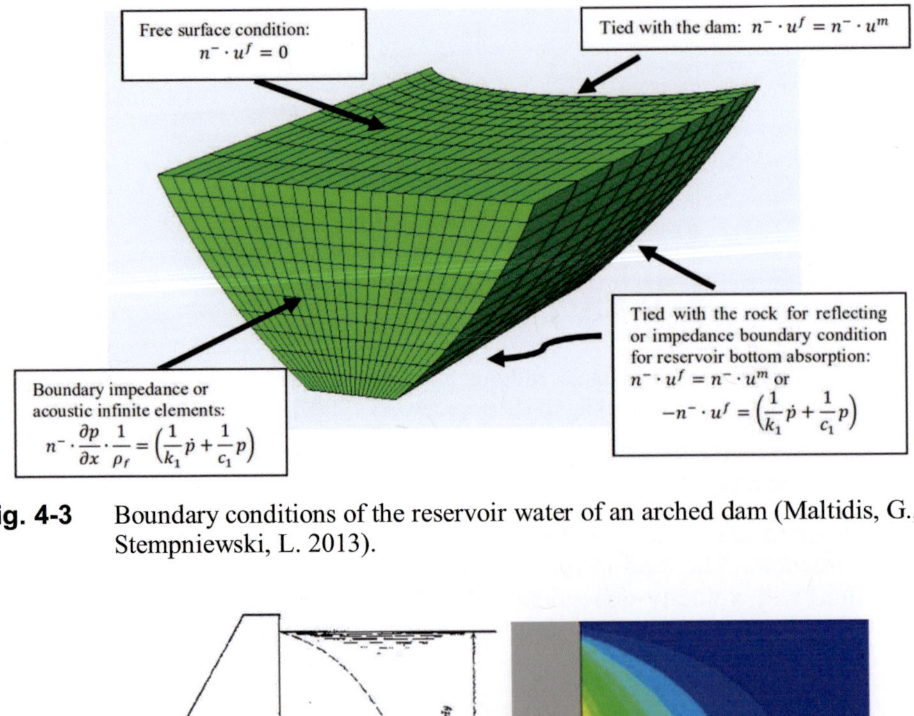

Fig. 4-3 Boundary conditions of the reservoir water of an arched dam (Maltidis, G., Stempniewski, L. 2013).

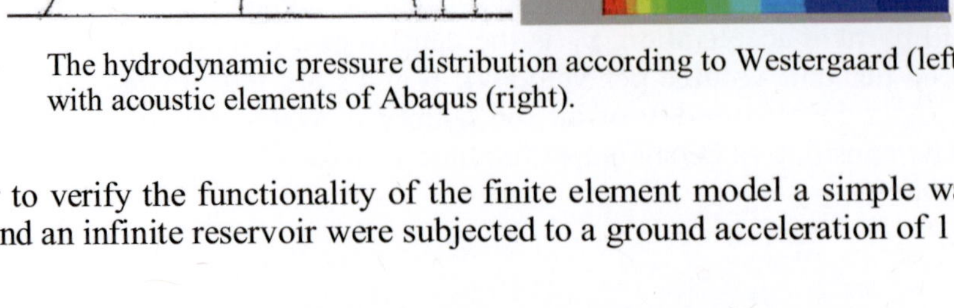

Fig. 4-4 The hydrodynamic pressure distribution according to Westergaard (left) and with acoustic elements of Abaqus (right).

In order to verify the functionality of the finite element model a simple wall of 8m height and an infinite reservoir were subjected to a ground acceleration of 1 m/s^2.

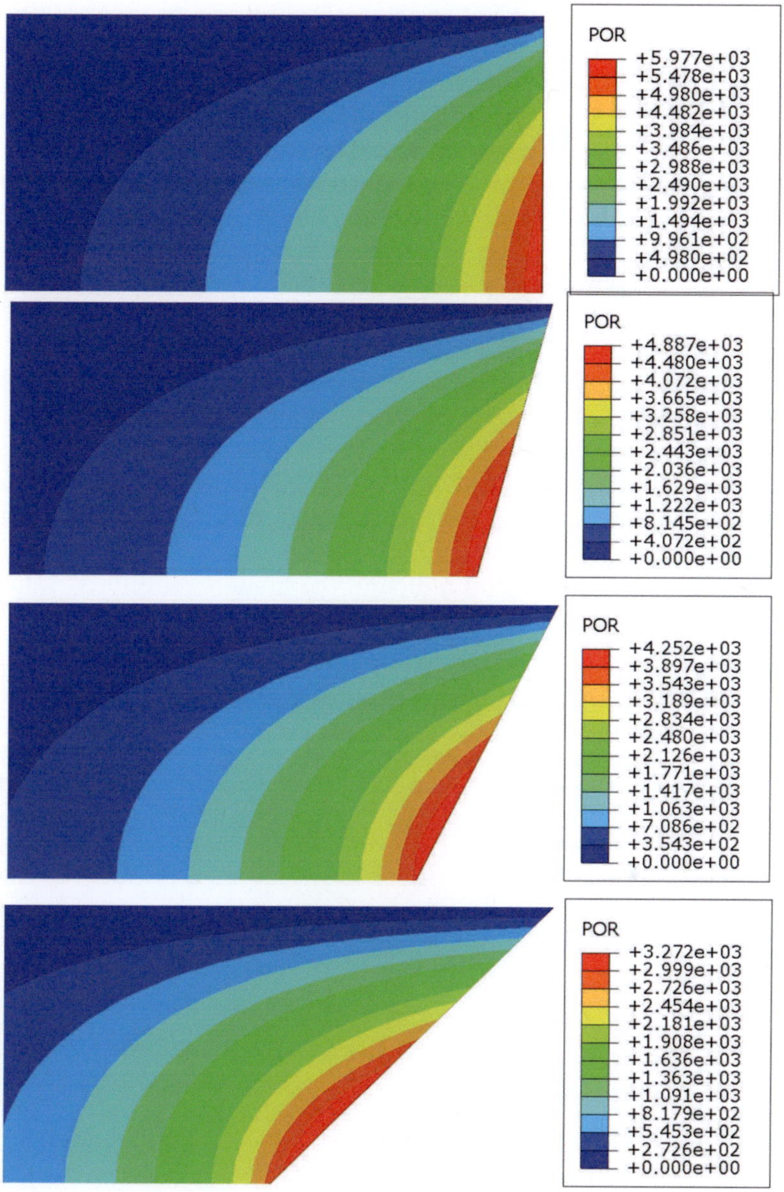

Fig. 4-5 The hydrodynamic pressure distribution and the maximum pressure values for different wall inclinations (90, 75, 60 and 45 degrees).

The wall is assumed rigid and fixed at its base. The formula of Westergaard provides a maximum hydrodynamic pressure of $0.743\times1 m/s^2\times8m\times10kN/m^3=5.944$ kN/m. The finite element analysis gives a maximum value for an infinite reservoir and rigid fixed wall of 5.977 kN/m, which is almost identical with the theoretical value. The theory of Zangar gives the reduction factors 0.85, 0.71 and 0.55 for an inclined wall with an angle of inclination of 75, 60 and 45 degrees respectively, delivering the values 5.05 kN/m, 4.22 kN/m and 3.27 kN/m. The finite element analyses give the values 4.89kN/m, 4.25 kN/m and 3.27 kN/m for the corresponding angles. The conformity of the values is obvious.

The reduction factors for the finite reservoir provided by (Brahtz, H. A., Heilbron, C. H. 1933; Sundquist, K. J., Werner, P. W. 1949; Newmark, N. M., Rosenblueth, E. 1971) were validated by a finite element model in which the reservoir has a height of 10 m. The walls are assumed to be rigid and fixed at their base and the water as incompressible. The L/H ratios take the values 1, 2, 3, 5 and 10. The following table gives the reduction factors and the theoretical and numerical values. The analysis performed is a time history analysis with sinusoidal excitation and ω=0.628 rad/sec in order to avoid any amplification although the compressibility of water was set $K=2.25\times10^{16}$ N/m^2 (practical incompressible).

Table 10 Effect of finite reservoir on hydrodynamic pressures

L/H	Theoretical value according to literature (kN/m)	Empirical formula (Eq. 2.10)	FE analysis (kN/m)
∞	7.43	-	
10	7.43	7.43	7.07
5	7.43	7.43	7.06
3	7.28	7.43	6.93
2	6.61	6.84	6.43
1	4.98	4.98	4.43

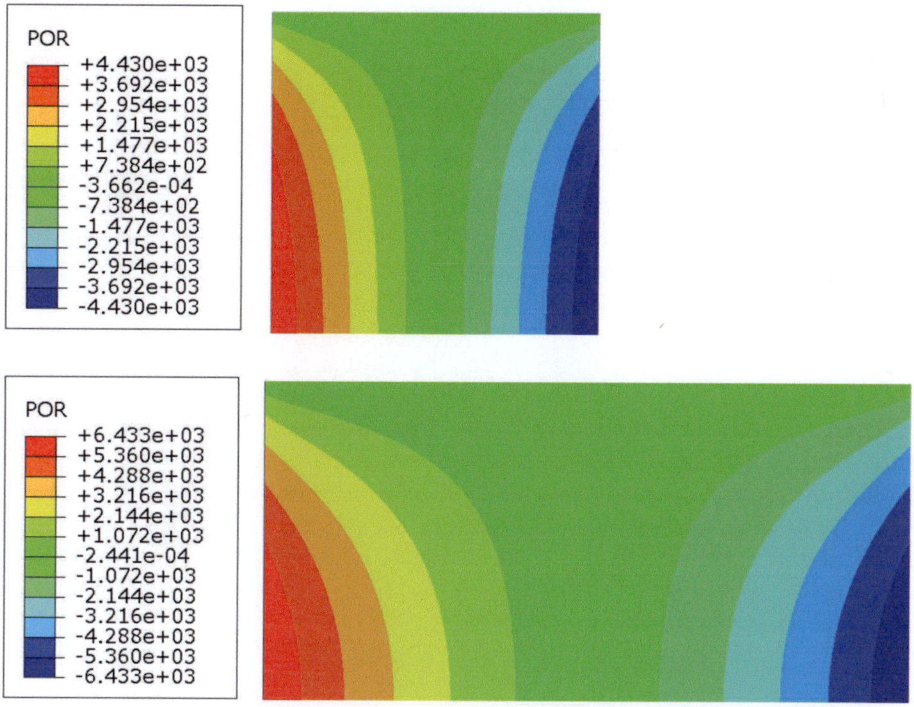

Fig. 4-6 The hydrodynamic pressure distribution and the maximum pressure values for $L/H = 1$ and $L/H = 2$.

The last verification of the eligibility of the acoustic elements for the investigation of the soil-structure interaction is a frequency extraction analysis. The following table gives the theoretical and calculated (with Abaqus) eigenfrequencies of one semi-infinite and two finite reservoirs with an L/H ratio equal to 1 and 10 respectively. The reservoirs have a depth of 8m. As it can be seen, there is a very good agreement between the theoretical and numerical results.

Table 11 Natural Frequencies calculated with Abaqus for reservoir depth of 8 m

	1st (Hz)		2nd (Hz)		3rd (Hz)	
L/H	Theoretical value	Abaqus	Theoretical value	Abaqus	Theoretical value	Abaqus
∞	44.94 (vertical)	46.83 (vertical)	134.81 (vertical)	139.83 (vertical)	224.69 (vertical)	228.40 (vertical)
10	44.94 (vertical)	46.83 (vertical)	45.8 (horizontal)	47.75 (horizontal)	46.7 (horizontal)	50.41 (horizontal)
1	44.94 (vertical)	46.83 (vertical)	100.5 (horizontal)	104.11 (horizontal)	134.8 (horizontal)	139.33 (horizontal)

4.2 Single wall-water system

Three main models were investigated; one in which the contained water extends to infinity; one in which two parallel walls bound the contained water and one in which the water is contained in a U-form section. For all of the models two dimensional and plain strain conditions are assumed. The walls are modelled with beam elements of type B21 (Euler-Bernoulli beam). Although it would be expected that the walls are modelled with continuum elements, this was not the case in this study in order to minimize the computational time (due to the bigger number of the nodes) and avoid a possible shear locking, which can increase the wall's stiffness. By comparing different element types of beam and continuum elements, the difficulty in achieving the precision of the analytical solution can be shown. Here, two beam elements (Euler-Bernoulli and Timoshenko beam elements) and eight continuum elements with different discretization refinement and integration order were compared under a horizontal load of 10 kN. The wall modelled with these elements has a height of 8 m, a width of 0.2 m, a Young modulus of 1 GPa, a Poisson's ratio of 0.2, and it is fixed at its base. The analytical solution gives a deflection of 2.56 m. Both beam element types give a value of 2.561 m and only the element type CPE4 with element dimensions 0.1 x 0.1 m gives a similar result (2.587 m), albeit with three times more nodes.

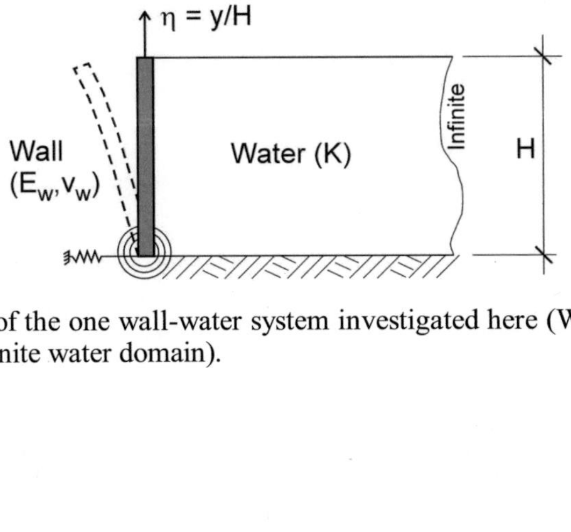

Fig. 4-7 Schema of the one wall-water system investigated here (Westergaard's model – semi-infinite water domain).

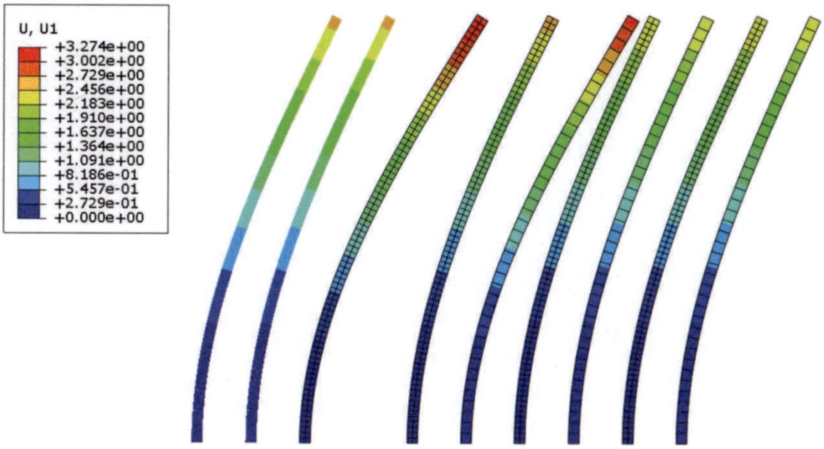

Fig. 4-8 From left to right: B21, B23, CPE4R, CPE4R (missing from the figure), CPE, CPE4, CP8, CP8, CP8R, CP8R.

Table 12 Deflection for different types of elements and discretization

Element type	Number of elements	Number of nodes	Deflection (m)
Analytical	-	-	2.560
Beam B21 (B-E)	40	41	2.561
Beam B23 (T)	40	81	2.561
CPE4R	80	123	3.274
CPE4R	40	82	819.1
CPE4	80	123	2.587
CPE4	40	82	3.072
CPE8	80	123	2.458
CPE8	40	82	2.455
CPE8R	80	123	2.458
CPE8R	40	82	2.455

The transverse dimension of the beam elements equals one meter in order to comply with the plane strain conditions of the water elements. The walls are constrained rotationally with a rotational spring, and a translational spring prevents the wall from moving transversely. The parametric study includes variation of four parameters for the one-wall model and five parameters for the two-wall model. These parameters are:

- Stiffness of the rotational spring (also complex stiffness)
- Stiffness of the translational spring (also complex stiffness)
- Stiffness of the wall
- Distance between the walls for the two-wall model
- Frequency of the excitation.

The water was modelled with acoustic elements available in Abaqus of type AC2D2. The effectiveness of this type of elements has been shown before (§4.1.1) and elsewhere (Dong J. et al. 2011). For the infinite domain of the reservoir an appropriate radiation (non-reflecting) boundary condition was applied (Sommerfeld boundary condition). The water elements are bonded with the beam elements of the wall. The acoustic elements convert the structure's accelerations to fluid pressures. At the free surface of the water zero pressure was given as a boundary condition so that water waves (sloshing) are neglected.

The bulk modulus of the water is taken equal to 2.25 GPa. It is also assumed that both walls are subjected to the same acceleration without phase, as the distances between the chamber walls of a navigation lock are smaller than 50 m and there is practically no phase at their movement. Three dimensionless parameters, following the concept of the next chapter and the notation of Veletsos and Younan for the wall-soil interaction, are used:

$$d_w = \frac{KH^3}{D_w} \quad \textbf{(4-4)}$$

$$d_\theta = \frac{KH^2}{R_\theta} \quad \textbf{(4-5)}$$

$$d_x = \frac{K}{K_x} \quad \textbf{(4-6)}$$

where K is the bulk modulus of the water, H the height of the wall, D_w the stiffness of the wall, and R_θ and K_x are the rotational and translational equivalent springs of the foundation. The dimensionless parameters d_w, d_θ and d_x correspond to the relative wall-water flexibility and foundation-water flexibility. The stiffness D_w of the wall can change by altering either the Young modulus of the wall's material or the section's width. In order to keep the mass of the wall constant and to avoid different dynamic mass forces of the walls, as the results have to be comparable, the Young modulus of the materials has been varied with the formula:

$$E_w = \frac{12(1-v_w^2)}{t_w^3} D_w = \frac{12(1-v_w^2)}{t_w^3} \frac{KH^3}{d_w} \quad \textbf{(4-7)}$$

where v_w is the Poisson's ratio of the wall, t_w the height of the wall section, E_w the Young modulus of the wall, G the shear modulus of the soil and H the height of the wall. As there is no shear wave propagation in the water the power applied to the height of the wall is not arising from a rigorous calculation but it is applied in order to obtain the dimensionless parameters following the concept of the next chapter.

The influence of the foundation flexibility and excitation frequency on the hydrodynamic pressures has been shown also by others (Chopra 1967, 1966; Chopra, A. K., Chakrabarti, P. 1973; Chopra, A. K., Fenves, G. 1983, 1984a, 1984b, 1985a, 1985b, 1985c; Papazafeiropoulos et al. 2011), but their field of investigation is restricted to dam-rock-reservoir interaction.

The damping of the walls was given in the form of Rayleigh damping for time integration analyses and in form of structural damping for modal dynamic and steady state analyses. In order to define the Rayleigh damping of the wall, the formula for a cantilever beam fixed at its one end was used:

$$\omega_i = \frac{\lambda_i^2}{H^2}\sqrt{\frac{D_w}{m_w}} \quad \textbf{(4-8)}$$

and the first and third eigenfrequencies (λ_1=1.8751, λ_3=7.8547) were used for the calculation of the parameters α and β of the Rayleigh damping. As the value of D_w does not remain constant with this parametric study (D_w is a function of E_w in order to model the different degrees of wall stiffness) the values for the Rayleigh damping take different values for each D_w analysed here. This was programed in the input file of Abaqus; however, given their large number, the values are not explicitly provided here.

The first model with the single wall has as unique source of damping the Sommerfeld radiation condition, which idealizes the infinite extent of the reservoir (no reflection of waves takes place). However, this condition is not enough to bound the response of the wall-reservoir system and the hydrodynamic pressures have enormous peaks in a frequency domain analysis. In order to mitigate this effect, the radiation damping and the hysteretic damping of the missing foundation as well as the damping of the wall itself must be considered. Because the radiation damping of the foundation of the wall/dam depends on many factors (the geometry and the embedment of the foundation, the modulus of elasticity and the mass of the foundation, Poisson's ratio and shear wave velocity), a small parametric study with constant damping values for all the frequency range was carried out. The modulus of elasticity is changing in order to keep the section of the wall constant and exclude the inertia effects in this analysis. The following table gives the modulus of elasticity used in this parametric analysis in conjunction with a section of 0.2m of the wall and the last column gives the corresponding section of the wall if we consider a concrete material with a Young modulus of 30 GPa.

Table 13 Values of E_w used in this analysis

d_w	E_w (Pa) (Wall section 0.2m)	Wall section (m) (E_w=30 GPa for concrete, wall height H=8m)
0.01	1.66E+17	35.0
25	6.64E+13	2.57
50	3.32E+13	2.04
100	1.66E+13	1.62

The values of the spring constant used here are in the range of real values of foundation impedances calculated for practical purposes. Real impedances can be calculated as suggested by (Mylonakis et al. 2006; Gazetas 1983).

Table 14 Values of R_θ and K_x used in this analysis

d_θ	R_θ (N/m)	d_x	K_x (N/m)
0.01	1.44E+13	0.01	1.80E+12
25	5.76E+09	3.125	7.20E+08
50	2.88E+09	6.25	3.60E+08
100	1.44E+09	12.5	1.80E+08

In order to investigate the influence of the damping of the foundation (radiation damping and hysteretic damping of the soil material) the given springs' constants were changed in a complex form $K\times(1+i\times\delta)$ assuming for convenience frequency independent damping. The damping given here as structural damping (δ) equals two times the viscous damping ratio (ξ).

Table 15 Values of hysteretic damping factor δ used at this analysis

δ
0.1
0.2
0.3
0.4

The reservoir bottom sediments, which mitigate the hydrodynamic pressures on the wall, are another source of damping. Because the navigation locks don't have or should not have dick stratums of sediments, this effect was not considered in this numerical investigation. In case one would like to do so, a simple impedance boundary condition could be added at the reservoir's bottom.

4.2.1 Influence of the wall's flexibility on the hydrodynamic pressures

As mentioned before, the Westergaard theory refers to rigid dams. However, often the navigation lock chamber walls are not compact enough to be considered rigid. A steady state analysis was carried out in order to determine whether the relative flexibility of the wall, expressed as d_w, can influence the hydrodynamic pressures on it. Here it has to be borne in mind that the reservoir's natural frequency depends on its depth, and that deep reservoirs have longer periods. However, this natural frequency refers to one-dimensional wave propagation of an infinite reservoir. It can be seen that the natural frequency changes when referring to a semi-infinite reservoir and depends also on the flexibility of the wall and the foundation. The following diagram shows that an increasing flexibility of the wall decreases the natural frequency of the system and can lead to bigger hydrodynamic pressures (here expressed as the total shear force at the base). For a quasi-rigid wall (d_w=0) the water-wall system has its resonance at $H/T \approx 360$ as also referred to by (Bustamante, J. I., Flores, A., E. Herrera, Rosenblueth I. 1963), which is ¼ of the speed of sound in water (1440 m/sec). This can be seen also in the next diagram. As the flexibility of the wall increases, the natural period T increases and the ratio H/T decreases. At frequencies reaching zero or with very large excitation periods (the ratio $H/T \to 0$) the total hydrodynamic pressures take the normalized value 0.55 as calculated by (Westergaard 1933) and (Karman 1933a). The pressure distribution varies also dramatically as the excitation frequency increases. The following figures show the dynamic water pressure distribution for different wall flexibilities for the two excitation frequencies f=10 and 35 Hz. The resonance frequency of the rigid wall found by Abaqus is 47 Hz (theoretical value is 45 Hz).

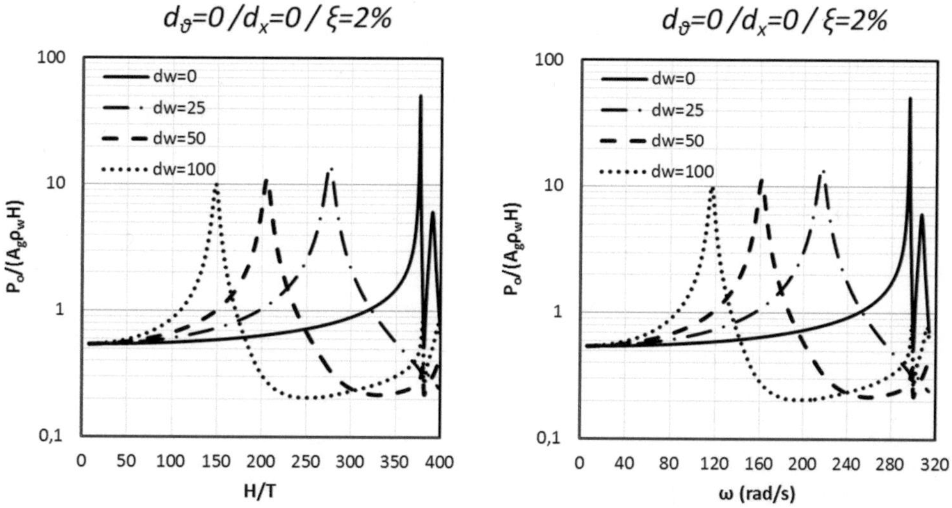

Fig. 4-9 Steady state response of the total hydrodynamic pressure for different wall flexibilities.

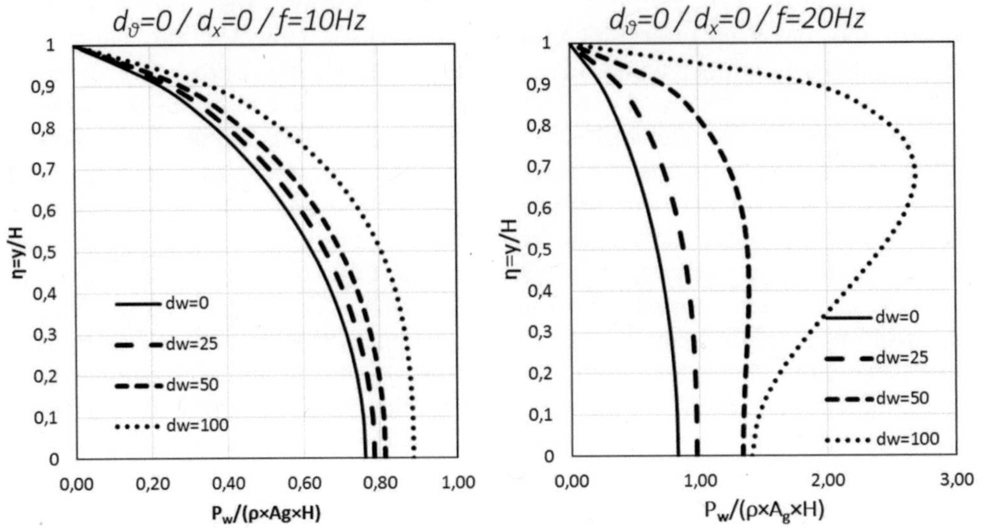

Fig. 4-10 Hydrodynamic pressure distribution for different wall flexibilities and excitation frequencies.

The maximum pressure for the low frequency (f=10Hz) at the rigid wall's base takes the value $0.75\times\rho\times Ag\times H$ whereas the theoretical value is $0.74\times\rho\times Ag\times H$ (Westergaard 1933).

4.2.2 Influence of the wall's damping on the hydrodynamic pressures

Although for rigid or semi-rigid structures such as dams or lock's monoliths a value of 2% of critical damping should be considered as justified, two other values, i.e. ξ=5% and 10%, were taken into account in order to see their influence on the hydrodynamic pressures. As was to be expected, the increasing damping reduces the hydrodynamic pressures and its influence is more obvious for flexible walls whereas it is negligible for rigid walls. The diagrams show the dependence of the total hydrodynamic pressure on the damping ratio of the wall for the resonance frequency.

4.2.3 Influence of the foundation's flexibility on the hydrodynamic pressures

The foundation's flexibility is expressed by means of the relative flexibility for the rotational flexibility of the foundation d_θ and the translational flexibility of the foundation d_x. The following figures give the amplification of the hydrodynamic pressures with increasing excitation frequency. The diagrams should be observed carefully. Although it is clear that the total hydrodynamic pressure at the resonance frequency is reduced dramatically as the foundation's flexibility increases, due to a shift of the natural frequency at lower frequencies a resonance is more likely to occur also for relatively small heights (the natural frequencies are shifted to the left in the diagrams). This can be understood better by making the following comparison of two foundation flexibilities, d_x=0 (or d_θ=0), which refers to a rigid base and d_x=10 (or d_θ=10), which refers to a flexible base. The total hydrodynamic pressure for the flexible foundation at its resonance frequency is much smaller than the total hydrodynamic pressure at the resonance frequency of the rigid base. However, if we compare the total hydrodynamic pressure at a smaller frequency, for example H/T=100, the hydrodynamic pressures of the flexible foundation are greater than the hydrodynamic pressures of the rigid foundation.

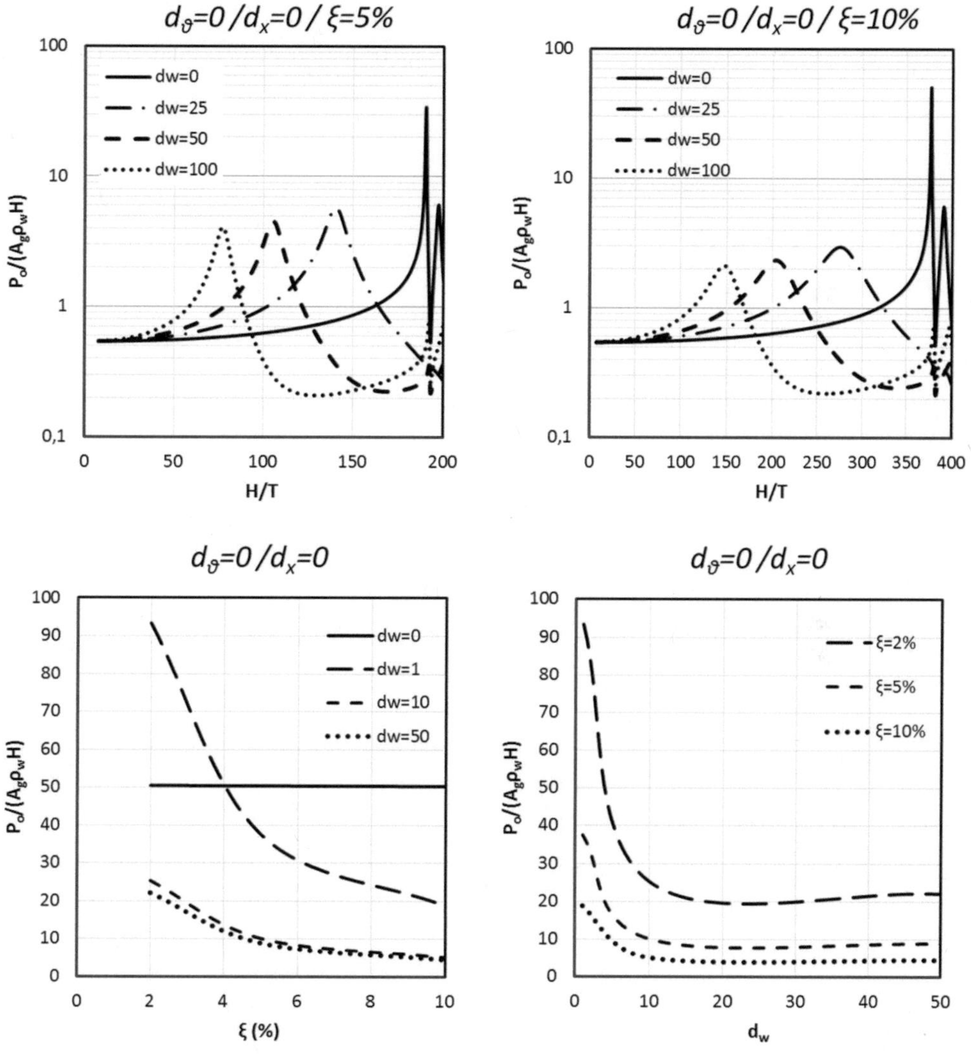

Fig. 4-11 Influence of the wall's damping on the hydrodynamic pressures at the resonance frequency.

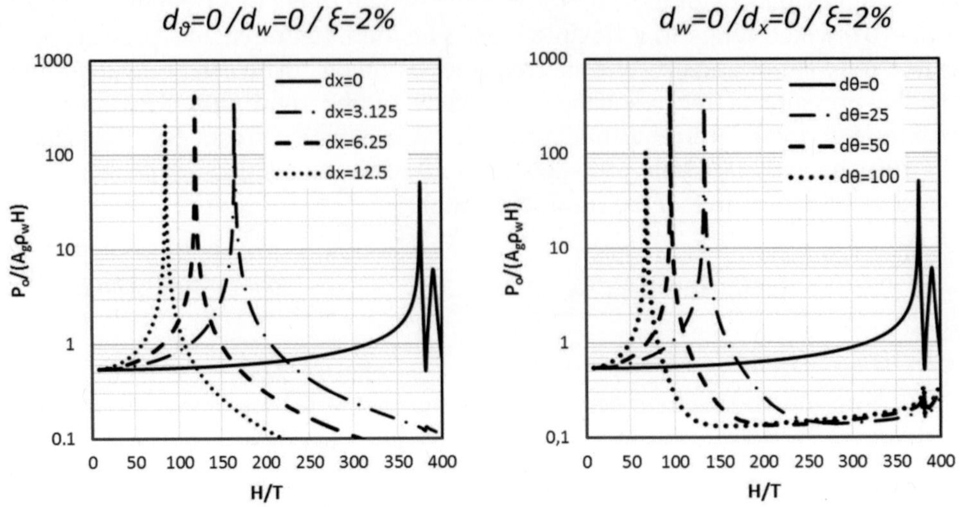

Fig. 4-12 Influence of the foundation's flexibility on the hydrodynamic pressures.

The latter can be seen better by comparing the hydrodynamic distribution for a frequency of 20 Hz, which for the numerical model of this study with a wall height of 8 m corresponds to a value of H/T=160.

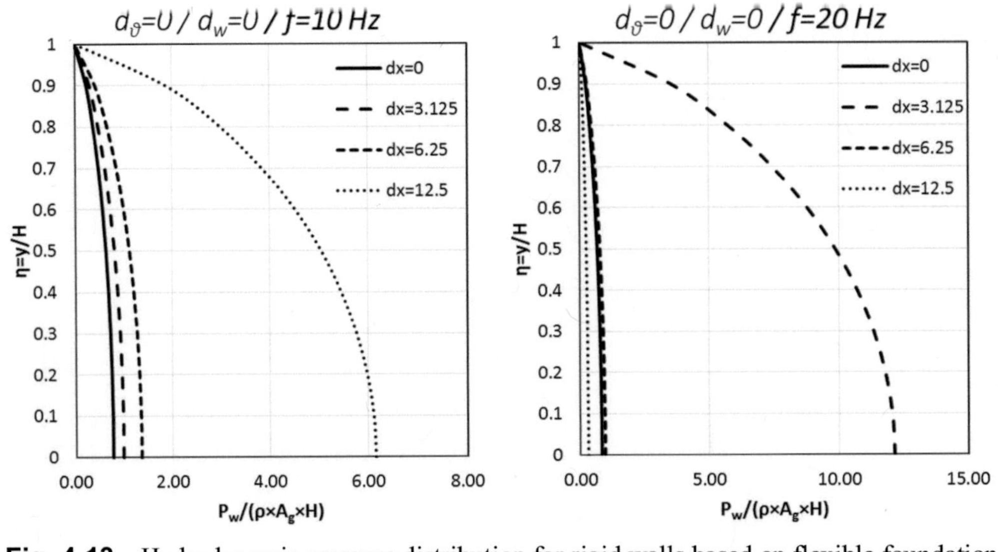

Fig. 4-13 Hydrodynamic pressure distribution for rigid walls based on flexible foundation for frequency excitations of 10 and 20 Hz (ξ=2%).

The same phenomenon is observed not only for an increasing foundation flexibility, but also for an increasing wall flexibility. This change in the hydrodynamic pressure distribution can be also observed at the height of application of the total hydrodynamic pressure. As the foundation becomes more flexible, the height of the total hydrodynamic force increases.

As the flexibility of the foundation increases, the natural frequency of the system becomes smaller. This reduction is not linear or constant due to the fluid-structure interaction and due to the effect of the base flexibility on the natural frequencies of the wall. The change of the natural frequency of the system is shown in the next figures.

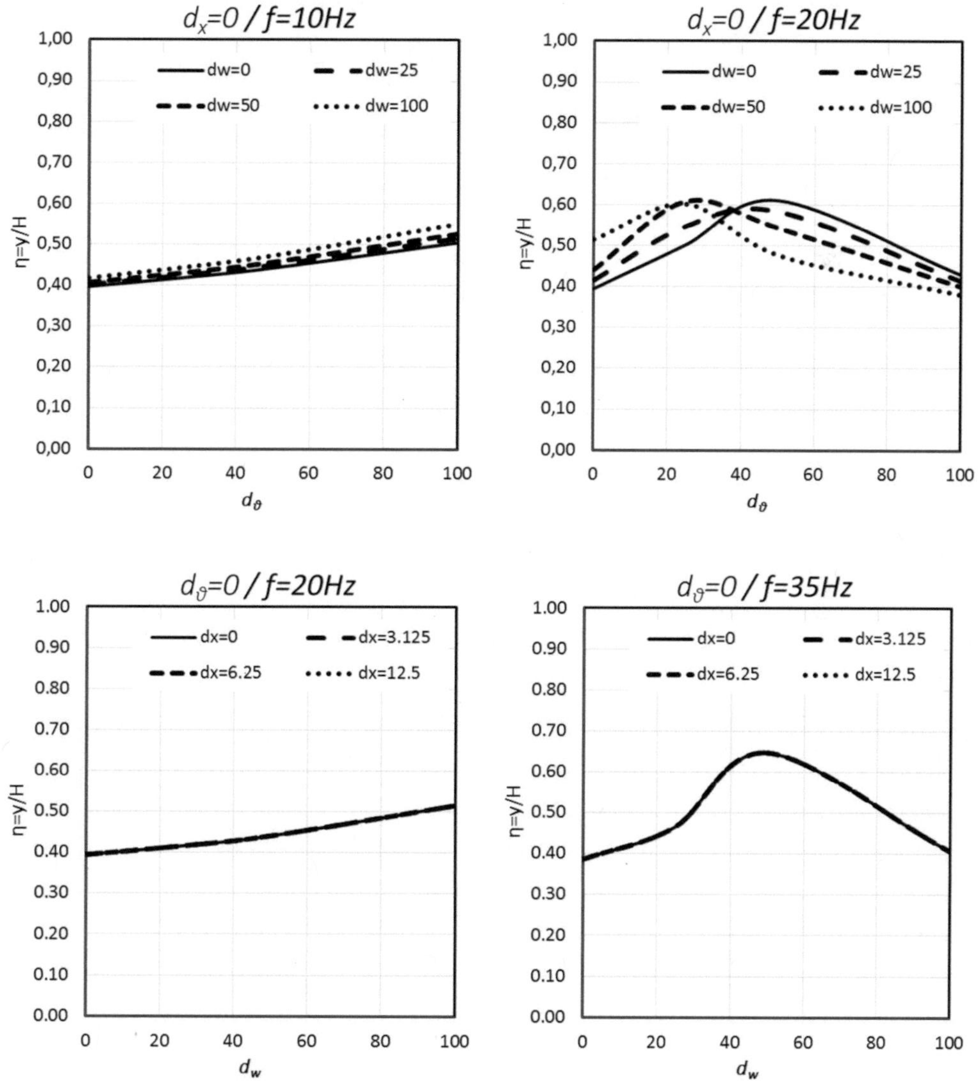

Fig. 4-14 Application height of the total hydrodynamic force for different base and wall flexibilities.

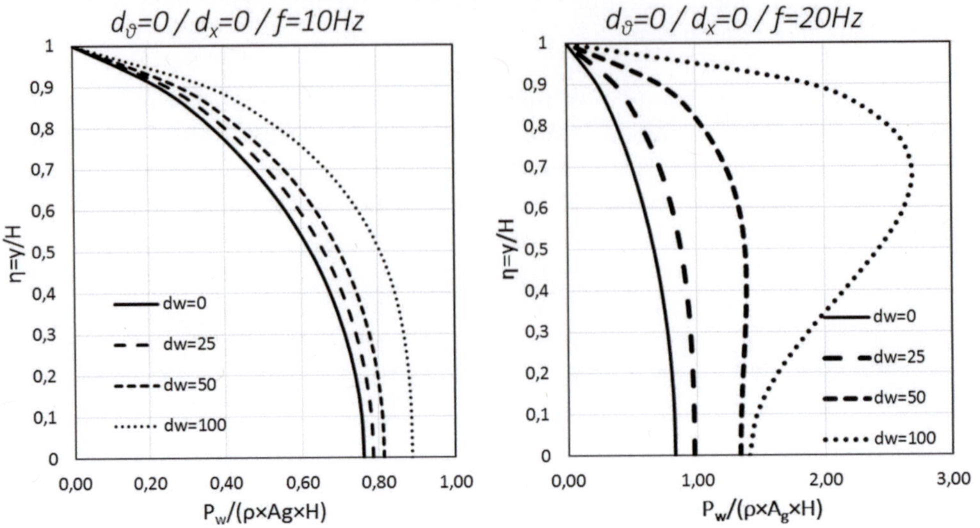

Fig. 4-15 Hydrodynamic pressure distribution for different wall flexibilities based on rigid foundation for two frequency excitations; left10 Hz and right 20 Hz.

The former diagrams show that the height of application of the hydrodynamic pressure is at 0.4H for "static" loading and for rigid systems. This value is also recommended at most legislations and its common practice to assume this height as point of application of the hydrodynamic pressure. As the system flexibility increases the application point of the hydrodynamic pressure also increases in height. The same is observed also for an increasing excitation frequency up to a point, where tension (positive pressures) develops and the height of the application point decreases rapidly. At frequencies for seismic design, which lie between 1.5 and 5.0 Hz, the value of 0.4H for the application point of the hydrodynamic pressures remains a very good estimation also for flexible systems.

4.2.4 Influence of the foundation's damping on the hydrodynamic pressures

The aforementioned diagrams refer to the case where the foundation is flexible but no damping takes place due to radiation and/or material damping. In order to investigate the influence of the foundation's damping on the hydrodynamic pressures, the springs were given in a complex form. Because the total damping (radiation and material damping) depends on mechanical characteristics of the soil (G-Modulus, Poisson's ratio), the foundation's geometry (width and embedment of the foundation), the existence of bedrock below the structure and the excitation frequency, only few discrete values were given here for the hysteretic damping ratio δ, which corresponds to several combinations of the aforementioned parameters. Moreover, damping remains constant with changing excitation frequency, which, however, is not realistic.

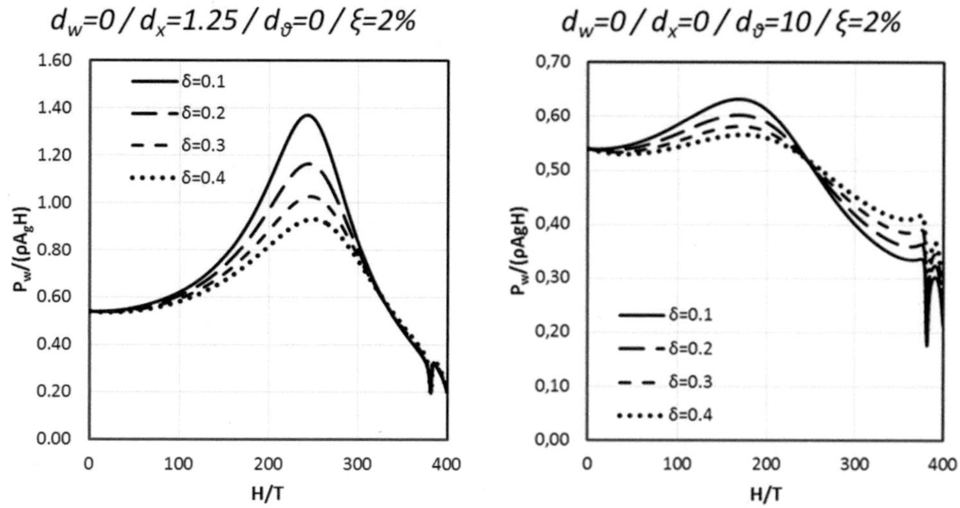

Fig. 4-16 Reduction of the maximum hydrodynamic pressure on the wall, based on flexible foundation as a function of the foundation's damping.

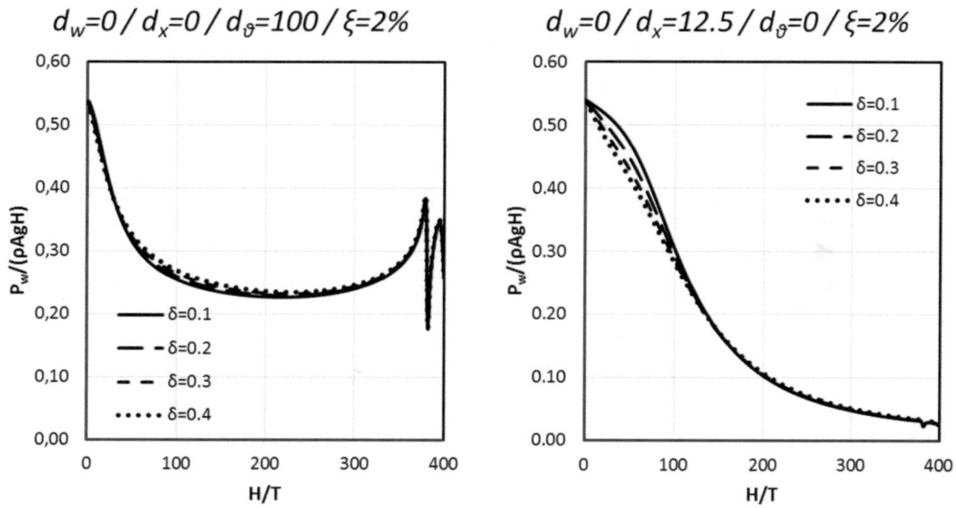

Fig. 4-17 Reduction of the maximum hydrodynamic pressure on a rigid wall based on flexible foundation as a function of the foundation's damping (here overdamped system).

4.2.5 Resonance frequency of fluid-structure system based on compliant base

As can be seen by observing the former diagrams the natural period of the water-wall system increases with increasing foundation flexibility. This effect can be better observed in the following diagrams, where the natural period of the system (taken as the value where the maximum hydrodynamic pressure occurs) is drawn versus the wall flexibility for different values of the relative foundation flexibility.

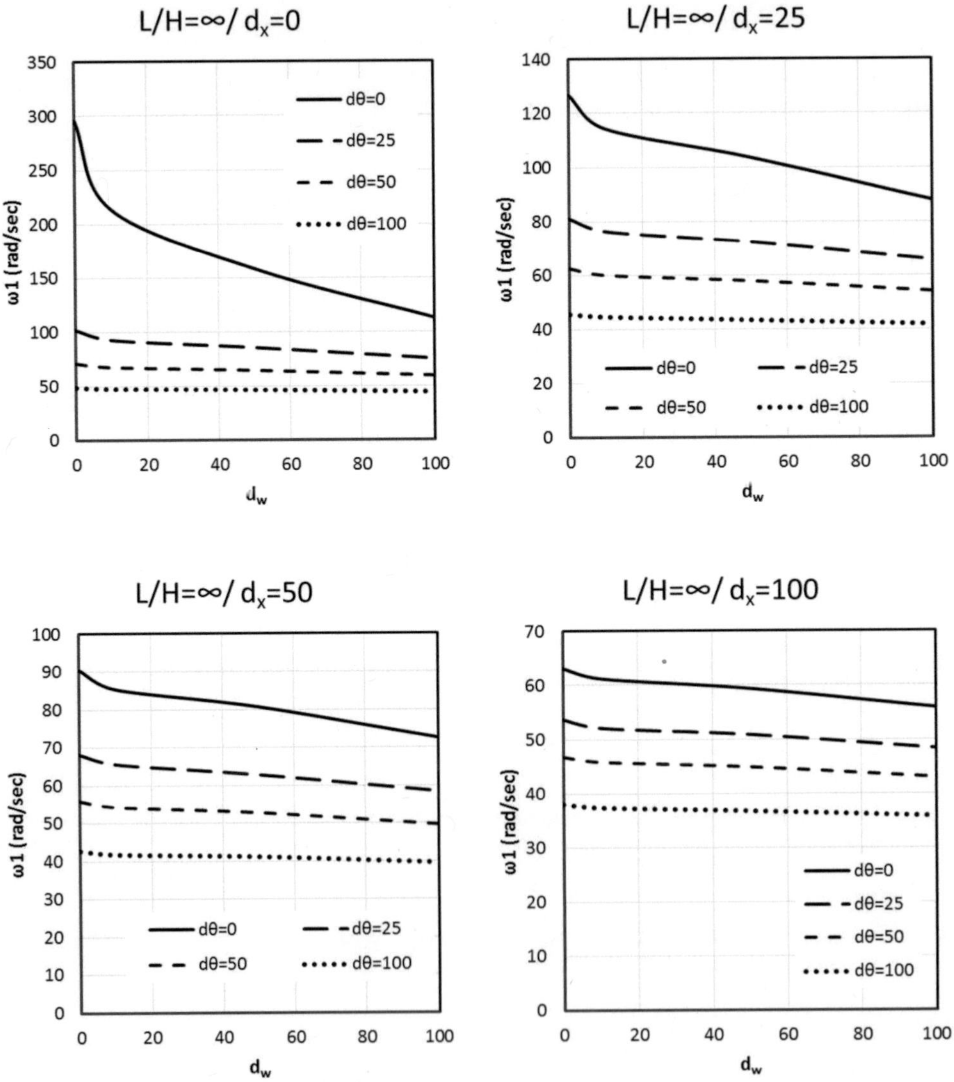

Fig. 4-18 Dependence of the natural frequency of water-wall systems on the relative wall and foundation flexibility (H=8.0 m).

Table 16 Values of the first natural frequency of the system for different wall and base flexibilities (H=8m)

L/H=∞	d_w	0	10	50	100	0	10	50	100
	d_θ	ω_1	ω_1	ω_1	ω_1	ω_1/ω_∞	ω_1/ω_∞	ω_1/ω_∞	ω_1/ω_∞
d_x=0	0	295.33	214.04	158.18	112.67	1.05	0.76	0.56	0.40
	25	101.69	92.58	85.37	75.01	0.36	0.33	0.30	0.27
	50	70.93	67.48	64.34	59.32	0.25	0.24	0.23	0.21
	100	48.65	47.39	46.14	43.94	0.17	0.17	0.16	0.16
d_x=3.125	0	126.79	114.24	103.57	87.88	0.45	0.41	0.37	0.31
	25	80.97	76.26	72.19	65.59	0.29	0.27	0.26	0.23
	50	62.46	59.94	57.75	53.98	0.22	0.21	0.20	0.19
	100	45.51	44.57	43.31	41.74	0.16	0.16	0.15	0.15
d_x=6.25	0	90.39	85.37	80.66	72.50	0.32	0.30	0.29	0.26
	25	68.10	65.59	62.77	58.38	0.24	0.23	0.22	0.21
	50	55.86	54.30	52.73	49.59	0.20	0.19	0.19	0.18
	100	42.68	41.74	41.11	39.55	0.15	0.15	0.15	0.14
d_x=12.5	0	63.08	61.20	59.32	55.86	0.22	0.22	0.21	0.20
	25	53.67	52.10	50.84	48.33	0.19	0.18	0.18	0.17
	50	46.76	45.82	44.88	43.00	0.17	0.16	0.16	0.15
	100	37.98	37.35	36.72	35.78	0.13	0.13	0.13	0.13

The natural periods found with Abaqus are a little bit higher than the theoretical values. The error is though only 5% for the rigid system.

This period lengthening due to the foundation's flexibility was also shown by (Chopra, A. K., Fenves, G. 1983, 1984a, 1984b, 1985a, 1985b, 1985c), who analytically derived relations for the dynamic behaviour of dam-reservoir-foundation systems. Their rigorously suggested formulas are, however, difficult to follow and difficult to apply in engineering practice, as they are based on a modal analysis with boundary elements formulation for the base flexibility. Moreover, their investigation is based on the response of the dam and not the water field, so the period lengthening in their approach refers to the lengthening of the dam's natural frequency.

4.3 Two-wall-water (bounded) system

The case of the one-wall-water system is interesting to investigate because there is a lot of literature available for comparison. However, this one-wall system finds little application in navigation locks, and it is carried out here only to investigate the influence of the wall's and the foundation's flexibility on the dynamic water pressures. It can be applied, however, for the design and analysis of quay walls or flood protection walls, which contrary to dams can be founded on soft soil and are generally much more flexible than gravity dams. For the navigation locks the presence of a second wall in a very small distance is of great interest. Two systems are investigated here; in the first, the walls are separated and based individually on a compliant base, and in the second system the two walls are monolithically connected with a rigid base plate and form a U-frame section based on a compliant base.

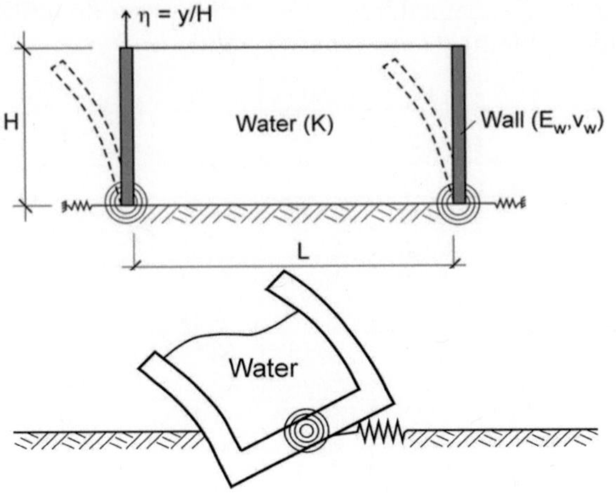

Fig. 4-19 Top: Schema of the pair of walls-water models (bounded water domain) investigated here (Brahtz and Heilbron/Bustamante and Flores/Werner and Sundquist model). Bottom: U-section navigation lock (tank formulation).

The numerical model has now walls with a height of 10.0 m and the reservoir has also a depth of 10.0 m. The dimensionless constants due to the new reservoir's depth take new values.

Table 17 Values of E_w used in this analysis

d_w	E_w (Pa) (Wall section 0.2m)	Wall section (m) (E_w=30 GPa for concrete, wall height H=10m)
0.01	3.24E+17	43.61
0.5	6.84E+15	11.84
1	3.24E+15	9.40
5	6.84E+14	5.49
10	3.24E+14	4.36
25	1.30E+14	3.21
50	6.84E+13	2.55
100	3.24E+14	2.02

The values of the spring constant used here are in the range of real values of foundation impedances calculated for practical purposes. Real impedances can be calculated as suggested by (Mylonakis et al. 2006; Gazetas 1983).

Table 18 Values of R_θ and K_x used in this analysis

d_θ	R_θ (N/m)	d_x	K_x (N/m)
0.01	2.25E+13	0.001	2.25E+12
0.5	4.50E+11	0.05	4.50E+10
1	2.25E+11	0.1	2.25E+10
5	4.50E+10	0.5	4.50E+9
10	2.25E+10	1	2.25E+9
25	9.00E+9	2.5	9.00E+8
50	4.50E+9	5	4.50E+8
100	2.25E+9	10	2.25E+8

The first natural period for all *L/H* ratios is theoretically the value $T_{11} = 4H/c \rightarrow H/T = 360$, where $c = 1440$ m/sec is the speed of sound in water. This value is the first natural frequency in horizontal direction (Bustamante, J. I., Flores, A., E. Herrera, Rosenblueth I. 1963) and in vertical direction. Abaqus finds this exact value but recognises it as vertical mode. This is why the following diagrams show as first natural period the theoretically second horizontal period of the system. As the ratio L/H increases, the natural frequency of the water domain decreases. Moreover, when the excitation frequency tends to zero, the total hydrodynamic pressure take the values analytically calculated by (Sundquist, K. J., Werner, P. W. 1949) and (Brahtz, H. A., Heilbron, C. H. 1933) for incompressible water.

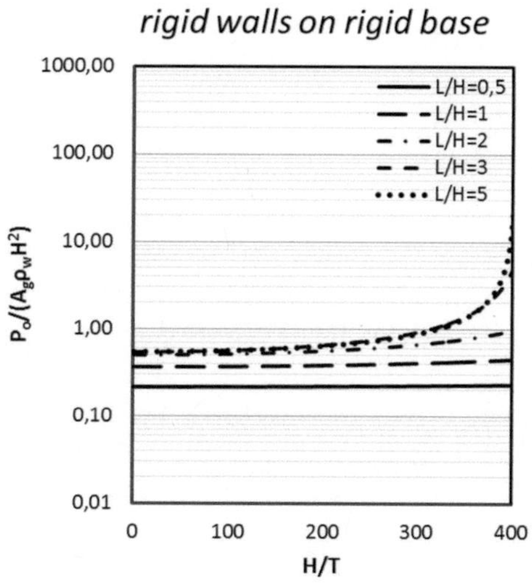

Fig. 4-20 Total hydrodynamic pressure vs H/T ratio for rigid systems and different L/H ratios for statically excited systems.

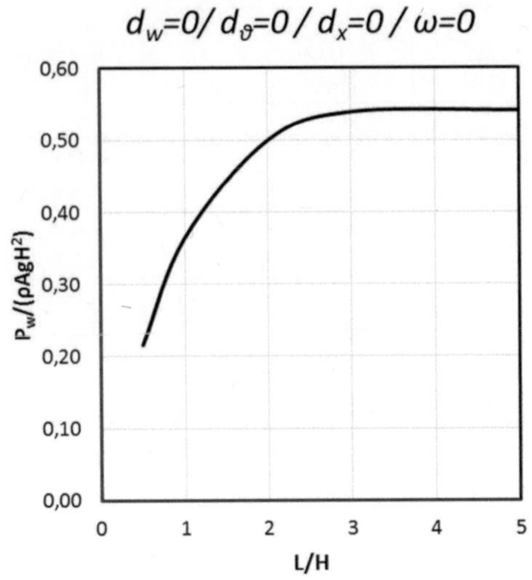

Fig. 4-21 Total hydrodynamic pressure vs L/H ratio for rigid walls based on a rigid base for statically excited systems.

A correction factor, which fits better the result of this analysis, is suggested here:

$$C_n = \begin{cases} 0.55 - 1.2\mathrm{Exp}\left(\dfrac{-L/H}{0.75}\right), & when\ L/H < 2.3 \\ 1, & when\ L/H \geq 2.3 \end{cases} \tag{4-9}$$

This correction factor serves as reduction factor of the total hydrodynamic pressure as the ratio *L/H* decreases and can be applied at the value calculated by (Westergaard 1933) for the total force of a semi-infinite water domain. This correction factor is another form of the factor suggested by (Sundquist, K. J., Werner, P. W. 1949) (see Equation 2.12 (Halabian 2015)). The diagrams are designed for a value of *H/T* up to 400, which corresponds to an excitation frequency of about 13 Hz for a wall height of 30 m. This frequency is about 2.5 times the usual predominant frequency of earthquakes, so a wide frequency range is covered.

4.3.1 Influence of the wall's and the foundation's flexibility on the hydrodynamic pressures

As before the same spring values were applied in order to account for a flexible foundation at the wall bases. Here it is once more noticed that the foundation's and the wall's flexibility influence the natural frequencies of the system and shift the first natural frequency to smaller frequencies, making a resonance more possible to occur. On the other hand, the flexibility of the wall and the foundation, which increase the possibility of resonance because they shift the natural frequency of the system to smaller values, should not be taken into consideration without the influence of damping. The more flexible the wall, the bigger is the influence of the damping on the pressures at resonance. The same holds also for the flexibility and damping of the foundation. In other words, the hazard of resonance because of the system's flexibility should be mitigated by the induced damping.

The natural frequencies of the two-wall-system were identified as the frequencies, where the maximum water pressures occur. Because the range of frequencies used here lies between 0.01 and 200 Hz, natural frequencies higher than the value of 200 Hz were not found. For rigid walls based on rigid foundations, these missing natural frequencies can be calculated with the formulas found in the literature. The next diagrams give a qualitative reproduction of the frequencies reduction as the flexibility of the system increases. The calculated values of the first natural frequencies can be found in Annex B. The L/H ratios of this investigation take the values 0.5, 1.0, 2.0, 3.0, 5.0 and 10.0 covering a wide range of cases.

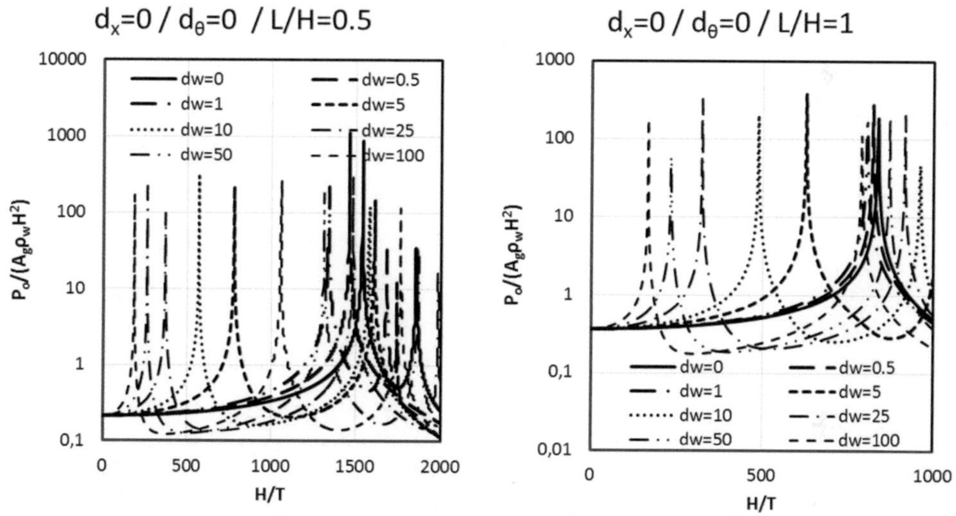

Fig. 4-22 Total hydrodynamic pressure for different wall and base flexibilities and for different L/H ratios (undamped system).

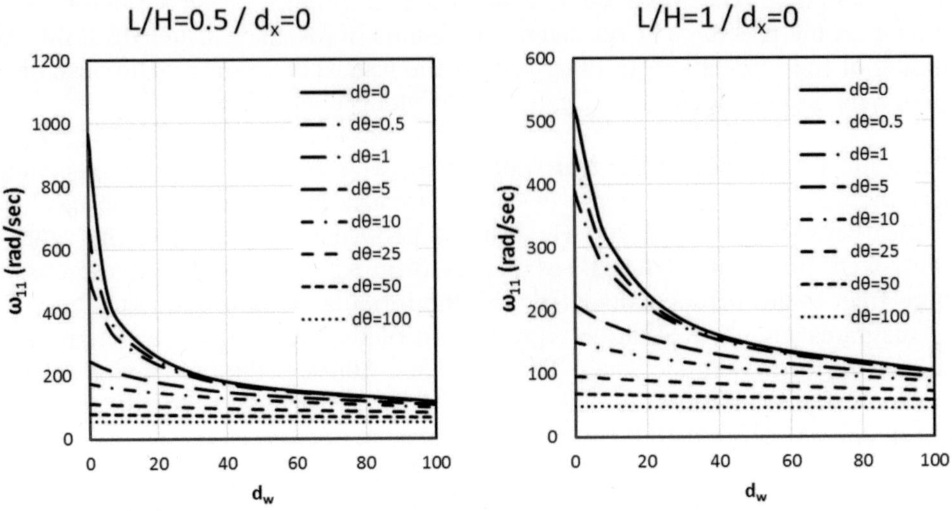

Fig. 4-23 Natural frequencies for different L/H ratios and wall and base flexibilities.

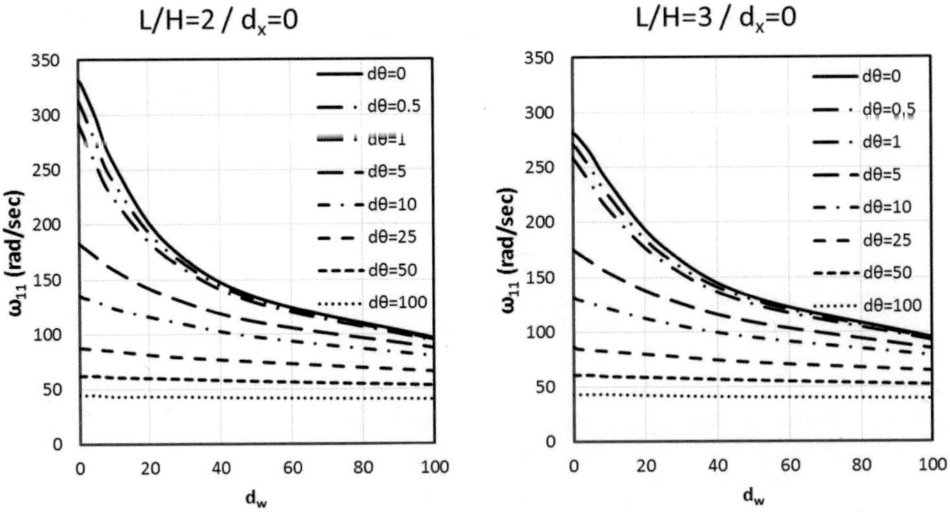

Fig. 4-24 Natural frequencies for different L/H ratios and wall and base flexibilities.

4.4 U-section-water system

Another system was also investigated. The lock section is now a U-section with monolithically connected walls and base plate. The base plate is assumed rigid and the walls can also be flexible. The soil springs are assigned at the middle of the base plate. This system has a different behaviour than the two-walls-water system. The natural frequencies of this system for the same L/H ratios are different from the natural frequencies of the two-walls system. The natural frequencies of this system and the ratio of the natural frequencies to the natural frequency of the unbounded reservoir are given in the Annex B. The following diagrams show the change in the natural frequency of this system as a function of the dimensionless parameters d_w, d_θ and d_x.

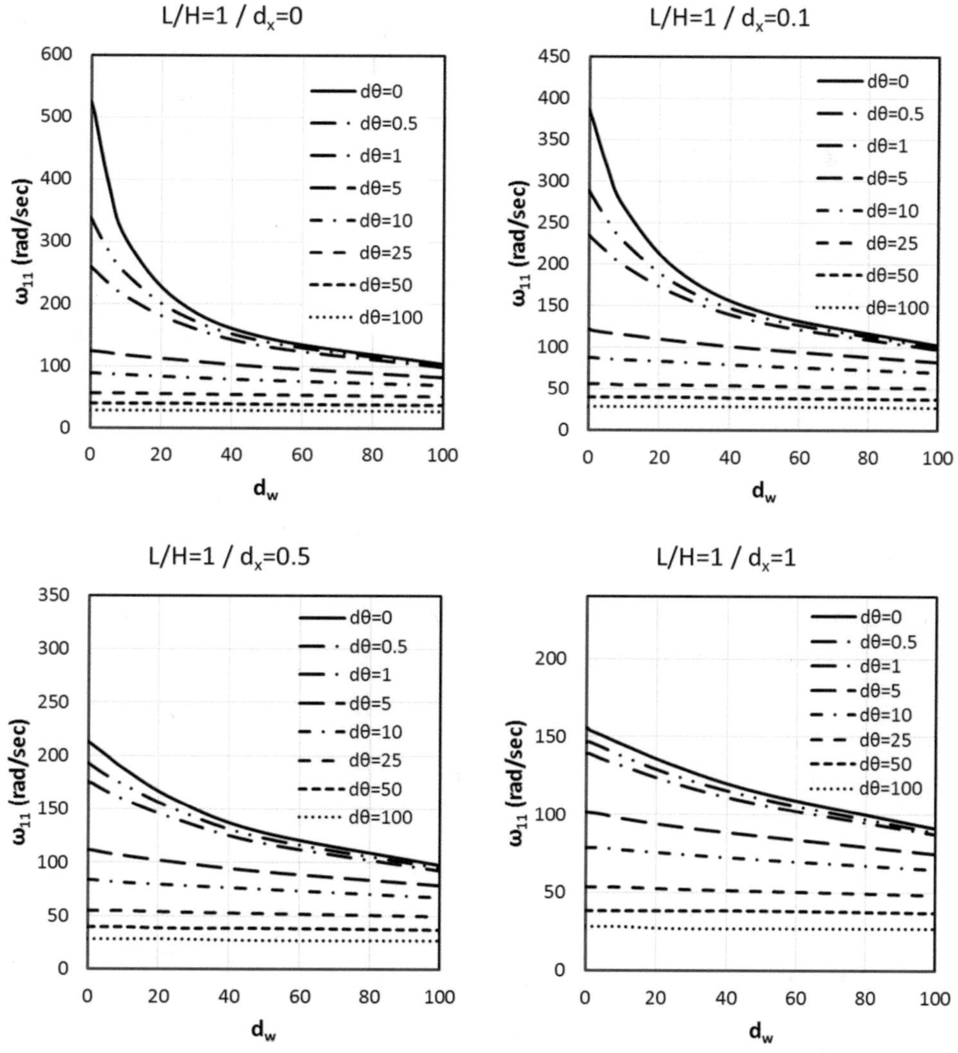

Fig. 4-25 Natural frequencies for different wall and base flexibilities (L/H=1).

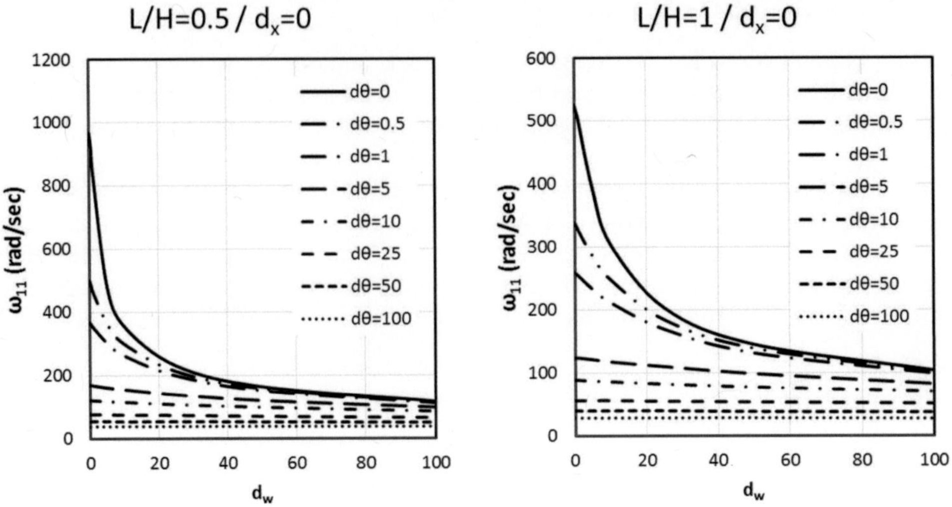

Fig. 4-26 Natural frequencies for different L/H ratios and wall and base flexibilities.

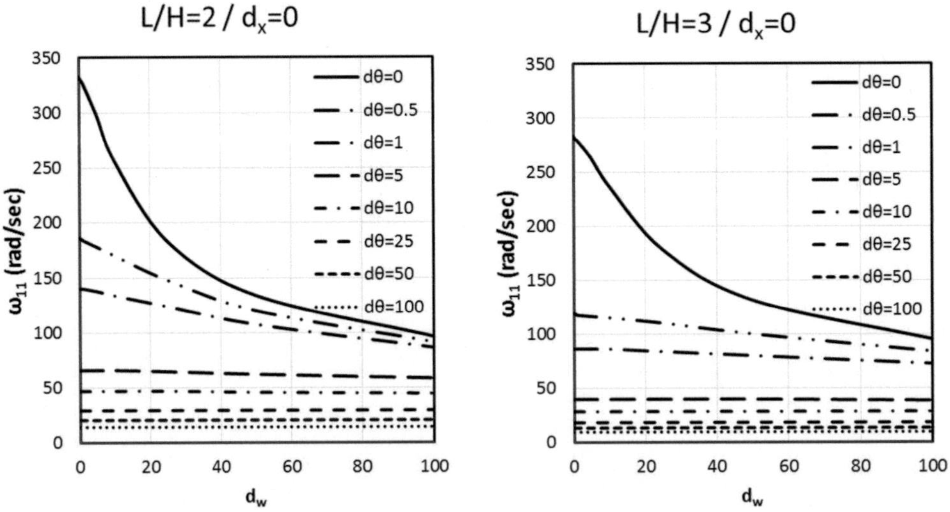

Fig. 4-27 Natural frequencies for different L/H ratios and wall and base flexibilities.

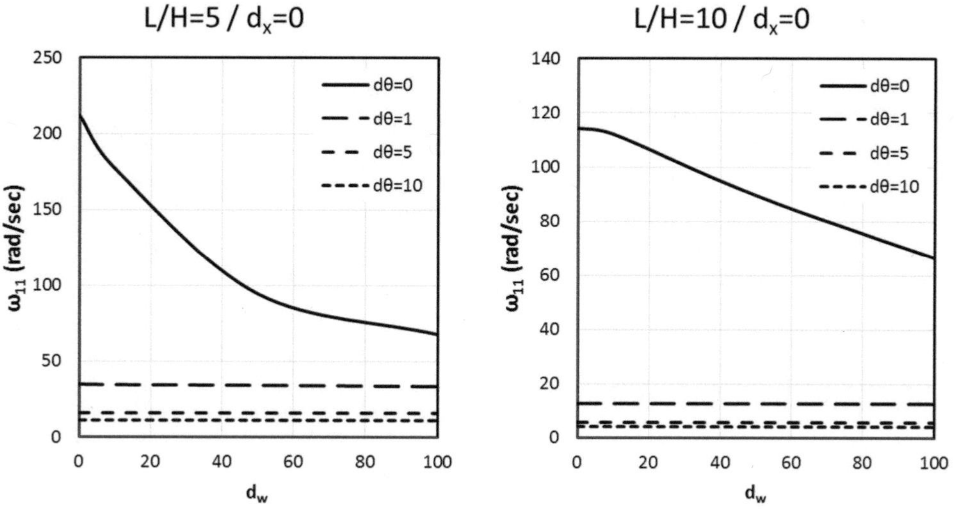

Fig. 4-28 Natural frequencies for different L/H ratios and wall and base flexibilities.

4.5 Conclusions

This numerical investigation has shown the influence of soil-structure and fluid-structure interactions on the hydrodynamic pressures. The added mass concept, which was suggested by (Westergaard 1933), has been long known to be inadequate for the design of hydraulic structures due to the assumptions made. The flexibility of the foundation and the structures reduces the eigenfrequency of the water domain and can lead to resonance. On the other hand the structural damping of the wall or dam and the foundation's damping (radiation and material damping of the rock/soil) significantly bound the total hydrodynamic pressure. The formula given by (Bustamante, J. I., Flores, A., E. Herrera, Rosenblueth I. 1963) for the calculation of the natural frequencies of bounded water systems applies only to rigid walls based on rigid foundations. Here values are given based on the dimensionless parameters for the relative flexibility of the wall and the foundation for flexible walls based on flexible foundation. It is noted that the two investigated systems have different eigenfrequencies as the applied boundary conditions are different.

Chapter 5
Soil-structure interaction of navigation locks

Summary

This chapter describes the analyses carried out at this study for the calculation of the dynamic soil pressures and presents the results of them.

At the beginning of the chapter the modelling parameters are described. Modelling aspects such as the type of finite elements used and the damping input form are discussed. Two main models are investigated following the concept of chapter 4; one with semi-infinite soil domain (one-wall-system) and one with finite soil domain (two-walls-system). For both systems a bonded contact (implying that the wall is attached to the soil and no relative movement takes place) and a smooth contact (the soil cannot separate but slide without friction along the wall) are investigated. The case of the soil-wall separation is investigated only for the case of a quasi-static excitation unsing a static analysis, where the gravitational force of the soil is applied towards the wall. For both systems the walls are assumed flexible and are based on elastic foundation, implying that they are able to rotate and elastically move horizontally (without sliding). The relative flexibilities are described by the dimensionless parameters d_w for the wall flexure, d_θ for the foundament rotation and d_x for the horizontal moevement oft he fundament. As analysis techniques both a time domain and a frequency domain analysis are carried out. For the time domain analysis, in order the results to be comparable to the ones of the frequency domain analysis, a harmonic sinuisodial excitation is given, with three different circular excitations frequencies ($\omega_1/6$, ω_1, $3\omega_1$, whith ω_1 the natural frequency of the soil stratum). These three frequencies tend to cover the case of a quasi-static excitation, the case of an excitation at the resonance frequency of the soil and the case of a high frequency excitation. For the frequency domain analysis a wider range of frequencies is applied. Regarding the soil, one homogeneous soil with constant shear modulus distribution along the wall height and two inhomogenous soils, with a parabolic and a linear distribution of the soil shear modulus, are assumed. The influence of the damping and Poisson's ratio of the soil is also investigated. Additionally, the influence of an inclined rigid wall on the dynamic soil pressures using wave propagation (linear analysis) is shown for first time. For the two-walls-system an additional parameter, the geometric ratio L/H is also considered. The chapter ends with a short investigation of other more general inhomogeneous soil profiles. The results are presented in form of diagrams and tables (Annexes C-G).

5.1 Single gravity or cantilever walls

5.1.1 Modelling parameters

The geometry of the model is identical with the one for the fluid-structure interaction model, but now the water domain is replaced by a linear soil. Both models have a wall height and soil depth of 8.0 m. The wall is restrained elastically in x-direction and rotationally and it is supposed to be flexible. The soil rests on bedrock, i.e. it is supported by hinges. The model investigated here is similar to that of Jung and Bobet (Bobet, A., Jung, C. 2008; Bobet, A., Fernández, G., Jung, C. 2010) and the mechanical characteristics were the same as those given by Psarropoulos et al. (Psarropoulos et al. 2005) and Veletsos and Younan (Veletsos, A. S., Younan, A. H. 1992, 1993, 1994a, 1997, 1998b, 1998a, 2000b).

The soil is adhered to the wall (no separation is allowed) and two different wall-soil contact properties were investigated. In the first case the wall is "bonded" to the soil and no separation and no sliding occurs, whereas in the second case there is a "smooth" contact, i.e. no separation occurs but frictionless sliding is allowed. Moreover, the tensile resistance of the contact interface was investigated. This was done in two ways: either by setting the tension forces to null and calculating for this soil pressure distribution the total shear force, moment and application height (for the one-wall model), or for the two-wall system, by allowing a separation between the soil and the wall and applying the seismic "1 g" forces as gravitational forces towards the one wall.

The infinite domain can be idealized with dashpots or infinite elements available in Abaqus. The finite elements cannot be assigned a user-defined material in order to give a depth dependent shear modulus of the soil. For this reason only dashpots were used in this study, whose properties were calculated in a table calculation depending on the soil characteristics. For the case of homogeneous soil, the modelling with dashpots and finite elements gives the same results.

The size of the finite elements of the soil domain as well as the size of the soil domain was also investigated. The results show that the finer elements give better results near the free surface boundary where a singularity exists for the bonded contact (upper 10% of the soil domain). The distance of the infinite boundary from the wall was also investigated. The results show that there is an error of 3% between the finer elements of 0.2 m size and the coarse ones of 0.8 m (at the free surface the error is significantly larger). The error for the distance between the wall and the infinite boundary is up to 15% for a distance of 50 m instead of 80 m. For the following analyses a soil domain of 80 m length is used, which corresponds to ten times the wall height. Two meshes were applied. An element size of 0.2 m for the whole soil domain was used for the homogeneous soil. A modified reduced mesh for computational efficiency with an element size of 0.2×0.2 m up to a distance of two times the wall height and then a gradually increasing element length up to 0.4 m is used. The reduced mesh is used for the time domain analyses. For the frequency domain analyses the soil domain has elements of 0.2×0.2 m size. In all analyses, the wall element size is 0.2 m so as to ensure that the soil and wall nodes coincide.

Two types of analyses were carried out: a series of time domain analyses and a series of frequency domain analyses. For the time domain analyses the system was excited at three different frequencies until steady state conditions were achieved. For this type of analysis, damping is given as Rayleigh damping. These analyses were very time consuming because of the size of the model, its discretization and the time needed to achieve the steady state conditions. It was carried out so as to compare the results of this study with the results provided by other researchers (Bobet, A., Fernández, G., Jung, C. 2010; Psarropoulos et al. 2005), and to prove the correctness of the models. The second type of analysis is a frequency domain analysis. The system is excited by a sinusoidal excitation with changing circular frequency within a predefined range. For this analysis, damping is given in form of structural damping. This analysis may need more computational capacity in terms of RAM but it is much faster and leaves aside uncertainties arising from modelling parameters such as the Rayleigh damping parameters and the "exact" statically exited frequency, which can be determined to be 0.01 Hz and not a fraction of f_1 as has been done for the time domain analysis (statically exited case is defined at $f_1/6$ or $f_1/8$, where f_1 is the natural frequency of the soil stratum). So a better estimation can also be made for the amplification factor (AF) between the resonance and the "static" case. The difficulty of the frequency domain analysis lies in the fact that the magnitude of the desired quantity is always positive as a result of the square root of the sum of squares of the real and the imaginary part of the quantity. So care must be taken in order to subtract the amplitude of the areas where tension forces develop. Because the identification of these areas is quite a complex task and prone to mistakes, the easiest way to have the total shear forces is to read the shear force at the base of the wall, for a wall with no damping and mass, in order to exclude additional damping and mass forces of the wall.

Different soil profiles were investigated in order to see the difference between the dynamic soil pressures. In order to investigate the influence of the relative soil-wall flexibility d_w, the Young modulus of the soil is kept constant and the Young modulus of the wall is changed according to the value of d_w with the following formula:

$$E_w = \frac{12(1-v_w^2)}{t_w^3} D_w = \frac{12(1-v_w^2)}{t_w^3} \frac{GH^3}{d_w} \rightarrow d_w = \frac{12(1-v_w^2)}{t_w^3} \frac{GH^3}{E_w} \quad \textbf{(5-1)}$$

where v_w is Poisson's ratio of the wall, t_w the height of the wall section, E_w the Young modulus of the wall, G the shear modulus of the soil and H the height of the wall.

The rotational flexibility of the wall at its base was modelled with a rotational spring. The stiffness of the rotational spring can be defined directly by:

$$R_\theta = \frac{GH^2}{d_\theta} \rightarrow d_\theta = \frac{GH^2}{R_\theta} \quad \textbf{(5-2)}$$

Here the wall was modelled as a beam with a constant width of 0.2 m and the values of R_θ depend only on d_θ. Real values of R_θ can be calculated for the formulas used for the machine foundation vibrations (see for example (Gazetas 1983; Mylonakis et al. 2006)).

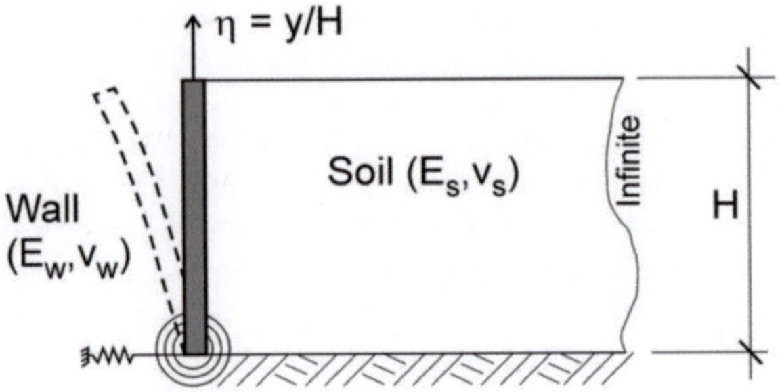

Fig. 5-1 Schema of the first model investigated here (similar to the model of Veletsos and Younan, Psarropoulos et al, Young and Bobet).

The translational flexibility of the wall at its base was modelled with a translational spring. The stiffness of the rotational spring can be defined directly by (Jung et al. 2010):

$$K_x = \frac{G}{d_x} \rightarrow d_x = \frac{G}{K_x} \tag{5-3}$$

For a non-homogeneous soil profile these parameters are rewritten:

$$d_w = \frac{12(1-v_w^2)}{t_w^3}\frac{\bar{G}H^3}{E_w}$$

$$d_\theta = \frac{\bar{G}H^2}{R_\theta} \tag{5-4}$$

$$d_x = \frac{\bar{G}}{K_x}$$

Where $\bar{G}$ is the average value of the shear modulus of the soil, and not the value of the shear modulus at mid-height of the wall ($\bar{G} \neq G^{h=0.5H}$).

5.1.2 Influence of the walls' modelling

In order to investigate the influence of the wall's modelling (apart from the comparison made in the former subchapter in terms of the type of finite elements) a quick parametric study consisting of the following models was done:

- Model A: the wall is massless and has no damping; the changing stiffness is given in terms of Young modulus.
- Model B: the wall is massless and has only stiffness proportional damping; the changing stiffness is given in terms of Young modulus.
- Model C: the wall has mass and both mass and stiffness proportional damping; the changing stiffness is given in terms of Young modulus.
- Model D: the wall has mass and both mass and stiffness proportional damping; the stiffness is given as a function of the section thickness in order to determine the influence of the inertia forces on these linear analyses.

All the models were excited by "static" excitation and at their resonance frequency. "Static" excitation refers to the excitation of the system at a very small fraction of its resonance frequency, here defined at $\omega_1/6$. The results show that there is no significant difference between the dynamic soil pressures except for the case of resonance for Model D. Model A was adopted for the further analyses in order for the results of this study to be directly comparable with the results found in the literature.

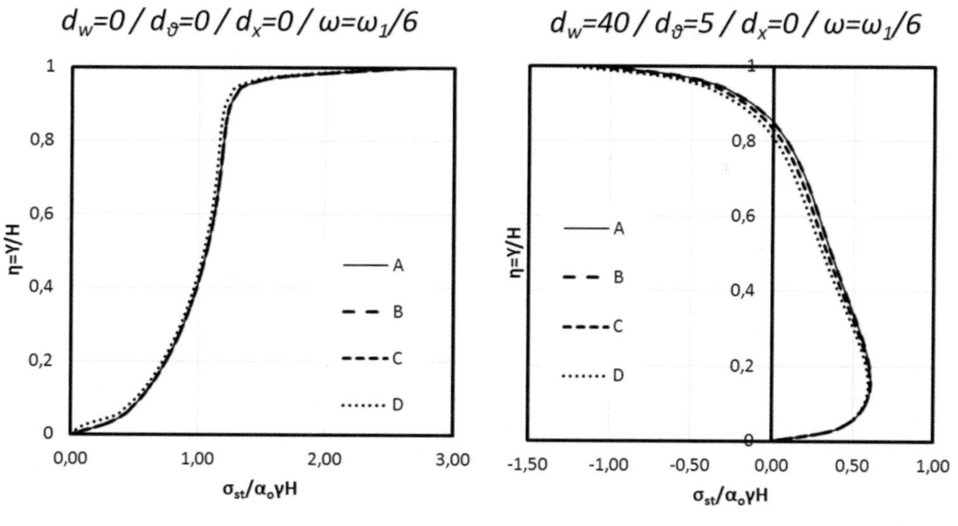

Fig. 5-2 Comparison of the pressure distribution of the different models of the wall for the quasi static case.

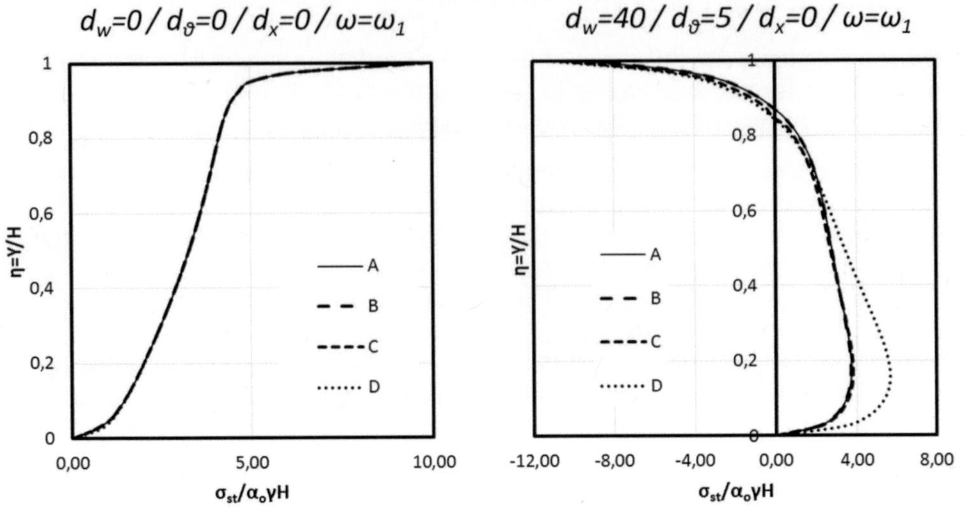

Fig. 5-3 Comparison of the pressure distribution of the different models of the wall for the resonance case.

5.1.3 Influence of the soil profile

A soil profile can vary as far as the shear modulus or the shear wave velocity is concerned. This variation can be expressed with the following formula (Di Laora 2015):

$$v(z) = v_H \left[b + (1-b)\frac{z}{H})\right]^n \qquad \textbf{(5-5)}$$

Where $v(z)$ is the shear wave of the soil in arbitrary location z; v_H the shear wave at depth z=H; and b and n are parameters controlling the shape of the stiffness distribution. For the investigation different soil profiles were assumed; a constant, a linear and a parabolic distribution of the shear modulus increasing with depth were assumed:

$$v(z) = \begin{cases} v_H \quad \textit{for the constant distribution} \\ v_H \left(\frac{z}{H}\right)^{0.5} \quad \textit{for the linear distribution} \\ v_H \left(\frac{z}{H}\right)^{0.25} \quad \textit{for the parabolic distribution} \end{cases}$$

This variation can be expressed also in the form of the shear modulus of the soil. The parabolic distribution is given by (Veletsos, A. S., Younan, A. H. 1997):

$$G(\eta) = G_o(1-\eta^2) \tag{5-6}$$

And the linear distribution of the shear modulus is given by:

$$G(\eta) = G_o(1-\eta) \tag{5-7}$$

Where η is the dimensionless depth parameter (η=y/H).

To enable a comparison of the influence of the different soil profiles on the dynamic soil pressures, the values of the shear modulus were fitted so that the mean shear wave velocity was equal to 100 m/s. The investigation follows the excitation characteristics of Psarropoulos et al. (Psarropoulos et al. 2005) and Jung and Bobet (Bobet, A., Jung, C. 2008), i.e. a harmonic sinus excitation with amplitude A_o = 1 m/s^2, and the natural frequency was chosen as f_1, $f_1/6$ and $3f_1$ with:

$$f_1 = \begin{cases} 0.25\dfrac{v_s}{H} & for\ the\ constant\ distribution \\ 0.19\dfrac{v_H}{H} & for\ the\ linear\ distribution \\ 0.223\dfrac{v_H}{H} & for\ the\ parabolic\ distribution \end{cases} \tag{5-8}$$

where v_H is the shear wave velocity at the bottom of the soil layer. These relations were taken by (Gazetas 1991). The distribution of the shear modulus for the linear and parabolic soil profile was adapted so that the average shear wave velocity of the soil stratum was equal to the one of the homogeneous soil deposit, i.e. 100 m/s. This results in v_H = 129 m/s for the parabolic profile and v_H =150 m/s for the linear soil profile.

Different values for the parameters α and β of the Rayleigh damping are necessary in order to achieve the same damping in all investigated frequencies. This can be done in several ways. One way is to define only mass proportional (α) or only stiffness proportional (β) damping, for which cases these parameters depend only on the target frequency. In case of full Rayleigh damping, where both parameters α and β have to be defined, a way is to define the investigated frequencies, i.e. ω_1 and $\omega_1/6$ or $3\omega_1$, as target frequencies. Other techniques suggest to set as second target frequency the next bigger odd number of ω/ω_1 as it has been done in (Vrettos., C., Feldbusch, A. 2016). For the soil medium the Rayleigh parameters take the values presented in Table 17. The values are similar to the values used by (Vrettos., C.,

Feldbusch, A. 2016) for homogeneous soil. The different damping techniques were investigated here and compared to each other and with the theoretical results. The comparison refers to the rigid wall, both statically excited systems and systems in resonance, homogeneous soil and includes the following damping alternatives

a) Mass proportional (α) damping only
b) Stiffness proportional (β) damping only
c) Rayleigh damping with target frequencies ω_1 and $\omega_1/6$
d) Rayleigh damping with target frequencies, the excitation frequency and the next bigger odd number of ω/ω_1
e) Modal dynamic analysis with structural damping ξ=5% (δ=0.1).

The values of the parameters of the above cases are presented in the table below.

Table 19 Values of α and β parameters for damping

	a	b	c		d		e
	α	β	α	β	α	β	
$\omega_1/6$	0.3272	0.0305	0.2805	0.0043	0.2152	0.0105	-
ω_1	1.9635	0.0051	0.2805	0.0043	0.9617	0.0026	-

The results are presented in the following diagrams.

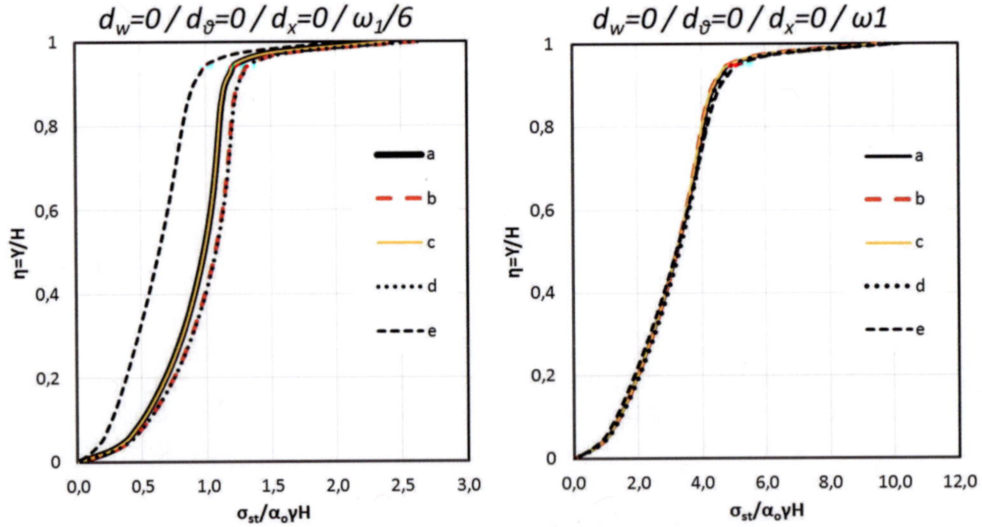

Fig. 5-4 Influence of different damping modelling techniques.

For the statically excited case the modal dynamic analysis with structural damping damps too much the response of the system and total soil pressure on the wall is too small. The mass proportional damping yields exactly the same results as the full Rayleigh damping. This is also a proof that only mass proportional damping damps more the small frequencies and influences in that way also the full Rayleigh damping. Where only stiffness proportional damping is used, the results are the same as with technique (d). For the resonance case all damping techniques give practically the same results. In the following table, these modelling techniques are also compared in terms of the total shear forces produced. It has to be mentioned here that the modal dynamic analysis equals exactly the theoretical value for the resonance case, and that it is much faster than the direct time integration analysis. However, for the statically excited system it delivers no satisfactory results and in fact no steady state response has been produced although a large number of cycles were run.

Table 20 Values of total shear force due to different damping techniques

	$P/(\rho AgH^2)$					
	a	b	c	d	e	Theoretical value[1]
$\omega_1/6$	0.92	1.00	0.92	1.00	0.63	0.94
ω_1	3.15	3.12	3.12	3.20	3.16	3.16
AF	3.42	3.12	3.39	3.20	5.01	3.36

For the further analyses conducted here, the technique (d) with the two target frequencies as described was used. The values of the parameters α and β are shown in the next table.

Table 21 Values of Rayleigh damping used in this study

	Constant		Linear		Parabolic	
	α	β	α	β	α	β
ω1/6	0.2152	0.0105	0.2339	0.00998	0.23544	0.00995
ω1	0.9617	0.0026	1.0229	0.00243	1.0276	0.0024
3ω1	2.0488	0.01107	1.6776	0.00117	1.6944	0.0011

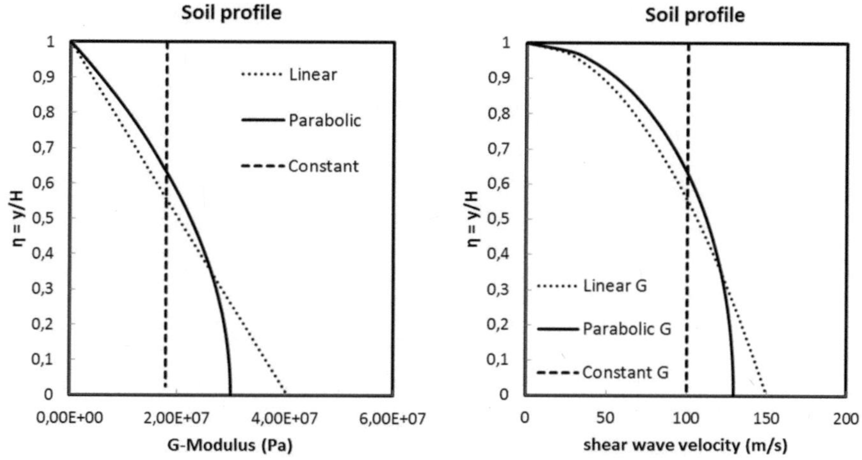

Fig. 5-5 Soil profiles investigated.

[1] Veletsos, A. S., Younan, A. H. 1997

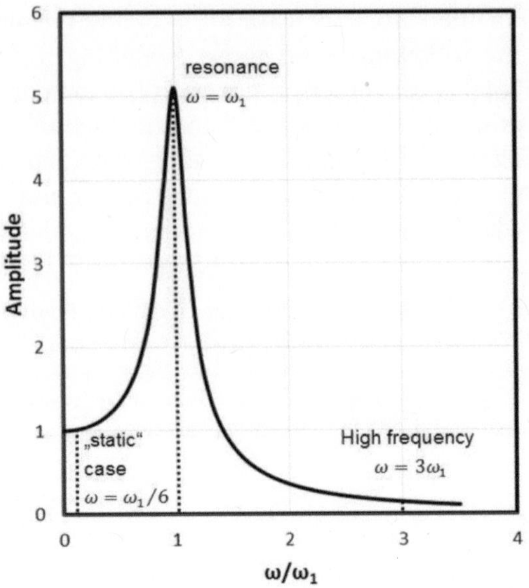

Fig. 5-6 The three frequencies investigated, where ω_1 corresponds to the first natural frequency of the soil stratum.

Table 22 Soil and wall mechanical characteristics

	Parameter	Value	Comment
Ground input motion	Peak acceleration a_g (m/s^2)	1.0	
	Angular frequency ω_1 (rad/sec)	19.63	Constant shear modulus distribution
		22.38	Linear -//-
		22.59	Parabolic -//-
Wall	Mass (kg)	2500	Or massless depending on the analysis
	E_w (MPa)	*	*Depends on the relative wall-soil flexibility
	v_w	0.2	
	H (m)	8.0	
	Damping ratio ξ_w	2%	Or no damping depending on the analysis
Soil	Mass (kg)	1800	
	E_s (MPa)	47.88	Homogeneous soil
	E_s (MPa)	107.73	Linear soil at η=0
	E_s (MPa)	79.81	Parabolic soil at η=0
	v_s	0.33	
	H (m)	8.0	
	Damping ratio ξ_s	5%	

Table 23 Values of E_w for different soil profiles

	Constant	Linear	Parabolic
d_w	E_w (Pa)	E_w (Pa)	E_w (Pa)
0.001	1.33E+16	1.49E+15	1.47E+15
1	1.33E+13	1.49E+13	1.47E+13
5	2.65E+12	2.99E+12	2.93E+12
10	1.33E+12	1.49E+12	1.47E+12
20	6.64E+11	7.46E+11	7.33E+11
30	4.42E+11	4.98E+11	4.89E+11
40	3.32E+11	3.73E+11	3.66E+11

Table 24 Values of R_θ for homogeneous soil

	Constant	Linear	Parabolic
d_θ	R_θ (Nm)	R_θ (Nm)	R_θ (Nm)
0.001	1.15E+11	1.30E+12	1.27E+11
0.5	2.30E+09	2.59E+09	2.54E+09
1	1.15E+09	1.30E+09	1.27E+09
2	5.76E+08.	6.48E+08	6.36E+08
3	3.84E+08	4.32E+08	4.24E+08
4	2.88E+08	3.24E+08	3.18E+08
5	2.30E+08	2,59E+08	2.54E+08

Table 25 Values of K_x for different soil profiles

	Constant	Linear	Parabolic
d_x	K_x (N/m)	K_x (N/m)	K_x (N/m)
0.001	1.80E+10	2.03E+10	1.99E+10
0.1	1.80E+08	2.03E+08	1.99E+08
0.5	3.60E+07	4.05E+07	3.98E+07
1	1.80E+07	2.03E+07	1.99E+07

Table 26 Element size for homogeneous soil

	$\omega_1/6$	ω_1	$3\omega_1$
Angular velocity (rad/sec)	3.27	19.63	58.86
Frequency (Hz)	0.52	3.12	9.37
Wave length λ (m)	192.15	32.02	10.67
$\lambda/10$	19.21	3.20	1.07
Element size (m)	0.2	0.2	0.2

Table 27 Element size for inhomogeneous soil (parabolic profile)

	$\omega_1/6$	ω_1	$3\omega_1$
Angular velocity (rad/sec)	3.76	22.59	67.75
Frequency (Hz)	0.59	3.59	10.77
Wave length λ (m)	169.49	27.85	9.28
λ/10	16.95	2.79	0.93
Element size (m)	0.2	0.2	0.2

Table 28 Element size for inhomogeneous soil (linear profile)

	$\omega_1/6$	ω_1	$3\omega_1$
Angular velocity (rad/sec)	3.73	22.37	67.1
Frequency (Hz)	0.59	3.56	10.68
Wave length λ (m)	169.49	28.09	9.36
λ/10	16.95	2.8	0.94
Element size (m)	0.2	0.2	0.2

The element size was chosen smaller than 1/10 (see (Kuhlemeyer R. L., Lysmer, J. 1973) of the length of the shear wave to ensure minimum frequency cut-off. With this element size a very high cut-off frequency (f=50 Hz) is achieved. The figure below gives the results of this study with finite elements and compares the graphs with the one calculated by Veletsos and Younan (Veletsos, A. S., Younan, A. H. 1997).On the vertical axis the normalized height of the wall is plotted, whereas on the horizontal axis the values of the soil pressures are shown, normalized with respect to $\alpha_o \gamma H$, where α_o is the maximum acceleration of the excitation, γ the unit weight of the retained soil and H the height of the retaining wall.

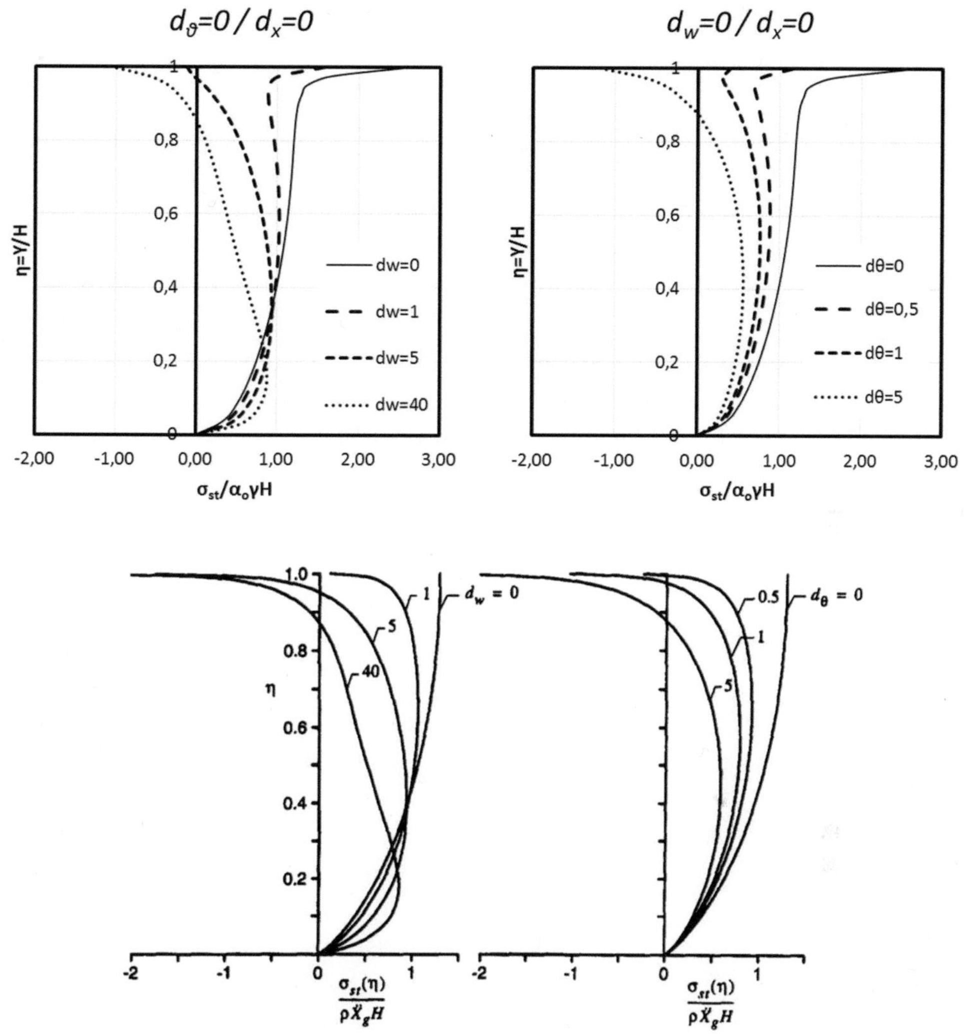

Fig. 5-7 Distributions of wall pressure for statically excited systems with different wall and base flexibilities (v = 1/3): for d_θ =0 (left); for d_w =0 (right) after this study with Abaqus (up); after (Veletsos, A. S., Younan, A. H. 1997)(down).

As it has been mentioned in (Wood 1973), the singularity at the top of the wall because of the bonded contact is typical of finite element calculations.

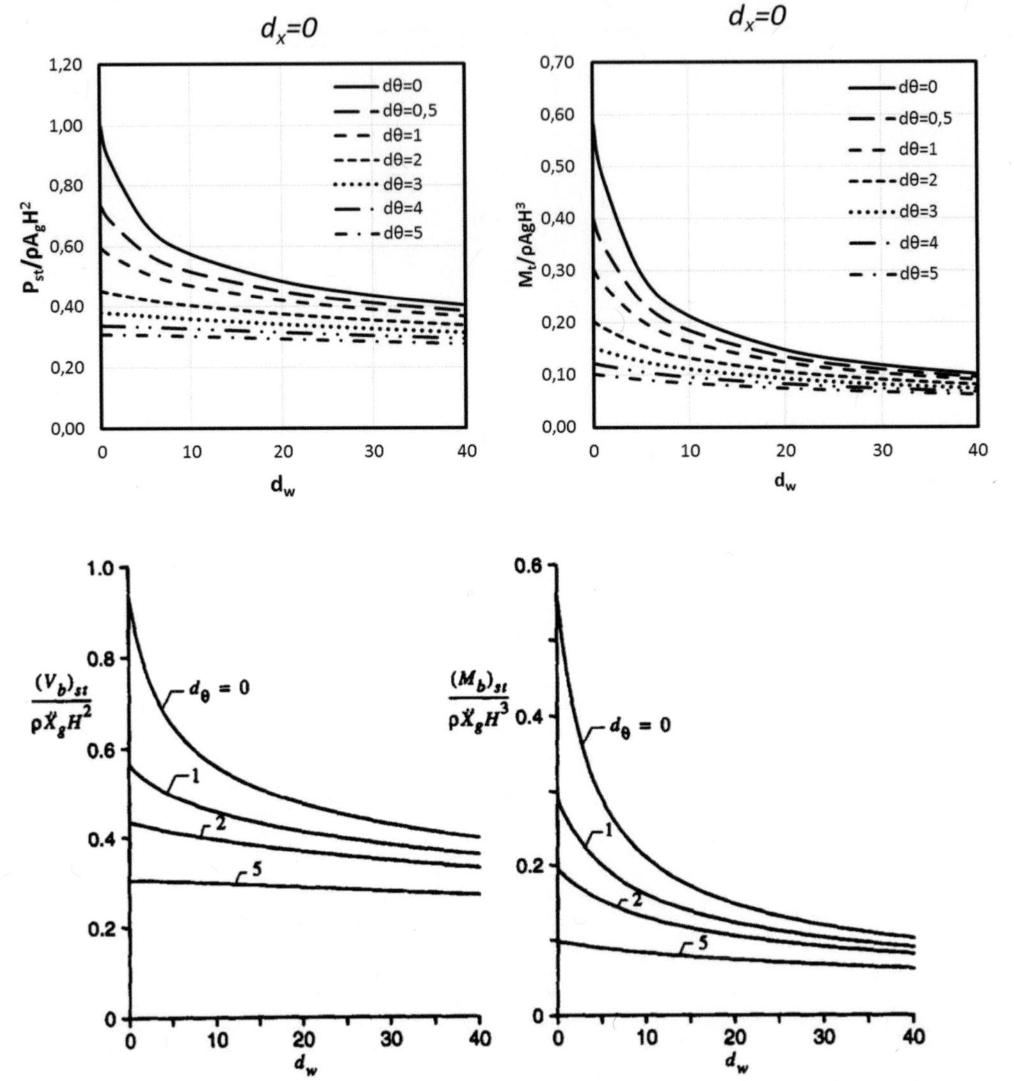

Fig. 5-8 Normalized values of base shear and moment in a wall of statically excited systems with different wall and base flexibilities (v =1/3): according to this study with Abaqus (up); according to (Veletsos, A. S., Younan, A. H. 1997) (down).

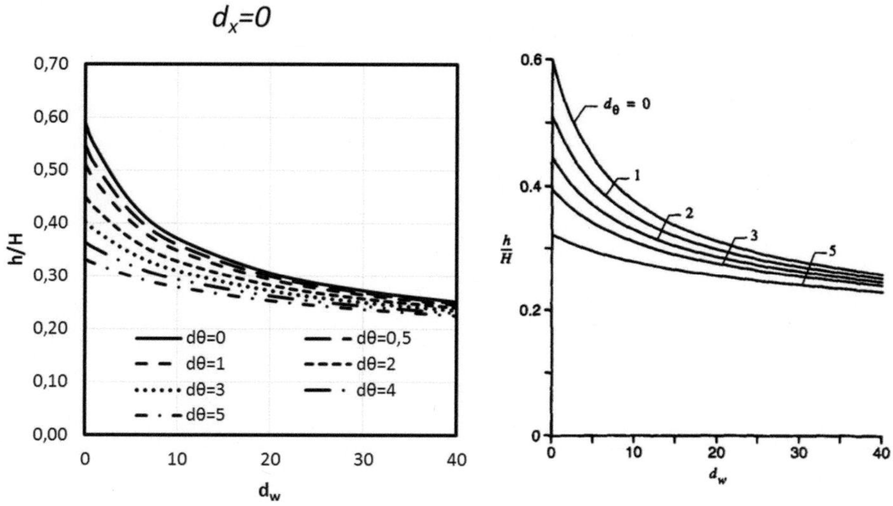

Fig. 5-9 Normalized effective heights of statically excited systems with different wall and base flexibilities (ν =1/3): according to this study with Abaqus (left); according to (Veletsos, A. S., Younan, A. H. 1997) (right).

Table 29 Comparison of the shear forces calculated by Veletsos and Younan (Veletsos, A. S., Younan, A. H. 1997) and this study

	$(V_b)_{st} / \alpha_0 \gamma H^2$				
d_w	V-Y[2]	This study (time domain)	Difference %	This study (frequency domain)	Difference %
0	0.940	1.003	6.28%	0.998	5.81%
1	0.838	0.886	5.42%	0.880	4.77%
5	0.653	0.674	3.12%	0.667	2.10%
40	0.397	0.404	1.73%	0.398	0.25%

[2] Veletsos, A. S., Younan, A. H. 1997

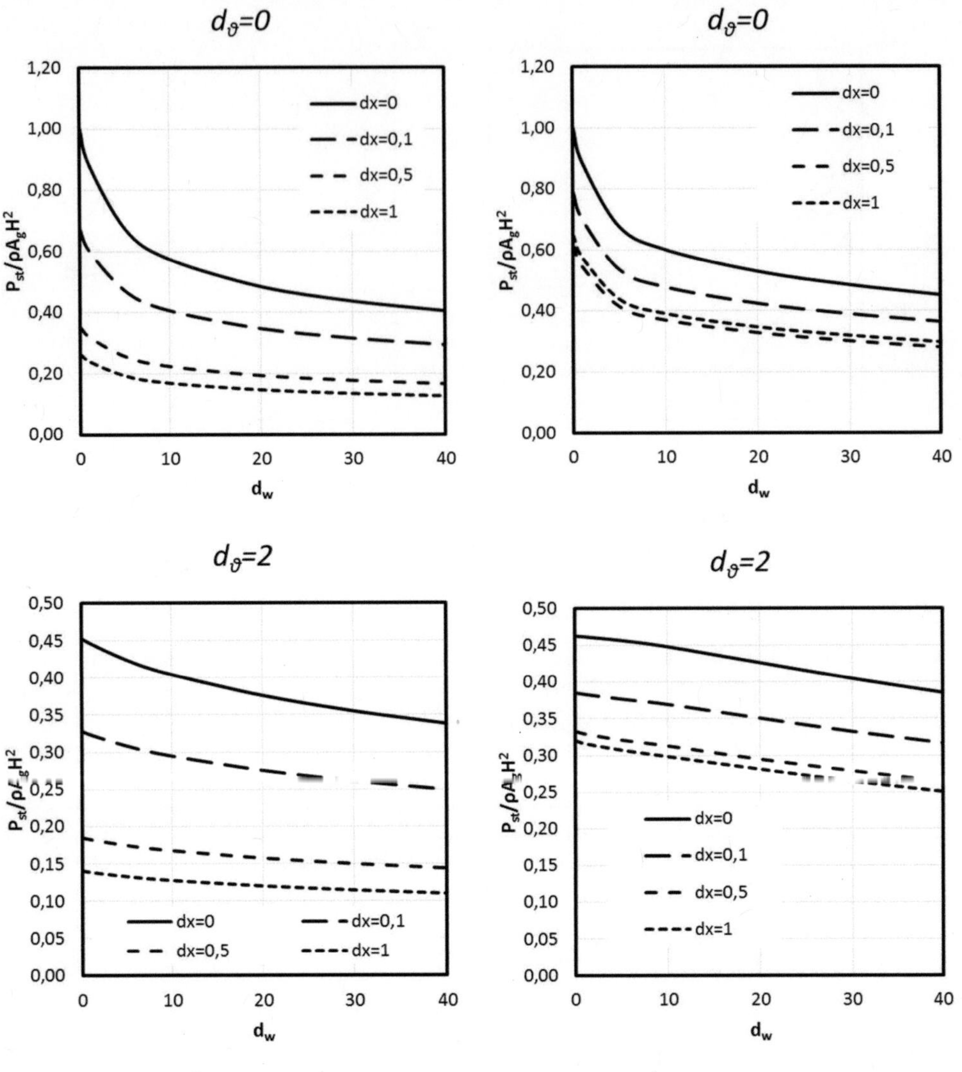

Fig. 5-10 Normalized base shears due to soil pressures statically excited systems with different wall and base flexibilities ($v = 1/3$): considering tension forces (left), setting tension forces to null (right).

Table 30 Comparison of the effective heights calculated by Veletsos and Younan (Veletsos, A. S., Younan, A. H. 1997) and this study

	h/H		
dw	V-Y	This study (time domain)	Difference %
0	0.599	0.588	-1.87%
1	0.553	0.546	-1.28%
5	0.444	0.439	-1.14%
40	0.257	0.252	-1.98%

It can be seen for the diagrams that for some values of wall and base flexibilities the soil pressures obtain negative values (because of the constraint of no separation between the two contact surfaces), i.e. the soil domain develops tension forces, which is unreal. Veletsos and Younan commented that a wall-soil separation will occur only when these tension pressures are bigger than the gravity pressures of the soil acting at this height. Psarropoulos et al. and Jung and Bobet assumed more realistically that these tension forces should not be taken into account when calculating the base shear and moment.

The following Figures give the dynamic soil pressure distribution for the other two profiles, i.e. the parabolic and the linear profiles.

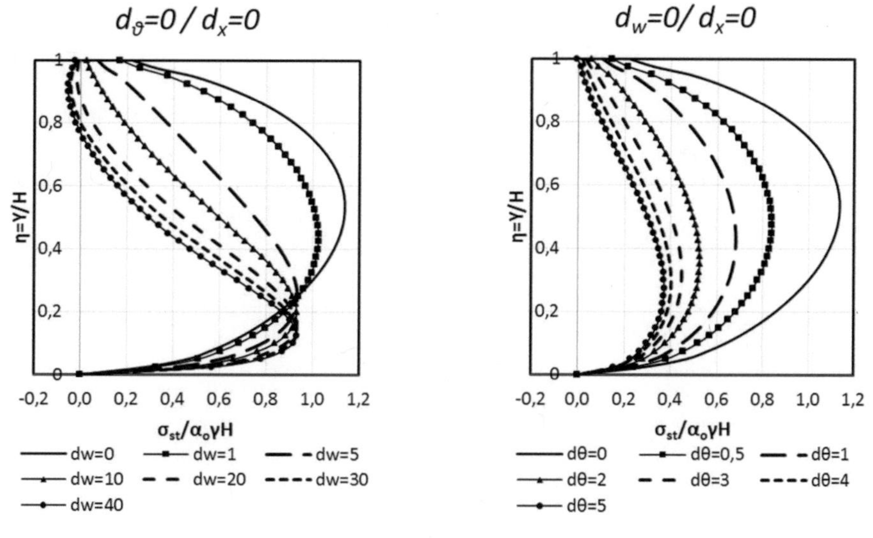

Fig. 5-11 Distributions of wall pressures for statically excited systems ($\omega_1/6$) with different wall and base flexibilities (v =1/3) for the parabolic soil profile.

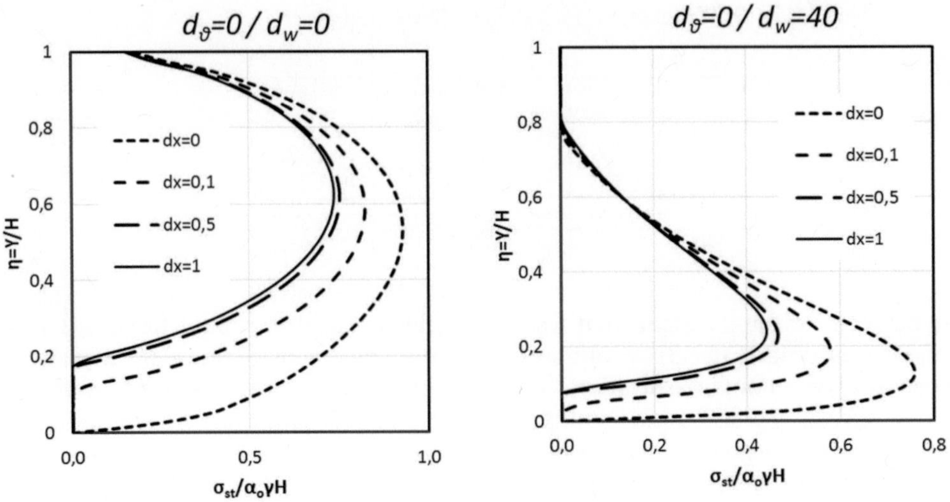

Fig. 5-12 Distributions of wall pressure for statically excited systems ($\omega_1/8$ instead of $\omega_1/6$) with different wall and base flexibilities (v =1/3) for the parabolic soil profile (the tension forces are set to null).

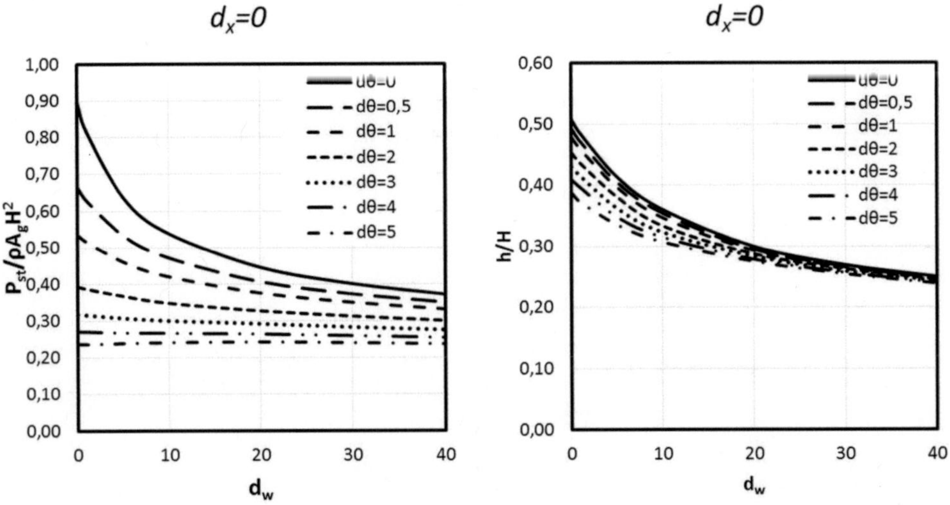

Fig. 5-13 Normalized values of base shear in the wall and normalized effective heights of statically excited systems ($\omega_1/6$) with different wall and base flexibilities (v=1/3) for the parabolic soil profile.

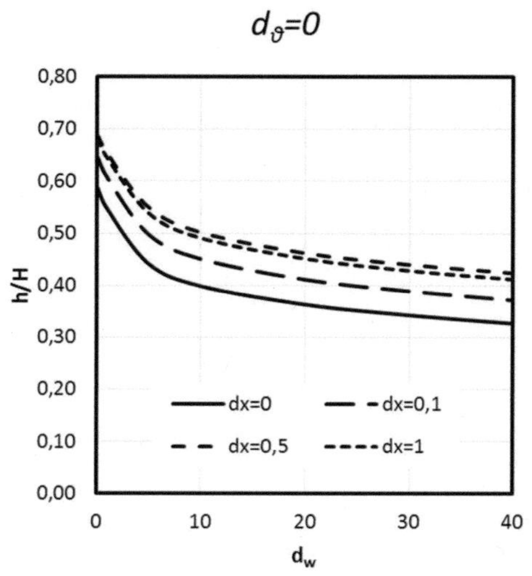

Fig. 5-14 Normalized effective heights of statically excited ($\omega_1/8$) systems with different wall and base flexibilities ($v = 1/3$) for the parabolic soil profile (the tension forces on the wall for $d_x > 0$ are set to null).

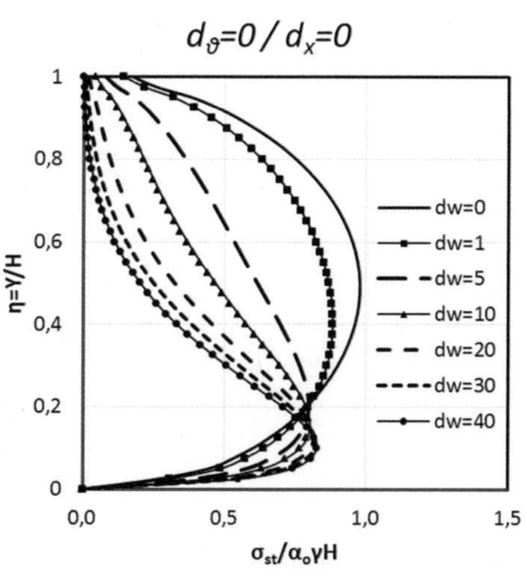

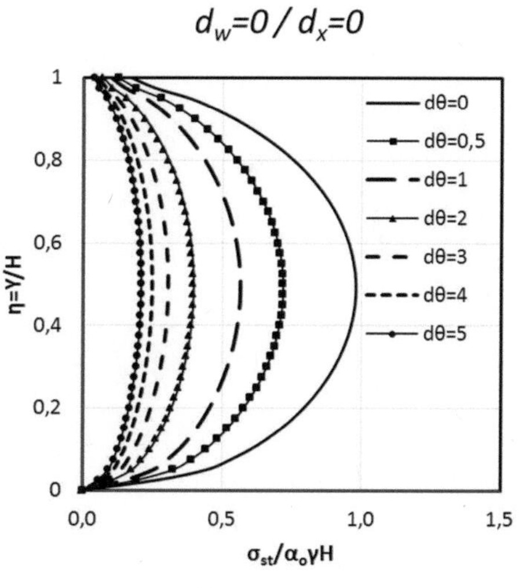

Fig. 5-15 Distributions of wall pressures for statically excited ($\omega_1/6$) systems with different wall and base flexibilities ($v = 1/3$) for the linear soil profile.

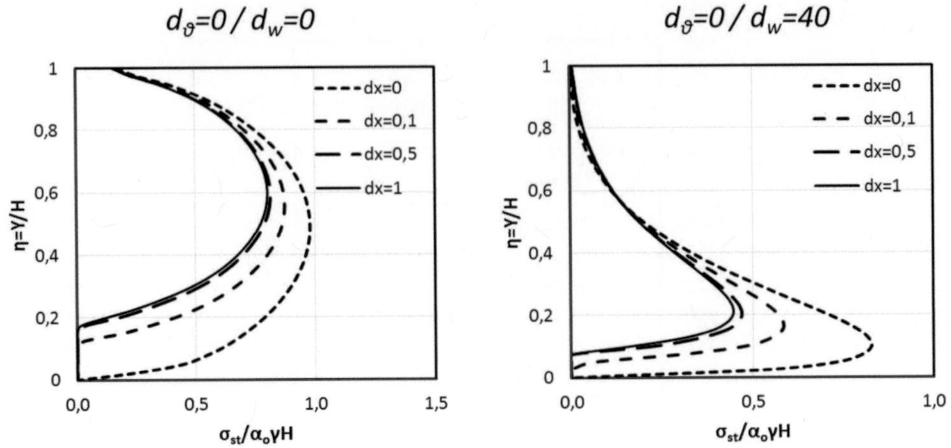

Fig. 5-16 Distributions of wall pressure for statically excited ($\omega_1/6$) systems with different wall and base flexibilities (v =1/3) for the linear soil profile (the tension forces are set to null).

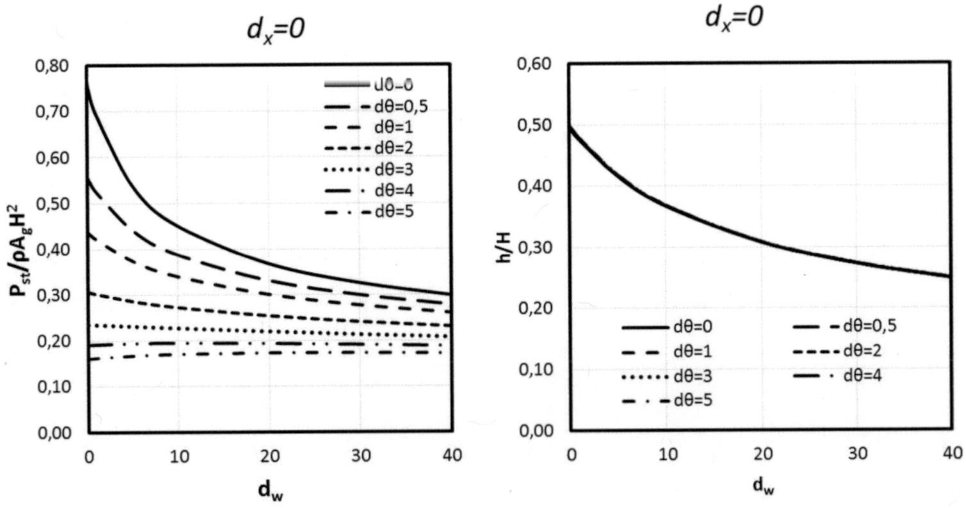

Fig. 5-17 Normalized values of base shear in the wall and normalized effective heights of statically excited ($\omega_1/6$) systems with different wall and base flexibilities (v=1/3) for the linear soil profile.

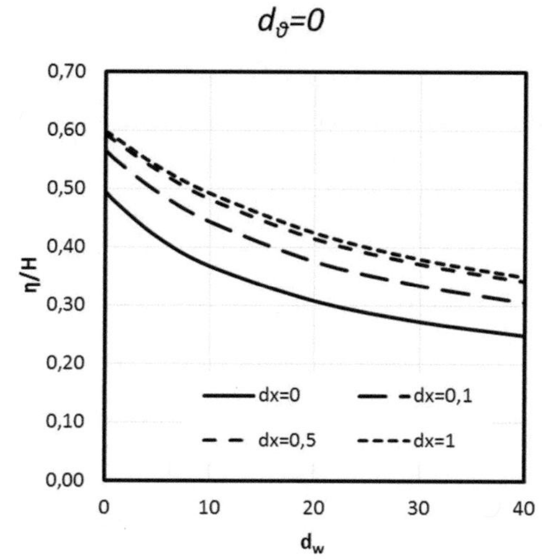

Fig. 5-18 Normalized effective heights of statically excited ($\omega_1/6$) systems with different wall and base flexibilities (v =1/3) for the Linear soil profile (the tension forces on the wall for d_x>0 are set to null).

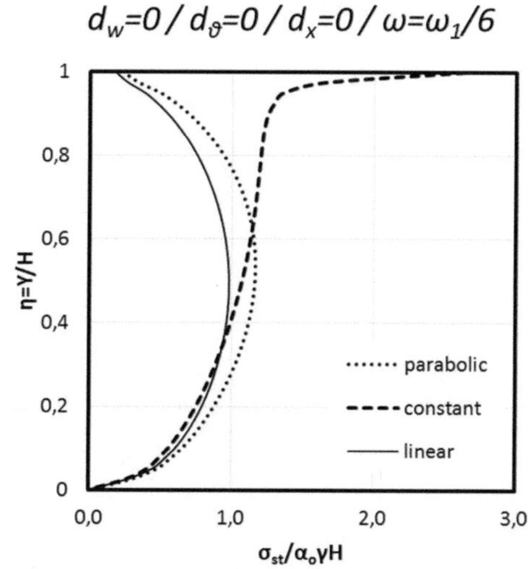

Fig. 5-19 Distributions of wall pressure for statically excited systems for different soil profiles (v =1/3, rigid system).

The examination of the influence of the horizontal spring on the effective height of the total force of the earth pressures shows that although the pressures decrease, the height of application increases. This paradox has been also been pointed out by (Bobet, A., Fernández, G., Jung, C. 2010) and indicates that the soil pressures become smaller near the wall base as the foundation's translational flexibility increases. It is very interesting to observe that the point of the resultant force varies between 0.3 and 0.6 times the wall's height. These are the values suggested by Mononobe (0.3*H*) for yielding walls and by Wood (0.6*H*), whereas Eurocode 8 has adopted the value of 0.5*H*, which corresponds to rigid walls on rigid foundations.

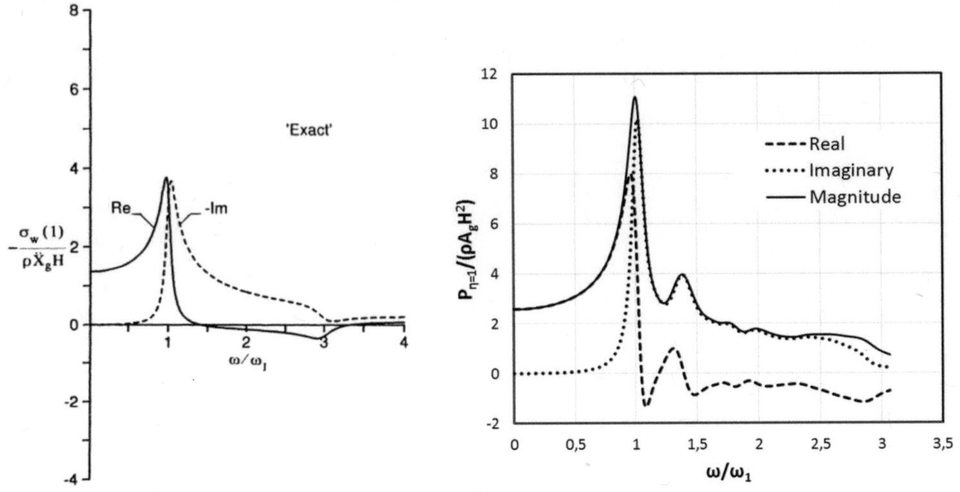

Fig. 5-20 The real, imaginary and magnitude part after (Veletsos, A. S., Younan, A. H. 1992) (left) and this study (right).

Approximate expressions for base shear and effective heights

Veletsos and Younan have provided tables with the normalized values of the base shear and the normalized effective heights for the case of the homogeneous soil profile, but not for the parabolic and the linear soil profiles. This paragraph gives formulas that approximate the values of the base shear and the effective heights calculated in the numerical analysis. For the homogeneous (constant) soil profile:

$$\frac{P_{st}}{\rho A_g H^2} = 1.05 - 0.44\sqrt[3]{d_\theta} - 0.19\sqrt[3]{d_w} + 0.11\sqrt[3]{d_\theta d_w} \quad \textbf{(5-9)}$$

$$\frac{h}{H} = 0.67 - 0.17\sqrt[3]{d_\theta} - 0.1\sqrt[3]{d_w} + 0.05\sqrt[3]{d_\theta d_w} \quad \textbf{(5-10)}$$

Table 31 Difference of the approximate equation for shear forces versus the values calculated by Veletsos and Younan (Veletsos, A. S., Younan, A. H. 1997) for the homogeneous soil profile

	$(V_b)_{st} / \alpha_o \gamma H^2$																	
d_θ	0			1			2			3			4			5		
d_w	Eq.	V-Y	Diff. %	Eq.	V-Y	Diff. %	Eq.	V-Y	Diff. %	Eq.	V-Y	Diff. %	Eq.	V-Y	Diff. %	Eq.	V-Y	Diff. %
0	1.050	0.94	10%	0.610	0.566	7%	0.496	0.437	**12%**	0.415	0.371	11%	0.352	0.331	6%	0.298	0.305	-2%
0.5	0.899	0.883	2%	0.547	0.556	-2%	0.455	0.434	5%	0.391	0.370	5%	0.339	0.331	2%	0.296	0.305	-3%
1	0.860	0.838	3%	0.53	0.547	-3%	0.444	0.431	3%	0.384	0.369	4%	0.336	0.331	1%	0.296	0.304	-3%
2	0.811	0.77	5%	0.509	0.531	-4%	0.431	0.426	1%	0.376	0.367	2%	0.332	0.330	1%	0.295	0.304	-3%
3	0.776	0.72	7%	0.495	0.517	-4%	0.421	0.421	0%	0.370	0.366	1%	0.329	0.329	0%	0.295	0.303	-3%
4	0.748	0.683	9%	0.483	0.506	-5%	0.414	0.417	-1%	0.366	0.364	1%	0.327	0.328	0%	0.295	0.303	-3%
5	0.725	0.653	10%	0.473	0.496	-5%	0.408	0.413	-1%	0.362	0.362	0%	0.325	0.327	-1%	0.294	0.302	-3%
6	0.705	0.628	11%	0.465	0.487	-5%	0.402	0.409	-2%	0.358	0.360	-1%	0.324	0.326	-1%	0.294	0.301	-2%
8	0.670	0.59	**12%**	0.450	0.471	-5%	0.393	0.402	-2%	0.353	0.356	-1%	0.321	0.324	-1%	0.294	0.299	-2%
10	0.641	0.561	**12%**	0.438	0.458	-5%	0.385	0.395	-3%	0.348	0.352	-1%	0.318	0.321	-1%	0.294	0.298	-1%
15	0.581	0.51	**12%**	0.413	0.433	-5%	0.369	0.38	-3%	0.338	0.343	-1%	0.314	0.315	0%	0.293	0.293	0%
20	0.534	0.475	11%	0.393	0.413	-5%	0.356	0.368	-3%	0.33	0.335	-2%	0.31	0.309	0%	0.292	0.288	1%
30	0.46	0.429	7%	0.361	0.382	-6%	0.336	0.347	-3%	0.318	0.319	0%	0.304	0.297	2%	0.292	0.279	4%
40	0.4	0.397	1%	0.336	0.360	**-7%**	0.32	0.33	-3%	0.308	0.307	0%	0.299	0.287	4%	0.291	0.270	7%

Table 32 Difference of the approximate equation for effective heights versus the values calculated by Veletsos and Younan (Veletsos, A. S., Younan, A. H. 1997) for the homogeneous soil profile

	h/H																	
d_θ	0			1			2			3			4			5		
d_w	Eq.	V-Y	Diff. %	Eq.	V-Y	Diff. %	Eq.	V-Y	Diff %	Eq.	V-Y	Diff. %	Eq.	V-Y	Diff. %	Eq.	V-Y	Diff. %
0	0.670	0.599	11%	0.500	0.512	-2%	0.456	0.447	2%	0.425	0.396	7%	0.400	0.356	11%	0.379	0.323	**15%**
0.5	0.575	0.575	0%	0.443	0.496	**-12%**	0.409	0.436	-7%	0.385	0.389	-1%	0.365	0.351	4%	0.349	0.320	8%
1	0.550	0.553	-1%	0.428	0.481	**-12%**	0.396	0.426	-8%	0.374	0.382	-2%	0.356	0.346	3%	0.341	0.316	7%
2	0.519	0.517	0%	0.409	0.456	-11%	0.381	0.408	-7%	0.361	0.369	-2%	0.345	0.337	2%	0.332	0.310	7%
3	0.497	0.488	2%	0.396	0.436	-10%	0.370	0.393	-6%	0.352	0.359	-2%	0.337	0.329	2%	0.325	0.305	6%
4	0.480	0.464	3%	0.386	0.418	-8%	0.361	0.381	-6%	0.344	0.349	-1%	0.331	0.323	2%	0.319	0.300	6%
5	0.465	0.444	5%	0.377	0.403	-7%	0.354	0.369	-4%	0.338	0.341	-1%	0.325	0.316	3%	0.314	0.295	6%
6	0.452	0.426	6%	0.369	0.390	-6%	0.348	0.360	-3%	0.333	0.333	0%	0.321	0.311	3%	0.310	0.291	6%
8	0.430	0.398	7%	0.356	0.368	-3%	0.337	0.343	-2%	0.323	0.321	1%	0.313	0.301	4%	0.303	0.284	6%
10	0.411	0.376	9%	0.345	0.351	-2%	0.328	0.329	0%	0.315	0.310	2%	0.306	0.293	4%	0.298	0.278	7%
15	0.374	0.337	10%	0.322	0.319	1%	0.309	0.304	2%	0.300	0.290	3%	0.292	0.277	5%	0.286	0.265	7%
20	0.344	0.311	10%	0.305	0.298	2%	0.294	0.286	3%	0.287	0.275	4%	0.281	0.265	6%	0.276	0.255	8%
30	0.297	0.278	6%	0.276	0.270	2%	0.271	0.262	3%	0.267	0.254	5%	0.264	0.247	6%	0.261	0.240	8%
40	0.260	0.257	1%	0.254	0.251	1%	0.252	0.245	3%	0.251	0.24	4%	0.25	0.234	6%	0.25	0.229	8%

As it can be seen, the errors are between +12% and -7% for the base shears and between 15% and -12% for the effective heights, good approximations for engineering purposes.

For the linear soil profile the approximate formulas are:

$$\frac{P_{st}}{\rho A_g H^2} = 0.7 - 0.25\sqrt{d_\theta} - 0.065\sqrt{d_w} + 0.03\sqrt{d_\theta d_w} \quad \textbf{(5-11)}$$

$$\frac{h}{H} = 0.5 - 0.04\sqrt{d_w} \quad \textbf{(5-12)}$$

These formulas deliver errors between +13% and -14% for the base shears and between +9% and -13% for the effective heights.

For the parabolic soil profile the approximate formulas are:

$$\frac{P_{st}}{\rho A_g H^2} = 0.84 - 0.28\sqrt{d_\theta} - 0.078\sqrt{d_w} + 0.038\sqrt{d_\theta d_w} \quad \textbf{(5-13)}$$

$$\frac{h}{H} = 0.52 - 0.053\sqrt{d_\theta} - 0.047\sqrt{d_w} + 0.008\sqrt{d_\theta d_w} \quad \textbf{(5-14)}$$

These formulas deliver errors between -15% and +10% for the base shears and between -3% and +9% for the effective heights.

It must be mentioned here that the suggested formulas are derived as best fit curves for the curves of the analysis and offer a rough estimation of the expected values. The error range indicates that the calculated values for the base shear or effective height derived from these formulas can lie at the next or previous value of d_w or d_θ in reality.

Table 33 Difference of the approximate equation for shear forces and effective heights versus the values of the numerical analysis for the linear soil profile. (Abq.=Abaqus analysis)

	$(V_b)_{st} / \alpha_o \gamma H^2$																	
d_w	0			5			10			20			30			40		
d_θ	Eq.	Abq.	Diff. %	Eq.	Abq.	Diff. %	Eq.	Abq.	Diff. %	Eq.	Abq.	Diff. %	Eq.	Abq.	Diff. %	Eq.	Abq.	Diff. %
0	0.71	0.70	2%	0.51	0.55	-8%	0.42	0.49	-13%	0.35	0.41	-15%	0.31	0.34	-10%	0.28	0.29	-3%
0.5	0.52	0.52	0%	0.42	0.43	-3%	0.37	0.38	-4%	0.31	0.33	-5%	0.28	0.28	1%	0.26	0.25	5%
1	0.41	0.45	-9%	0.35	0.37	-5%	0.32	0.34	-6%	0.28	0.29	-2%	0.26	0.26	0%	0.25	0.23	7%
2	0.29	0.35	**-18%**	0.27	0.30	-10%	0.26	0.28	-8%	0.24	0.25	-4%	0.23	0.22	3%	0.22	0.20	**9%**
3	0.22	0.27	**-18%**	0.22	0.24	-9%	0.21	0.23	-7%	0.21	0.21	0%	0.20	0.20	0%	0.20	0.18	**9%**
4	0.18	0.20	-10%	0.18	0.19	-3%	0.18	0.18	0%	0.18	0.18	0%	0.18	0.17	6%	0.18	0.17	4%
5	0.15	0.14	8%	0.16	0.15	5%	0.16	0.15	8%	0.16	0.15	9%	0.16	0.15	9%	0.16	0.15	8%
	h/H																	
0	0.50	0.50	0%	0.42	0.41	**2%**	0.37	0.37	0%	0.31	0.32	-3%	0.27	0.28	**-4%**	0.25	0.25	0%
0.5	0.50	0.50	0%	0.42	0.41	**2%**	0.37	0.37	0%	0.31	0.32	-3%	0.27	0.28	**-4%**	0.25	0.25	0%
1	0.50	0.50	0%	0.42	0.41	**2%**	0.37	0.37	0%	0.31	0.32	-3%	0.27	0.28	**-4%**	0.25	0.25	0%
2	0.50	0.50	0%	0.42	0.41	**2%**	0.37	0.37	0%	0.31	0.32	-3%	0.27	0.28	**-4%**	0.25	0.25	0%
3	0.50	0.50	0%	0.42	0.41	**2%**	0.37	0.37	0%	0.31	0.32	-3%	0.27	0.28	-2%	0.25	0.25	0%
4	0.50	0.50	0%	0.42	0.41	**2%**	0.37	0.37	0%	0.31	0.32	-3%	0.27	0.28	-2%	0.25	0.25	0%
5	0.50	0.50	0%	0.42	0.41	**2%**	0.37	0.37	0%	0.31	0.32	-3%	0.27	0.28	-2%	0.25	0.25	0%

Table 34 Difference of the approximate equation for shear forces and effective heights versus the values of the numerical analysis for the parabolic soil profile. (Abq.=Abaqus analysis)

	$(V_b)_{st} / \alpha_o \gamma H^2$																	
d_w	0			5			10			20			30			40		
d_θ	Eq.	Abq.	Diff. %	Eq.	Abq.	Diff. %	Eq.	Abq.	Diff. %	Eq.	Abq.	Diff. %	Eq.	Abq.	Diff. %	Eq.	Abq.	Diff. %
0	0.88	0.84	5%	0.62	0.67	-7%	0.52	0.59	-12%	0.43	0.49	-12%	0.39	0.41	-5%	0.36	0.35	3%
0.5	0.65	0.64	2%	0.52	0.53	-2%	0.46	0.48	-4%	0.39	0.41	5%	0.36	0.36	0%	0.34	0.32	6%
1	0.52	0.56	-7%	0.45	0.47	-4%	0.40	0.43	-7%	0.36	0.38	5%	0.34	0.34	0%	0.32	0.31	3%
2	0.38	0.44	-14%	0.36	0.39	-8%	0.34	0.37	-8%	0.32	0.34	6%	0.30	0.31	-3%	0.29	0.29	0%
3	0.31	0.36	-14%	0.30	0.33	-9%	0.29	0.32	-9%	0.28	0.30	7%	0.27	0.29	-7%	0.26	0.28	-7%
4	0.26	0.28	-7%	0.26	0.28	-7%	0.26	0.27	-4%	0.25	0.27	7%	0.25	0.27	-7%	0.24	0.27	-11%
5	0.23	0.21	**10%**	0.23	0.23	0%	0.23	0.24	-4%	0.23	0.25	8%	0.23	0.25	-8%	0.22	0.26	**-15%**
	h/H																	
0	0.51	0.52	-2%	0.41	0.41	0%	0.36	0.37	-3%	0.30	0.31	**-3%**	0.26	0.26	0%	0.24	0.22	**9%**
0.5	0.49	0.48	2%	0.40	0.39	3%	0.35	0.35	0%	0.29	0.30	**-3%**	0.26	0.26	0%	0.24	0.22	**9%**
1	0.48	0.47	2%	0.39	0.38	3%	0.34	0.34	0%	0.29	0.29	0%	0.26	0.25	4%	0.24	0.22	**9%**
2	0.45	0.45	0%	0.38	0.37	3%	0.33	0.33	0%	0.28	0.29	**-3%**	0.25	0.25	0%	0.23	0.22	5%
3	0.43	0.43	0%	0.36	0.35	2%	0.32	0.32	0%	0.28	0.28	-2%	0.25	0.25	0%	0.23	0.22	4%
4	0.41	0.41	0%	0.34	0.34	0%	0.31	0.32	-2%	0.27	0.28	**-3%**	0.25	0.24	2%	0.23	0.22	3%
5	0.39	0.40	**-3%**	0.33	0.34	-3%	0.30	0.31	-2%	0.26	0.27	-2%	0.24	0.24	1%	0.22	0.22	2%

5.1.4 Influence of the excitation frequency

As the computed dynamic soil pressures are referred to as "static" (static refers not to the pressures caused by gravitational forces, but to dynamic pressures with an angular excitation velocity $\omega \rightarrow 0$), it is of interest to see the influence of other frequencies of excitation. The following figures give the dynamic soil pressures for a rigid and constraint wall for the resonance frequency of the soil stratum and a higher frequency (three times the resonance frequency) of the soil stratum for two soil profiles (constant and parabolic). As it can be seen, the frequency of the excitation influences the dynamic soil pressures. An excitation at resonance frequency of the soil stratum can can lead to shear force up to 3.5 times bigger in the case of a rigid wall and up to 7 times bigger in the case of a flexible wall. These values are the results of the time history analysis (until steady state response is achieved in time domain) and they are a slightly bigger than the values calculated by (Veletsos, A. S., Younan, A. H. 1997).

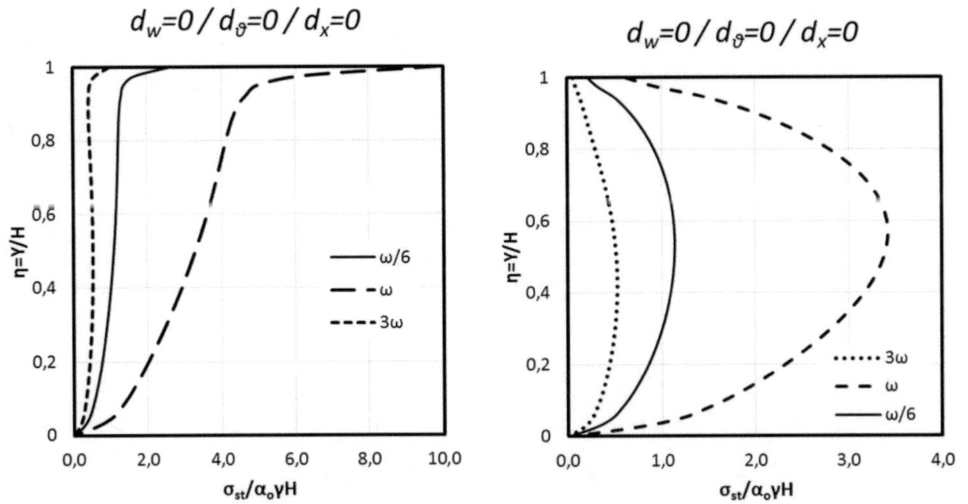

Fig. 5-21 Dynamic soil pressure distribution for the constant (left) and parabolic (right) soil profiles for the three excitation frequencies.

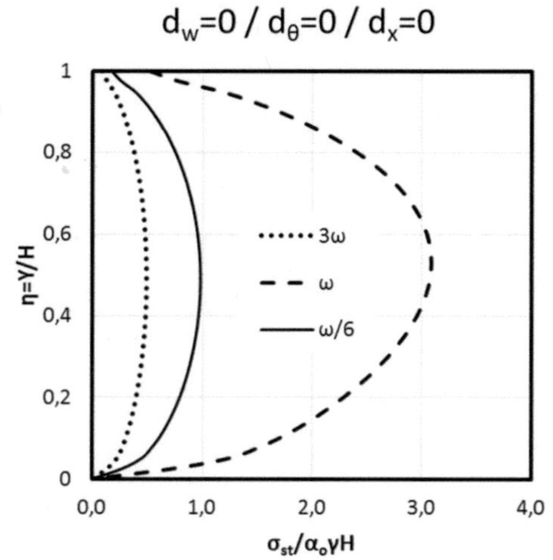

Fig. 5-22 Dynamic soil pressure distribution for linear soil profile for the three excitation frequencies.

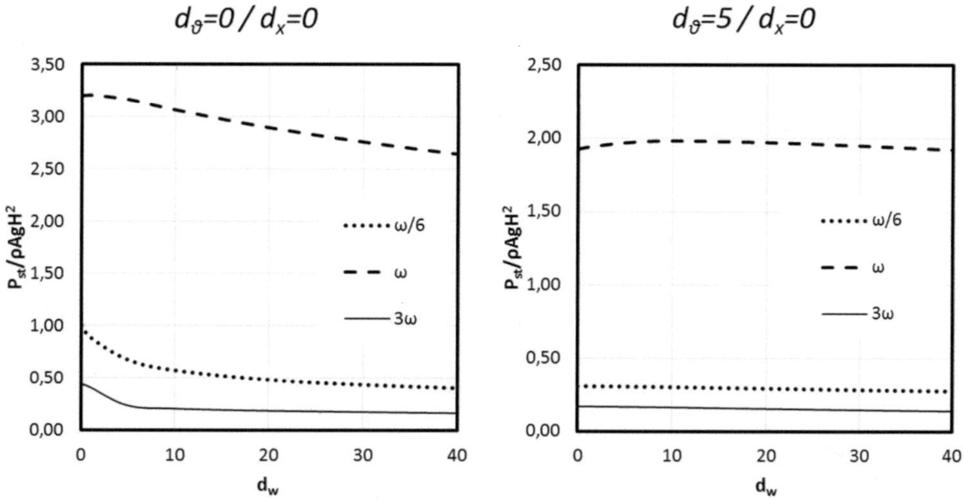

Fig. 5-23 Normalized shear forces for homogeneous (constant) soil profile for the three excitation frequencies.

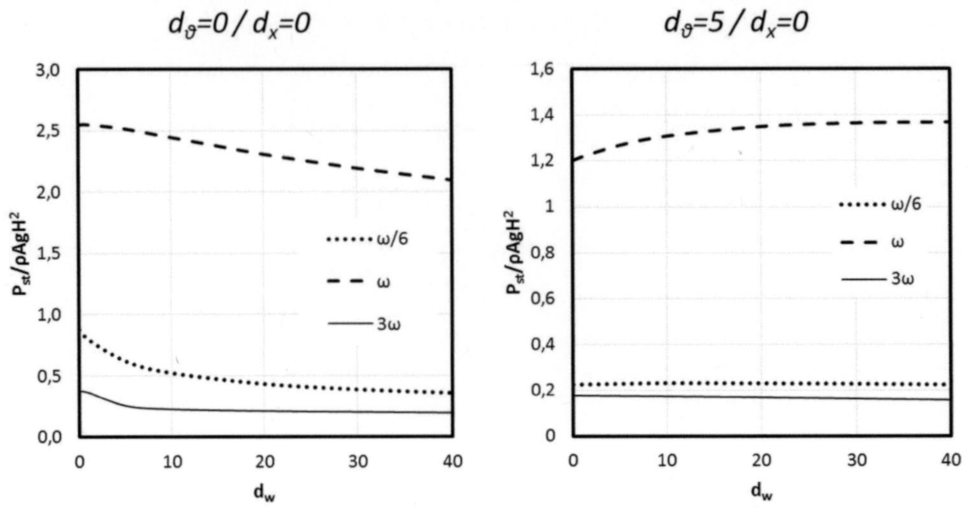

Fig. 5-24 Normalized shear forces for parabolic soil profile for the three excitation frequencies.

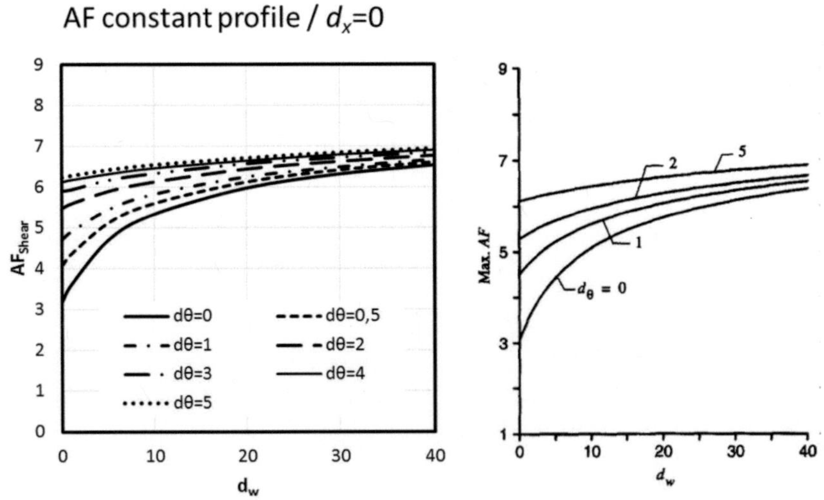

Fig. 5-25 Amplification factors of the base shear of this numerical study and the analytical study conducted by (Veletsos, A. S., Younan, A. H. 1997) for homogeneous soil.

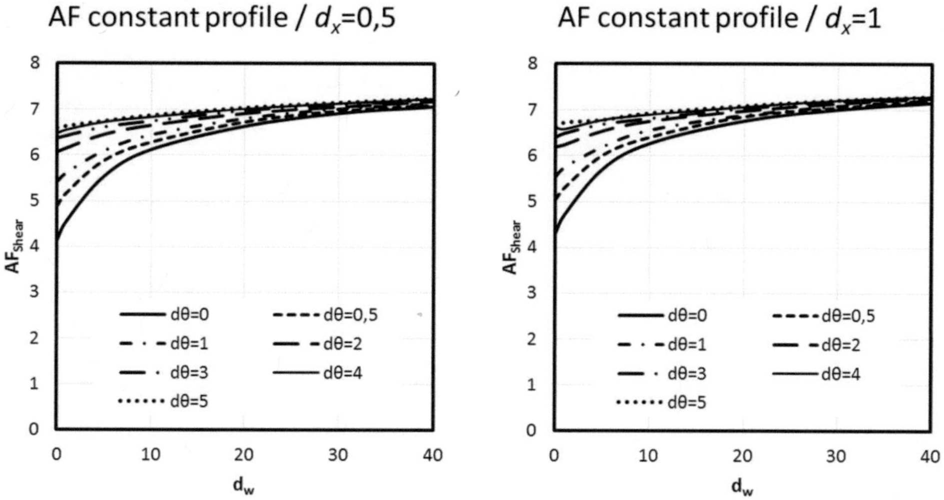

Fig. 5-26 Amplification factors of the base shear for different base flexibilities according to this study.

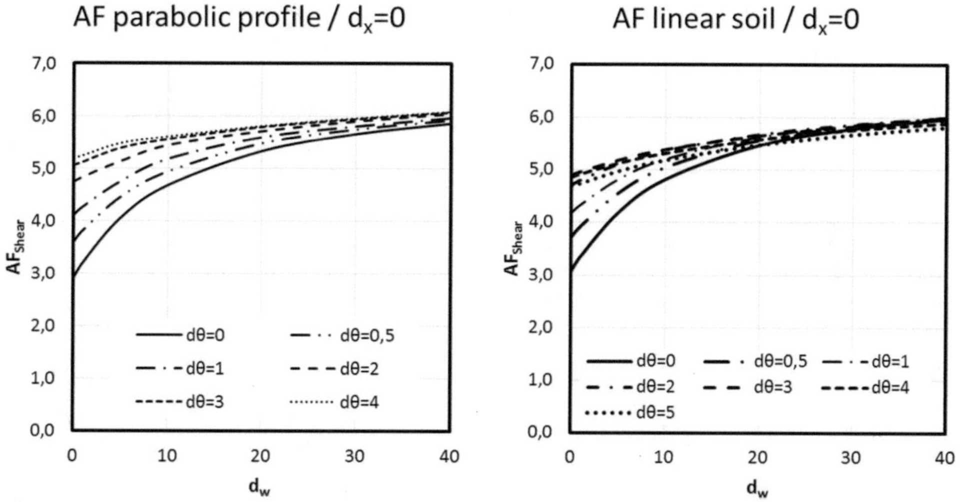

Fig. 5-27 Amplification factors of the base shear for the two inhomogeneous soil profiles of this study.

The increase in the base flexibility by means of a translational spring leads to a small (practically negligible) increase in the amplification factors of the base shear for the homogeneous soil.

As it can be seen from the next figures, the amplification factors are reduced for the more realistic soil profiles with parabolic and linear shear modulus.

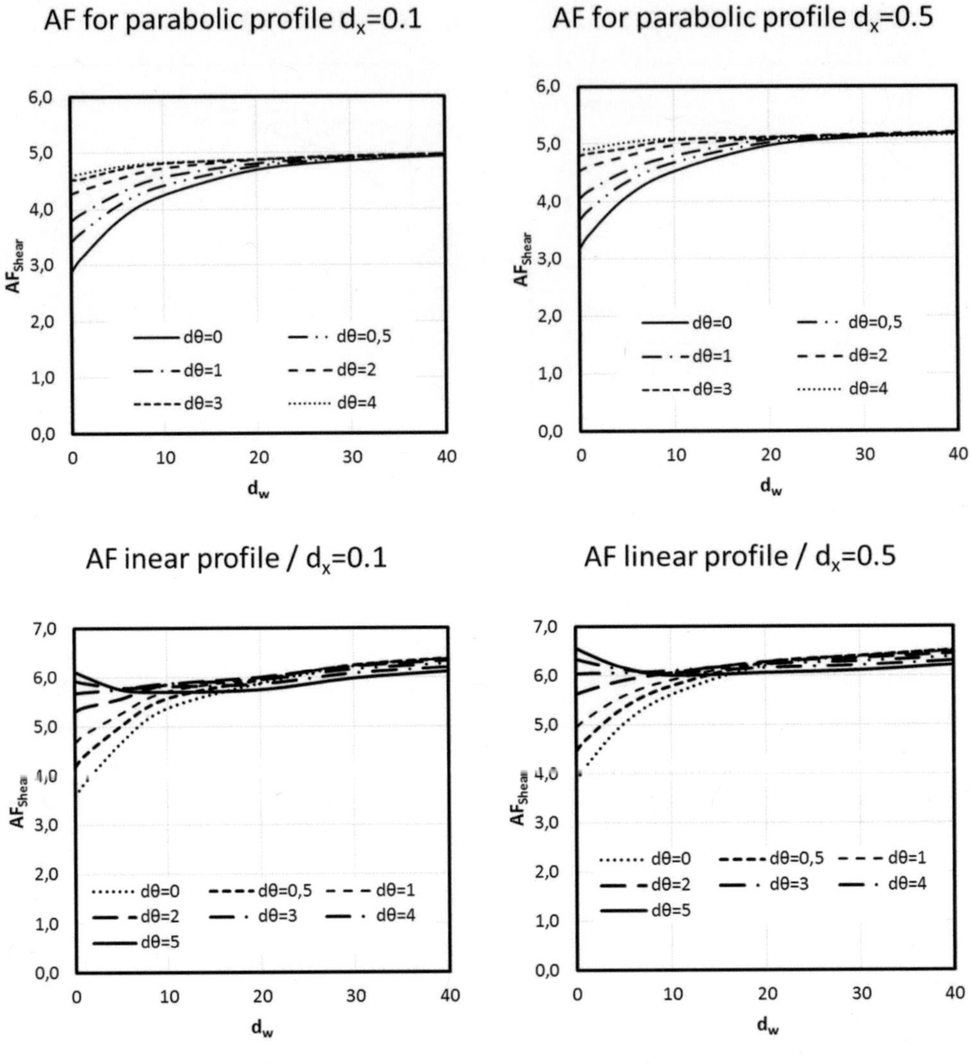

Fig. 5-28 Amplification factors of the base shear for the two inhomogeneous soil profiles including horizontal base flexibility (tension forces set to null).

The influence of the excitation frequency has also been investigated by means of a steady state analysis (frequency domain) available in Abaqus. This analysis offers a better overview of the resonance of the soil pressures, which depend on the relative wall and base flexibility and the soil profile. For this analysis the same values of d_w and d_θ as before were investigated (i.e. d_w=0/1/5/10/20/30/40 and d_θ=0/0.5/1/2/3/4/5) but the influence of the translational spring was excluded from the analysis. The frequency range examined was from 0.01 to 16 Hz which corresponds to ω values of 0.0628 (quasi static loading) to about 100 rad/sec or to period T values of 0.0625 to 10 sec.

However, despite the greater difficulty in defining the damping characteristics (Rayleigh damping) properly and the possible error at the excitation frequency, the results of the time domain analysis are in better agreement with the theoretical results provided by other researchers (Veletsos, A. S., Younan, A. H. 1992, 1993, 1994a, 1997, 2000b). The amplification factor of the soil pressures of the frequency domain analysis fluctuates between 3.75 for the rigid wall on a rigid base (the time domain analysis gave an amplification factor of 3.19) and 7.78 for a very flexible system (instead of the factor 6.95 in the time domain analysis). According to Veletsos and Younan (Veletsos, A. S., Younan, A. H. 1993) the maximum values in the resonance period for rigid walls take a value of $1/\sqrt{\delta}$ (=$1/\sqrt{0.1} = 3.16$) which tends to the value of the single degree of freedom oscillator $1/\delta$ (=$1/0.1 = 10$) for the more flexible systems.

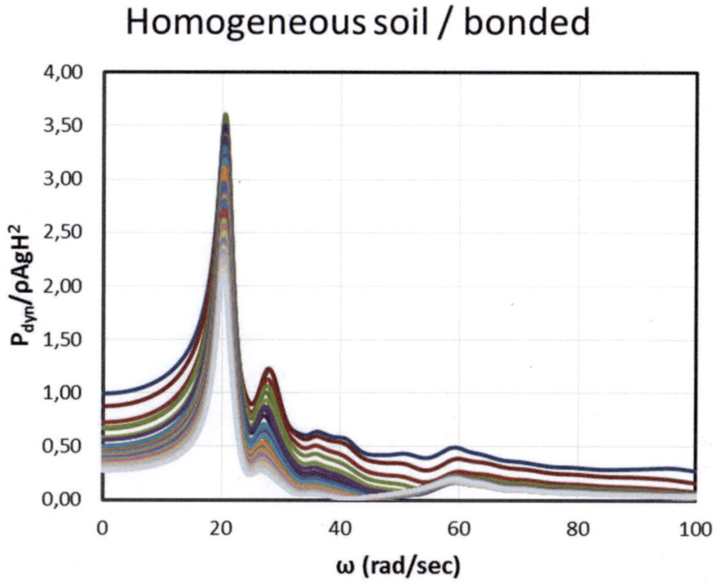

Fig. 5-29 Steady state response of soil pressures of homogeneous soil for all investigated relative wall and base flexibilities (bonded contact).

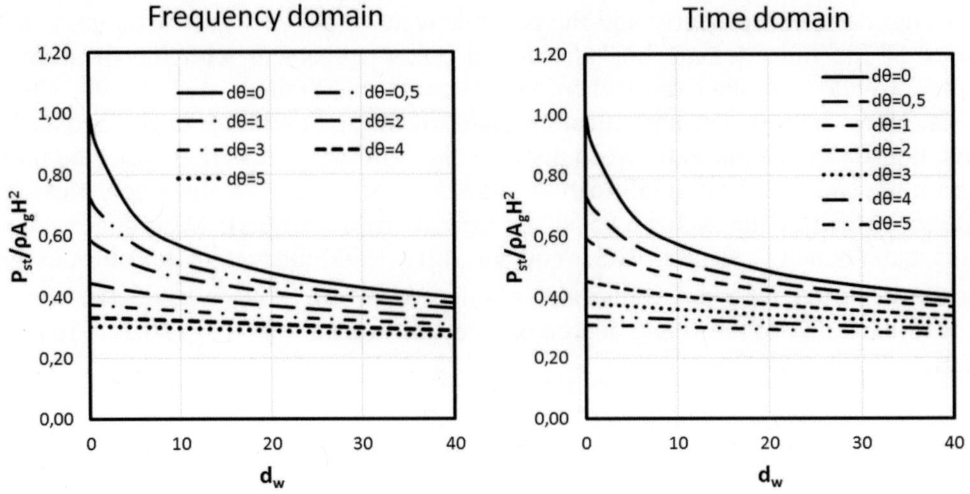

Fig. 5-30 Comparison of the normalized base shear forces: frequency domain vs time domain analysis (ξ=5%, bonded contact).

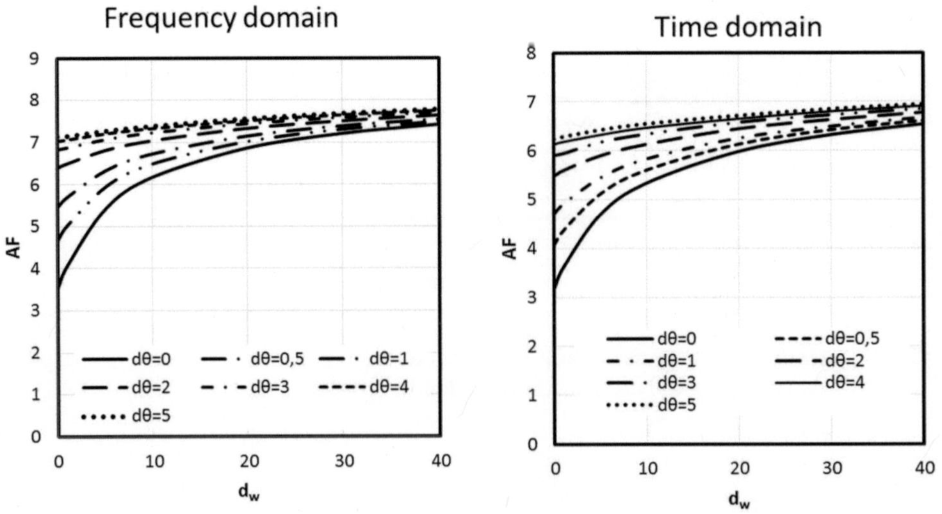

Fig. 5-31 Comparison of the amplification factors for the base shear forces for homogeneous soil: frequency domain vs time domain analysis (ξ=5%, bonded contact).

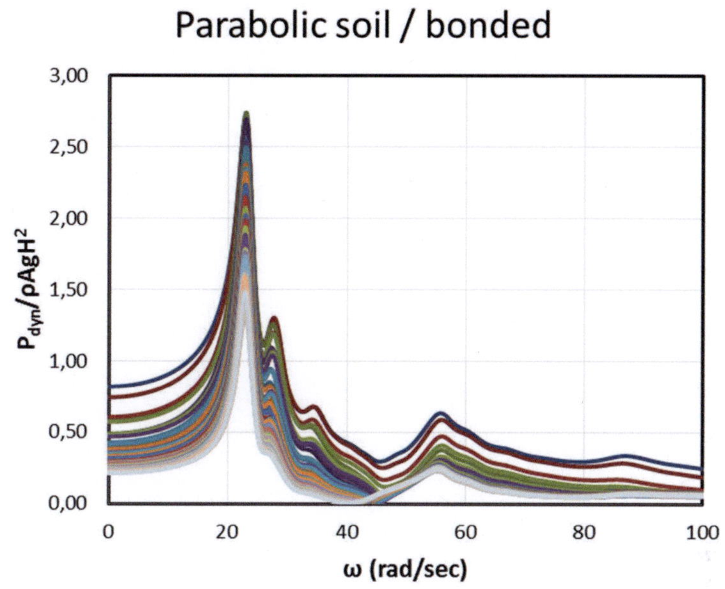

Fig. 5-32 Steady state response of soil pressures of soil with parabolic profile for all investigated relative wall and base flexibilities (bonded contact).

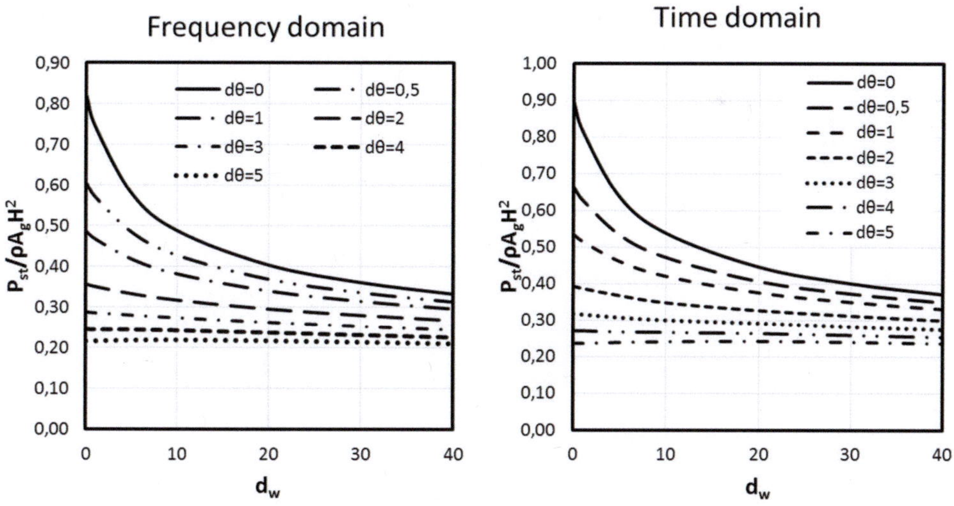

Fig. 5-33 Comparison of the normalized base shear forces for parabolic soil profile: frequency domain vs time domain analysis (ξ=5%, bonded contact).

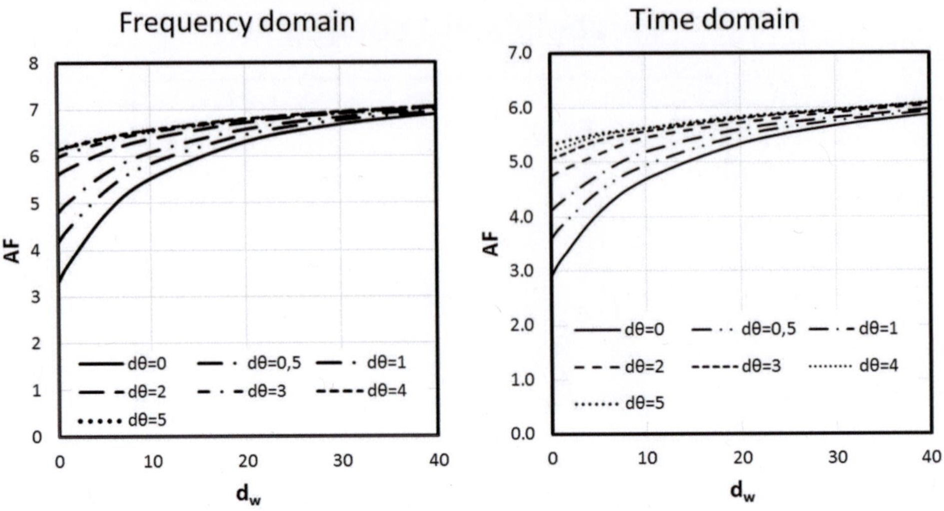

Fig. 5-34 Comparison of the amplification factors for the base shear forces for parabolic soil profile: frequency domain vs time domain analysis (ξ=5%, bonded contact).

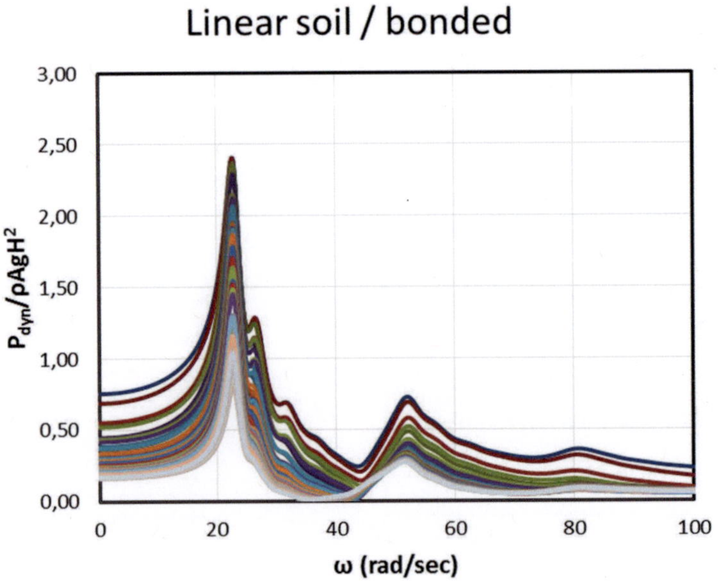

Fig. 5-35 Steady state response of soil pressures of soil with linear profile for all investigated relative wall and base flexibilities.

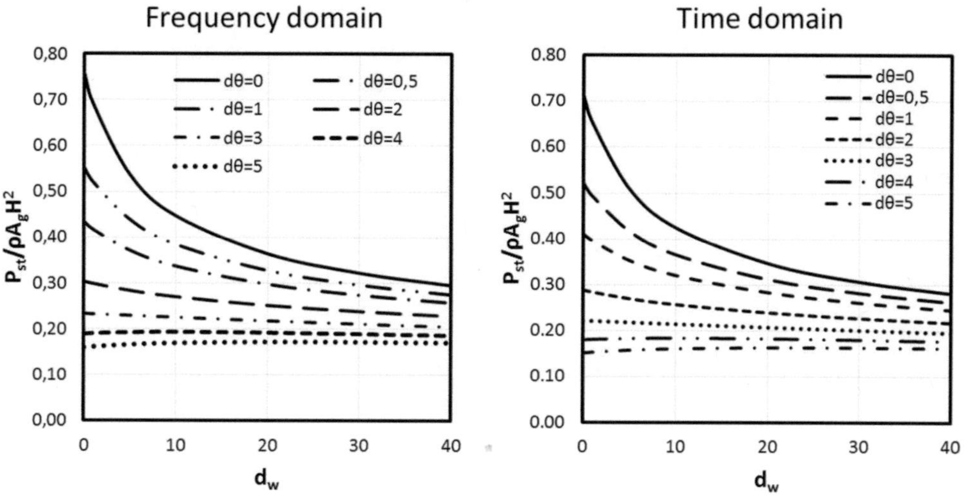

Fig. 5-36 Comparison of the normalized base shear forces for linear soil profile: frequency domain vs time domain analysis (ξ=5%, bonded contact) for statically excited system.

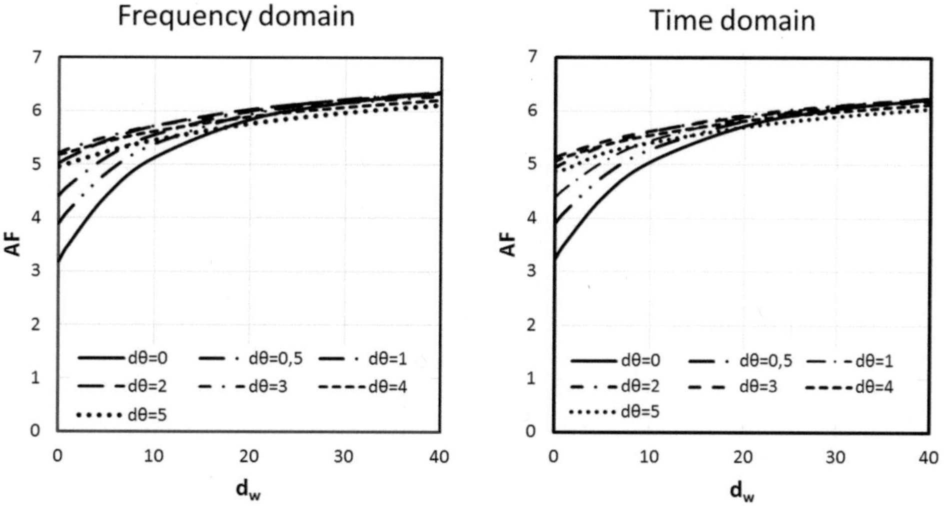

Fig. 5-37 Comparison of the amplification factors for the base shear forces for linear soil profile: frequency domain vs time domain analysis (ξ=5%, bonded contact).

5.1.5 Influence of the contact modelling

(Wood 1973) had calculated dynamic soil pressures with the assumption of both a bonded and a smooth contact. Bonded contact means that the wall is attached to the soil and no relative movement occurs. Smooth contact means that the wall is attached to the soil (no separation occurs) but the soil is able to slide along the soil-wall interface. For the case investigated by (Wood 1973) of two rigid walls containing soil (see next chapter) this different contact modelling leads to different eigenfrequencies of the two systems but similar static pressures. For the case of the bonded contact the soil pressures have a singularity at $\eta=1$, i.e. at the soil's free surface. Both the bonded and the smooth contact were investigated here by means of a frequency domain analysis. The smooth contact results in higher soil pressures and shear forces at the wall. The amplification factor AF is also higher for the smooth contact. Moreover, a non-bonded contact was also investigated in means of a frequency domain (steady state analysis). The results of this type of contact are identical to the results of the smooth contact.

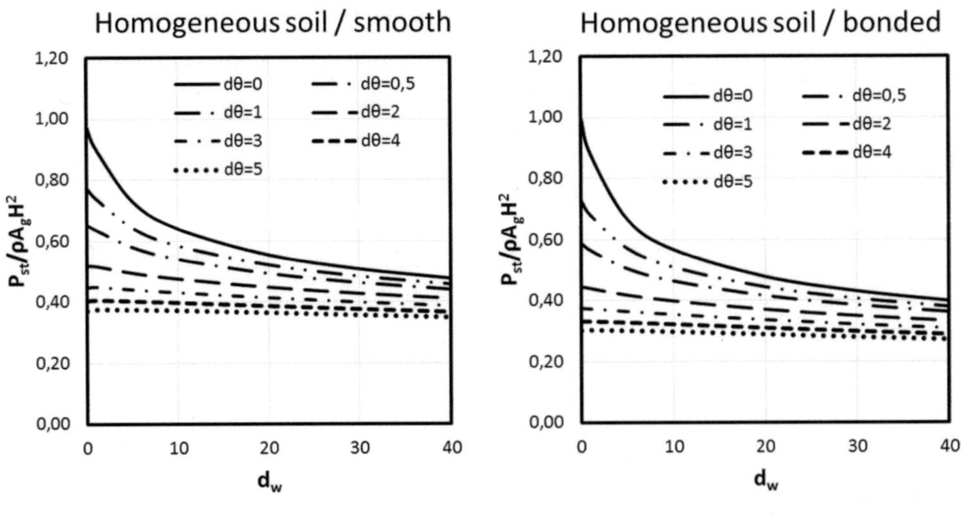

Fig. 5-38 Comparison of the normalized shear forces between bonded and smooth contact for the homogeneous soil profile (ξ=5%).

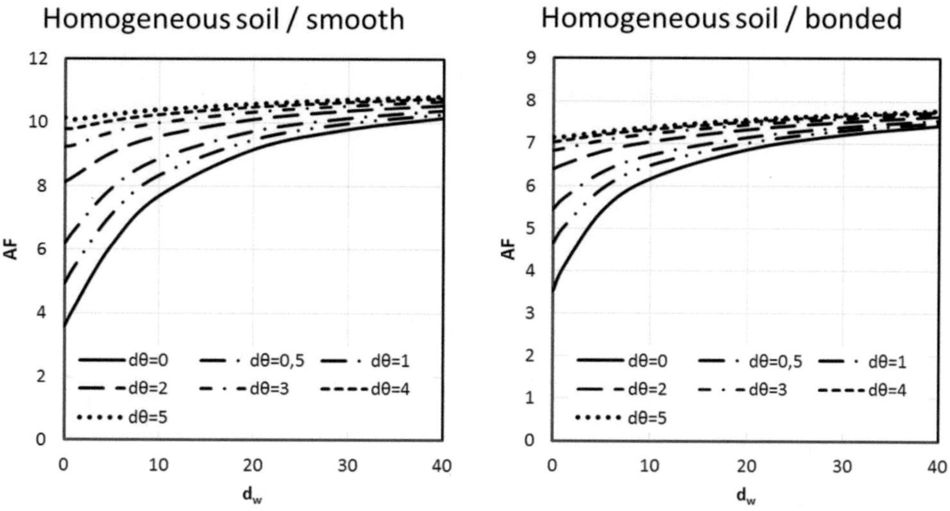

Fig. 5-39 Comparison of the amplification factors between bonded and smooth contact for the homogeneous soil profile (ξ=5%).

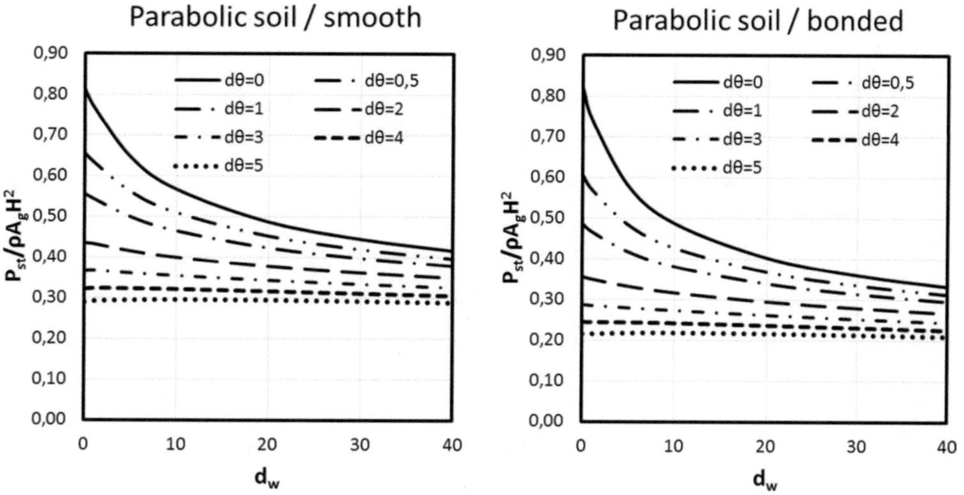

Fig. 5-40 Comparison of the normalized shear forces between bonded and smooth contact for the parabolic soil profile (ξ=5%).

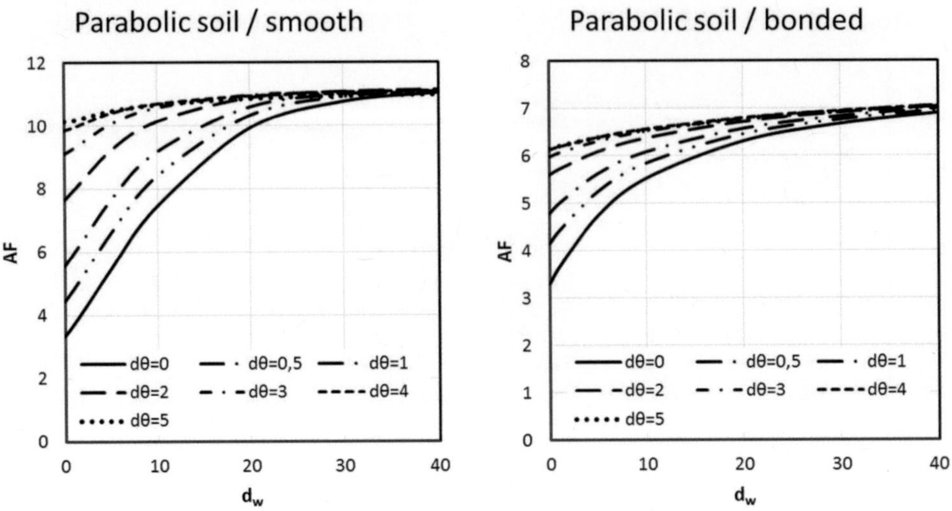

Fig. 5-41 Comparison of the amplification factors between bonded and smooth contact for the parabolic soil profile (ξ=5%).

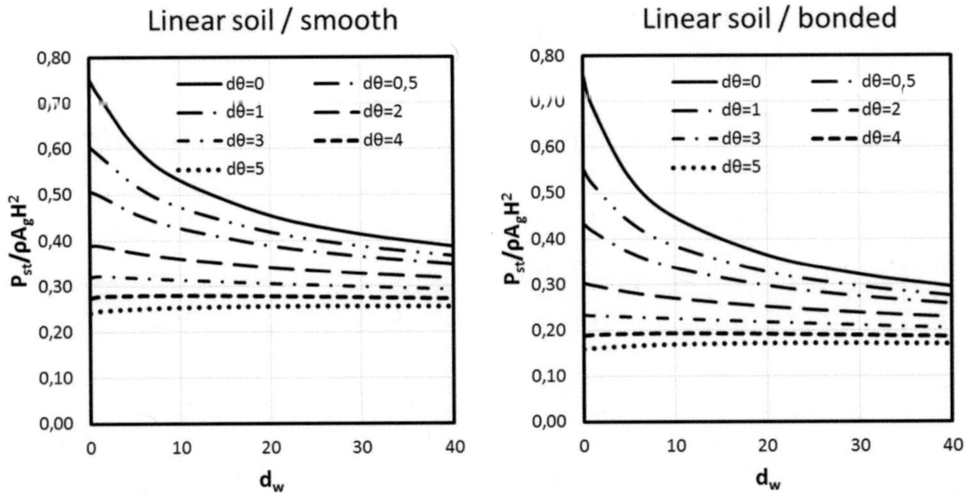

Fig. 5-42 Comparison of the normalized shear forces between bonded and smooth contact for the linear soil profile (ξ=5%).

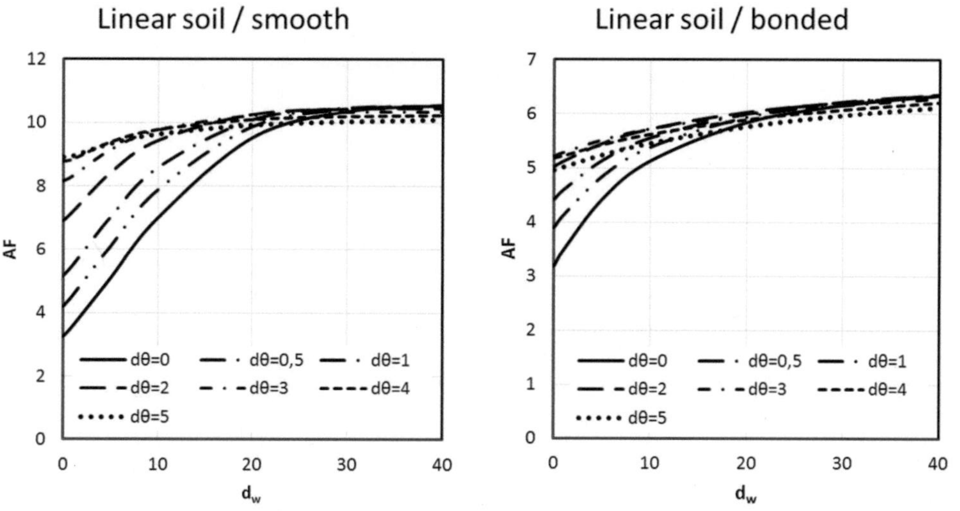

Fig. 5-43 Comparison of the amplification factors between bonded and smooth contact for the linear soil profile (ξ=5%).

5.1.6 Influence of the wall's inclination

The semi-infinite system was also investigated for the case of an inclined wall. For this investigation the wall is assumed to be rigid (d_w=0) and to lie on a rigid foundation (d_θ=0, d_x=0). The following diagram shows the dynamic soil pressure distribution for the static case (ω=3.27 rad/sec) for different wall inclination angles. As it can be seen from the following diagrams, the total force and its height of application decrease linearly with a decreasing wall inclination. Based on the results the following formulas for the correction factors can be proposed:

$$C_{P,incl} = 0.009(\theta^o) + 0.18 \quad \textbf{(5-15)}$$

$$C_{h,incl} = 0.0024(\theta^o) + 0.37 \quad \textbf{(5-16)}$$

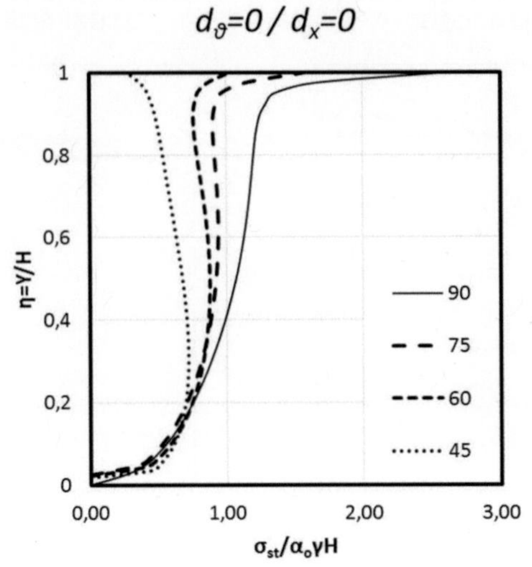

Fig. 5-44 Soil pressure distribution for statically excited system for wall inclinations of 45, 60, 75 and 90 degrees (homogeneous soil).

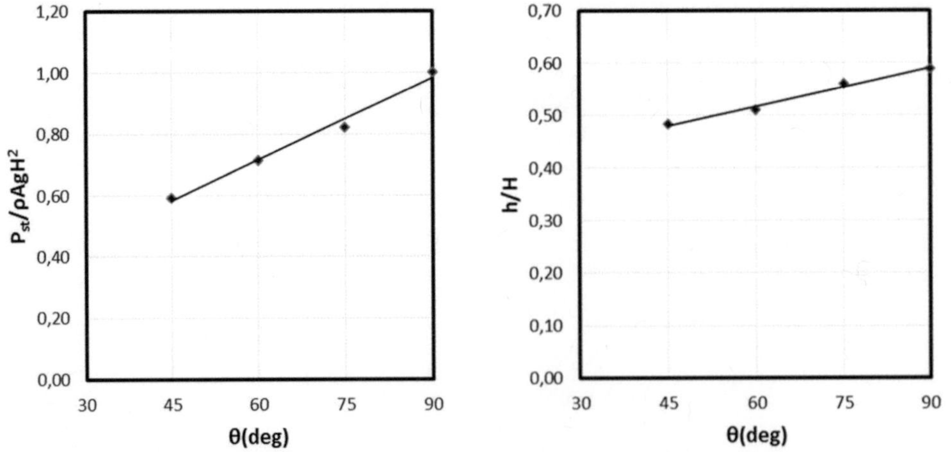

Fig. 5-45 Normalized values of base shear and effective heights of statically excited systems ($v=1/3$, $d_\theta=0$, $d_x=0$, $d_w=0$) for different wall inclinations of 45, 60, 75 and 90 degrees (homogeneous soil).

5.1.7 Influence of Poisson's ratio of soil

The influence of Poisson's ratio was also investigated. The numerical analysis showed that the soil pressures increase with an increasing Poisson's ratio. The height of application of the resultant force remains constant at $0.59H$ for values of Poisson's ratio between 0.1 and 0.4 and decreases to 0.55 for the incompressible soil.

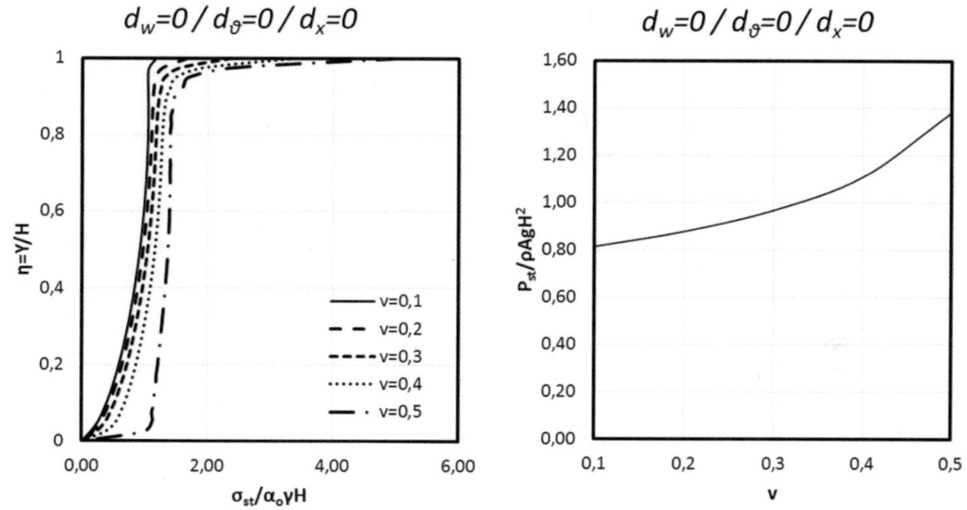

Fig. 5-46 Soil pressure distribution for different values of Poisson's ratio(homogeneous soil) and the normalized shear force versus Poisson's ratio (statically excited system).

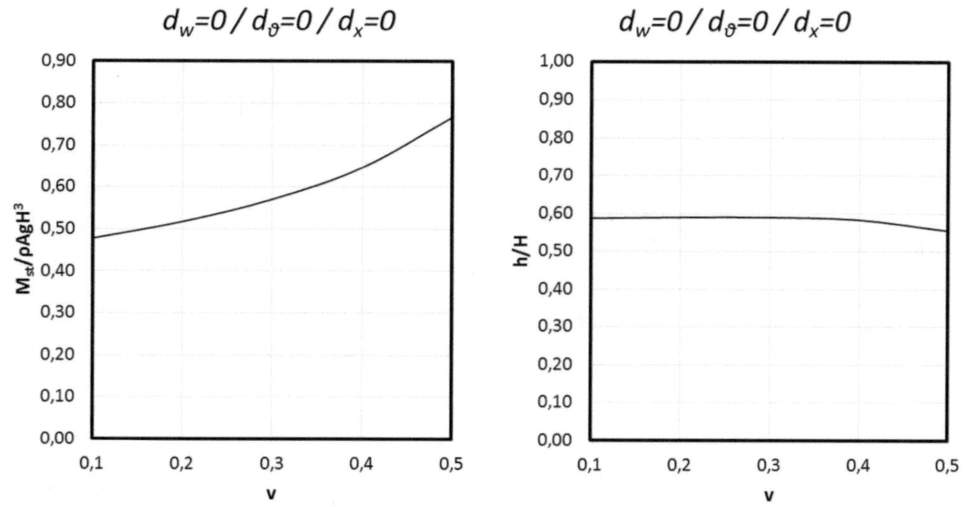

Fig. 5-47 The normalized base moment (left) and the effective height (right) versus Poisson's ratio (homogeneous soil - statically excited system).

Assuming as reference a Poisson's ratio of ν=0.33, which in this numerical study delivers a normalized shear force equal to unity for the rigid wall, the following simplified expression can be used to multiply the soil pressures and/or the normalized shear force and moment, which respond to Poisson's ratio 0.33 in order to obtain the soil pressures of other Poisson's ratios.

$$Bonded\ contact\colon C_v = \frac{0.8 + (10v)^v}{2.2} \tag{5-17}$$

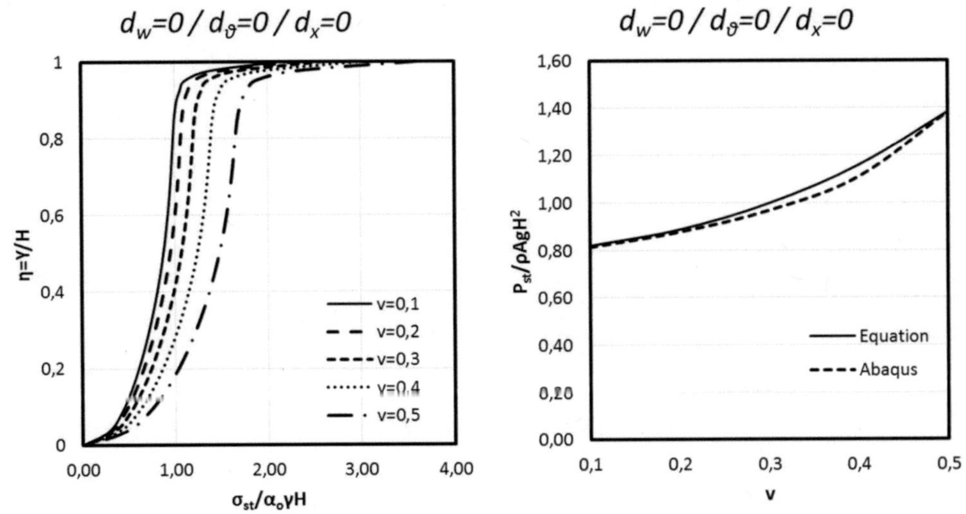

Fig. 5-48 Soil pressure distribution (left) and normalized shear pressures (right) calculated with the proposed formula (bonded contact – rigid system).

When a smooth contact is applied, the pressures distribution is different than the one as in the bonded contact. For smooth contact the next formula should apply:

$$Smooth\ contact\colon C_v = \frac{0.81 + (6v)^v}{2.05} \tag{5-18}$$

These formulas hold for homogeneous soil, rigid wall and rigid foundation.

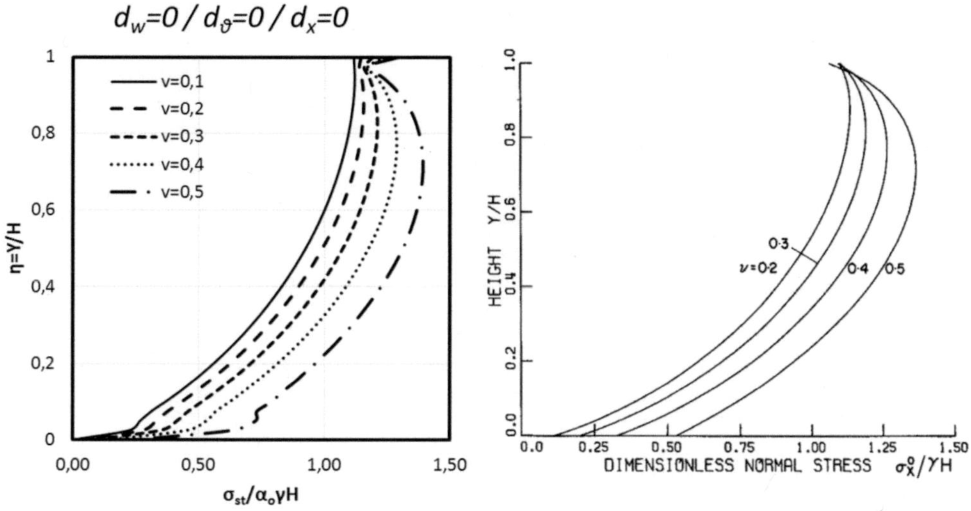

Fig. 5-49 Soil pressure distribution according to this study for smooth wall (left) and according to (Wood 1973) (right).

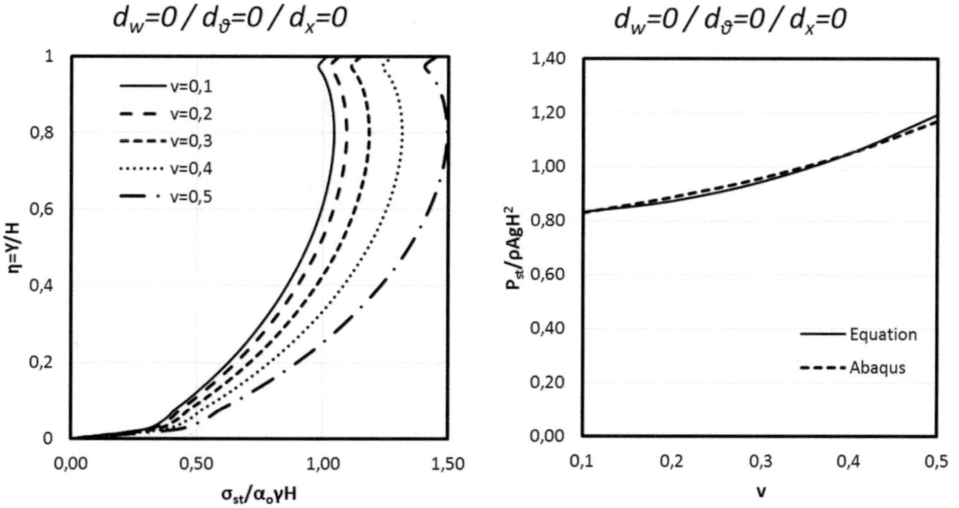

Fig. 5-50 Soil pressure distribution (left) and normalized shear pressures (right) calculated with the proposed formula (smooth contact – rigid system).

5.1.8 Influence of soil damping

In all former analyses damping was given either as Rayleigh damping for the time domain analyses or as structural damping for the frequency domain analyses, being kept constant at 5% of the critical threshold for the soil material. In reality the hysteric damping of the soil depends on the amplitude of the excitation. In order to see the influence of the hysteretic damping of the soil on the dynamic soil pressures, different damping values were used for frequency domain analyses (steady state response). The values chosen here are equal to 5%, 7.5%, 10%, 15% and 20% of the critical damping. It must be clarified that the impedance of the foundation was neglected in all of the analyses.

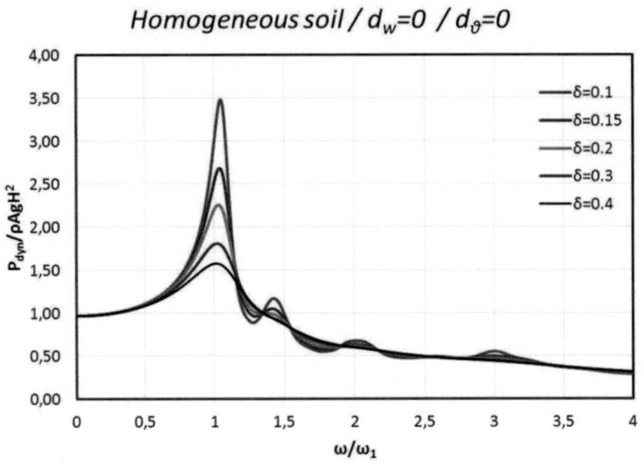

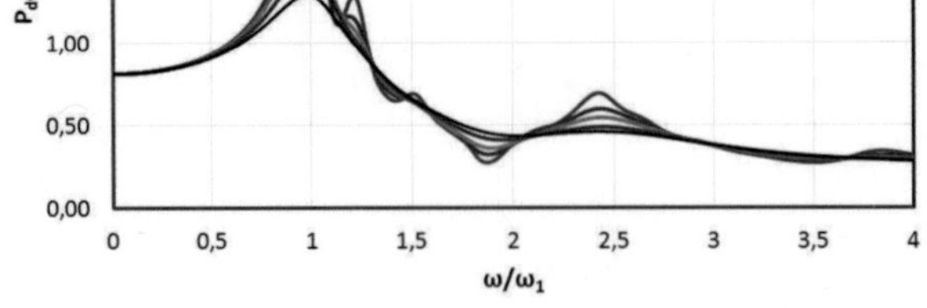

Fig. 5-51 Normalized shear forces for different soil profiles and damping ratios ($\delta=2\xi$) for a rigid wall on a rigid base (smooth contact).

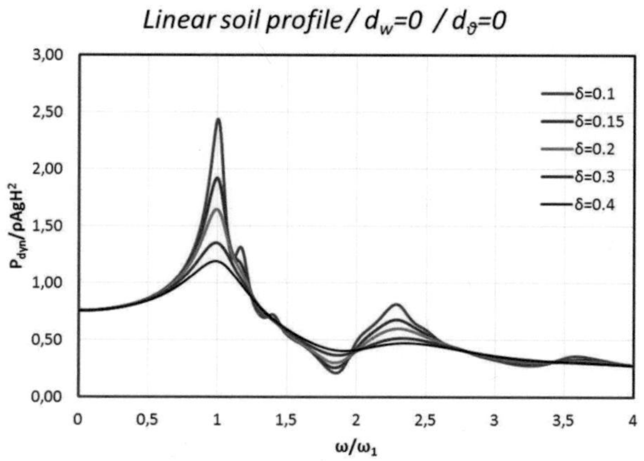

Fig. 5-52 Normalized shear forces for different soil profiles and damping ratios (δ=2ξ) for a rigid wall on a rigid base (smooth contact).

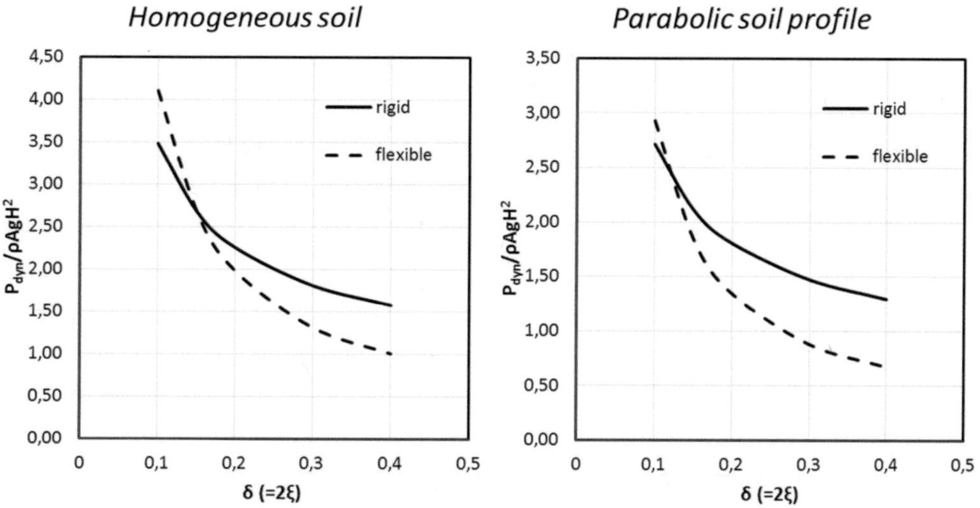

Fig. 5-53 Dependence of the normalized shear force on the damping ratio for different soil profiles, resonance frequency and the two extreme cases: rigid and flexible wall (smooth contact).

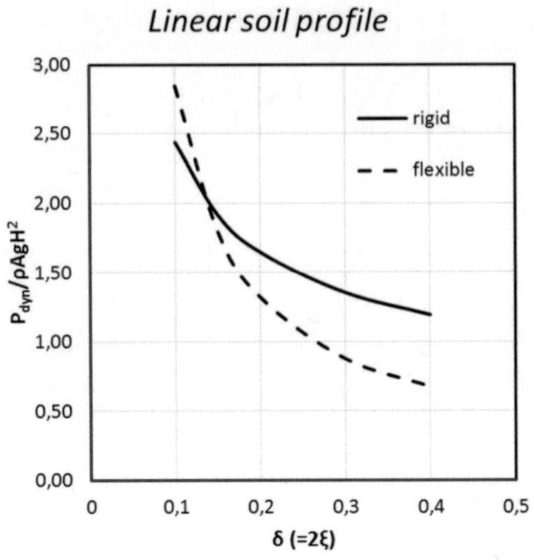

Fig. 5-54 Dependence of the normalized shear force on the damping ratio for different soil profiles, resonance frequency and the two extreme cases: rigid and flexible wall (smooth contact).

Since the damping ratio plays no role for the statically excited system, the results presented here refer to the resonance response. This is why the shear force of the flexible wall appears to be bigger than that of the rigid wall. The influence of the damping ratio of the soil on the dynamic soil pressures is not the same for rigid and flexible walls. For flexible systems the influence of damping is bigger and the reduction of the peak soil pressures at the resonance frequency is greater than the reduction for rigid systems. This can be observed also in the following figures.

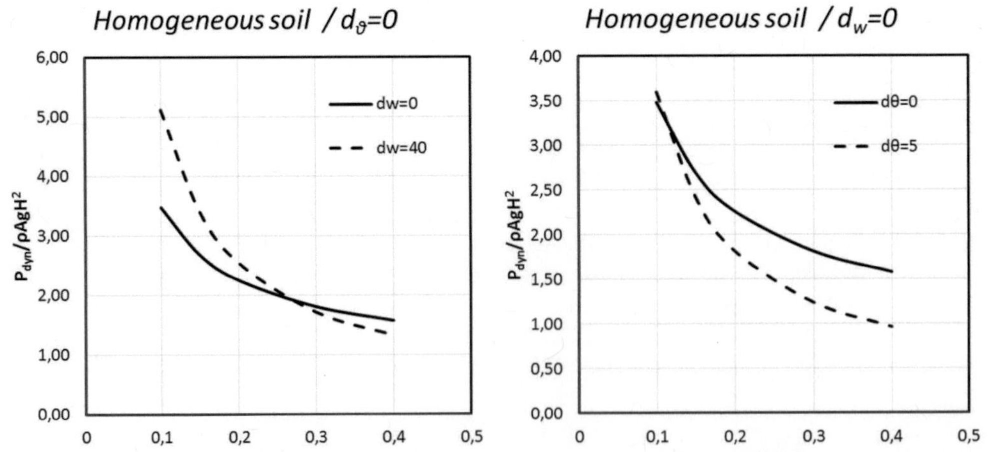

Fig. 5-55 Influence of the soil damping ratio on the total soil pressures for different wall and base flexibilities.

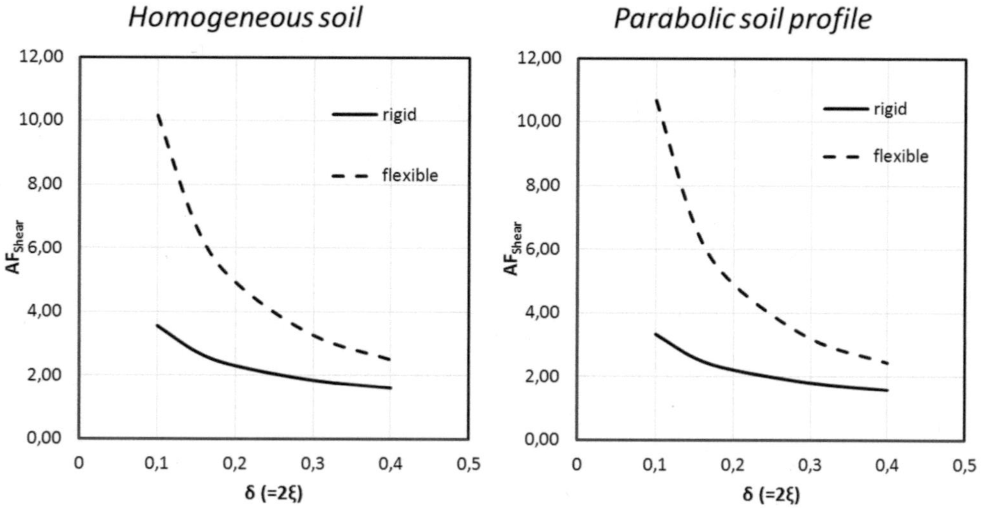

Fig. 5-56 Amplification factor vs the damping of the soil.

The previous figures show the relation between the amplification factor and the damping of the soil. It is interesting to observe that for the rigid wall based on a rigid foundation the amplification factor is given by the relation $1/\sqrt{\delta}$ and that for the flexible system by the relation $1/\delta$, which corresponds to an SDOF system. This was also observed by (Arias, A., Sanchez-Sesma, F. J., Ovando-Shelley, E. 1981) and (Veletsos, A. S., Younan, A. H. 1994a).

5.1.9 Influence of shear strain-dependent soil stiffness

In reality the soil stiffness depends on shear strain and does not remain constant. This fact has an influence on the soil pressures near the resonance frequency, where shear strains increase dramatically. In order to investigate shear strain-dependent soil stiffness the following dependence of the soil shear modulus on the shear strain was assumed (Seed et al. 1970; Vucetic und Dobry 1991).

As the stiffness of the soil decreases with increasing shear strain, relative wall-soil stiffness normally should also change. In this study relative wall-soil stiffness remains constant and equal to the one which corresponds to null shear strain (initial stiffness). In this way not only the influence of the reduction of soil stiffness can be observed, but it is at the same time meaningful and realistic to keep the wall stiffness constant as it is not going to change during a dynamic phenomenon. The analysis carried out is a frequency domain (steady state response). As it can be seen from the diagrams a strain-dependent stiffness of the soil reduces significantly the soil pressures near the resonance frequency (where the soil strains are also the biggest) to about 62% of a perfectly linear soil.

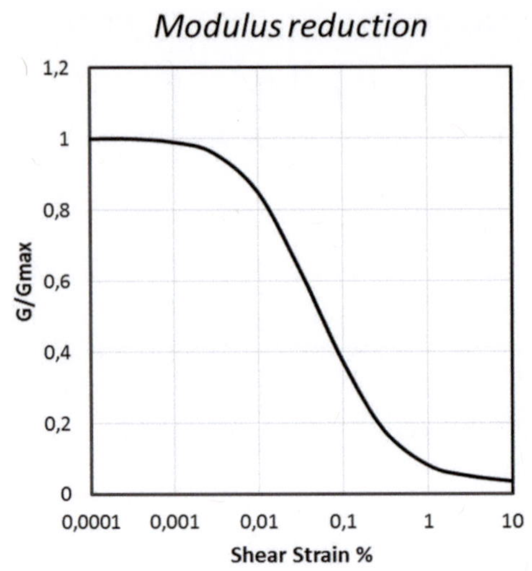

Fig. 5-57 Strain-dependent shear modulus used in this problem (PI=20); taken by (Vucetic und Dobry 1991).

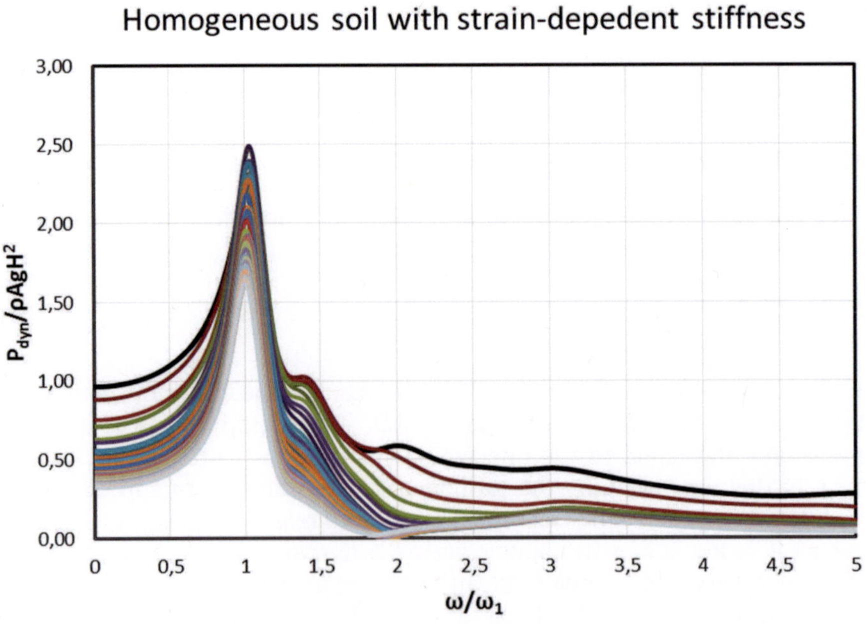

Fig. 5-58 Steady state response of the normalized shear force due to soil pressures of homogeneous soil with strain-dependent stiffness for all investigated relative wall and base flexibilities.

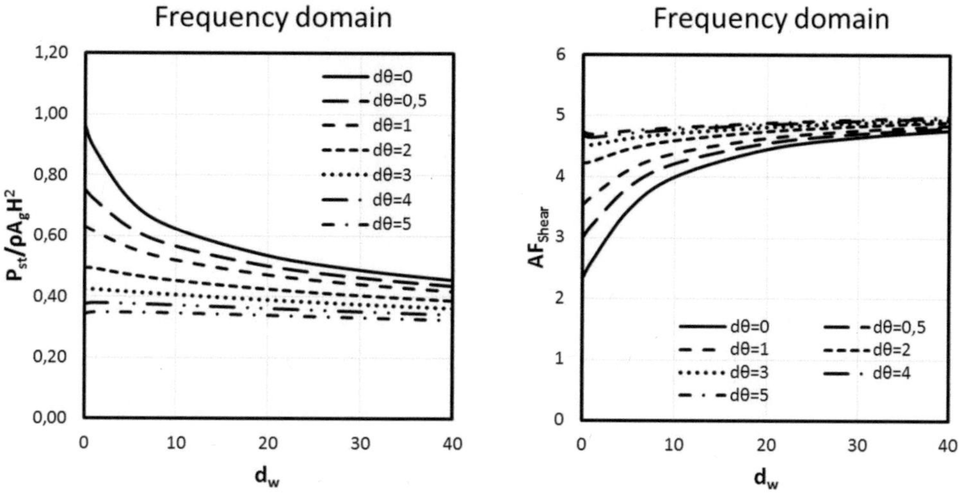

Fig. 5-59 Shear force and amplification factor of the soil pressures of a homogeneous soil with shear-dependent stiffness.

5.2 Finite element model for a pair of walls (bounded system)

Apart from the case where the soil extends to infinity (semi-infinite or semi-bounded domain) another interesting case is the one of two twin chambers placed at a small distance from each other. In this case the soil is bounded by the two chambers and it is interesting to additionally investigate the influence of the distance between these two chambers walls on the dynamic soil pressures. The two walls are able to elastically move in horizontal direction and to rotate without phase difference. The two walls are also flexible. The effect to be investigated here is similar to the silo effect described in the statutory provisions (for example (DIN 4085:2011-05)) and it is known that the soil pressures generally decrease as the L/H ratio decreases. For the dynamic case the amplification of the soil pressures is also investigated.

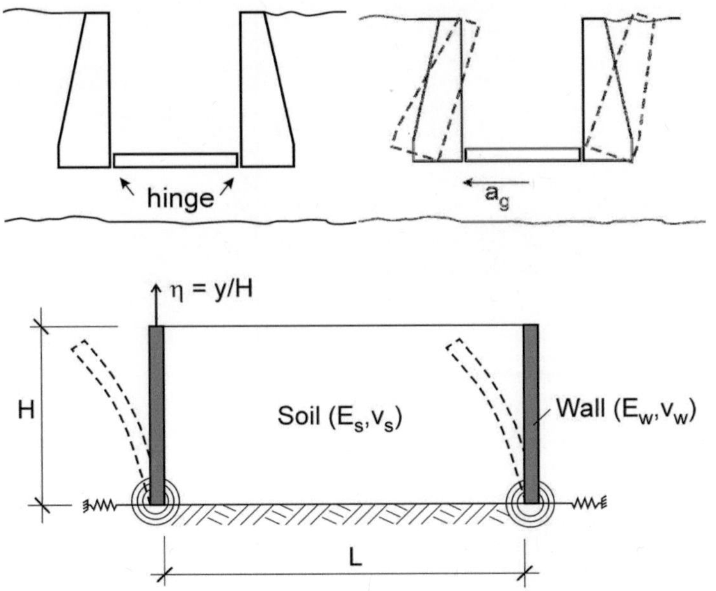

Fig. 5-60 Schema of the second model investigated here (similar to the model of Wood, Papazafeiropoulos and Psarropoulos, Vrettos et al., Theodorakopoulos et al.).

The same soil profiles as before are investigated as well as the same values for the horizontal and rotational springs and the wall flexibility are applied. As the system is now bounded, the frequencies of the system are not the same as in the semi-infinite model (unbounded soil shear column). The first frequencies are taken using the formula provided by (Parikh, V. H., Veletsos, A. S., Younan, A. H. 1995) and are given in the table below. These values are also in agreement with the ones provided by (Wood 1973):

$$\omega_{m,n} = \frac{\pi V_s}{2H}\sqrt{(2n-1)^2 + 4(2m-1)^2\frac{2}{1-v}\left(\frac{H}{L}\right)^2} = \omega_\infty\sqrt{(2n-1)^2 + 4(2m-1)^2\frac{2}{1-v}\left(\frac{H}{L}\right)^2} \quad \textbf{(5-19)}$$

Where *m*, *n* are the *m*th horizontal and the *n*th vertical participating mode of the system. For the case of non–homogeneous soil, Wu and Finn (Wu und Liam Finn 1999) suggested to replace the shear modulus with an effective modulus as:

$$V_s = \sqrt{\frac{G}{\rho}} = \sqrt{\frac{\sum G_i h_i}{\rho H}} \tag{5-20}$$

Table 35 Ratio of the first horizontal natural frequencies for different values of L/H to the first natural frequency of the unbounded domain for homogeneous soil with bonded and smooth soil-wall (Wood 1973)

	bonded	smooth
L/H	$\omega_{11}/\omega_\infty$	$\omega_{11}/\omega_\infty$
1	3.60	2.18
2	2.00	1.81
3	1.53	1.51
5	1.22	1.21
10	1.06	1.06

5.2.1 Influence of the L/H ratio

As it can be seen, the first natural frequency of the system is not constant for all the ratios *L/H*. Because of this fact, it is quite difficult to determine a quasi-static excitation frequency for all systems. Although a very small value of the circular frequency could be adopted, even small deviations from the absolute value ω=0 influence the response and cause dissimilar amplifications. For this reason, another strategy was followed for the quasi static response of the walls, by simply applying the gravitational force of the soil towards the wall. The results are in good agreement with the ones given by other researchers (Wood 1973; Vrettos et al. 2016) as it can be seen in the following diagrams.

Analogously to the former chapter on the hydrodynamic pressures of bounded basins the following reduction factor for the soil pressures can be applied (two proposed formulas):

$$A.\ C_{wd} = \begin{cases} 1 - 1.2 * \text{Exp}\left(\frac{-L/H}{1.4}\right), & when\ L/H < 5 \\ 1, & when\ L/H \geq 5 \end{cases} \tag{5-21}$$

$$B.\ C_{wd} = \begin{cases} \frac{6}{5}\left(\frac{L/H}{1 + L/H}\right), & when\ L/H < 5 \\ 1, & when\ L/H \geq 5 \end{cases} \tag{5-22}$$

The equations 5.20 and 5.21 are plotted versus the computed values in the following chart in the form of the resulting shear force due to the soil pressures.

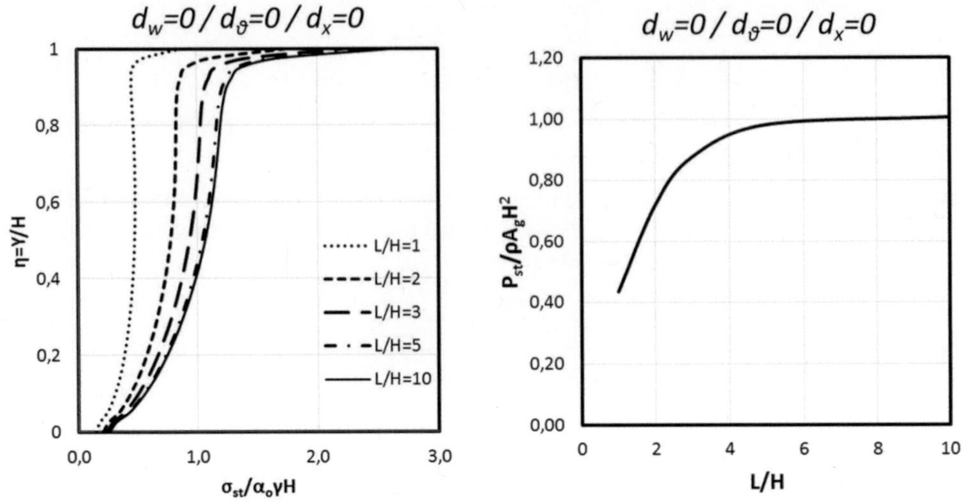

Fig. 5-61 Soil pressure distribution on a rigid wall for different ratios of L/H and the normalized base shear versus the L/H ratio (statically excited system).

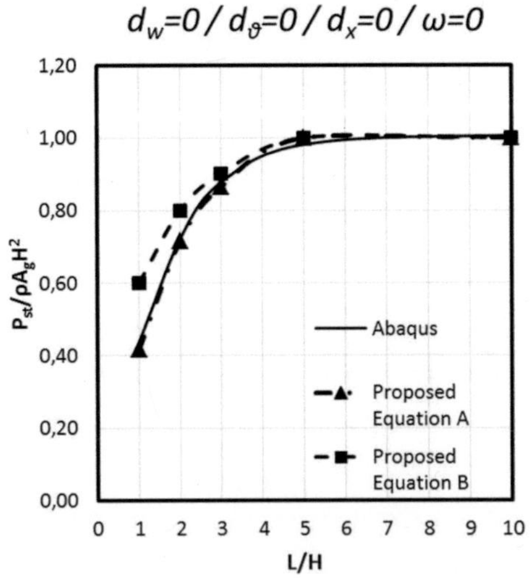

Fig. 5-62 The computed values of the numerical analysis and the approximate equation 5.21 and 5.22 in the form of shear forces (statically excited system).

For the two other inhomogeneous soil profiles the following formulas for the calculation of the shear forces of the statically excited system and rigid wall are proposed:

$$Parabolic\ soil:\quad C_{wd} = 0.81 - \mathrm{Exp}(-L/H) \qquad \textbf{(5-23)}$$

$$Linear\ soil:\quad C_{wd} = 0.75 - 0.9\ \mathrm{Exp}\left(\frac{-L/H}{0.95}\right) \qquad \textbf{(5-24)}$$

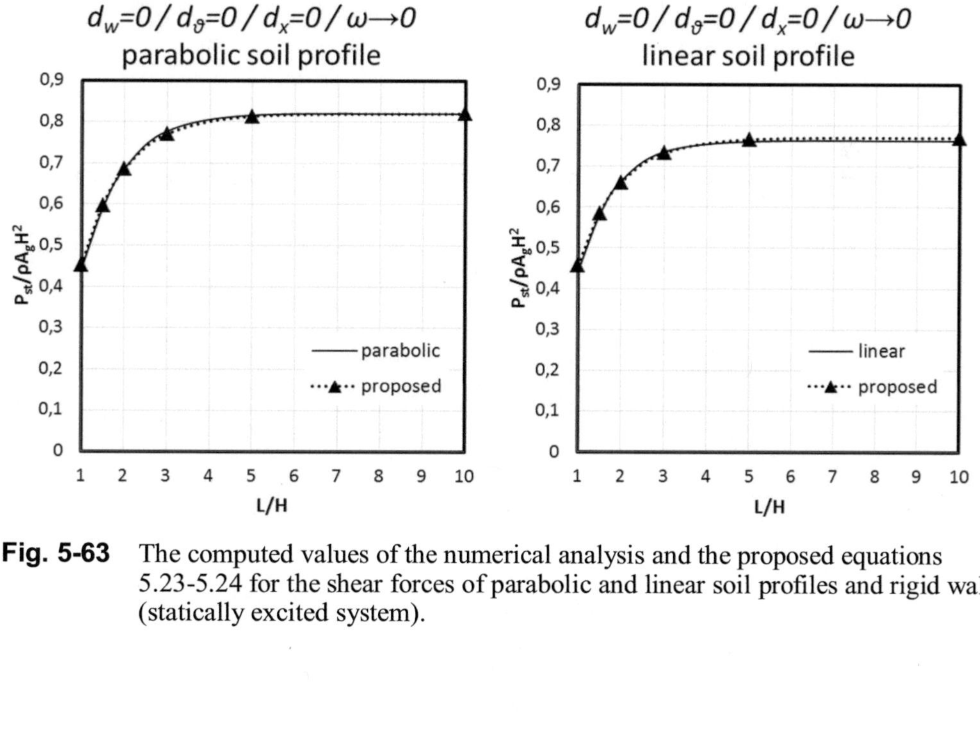

Fig. 5-63 The computed values of the numerical analysis and the proposed equations 5.23-5.24 for the shear forces of parabolic and linear soil profiles and rigid wall (statically excited system).

5.2.2 Influence of the flexibility of the wall and the foundation

As before, the soil-wall system was investigated for the static case for different wall and foundation flexibilities. The values of the relative flexibilities are the same as in the previous subchapter. The analyses were static 1g analyses, with the gravitational force of the soil applied to the one wall.

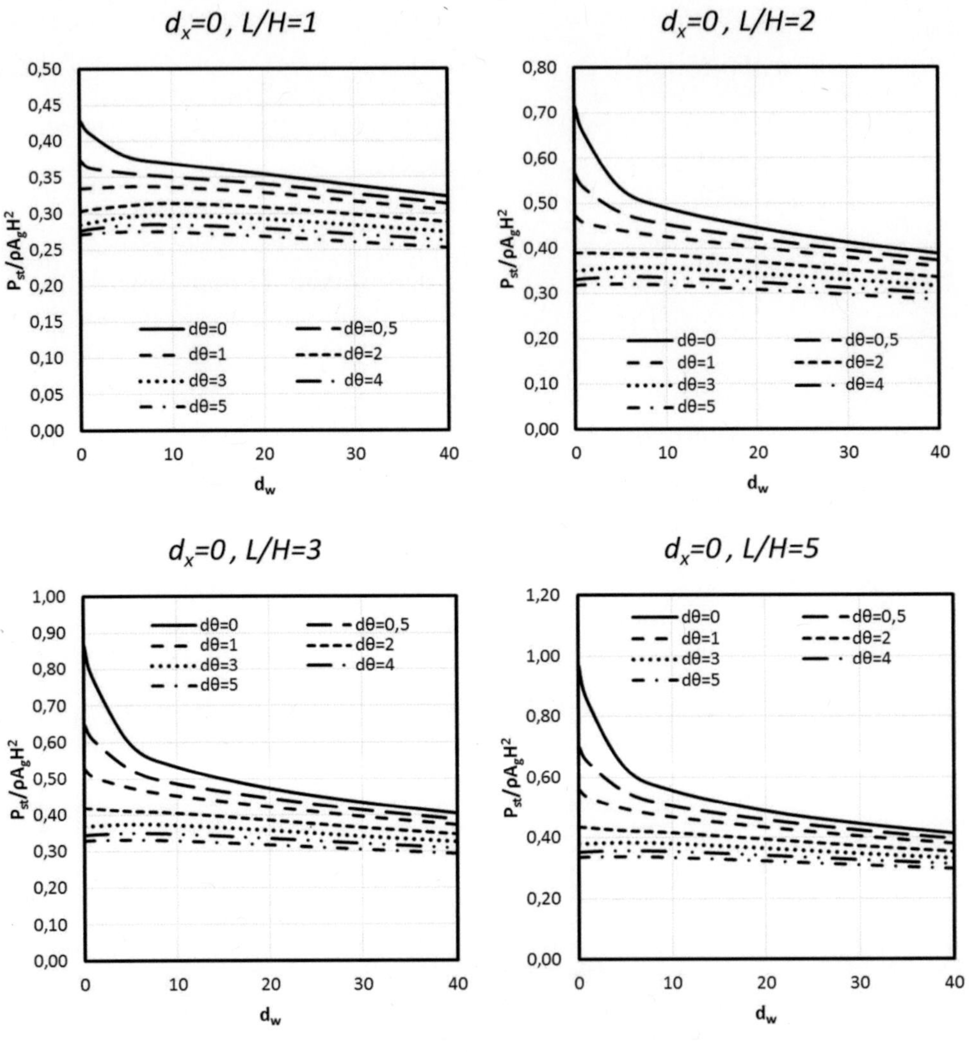

Fig. 5-64 Values of the normalized base shear for different values of d_w, d_θ and L/H (statically excited system).

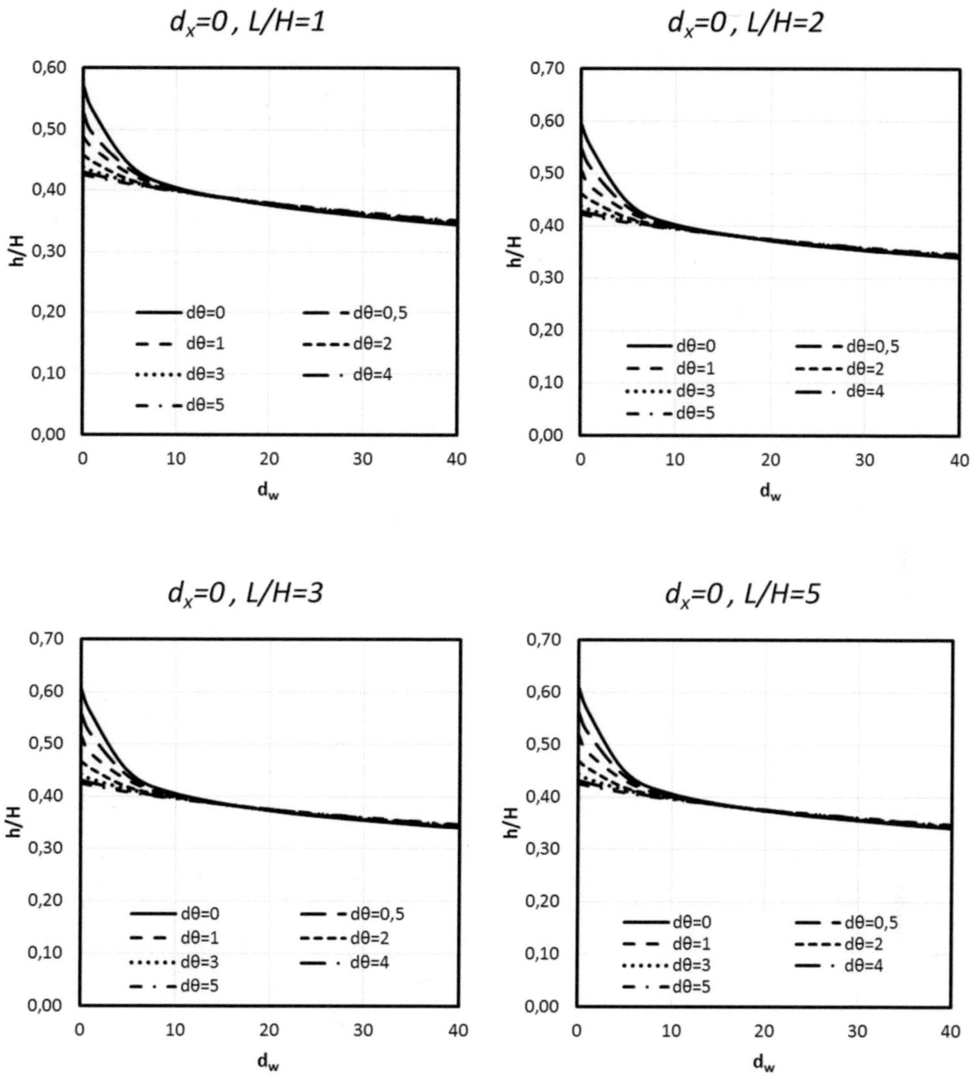

Fig. 5-65 Values of the effective height for different values of d_w, d_θ and L/H (statically excited system).

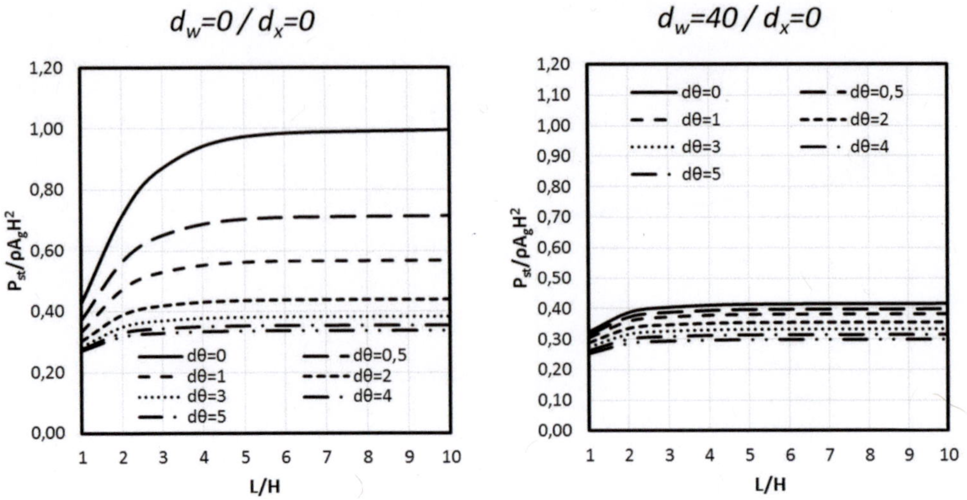

Fig. 5-66 Change in the normalized base shear for different values of d_θ and L/H (homogeneous soil, statically excited system).

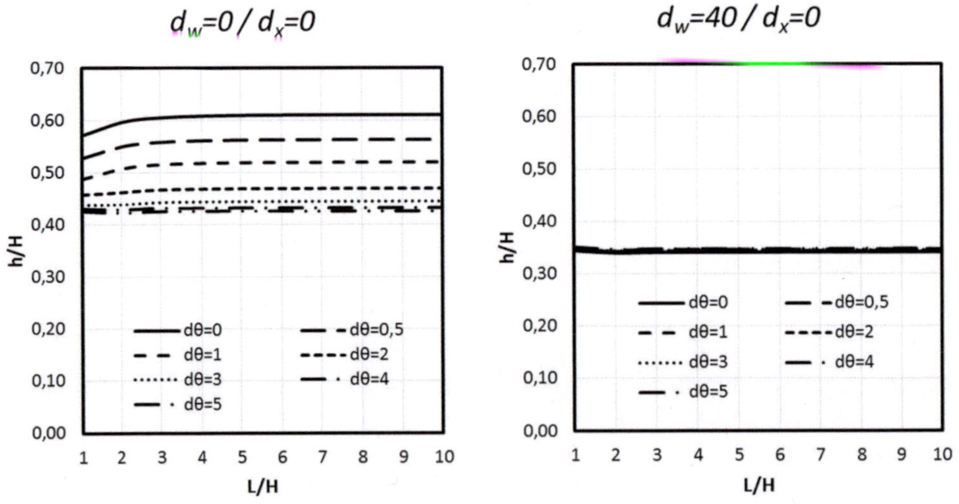

Fig. 5-67 Change in the effective height for different values of d_θ and L/H (homogeneous soil, statically excited system).

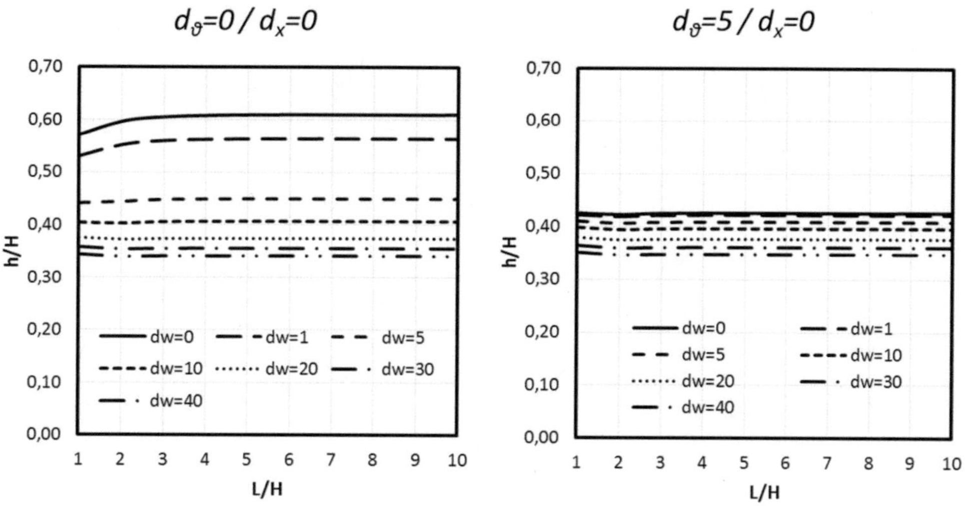

Fig. 5-68 Change in the effective height for different values of d_w and L/H (homogeneous soil, statically excited system).

5.2.3 Influence of the soil profile

Three soil profiles are investigated here: a constant shear modulus (homogeneous soil), a linear and a parabolic shear modulus distribution. The results are similar with the one derived for the semi-infinite domain, but with the soil pressures reducing as the *L/H* ratio also reduces. The total soil pressures on the wall reduce as the flexibility of the wall or its foundation increases. For the diagrams the following observations can be made:

- For a given wall flexibility d_w, the effective height remains practically constant and is independent from the value of the base rotation d_θ and the *L/H* ratio.
- For the constant soil profile the value of the effective height fluctuates between 0.6*H* for a rigid wall and 0.33*H* for a flexible wall.
- For the parabolic soil profile the value of the effective height fluctuates between 0.52*H* for a rigid wall and 0.27*H* for a flexible wall.
- For the linear soil profile the value of the effective height fluctuates between 0.51*H* for a rigid wall and 0.27*H* for a flexible wall.
- The total shear force decreases with decreasing *L/H* ratio and with increasing rotational flexibility of the wall and the foundation.

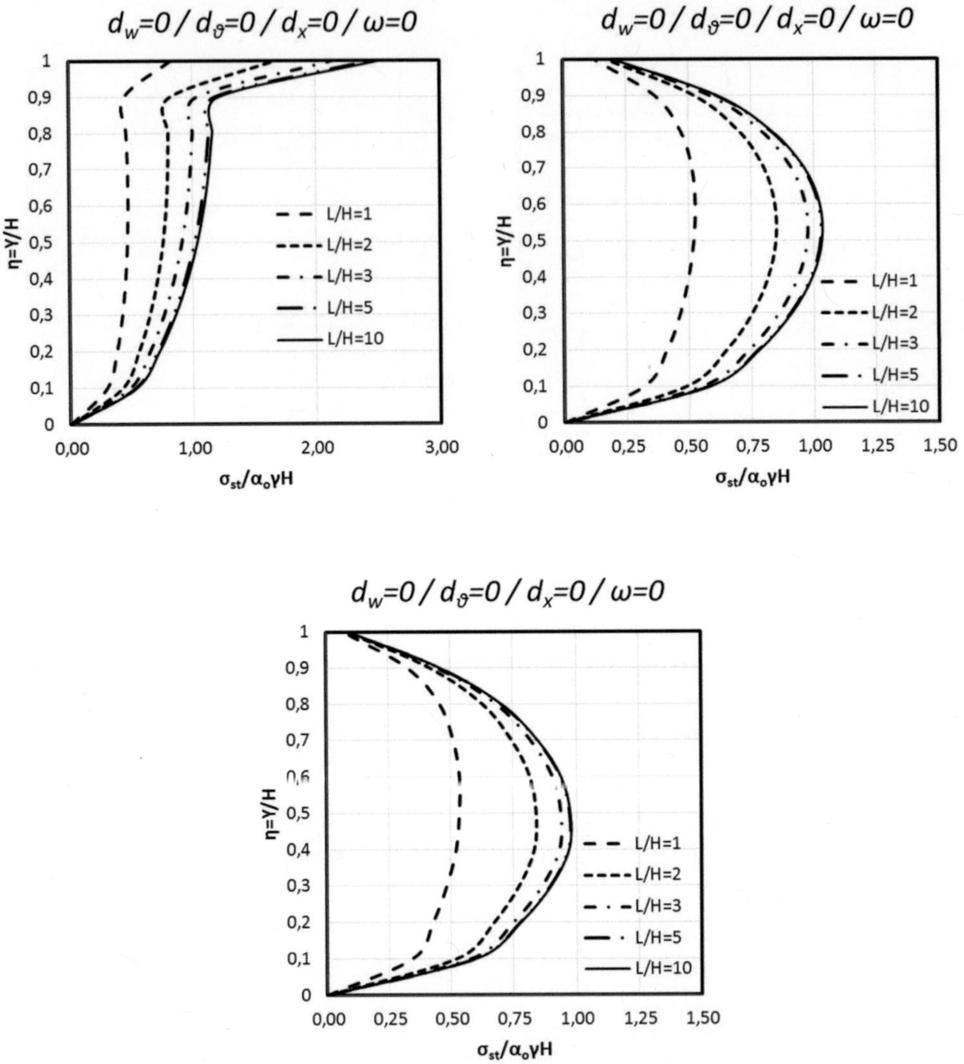

Fig. 5-69 Soil pressure distribution for the static case for different ratios of L/H and soil profiles (top left: constant; top right: parabolic; bottom: linear).

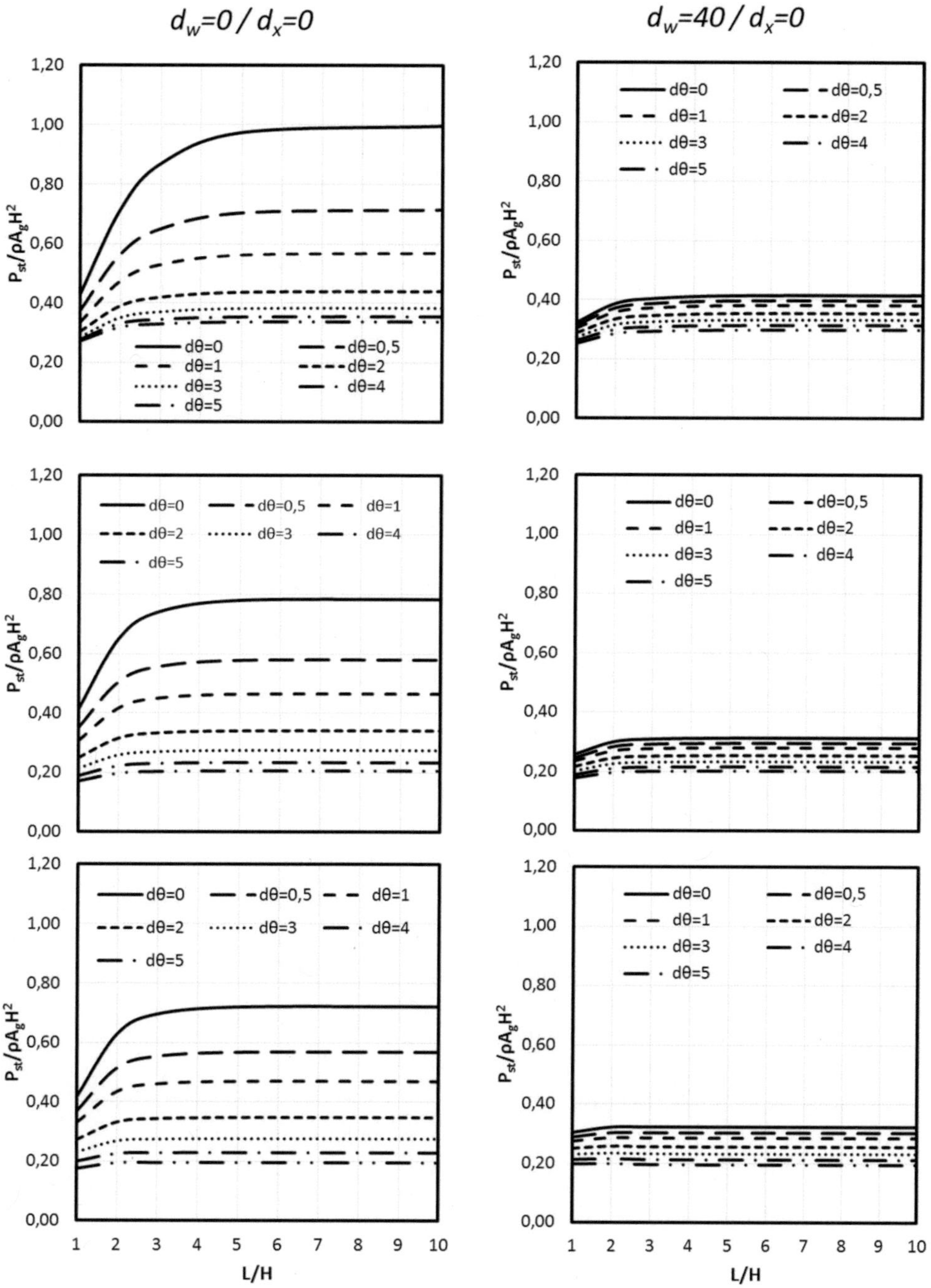

Fig. 5-70 Change of the normalized shear force for different wall and foundation flexibilities and different soil profiles (top: constant; middle: parabolic; bottom: linear soil profile and statically excited system).

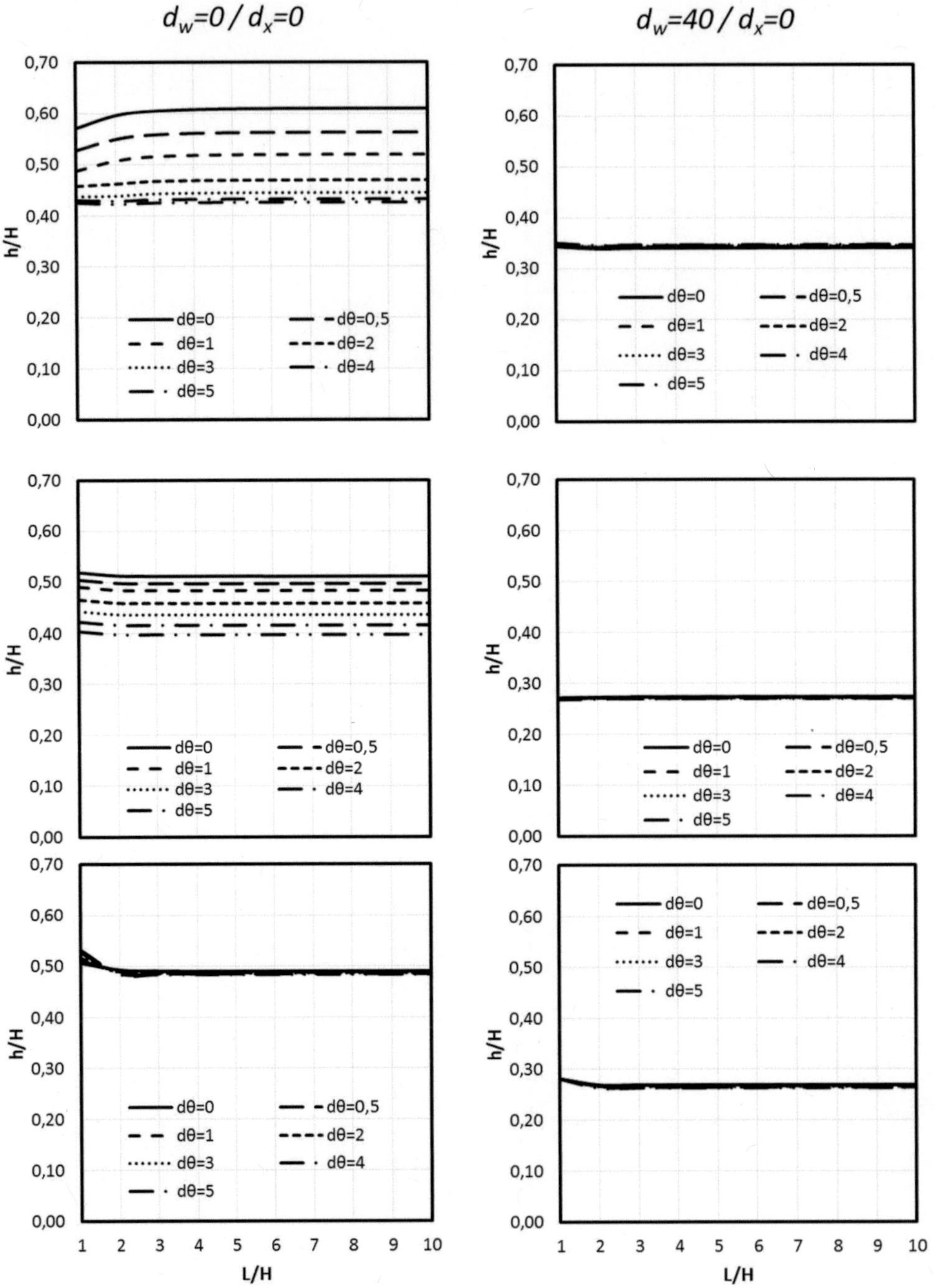

Fig. 5-71 Change in effective height for different wall and foundation flexibilities and different soil profiles for a statically excited system (top: constant; middle: parabolic; bottom: linear soil profile).

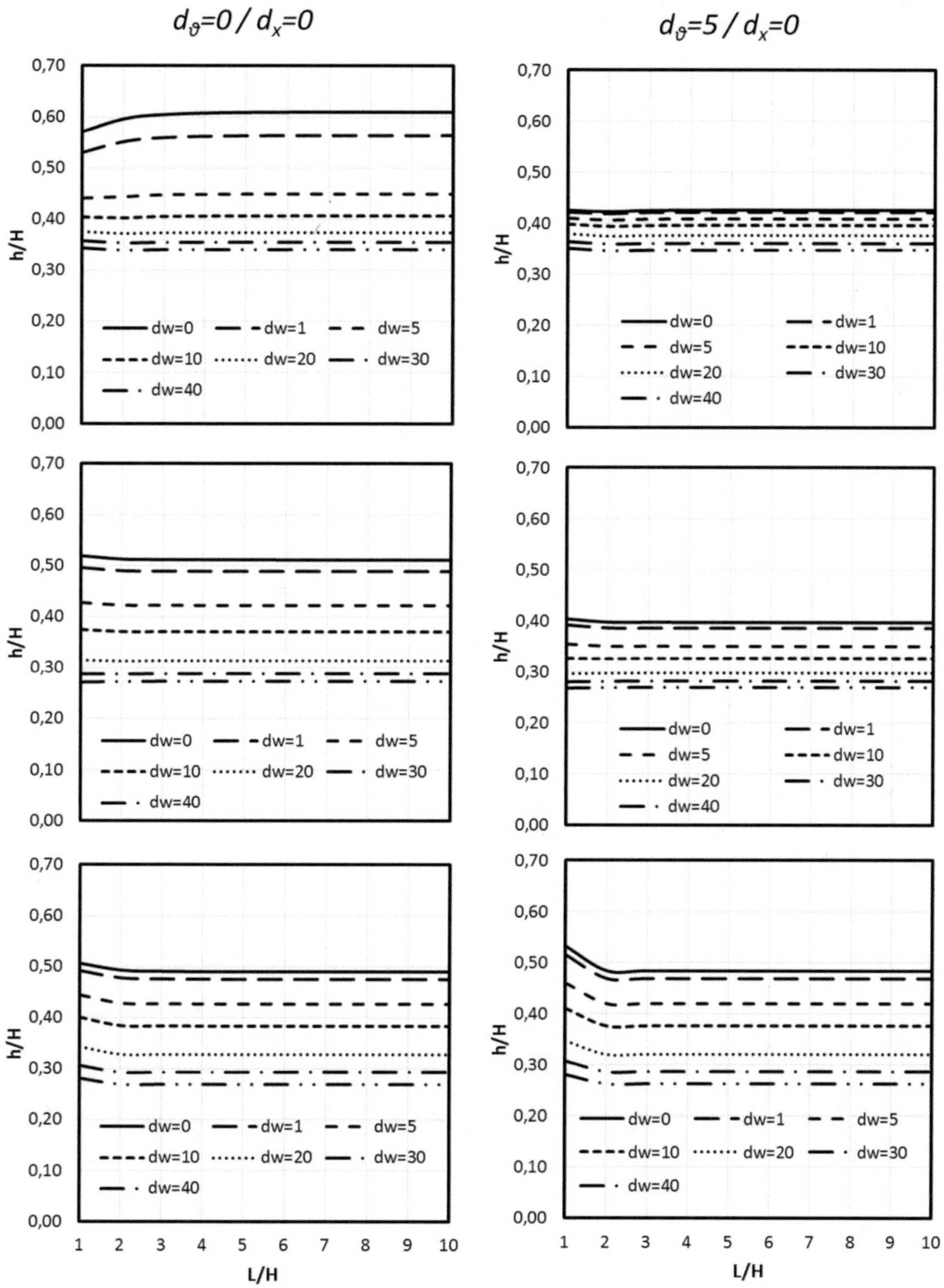

Fig. 5-72 Change in effective height for different wall and foundation flexibilities and different soil profiles (top: constant; middle: parabolic; bottom: linear statically excited system).

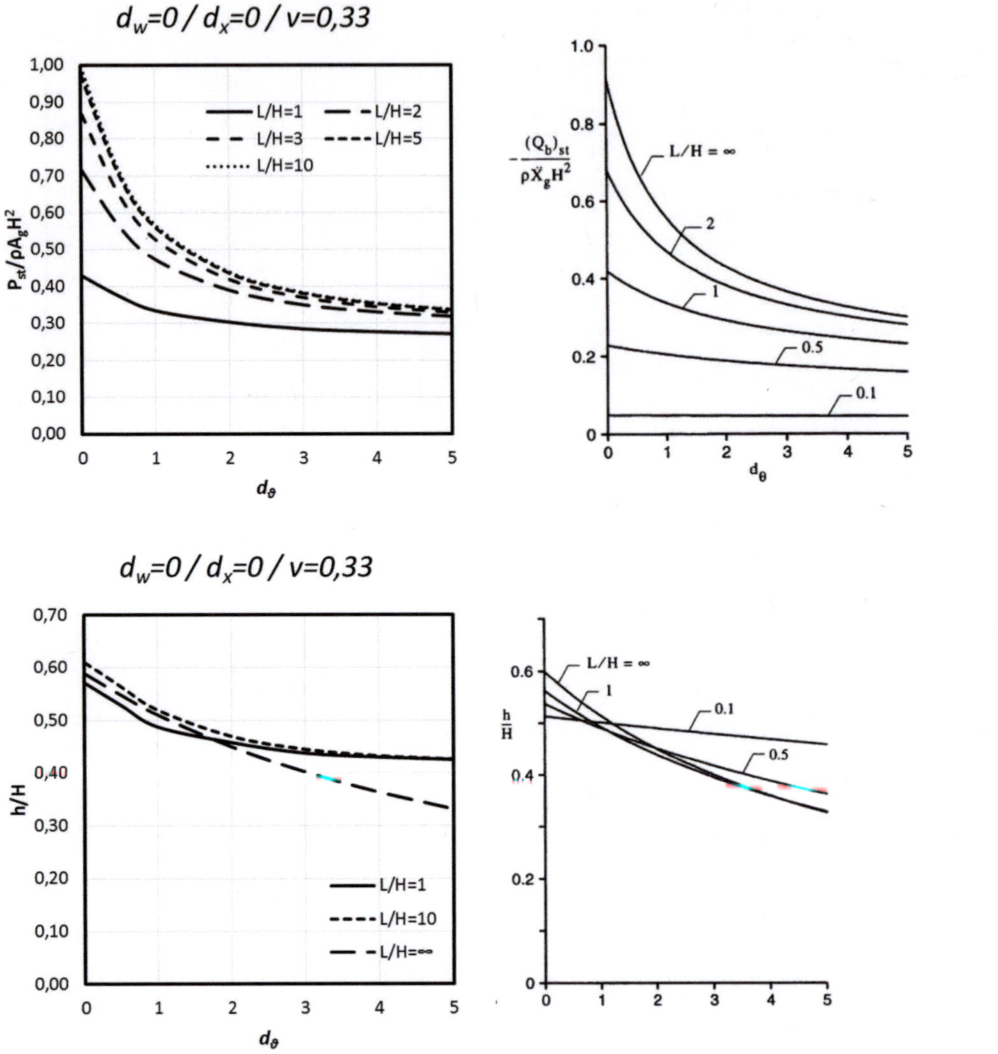

Fig. 5-73 Results of this numerical study versus the analytical solution provided by (Parikh, V. H., Veletsos, A. S., Younan, A. H. 1995) but for different Poison number values (0,33 at this study instead of 0,3) for a statically excited system.

The investigation of the translational movement of the wall delivers similar results with the semi-infinite model. Although total soil pressure decreases with increasing translational flexibility, the compression forces are gathered in the upper part of the wall. This fact leads to bigger overturning moments and, in turn, to greater effective heights. As mentioned before, the same observations were made by (Bobet, A., Jung, C. 2008; Bobet, A., Fernández, G., Jung, C. 2010) and these are due to the contact modelling (soil is attached to the wall).

5.2.4 Influence of Poisson's ratio of the soil in bounded systems

The change of Poisson's ratio in bounded systems was first investigated by (Wood 1973). Here, in order to verify the numerical model a similar static analysis was carried out. The results show that an increasing Poisson's ratio increases the soil pressures on the wall, but the height of application of the resultant force decreases due to disproportionally increasing moments (the initially parabolic pressure distribution becomes almost rectangular).

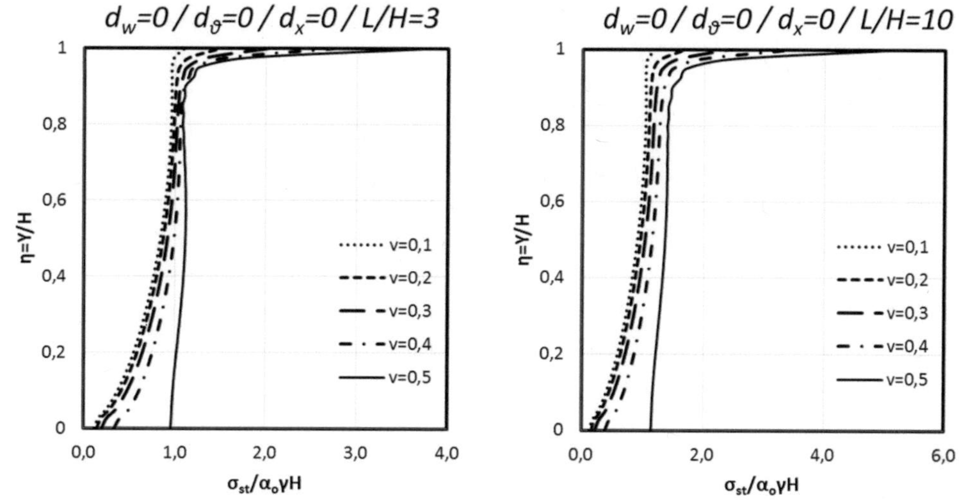

Fig. 5-74 Soil pressure distribution for the static case for different L/H and Poisson's ratios (bonded contact).

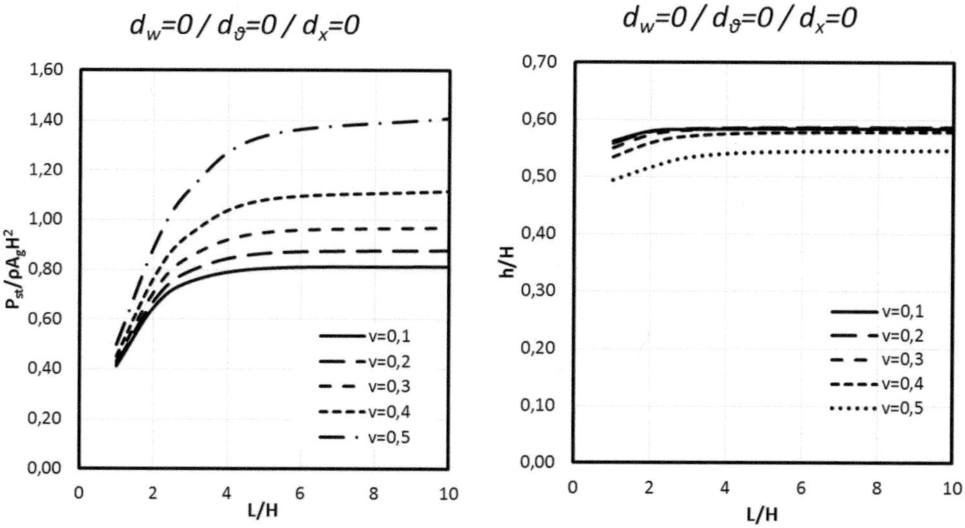

Fig. 5-75 Influence of Poisson's ratio of the soil on normalized shear pressures and effective height (statically excited system – bonded contact).

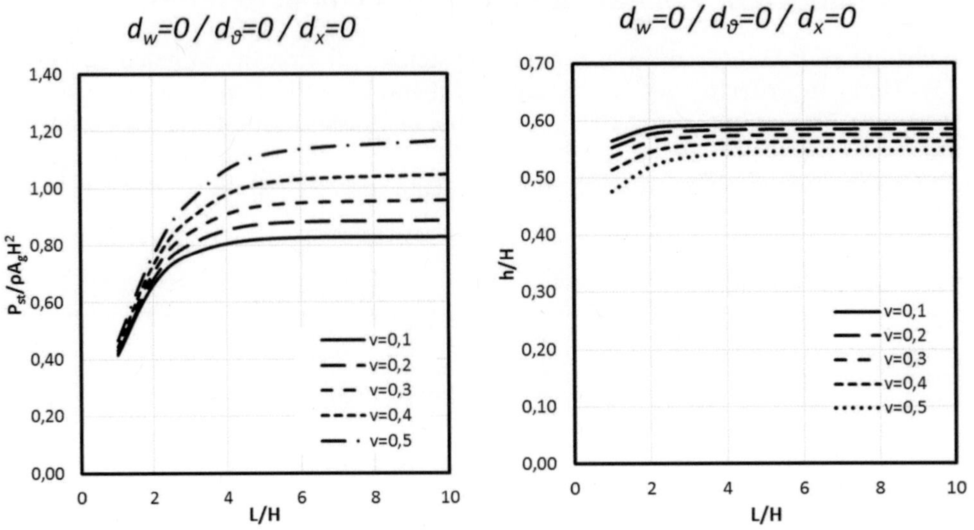

Fig. 5-76 Influence of Poisson's ratio of the soil on normalized shear pressures and effective height (statically excited system – smooth contact).

5.2.5 Influence of the excitation frequency in bounded systems

The influence of the excitation frequency in bounded soil-wall systems is a slightly more time-consuming task than the investigation of the semi-infinite model. This is because of the different resonance frequencies for each *L/H* ratio and the defined damping parameters. In order to avoid a thorough investigation of the Rayleigh parameters and a direct integration time history analysis, the steady state dynamics analysis was chosen. With the steady state analysis, damping can be defined as modal or structural damping A comparison of the analytical and exact solutions of (Papazafeiropoulos, G., Psarropoulos, P. N. 2010) and the results of the steady state response of the finite element analysis shows that the dynamic pressures are in complete agreement. The peaks of the total shear force acting on a wall appear at the first natural frequencies of the bounded system. The first natural frequencies calculated with Abaqus and (Papazafeiropoulos, G., Psarropoulos, P. N. 2010) are in good agreement. The only difference is observed for the size of the total shear force for the *L/H* ratio equal to 2. The finite element analysis delivers a slightly smaller amplitude for this ratio.

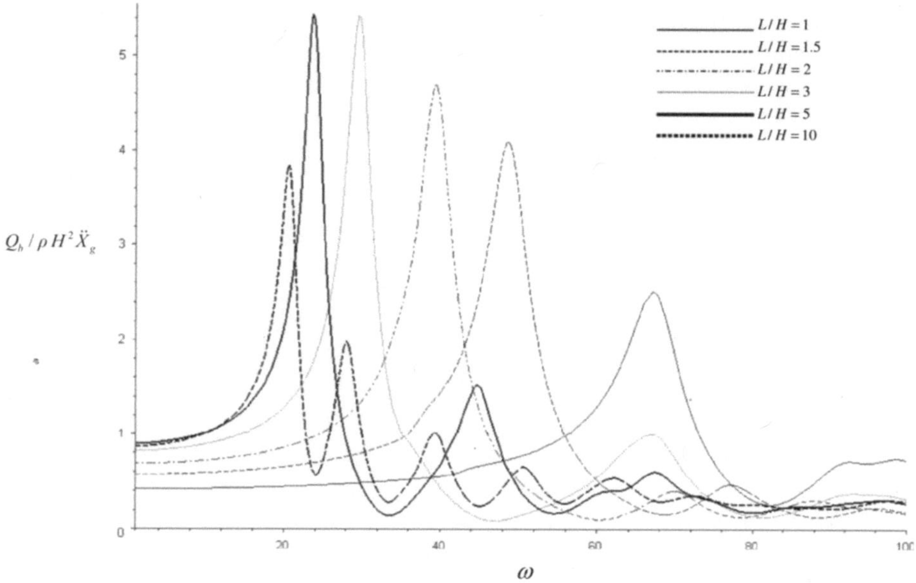

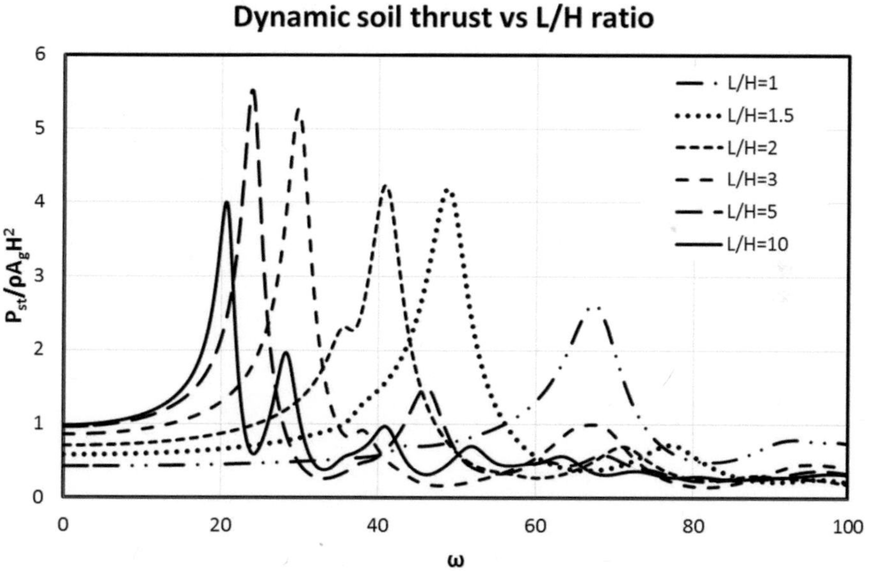

Fig. 5-77 Top: steady state response of rigid wall-soil systems with different L/H ratios, ν=0.3, δ=0.1 (→ξ=5%) according to (Papazafeiropoulos, G., Psarropoulos, P. N. 2010) and this numerical study; bottom: results of this numerical study (ν=0.33).

Wu and Finn (Wu und Liam Finn 1999) after performing a series of finite element analysis for different values of *L/H* conclude that: "*Dynamic amplification of seismic thrusts increases as the L/H ratio decreases from 5.0 to 1.5. The peak dynamic thrusts close to resonance (ω/ω_{11} = 0.7 to 1.0) are greater for backfills with L/H = 1.5 than for backfills with L/H = 5.0, although the dynamic thrusts at a very low frequency ratio ($\omega/\omega_{11} \leq 0.1$) are lower for backfills with L/H = 1.5 than for backfills with L/H = 5.0.*" In the analysis of (Wu und Liam Finn 1999) L/H is the ratio of the half-length of the domain as they performed an axisymmetric analysis (*L/H*=5 corresponds to *L/H*=10 in this analysis). Moreover, they pointed out that the amplification factors for wall-soil systems with finite backfills are larger than the amplification factors of wall-soil systems with semi-infinite backfills. Some of these results are proved analytically by (Papazafeiropoulos, G., Psarropoulos, P. N. 2010). (Papazafeiropoulos, G., Psarropoulos, P. N. 2010) provided the normalized shear forces for different L/H ratios and frequencies, but not directly the amplification factors. For *L/H*=3 the analytical solution of (Papazafeiropoulos, G., Psarropoulos, P. N. 2010) gives also a bigger value for soil thrust than a ratio of *L/H* equal to 10. Unfortunately (Wu und Liam Finn 1999) did not perform analyses for other *L/H* ratio values in order to compare the numerical with the analytical results. The difference in the amplitude of the total soil thrust of these two analyses is due to the different damping and Poisson's ratio values used (5% critical damping and ν=0.3 for the analytical solution of (Papazafeiropoulos, G., Psarropoulos, P. N. 2010) and 10% damping and ν=0.4 for the numerical analysis of (Wu und Liam Finn 1999)).

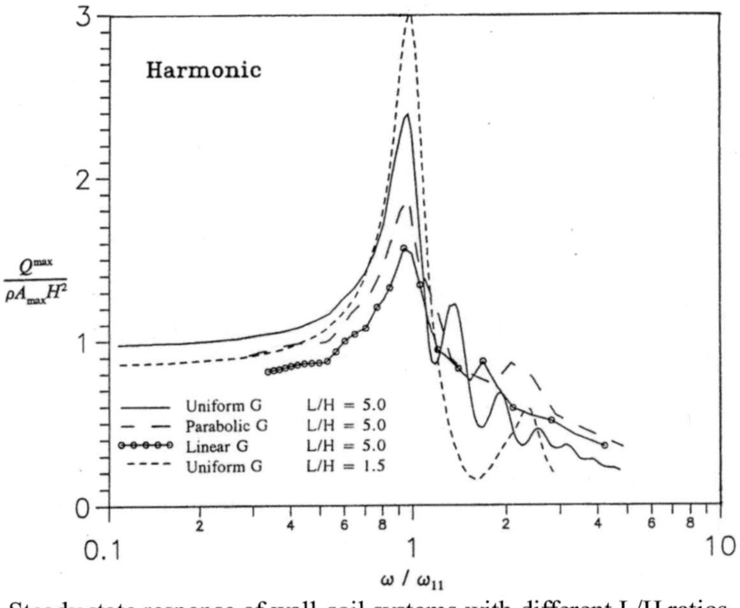

Fig. 5-78 Steady state response of wall-soil systems with different L/H ratios, v=0.4, λ=10% according to (Wu und Liam Finn 1999).

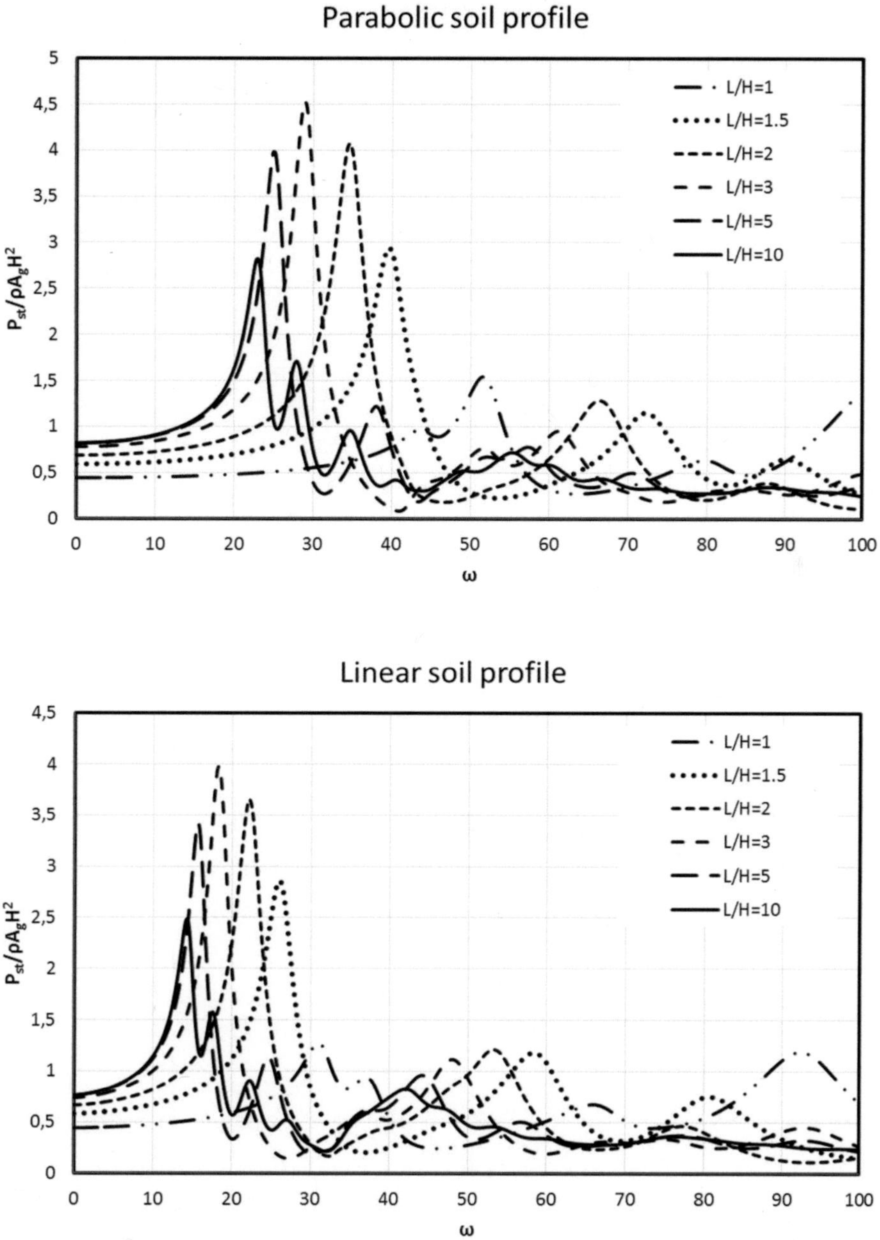

Fig. 5-79 Steady state response of rigid wall-soil systems with different L/H ratios for parabolic and linear soil profiles according to this numerical study (ν=0.33).

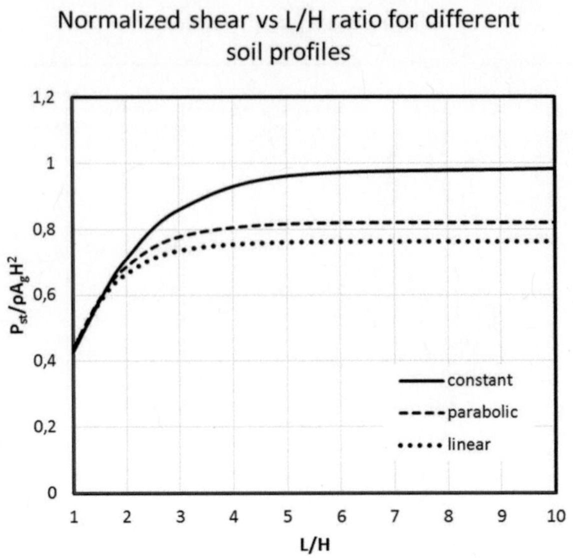

Fig. 5-80 Normalized shear due to soil pressures for the statically excited system for different soil profiles as result of the steady state response analysis.

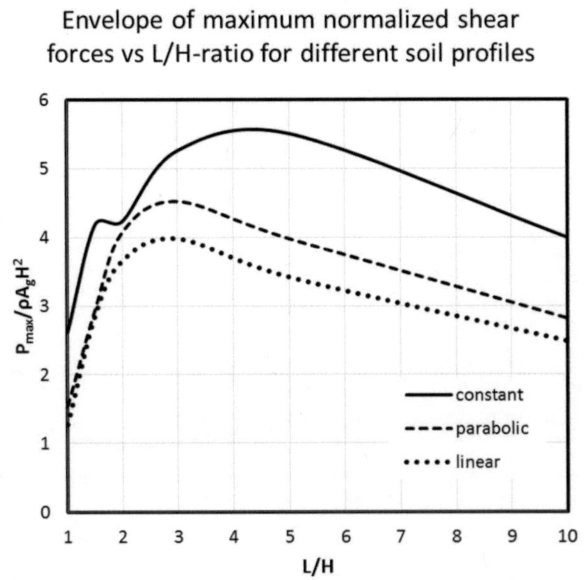

Fig. 5-81 Envelope of the maximum (resonance) normalized shear forces for different L/H ratios for rigid walls according to this numerical study.

As it can be seen, as the shear modulus decreases in the vicinity of the free surface, the total soil thrust also decreases and the peaks for the different *L/H* ratios are gathered at a narrower range of frequencies (15-30 rad/sec for the linear soil, 25-55 rad/sec for the parabolic soil and 20-70 rad/sec for the homogeneous soil). The total soil thrust and the AF of the total shear force also decrease as the shear modulus of the soil decreases near the free surface. For a pair of rigid walls and homogeneous soil the AF lies between 4 and 7, for a parabolic soil profile between 3.5 and 6, and for the linear soil profile between 2.8 and 5.5.

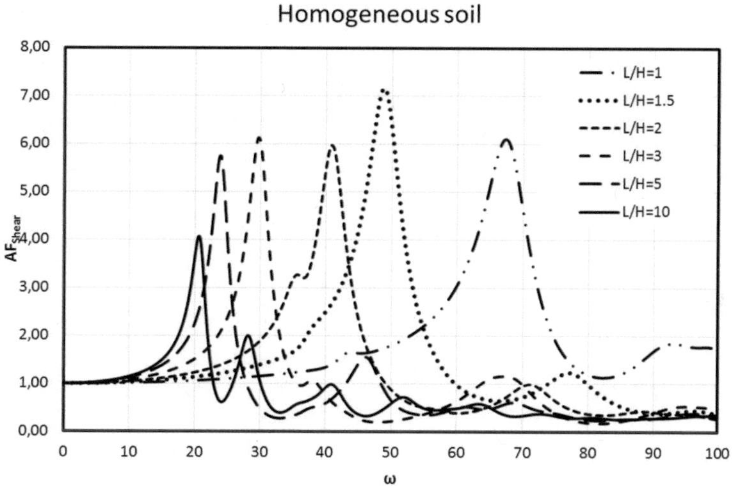

Fig. 5-82 Amplification factors of the base shear forces for rigid wall systems with different L/H ratios and for a homogeneous soil profile.

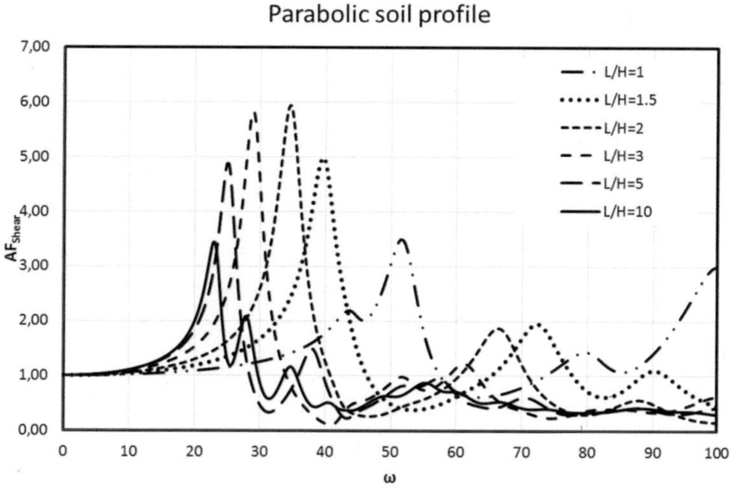

Fig. 5-83 Amplification factors of the base shear forces for rigid wall systems with different L/H ratios and for a parabolic soil profile.

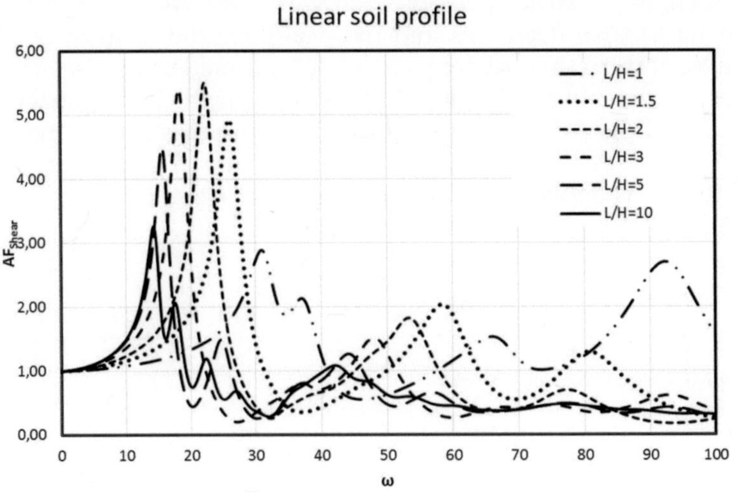

Fig. 5-84 Amplification factors of the base shear forces for rigid wall systems with different L/H ratios and for a linear soil profile.

The literature offers results only for the case of rigid walls for different excitation frequencies and different *L/H* ratios. The results for the rigid walls are in good agreement with the ones found in the literature. At this point it is interesting to investigate the influence of the excitation frequency on flexible systems (flexible walls on flexible foundations). This has been done in the same way as before, with the same values for the relative wall-soil flexibilities and for the three soil profiles investigated here. It has to be mentioned here that the amplification factors for the flexible wall-soil bounded systems are much bigger than the amplification factors for the rigid wall systems. This was also the case for the semi-infinite soil-wall systems. Moreover, the AFs are bigger for small values of the *L/H* ratio. The analysis procedure followed here is the steady state dynamics available in Abaqus.

Table 36 Frequencies with the maximum contribution to the soil pressures of the bounded domain for homogeneous soil with smooth and bonded soil-wall contact after this study

	Smooth contact		Bonded contact	
	Wood[3] (v=0.3)	This study (v=0.33)	Wood (v=0.3)	This study (v=0.33)
L/H=1	$\omega_{11}/\omega_\infty$	$\omega_{11}/\omega_\infty$	ω_1/ω_∞	ω_1/ω_∞
1	3.45	3.38	3.63	3.60
2	1.81	2.06	2.00	2.00
3	1.51	1.50	1.53	1.53
5	1.21	1.21	1.22	1.21
10	1.06	1.05	1.06	1.05

[3] Wood 1973

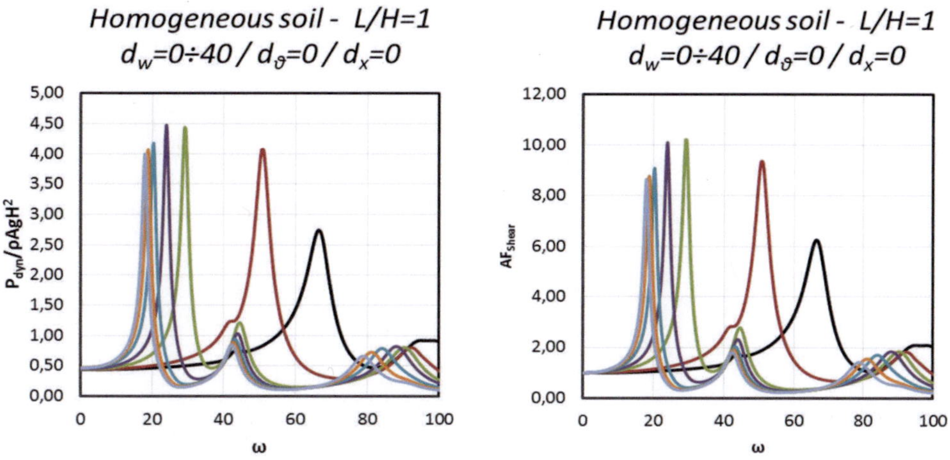

Fig. 5-85 Normalized base shears and amplification factors of the base shear forces for only the flexure relative flexibilities (d_θ=0, d_x=0) for homogeneous soil and L/H=1.

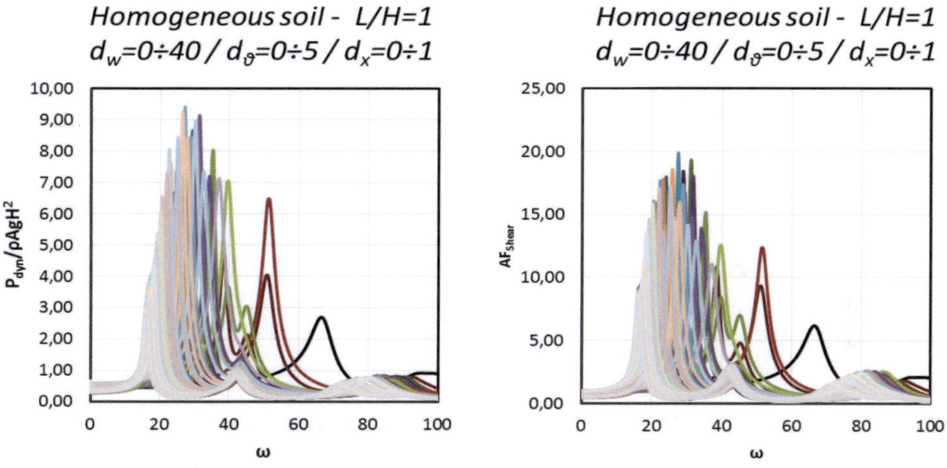

Fig. 5-86 Normalized base shears and amplification factors of the base shear forces for all the relative flexibilities investigated here for homogeneous soil and L/H=1.

5.2.6 Influence of soil damping

As for the one-wall system, the influence of the damping ratio is also investigated here. Damping is given as structural damping and remains constant for all the frequency range. As expected, an increasing damping ratio reduces the dynamic soil pressures. This can be clearly seen for the resonance case as well as for the statically excited case: there is no damping influence on the system. The following diagrams show the influence of damping for different L/H ratios and soil profiles for a rigid (*d_w=0, d_θ=0, d_x=0*) and a flexible wall (*d_w=40, d_θ=5, d_x=1*).

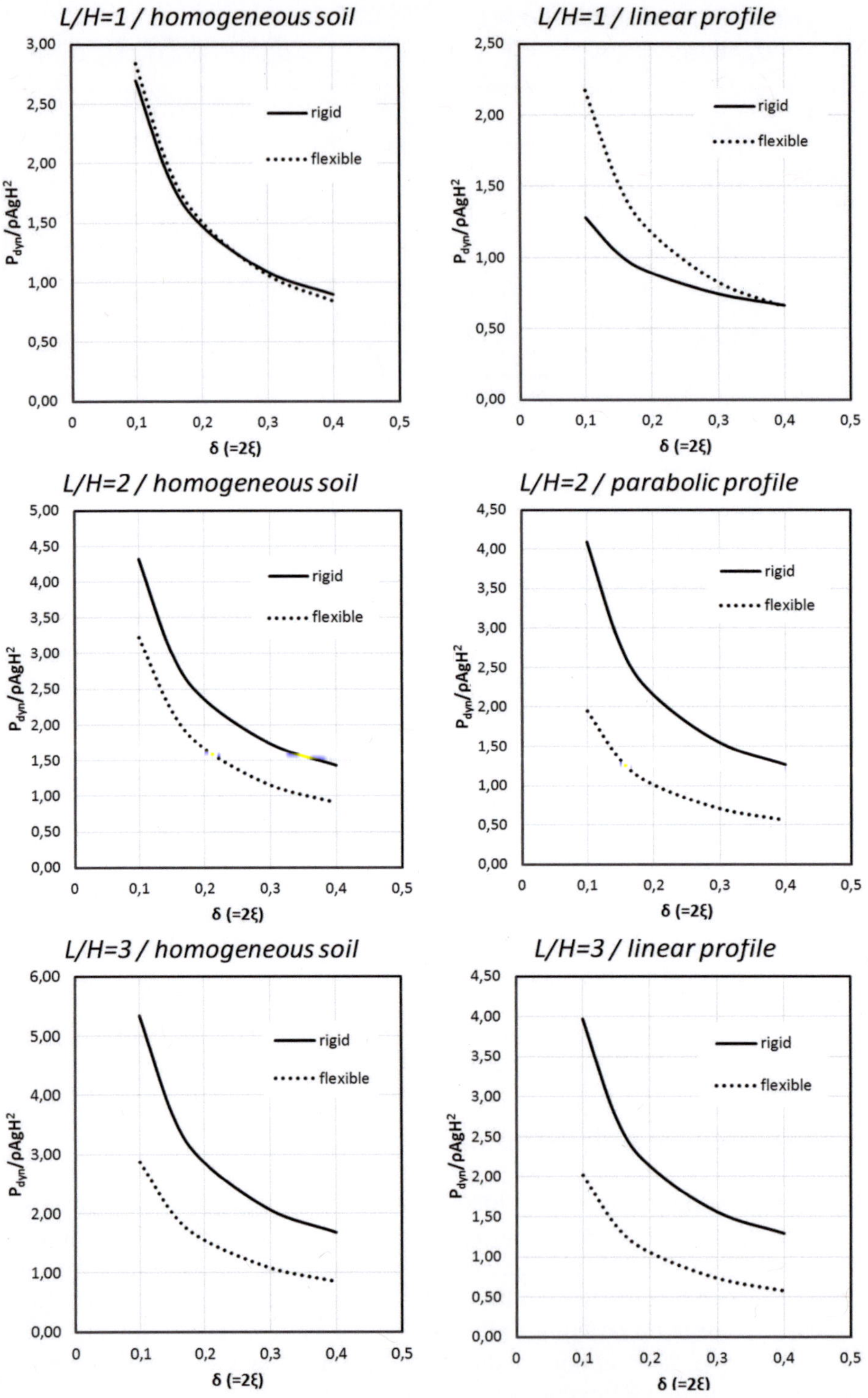

Fig. 5-87 Dependence of the normalized shear force on the structural damping ratio for different soil profiles and L/H ratios.

5.2.7 Influence of the soil-wall interface

For the "static" case (i.e. ω=0), the mass forces of the soil can be replaced by a static gravitational force acting in x-direction (horizontally) towards the wall. The previous analyses assumed that both walls are connected to the soil. If we consider that there is no tensile resistance between the soil and the wall, separation can occur. This fact has a double influence on the results presented before. As it can be seen from the following diagrams, the soil pressures on a wall without bonding are significantly greater than for the case of complete bonding when the *L/H* ratio equals 1. From a value $L/H \geq 2$, total soil pressure and base shear remain practically constant. For the case of complete bonding this occurs for values of $L/H \geq 5$. This fact indicates that the stiffness of the wall with bonding and its base compliance play a role because due to the tensile resistance of the interface the mass forces are divided between the two walls instead of the total mass forces being applied on the one wall. This fact can be seen for the case where there is no tensile resistance at the wall-soil interface, as total soil pressure on the wall is much greater for a ratio *L/H*=1. The higher soil pressures on a wall without tensile resistance were indicated also by (Veletsos, A. S., Younan, A. H. 1997), albeit without further explanation. The more realistic case of no tensile resistance leads to greater base shear in comparison to the complete bonding case because the flexibility of the wall or its base increases. This series of analyses was carried out as a static loading of 1g towards the one wall.

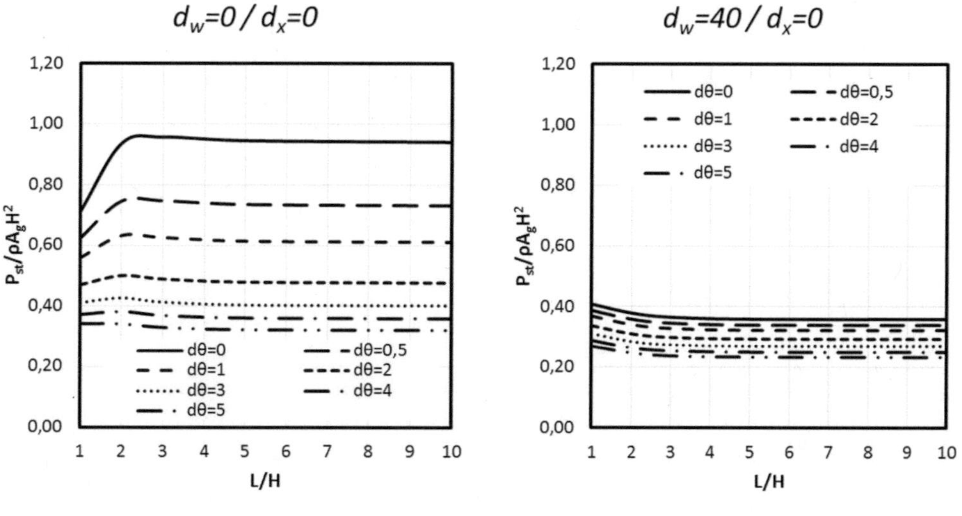

Fig. 5-88 Normalized base shears for the static case for different wall and base flexibilities for no tensile resistance of the soil-wall interface.

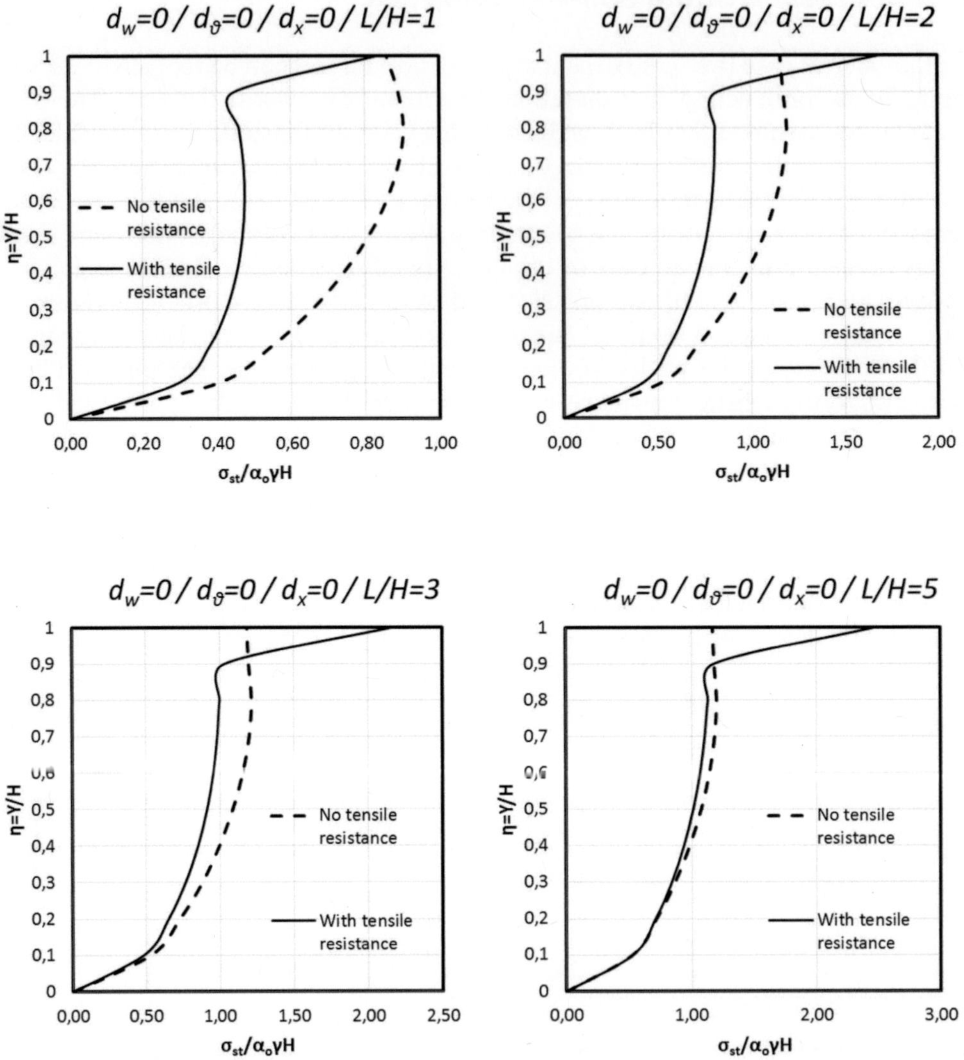

Fig. 5-89 Comparison of the distribution of the soil pressures of a homogeneous soil on a rigid wall for the static case with and without tensile resistance of the soil-wall interface for different L/H ratios.

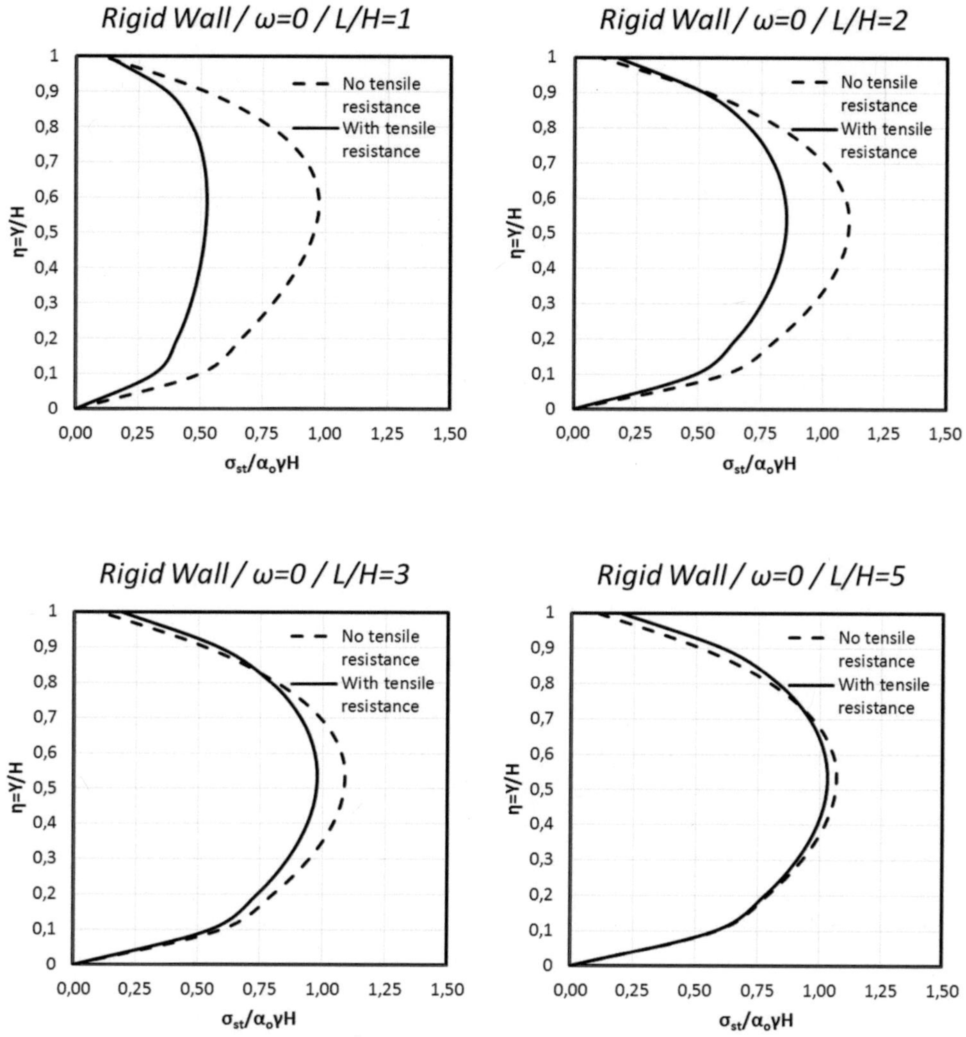

Fig. 5-90 Comparison of the distribution of the soil pressures of a parabolic soil profile on a rigid wall for the static case with and without tensile resistance of the soil-wall interface for different L/H ratios.

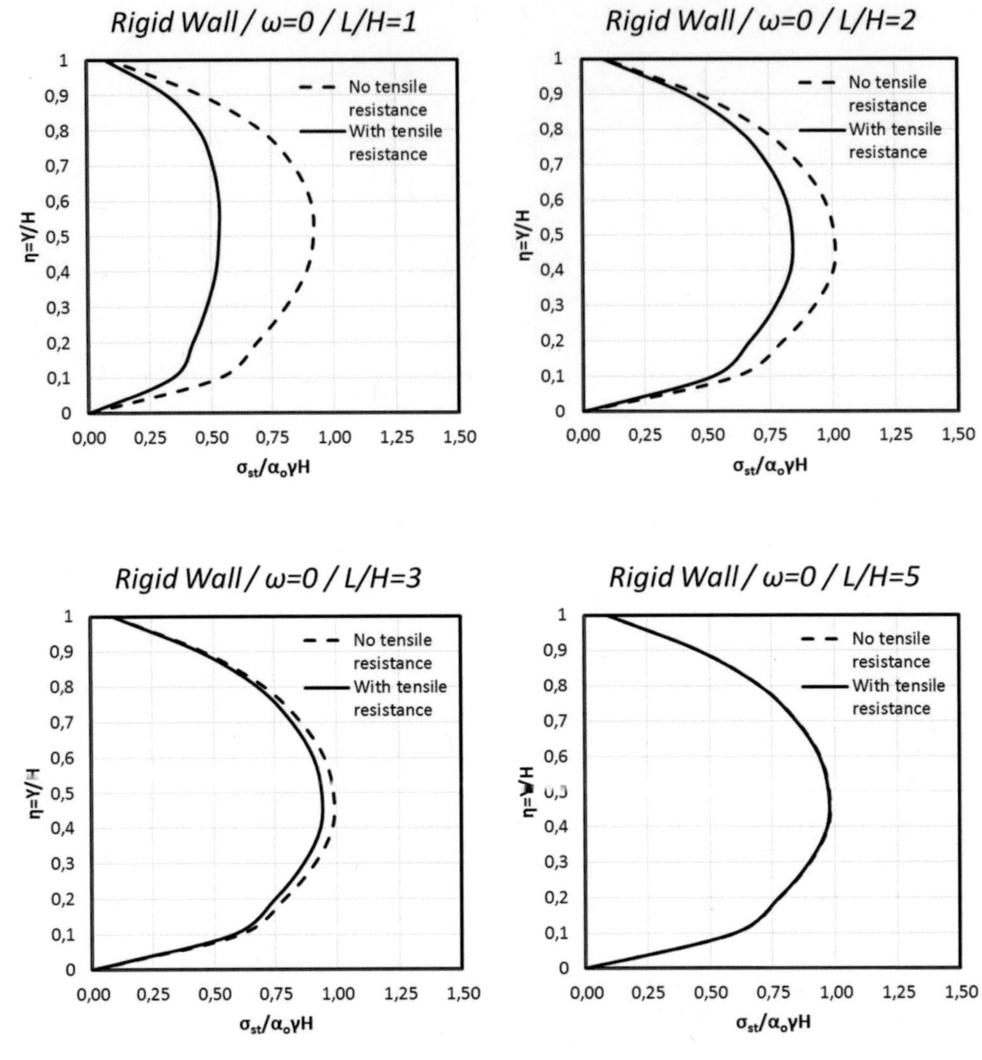

Fig. 5-91 Comparison of the distribution of the soil pressures of a linear soil profile on a rigid wall for the static case with and without tensile resistance of the soil-wall interface for different L/H ratios.

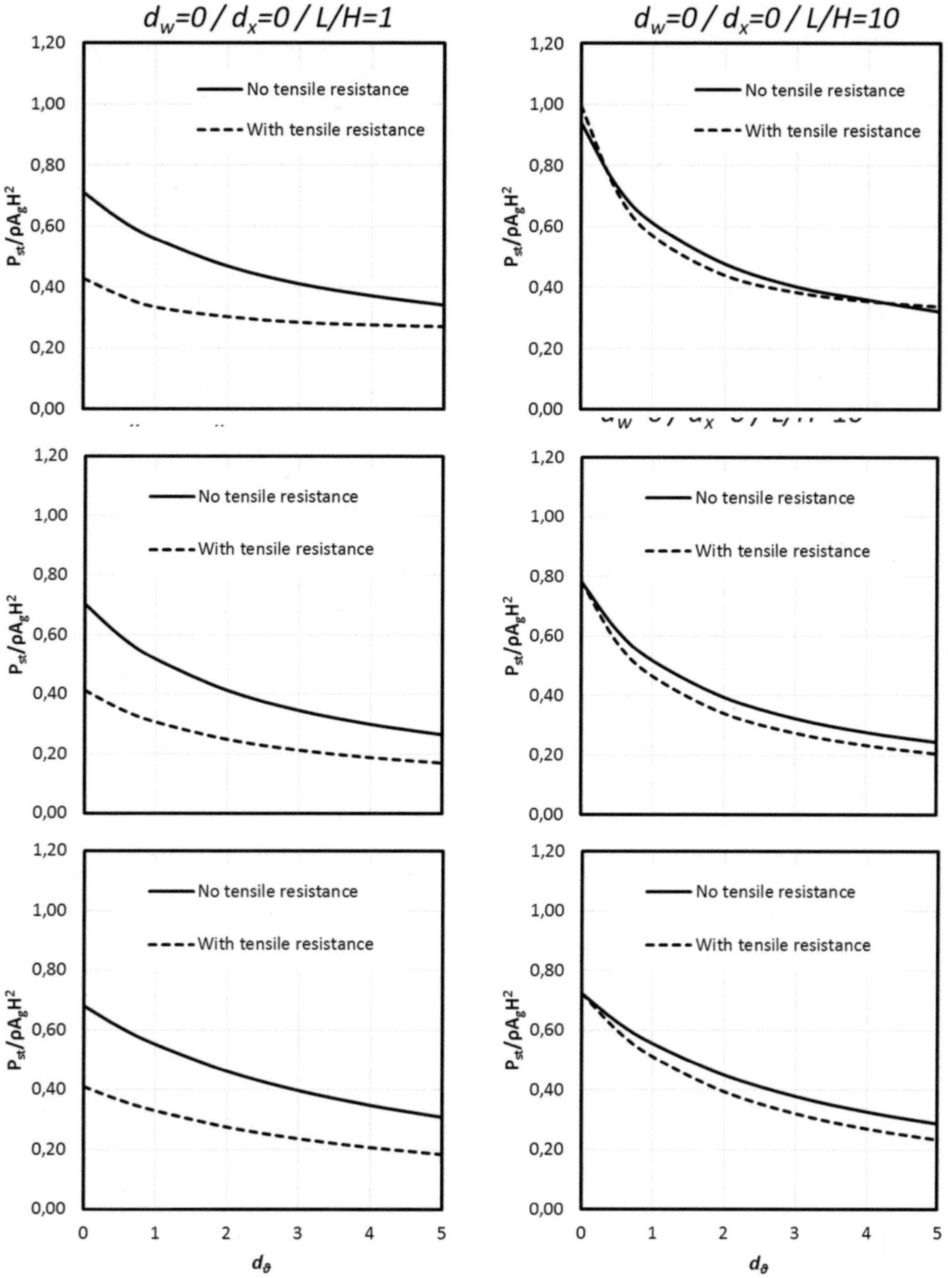

Fig. 5-92 Comparison of the variation of the base shear for the static case with and without tensile resistance of the soil-wall interface for homogeneous (top), parabolic (middle) and linear (bottom) soil profiles.

5.2.8 Other soil profiles

The investigations made until now refer to three "typical" soil profiles, i.e. soil with a constant or homogeneous shear modulus distribution, soil with a parabolic and soil with a linear or triangular shear stiffness. The two non-homogeneous soil profiles have zero shear stiffness at the free surface. However, these soils are often compacted, and have a shear stiffness greater than null also at the free surface. The distribution of the shear stiffness of such soil profiles along the depth can be described by the expression (Vrettos 1988):

$$G(z) = G_o + (G_\infty - G_o)(1 - e^{-\alpha z}) \tag{5-25}$$

where α is the non-homogeneity (constant with dimension of inverse length) and G_o and G_∞ are the shear moduli at the surface and at infinite depth. The non-homogeneity parameter is defined:

$$\Xi_o = 1 - \frac{G_o}{G_\infty} \tag{5-26}$$

and by introducing a new variable ξ instead of z in the form:

$$\xi = \Xi_o e^{-\alpha z} \tag{5-27}$$

Equation 5-24 becomes:

$$G(\xi) = G_\infty(1 - \xi) \tag{5-28}$$

And η a parameter which normalizes the wall height H with respect to the constant α:

$$\eta = \alpha H \tag{5-29}$$

The non-homogeneity constant α influences the curve of the distribution and the non-homogeneity parameter Ξ_o defines the value of the shear modulus at a specific depth as a fraction of the shear modulus of the surface. A non-homogeneity parameter Ξ_o equal to zero refers to a homogeneous soil profile. The influence of these two parameters is shown in the following figures. For rigid walls a frequency domain analysis was carried out in order to compare the results with (Vrettos et al. 2016). The frequency range used is the same as before ($f \in [0.01 \div 16]$ Hz) to cover the statically excited systems and high frequency excited systems.

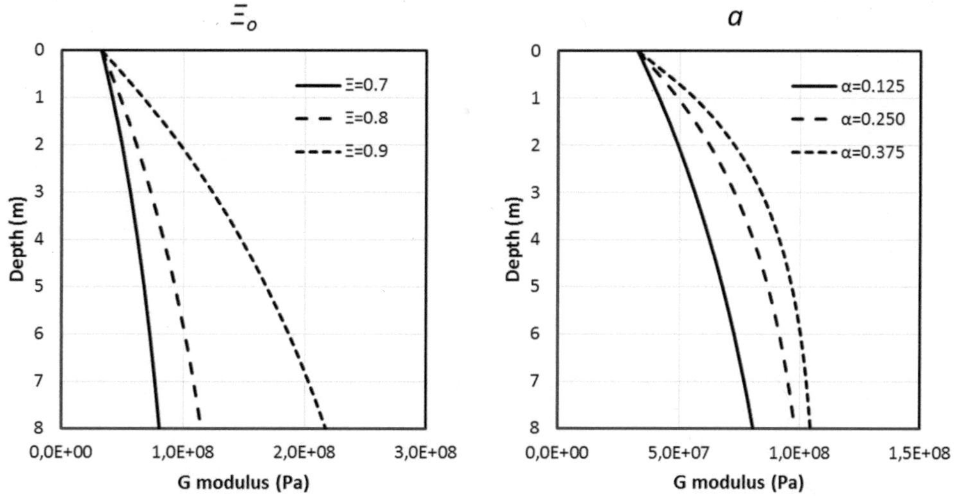

Fig. 5-93 Variation of the non-homogeneity parameter Ξ_o (left) for α=0.125 and variation of the non-homogeneity constant α (right) for Ξ_o=0.7.

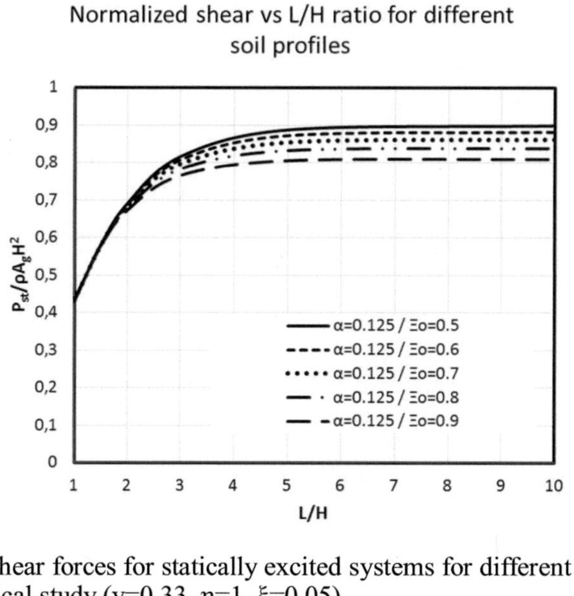

Fig. 5-94 Total shear forces for statically excited systems for different values of Ξ_o in this numerical study (v=0.33, η=1, ξ=0.05).

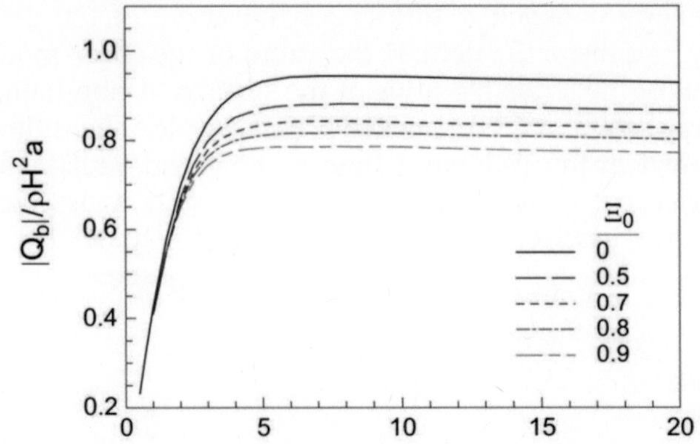

Fig. 5-95 Total shear forces for statically excited systems for different values of Ξ_o (v=0.3, η=1, ξ=0.05) (Vrettos et al. 2016).

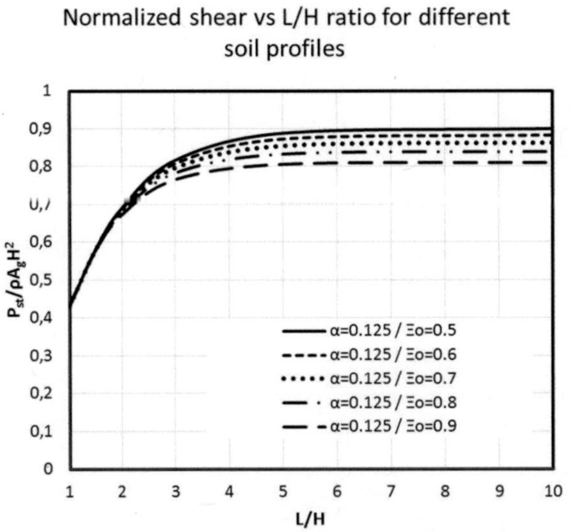

Fig. 5-96 Total shear forces for statically excited systems for different values of Ξ_o in this numerical study (v=0.33, η=2, ξ=0.05).

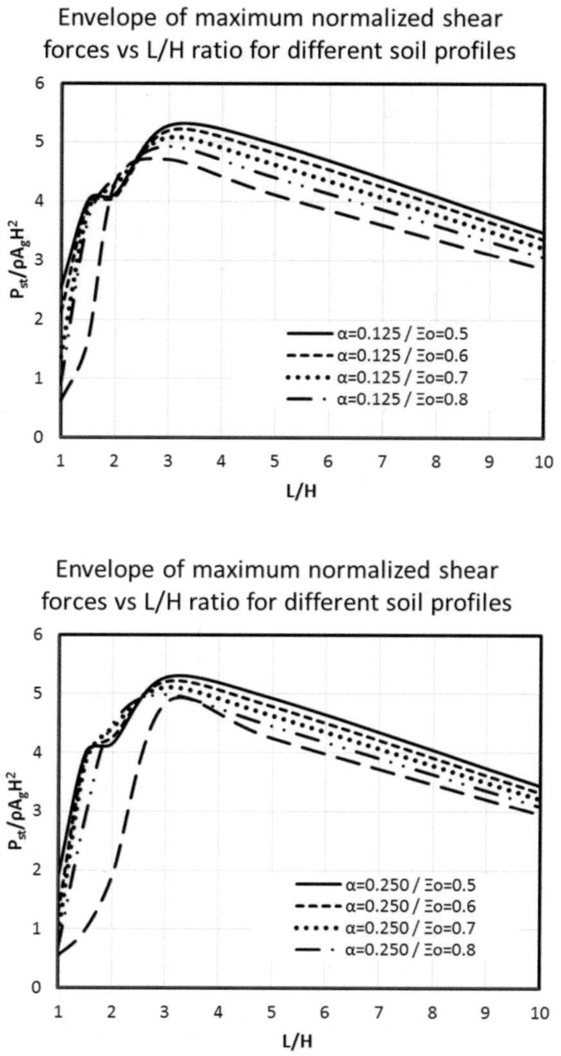

Fig. 5-97 Envelope of the maximum (resonance) normalized shear forces for different L/H ratios for rigid walls according to this numerical study (ν=0.33, ξ=0.05).

The normalized total soil pressure and amplification factors for the investigated parameters α and Ξ_o are presented in Annex D. The small differences between the results of this study and the results of (Vrettos et al. 2016) are due to the different Poisson's ratio values used in the studies (here 0.33 instead of 0.3 as in (Vrettos et al. 2016)).

Psarropoulos et al. (Psarropoulos et al. 2005) include in their investigation a soil profile according to (Gazetas, G., Travasarou, T. 2004). This soil profile is described by the expression for the shear wave velocity:

$$v_s = m(H - y)^{2/3} \tag{5-30}$$

Where m is a parameter chosen so as to determine the mean shear wave velocity of the soil stratum to be 100 m/sec and to ensure comparability of the results with the other soil profiles. For a wall height equal to 8.0 m and a target mean wave velocity of 100 m/sec the parameter m takes the value 41.67. The following figures compare the distribution of the shear wave velocity and the shear modulus of the relation proposed by (Gazetas, G., Travasarou, T. 2004) with the linear (triangular) shear modulus distribution. As it can be seen, the distributions are similar and this is also evident from comparing the results for the statically excited case and at the resonance frequency. In order to provide a name for this soil profile the term "quasi-linear" is adopted because of the linear-like distribution of the shear wave velocity along the depth.

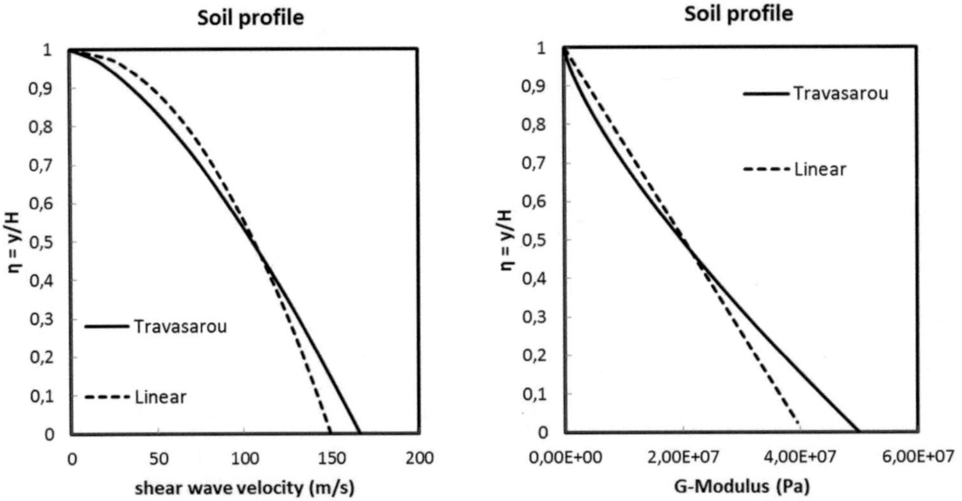

Fig. 5-98 Shear wave and shear modulus distribution according to (Gazetas, G., Travasarou, T. 2004).

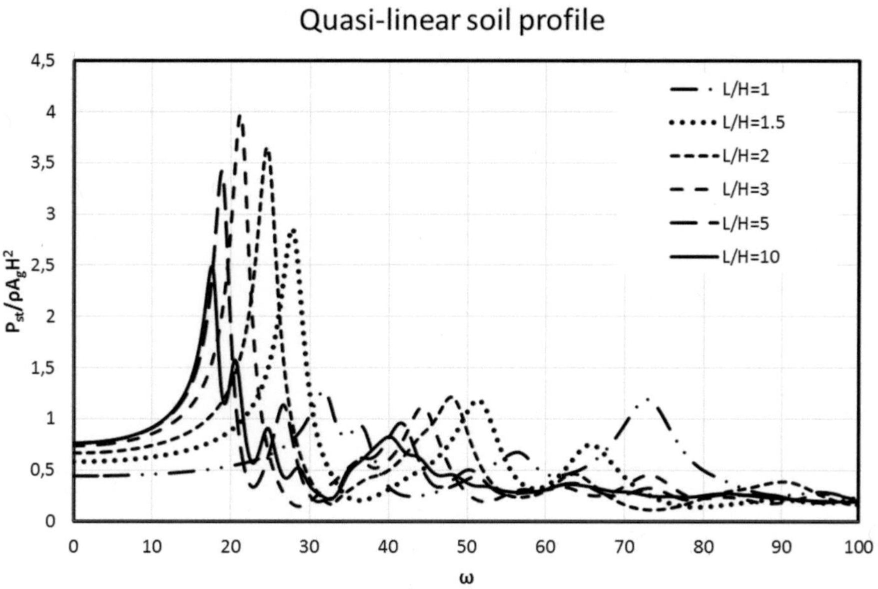

Fig. 5-99 Steady state response of rigid wall-soil systems with different L/H ratios for the quasi-linear soil profile according to this numerical study (ν=0.33).

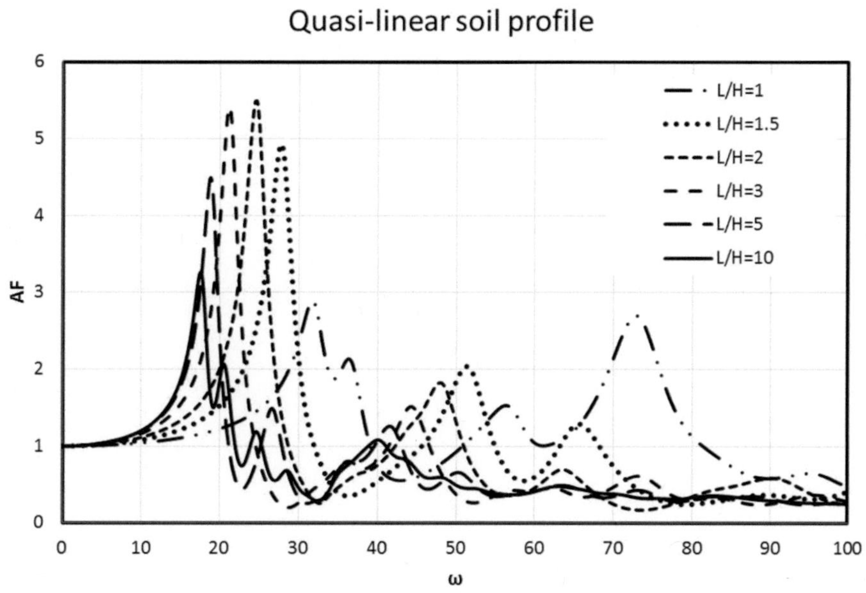

Fig. 5-100 Amplification factors of the total shear force for the quasi-linear soil profile.

Chapter 6
Comparison of boundaries modelling for dynamic soil-structure interaction

Summary

This chapter investigates different modelling techniques of boundary conditions used in a finite element model in order to have the minimum influence of the boundaries on the wave propagation in the soil domain. This problem is very common as it is state of practice to use a deconvolution analysis of the target time history of accelerations at a soil level, which corresponds to the bottom of the finite element model. The deconvolution analysis is made using one dimensional wave propagation, whereas at a finite element model a two or three dimensional domain exists. Twelve numerical models with some additional modifications are investigated. The analyses are carried out in the time domain with the finite element program Abaqus. The resulted time history acceleration of each numerical model is compared with the desired target acceleration in terms of response spectra. .

6.1 Description of the numerical model

A common problem in earthquake analysis of structures is to assign a predefined time history to the finite element model. Several problems can arise: assigning the seismic motion as "outcrop" or "within", the damping of the soil material, the finite element model's boundaries, to mention but a few. For the first issue referred to, the reader can find information elsewhere (Kwok et al. 2007), (Dawson, E. M., Mejia, L. H. 2006). The second issue has been investigated by others and information can also be found elsewhere ((Kwok et al. 2007), (LAM et al. 2007). The third issue regarding the finite element model has been investigated by (LAM et al. 2007). Although in their investigation they analysed many aspects of the modelling, they did not investigate thoroughly the case of transmitted boundaries, which is the most common one for dynamic analyses. Instead, they concentrated only on the case of dashpots as transmitted boundaries as they were implemented by (Kuhlemeyer R. L., Lysmer, J. 1969). An alternative for modelling transmitted boundaries is using the infinite elements (Bettess, P., Emson, C., Zienkiewicz, O. C. 1983) available in Abaqus and perfectly matched layers (Basu, U., Chopra, A. K. 2004; Govindjee, S., Sagiyama K., Persson P. O. 2014). These techniques are investigated among others in this thesis.

When performing a seismic analysis, the engineer has to use many times code-specific PGAs and response spectra. For a given PGA and response spectrum and with the use of specific programs artificial time histories can be produced, whose response spectra fit the design response spectrum. As the finite element model cannot have infinite dimensions, the boundaries of the soil domain extend to 2-3 times the height and/or width of the embedded structure according to some provisions (Hettler 2012). The artificial time histories, which have been produced for the free surface, have to be transformed in order to be applied at the model's base. This transformation is known as deconvolution and it can be performed with the use of several programs (for example SHAKE 91, SHAKE 04). The most common program for this procedure is SHAKE91 (Schnabel, P. B., Lysmer, J., Seed, H. B.). Shake is a computer program that uses the Fast Fourier Transform in order to estimate one-dimensional wave propagation through different soil layers and to give estimated accelerations and stresses at different soil levels using the equivalent linear method for the hysteretic damping of the soil. Since SHAKE 91 uses the shear beam approach for the one-dimensional wave propagation, it is expected that a continuum finite element model with different types of boundaries will not deliver exactly the same results because the propagating waves are reflected by the free surfaces and the model's boundaries. A sensitivity analysis of the different modelling of boundaries is conducted here. An artificial earthquake motion deconvoluted with SHAKE91 is given as input motion in the several finite element models. Another important parameter is the predefined damping ratio both in the SHAKE91 program and the finite element program. When the bedrock is at a finite depth below the surface and soft soil lies above it, the bedrock can be approximated with common boundary conditions in the finite element model. When the bedrock is at a very great depth and we practically have a half space of soil, then a part of the soil is modelled and absorbing boundaries have to be considered in the model. In the first case, when the bedrock is modelled with common boundaries, the deconvoluted seismic motion can be given as prescribed accelerations of displacements at the base of the FE model. In the latter case, the time histories of accelerations have to be transformed into time histories of velocities and then the appropriate nodal or traction forces can be computed and given at the base of the FE model. The traction forces have the advantage compared to the nodal forces that no specific effective nodal area needs to be computed, which can be tedious for irregular meshes. The traction forces are given at the whole bottom surface as a surface load.

If a half-space is modelled and infinite elements or dashpot elements are present, then the seismic input has to be given in the form of forces or tractions. The damping at the boundaries is proportional to the density of the medium and the wave velocities:

$$d = \rho C_s \tag{6-1}$$

where d represents the damping factor, ρ the density of the medium and C_s the velocity of shear waves.

However, the wave absorption has to be taken into account for the seismic input. The accelerations are converted to time history velocities and then multiplied with the shear wave velocity, the density of the medium and the contributing area around the node in order to obtain forces. The factor 2 accounts for the damping at the boundary (see for derivation Annex I).

$$F_S(t) = 2\rho C_S V_S(t) dA \qquad \textbf{(6-2)}$$

$\sigma_s(t)$ means the shear traction as a time history, ρ the density of the medium and C_s the shear velocity of wave propagation in the medium, $V_s(t)$ the earthquake velocity time history, and dA the contributing area around the node. In terms of tractions along the surface of the basis:

$$\sigma_S(t) = 2\rho C_S V_S(t) \qquad \textbf{(6-3)}$$

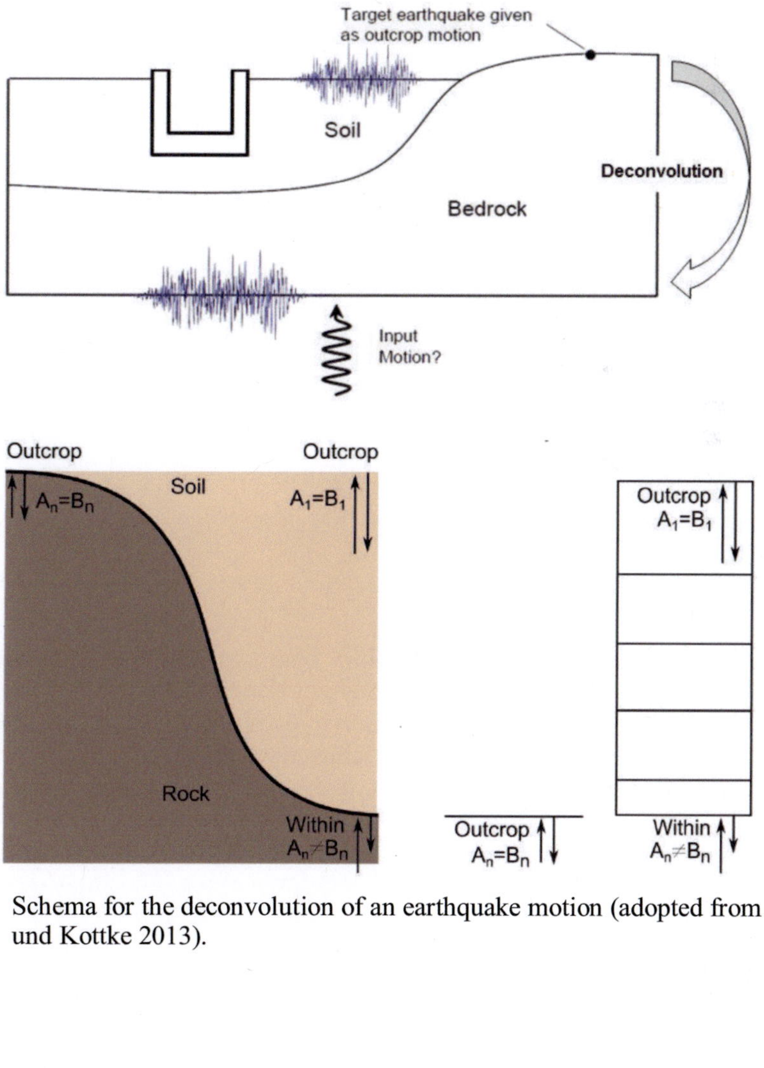

Fig. 6-1 Schema for the deconvolution of an earthquake motion (adopted from (Rathje und Kottke 2013).

Similar studies have been carried out by other researchers (Lam, I.P., Law, H., Yang, Ch. 2004). They compared the one-dimensional soil response in the frequency domain of (Schnabel, P. B., Lysmer, J., Seed, H. B.) to the numerical results in the time domain of the finite element program ADINA. They concluded that the most appropriate boundaries of the finite element model are the slave left-right boundaries (Case (d) of their study and Model 4 of this study). Moreover, they showed that in the case of the model with two edge columns, the best results are delivered by the model with a column width equal the soil's depth $H=D$. They further investigated the influence of the time integration scheme and the type of damping. They compared three time integration schemes, i.e. Wilson's θ Method, Newmark Method (δ= 0.5 and α = 0.25) and Newmark Method (δ = 0.65 and α = 0.331) and they pointed out that different time integration schemes give different results at high frequencies but are all still acceptable. Regarding damping they used three types of Rayleigh damping; mass proportional damping only (α), stiffness proportional damping only (β) and both mass and stiffness proportional damping (α and β). They concluded that all the three methods of applying the Rayleigh damping are appropriate with the best results given by mass proportional damping only.

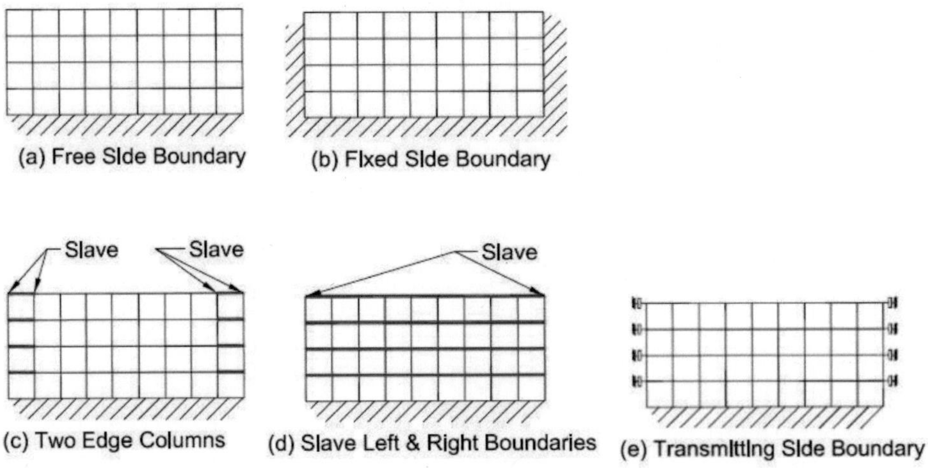

Fig. 6-2 Models investigated by (Lam, I.P., Law, H., Yang, Ch. 2004).

In this study a very stiff soil with rock-like characteristics (v_s=800 m/s) is used. Although such a high shear wave velocity is not common for soil-structure interaction problems, it is common in dam engineering, where dam-rock interaction takes place. The high shear wave velocity was chosen so as to avoid problems such as hysteretic damping because of big strains of the soil and frequency cut-offs of the seismic excitation and to concentrate more on the investigated issue of boundaries modelling. As the soil stratum investigated here is very stiff, little shear strains are expected and the hysteretic damping remains at low values. Thus no extra care needs to be taken to model a shear-dependent damping of the soil at the finite element code.

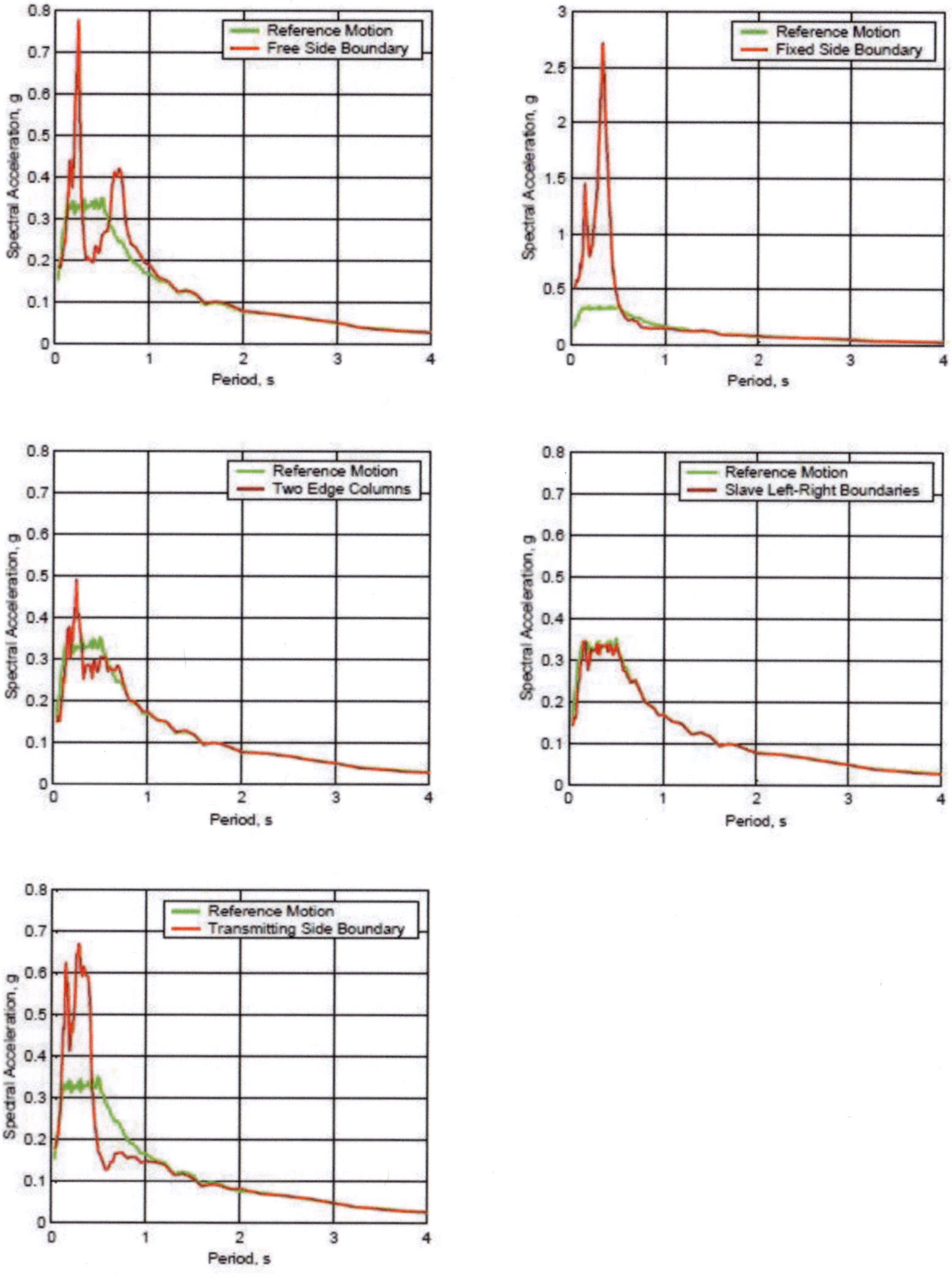

Fig. 6-3 Results of the different boundary modelling provided by (Lam, I.P., Law, H., Yang, Ch. 2004).

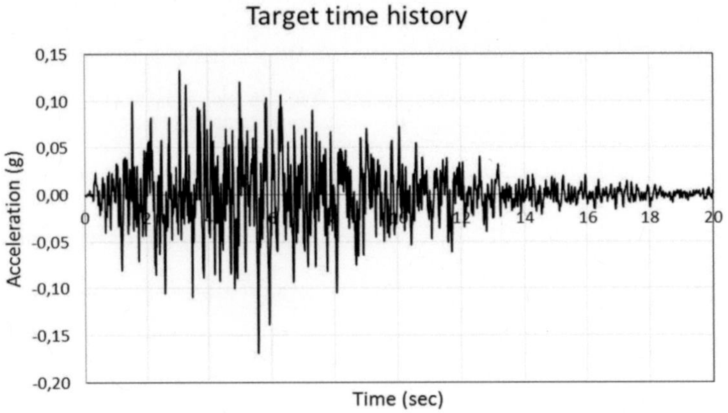

Fig. 6-4 The artificial target time history of acceleration.

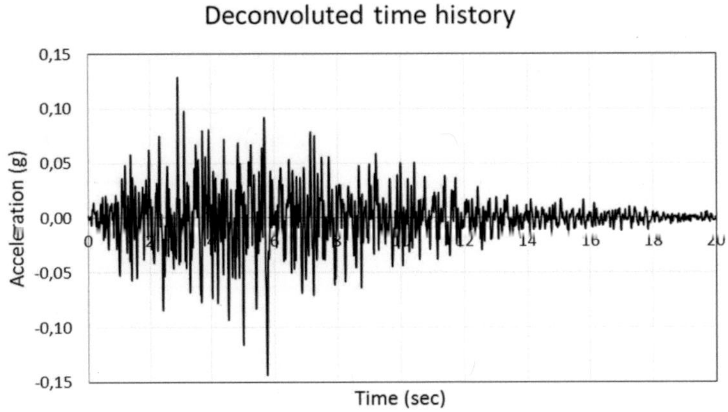

Fig. 6-5 The deconvoluted time history of acceleration.

All the finite element models have the same size and the same material for the soil as well as the same discretization. The soil domain is 30 m deep and 105 m wide, the damping ratio is 5%, the shear wave velocity of the soil is 800 m/s and its density is ρ=2194 kg. The material damping is given in terms of Rayleigh damping. Information on the influence of the Rayleigh damping on the analysis results can be found elsewhere (Kwok et al. 2007), (LAM et al. 2007). It was stated in (Kwok et al. 2007) that along the simple and full Rayleigh damping the best results, which fit the frequency domain solution of SHAKE91, are given for the calculation of the α and β parameters for the first natural frequency of the system and a frequency three times higher than the first one. The first natural period of the soil stratum is 0.15 sec (6.67 Hz).

The size of the finite elements is 1 m, which is exactly one tenth of the wavelength of 10 m for a cut-off frequency of 80 Hz. This element size is assumed to be adequate as most of the earthquakes have a frequency content between 1.5-5 Hz (Bray,

J. D., Faraj, F., Rathje, E. M., Russell, S. 2004). The finite element models with the different boundary conditions are described sequentially:

Model 1: In Model 1 both the base and the sides are supported. The acceleration is given at all boundaries as the same time history. This model is investigated here only to indicate that the commonly used boundaries for static analyses are not appropriate for dynamic analyses. The response spectrum of the acceleration of the free surface is, as expected, several times bigger than the desired one, due to the reflexions at the boundaries. A modification of Model 1, referred to as Model 1a, is to constrain the side nodes so that they move only horizontally but not vertically. This will impose a shear beam movement of the soil domain. However, this modification is not applicable for vertical components of a time history.

Model 2: It is the same as Model 1, but massless. This model is often used in dam engineering, where the elasticity of the underlying rock is important for the dynamic analysis but the real stress field of the underlying rock can be neglected. In this model wave propagation is neglected and damping can be given only in terms of the stiffness matrix (parameter β of the Rayleigh damping). This approach is applicable only to implicit codes (by assigning zero value to the density) as density is required for explicit codes (very small values of density lead to an extremely small stable analysis time step) (U.S. Department of the Interior, Bureau of Reclamation 2006)

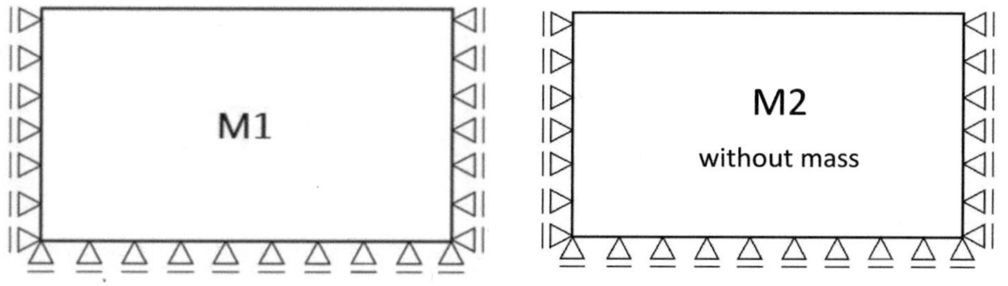

Fig. 6-6 The boundaries of Models 1 and 2.

Model 3: Model 3 is supported only at its base forming a soil shear beam. This model is expected to give the desired results as it reproduces the shear beam of the one-dimensional propagation analysis of SHAKE and similar programs. Although the input motion should cause only propagation of shear waves in the model, reflection at the side boundaries of the finite element model cannot be excluded, which can affect the free surface motion.

Model 4: This Model is the same as Model 3 with the difference that the side nodes are coupled together. This "slaving" of the side nodes is to enforce a better shear beam behaviour.

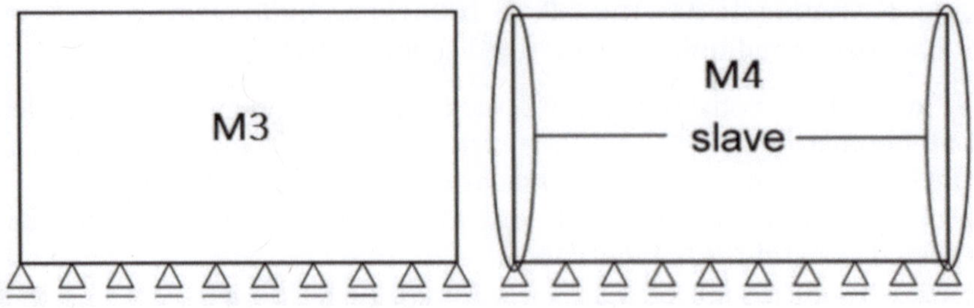

Fig. 6-7 The boundaries of Models 3 and 4.

Model 5: In this model two soil columns extend at the left and right sides of the model. The side nodes of each column are coupled together to enforce a shear beam behaviour, which in turn should enforce a shear beam behaviour in the main soil domain. In order to investigate the influence of the width of the columns, two column widths are investigated here, i.e. $H/3$ (Model 5) and $2H/3$ (Model 5a) where H represents the depth of the soil domain. The advantage of this model is that the soil domain to the left and right of the investigated structure is not required to be at the same level for the coupling of the opposite nodes to be feasible. This is the case in many geotechnical problems where the behaviour of a gravity wall has to be investigated.

Model 6: This model is similar to Model 5 with the following difference: the side nodes are not coupled together and the column is divided into thinner columns that have been assigned the same material properties but have increasing damping as we move to the sides. These thin soil columns, with an increasing damping ratio reaching 100% of the critical damping at the outer columns, reproduce in a way the consistent wave-absorbing boundary condition (Basu, U., Chopra, A. K. 2004; Govindjee, S., Sagiyama K., Persson P. O. 2014). This model has the same advantage as Model 5 relating to the soil level at the boundaries. A modification of this model (Model 6a) is to constrain the side nodes so that they only move horizontally. As the damping of the outer column is 100% of the critical damping, the wave reflections at the boundaries should not be a problem, and these constrains shall force a shear beam movement as in Model 1a. A further modification of this model, referred to as Model 6b, is to enforce coupling of the outer nodes as in Model 4.

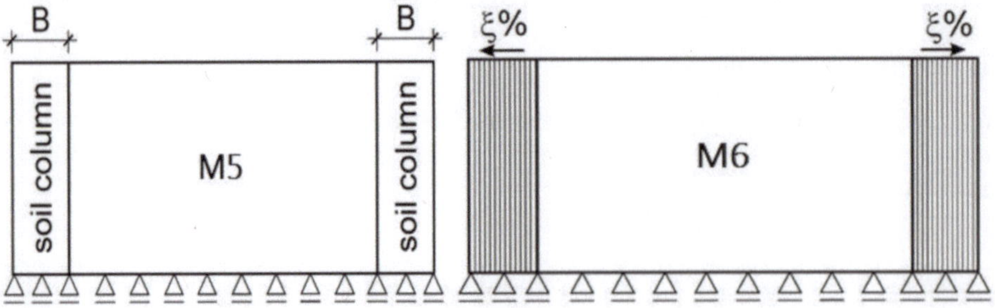

Fig. 6-8 The boundaries of Models 5 and 6.

Model 7: This Model has ordinary dashpots at the side boundaries in order to absorb any reflecting waves.

Model 8: The same as Model 7, but the side nodes are additionally coupled as in Model 4.

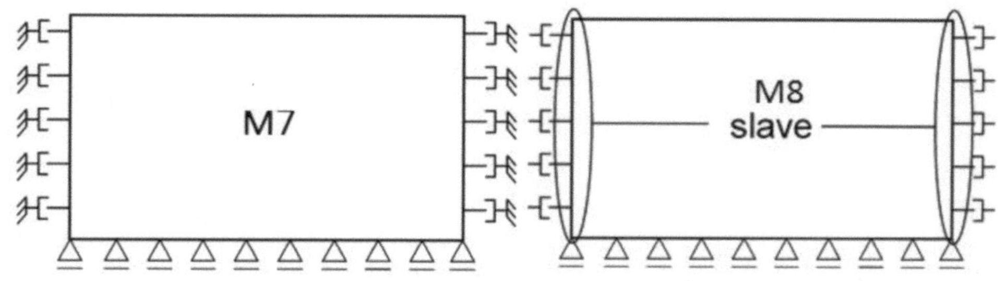

Fig. 6-9 The boundaries of Models 7 and 8.

Model 9: Instead of ordinary dashpots, this model has the infinite elements available in Abaqus. The infinite elements enforce the non-absorbing boundary condition for dynamic problems (Kuhlemeyer R. L., Lysmer, J. 1969), (Bettess, P., Emson, C., Zienkiewicz, O. C. 1983).

Model 10: The same as Model 9, but with additionally coupled side nodes.

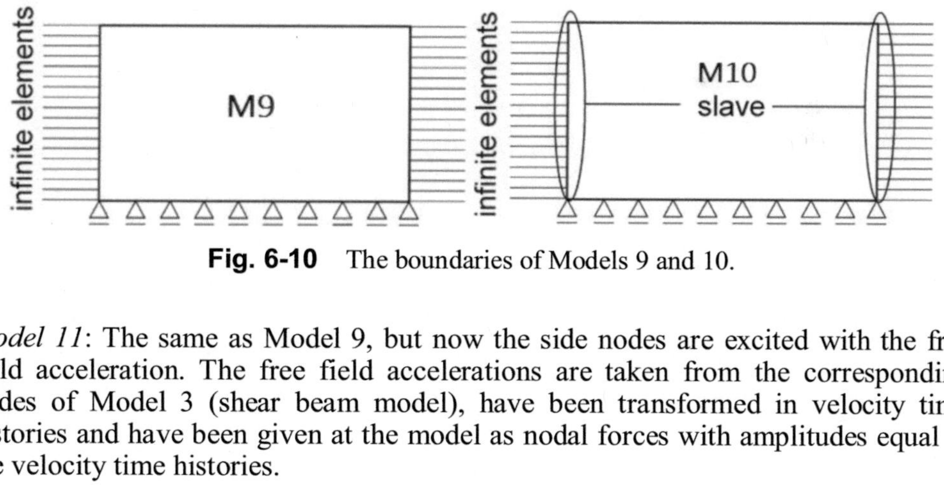

Fig. 6-10 The boundaries of Models 9 and 10.

Model 11: The same as Model 9, but now the side nodes are excited with the free field acceleration. The free field accelerations are taken from the corresponding nodes of Model 3 (shear beam model), have been transformed in velocity time histories and have been given at the model as nodal forces with amplitudes equal to the velocity time histories.

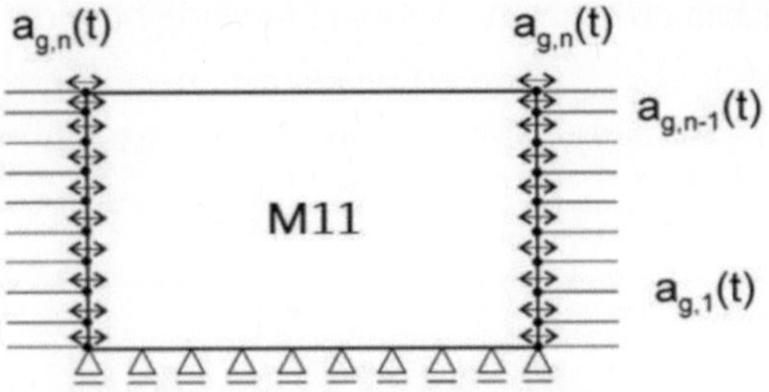

Fig. 6-11 The boundaries of Model 11.

Model 12: The same as Model 7, but now the dashpots are not connected to the ground but to the nodes, to which the accelerations of the corresponding nodes of Model 3 (shear beam model) are given as nodal forces at the side nodes.

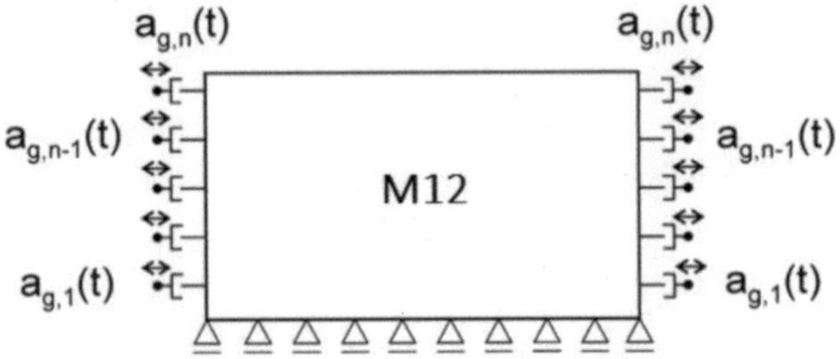

Fig 6-12 The boundaries of Model 12.

6.1.1 Results of the different models

In order to compare the results, the outcrop motion at the surface of the finite element model is written down and its response spectrum is compared with the response spectrum of the target acceleration. The results are shown in the following response spectra.

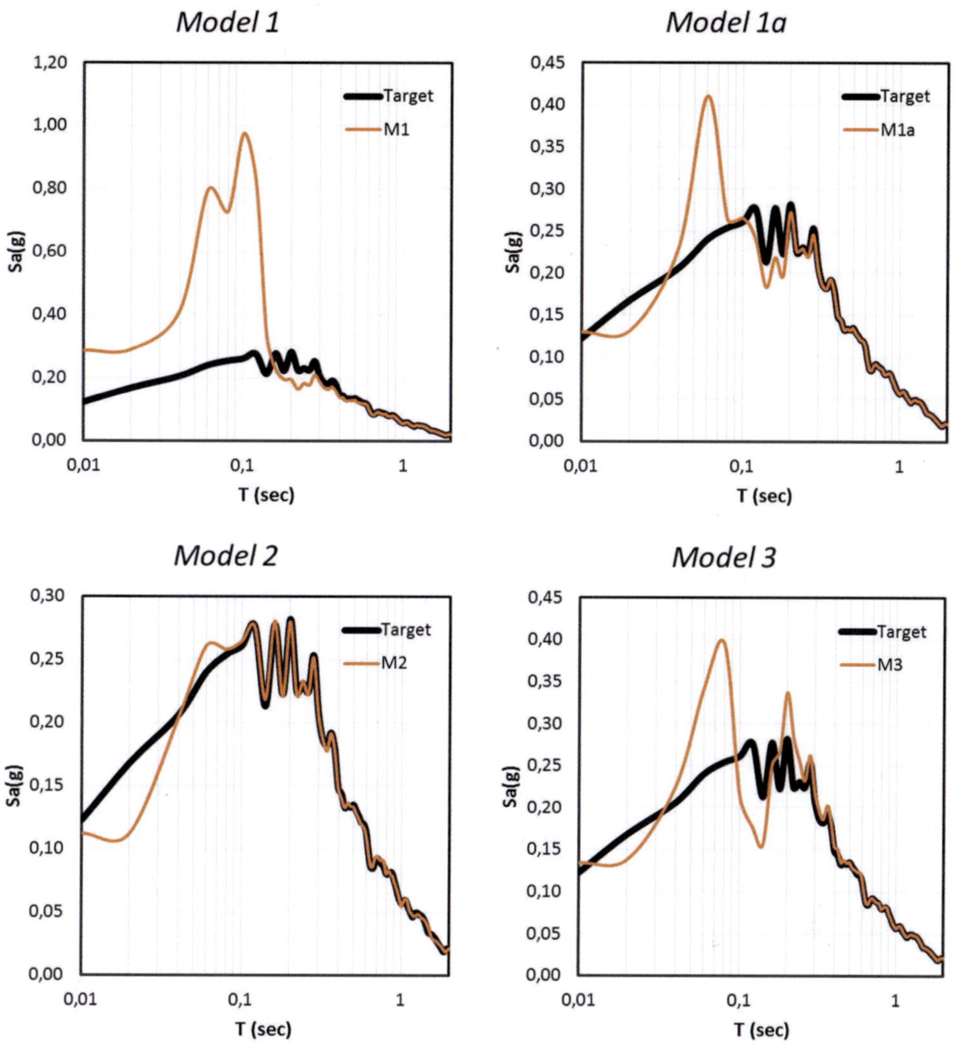

Fig. 6-13 Response spectra of the free field acceleration of the models.

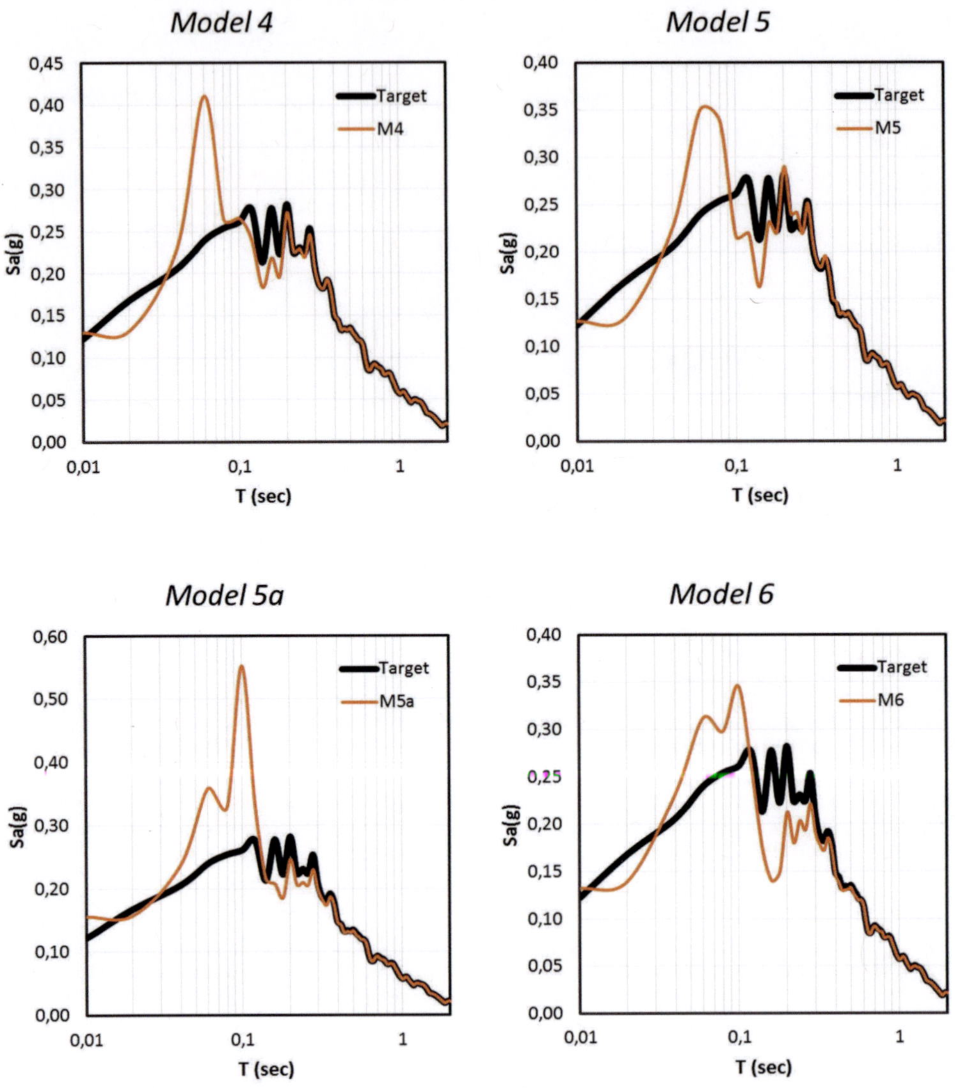

Fig. 6-14 Response spectra of the free field acceleration of the models (cont'd).

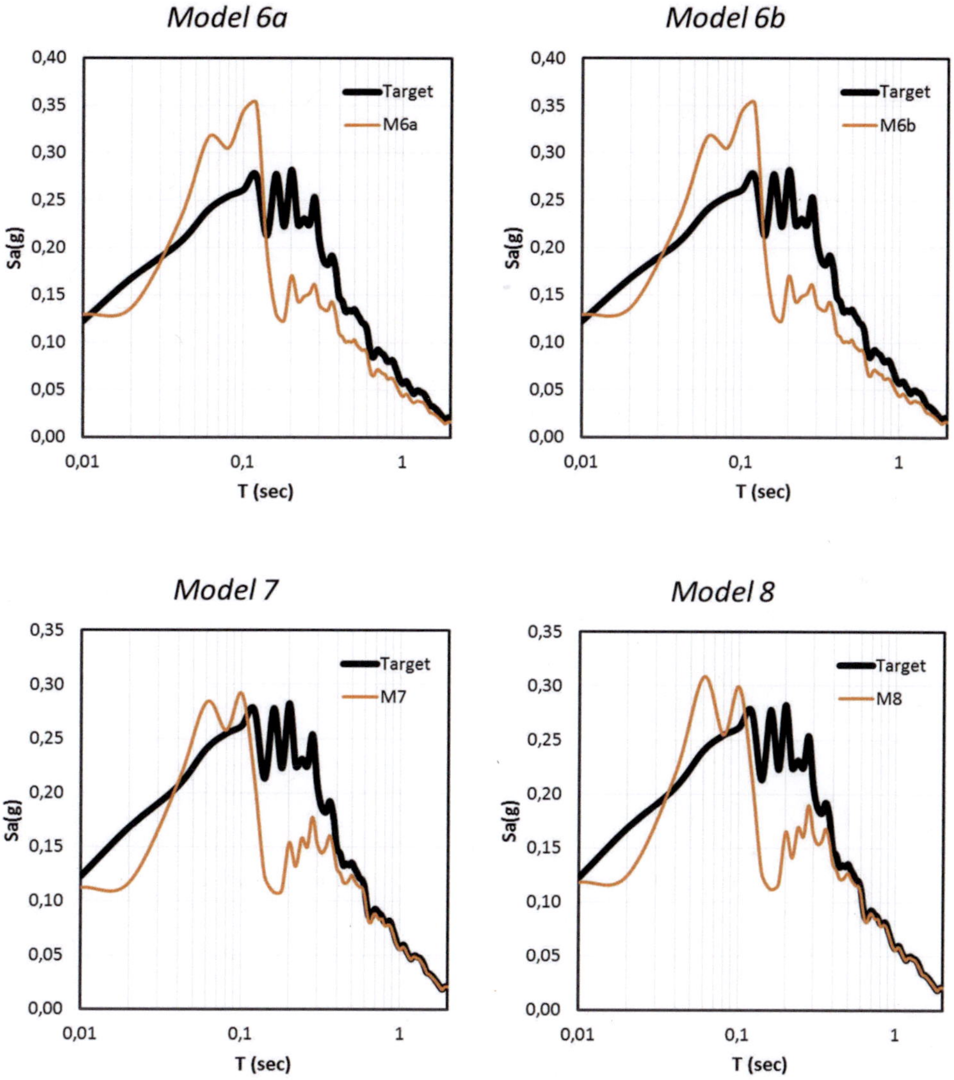

Fig. 6-15 Response spectra of the free field acceleration of the models (cont'd).

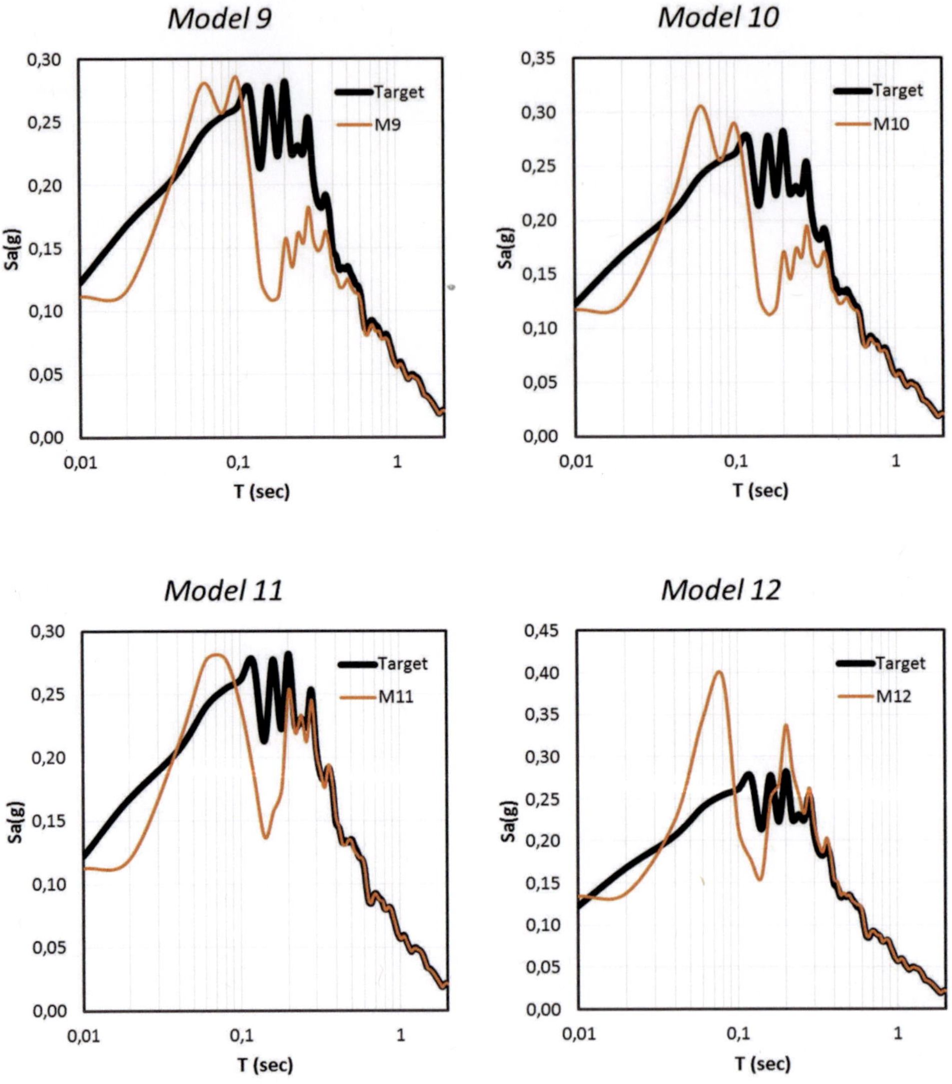

Fig. 6-16 Response spectra of the free field acceleration of the models (cont'd).

Model 6 was further investigated as it is the easiest to implement in common finite element programs, which do not have the ability of modelling dashpots or infinite elements. As a modification, there is no linear increase in damping as before, but an exponential and parabolic increase. A third, more simple modification is to have only the outer soil column with 100% damping. The parabolic increase used here has the function $d(\%) = 5\% + 0.0001x^6$, where x is the distance from the soil domain with 5% damping as far as the outer boundary. The best results regarding these three modifications are offered by the parabolic increase of damping, although the other two modifications are satisfactory for engineering purposes.

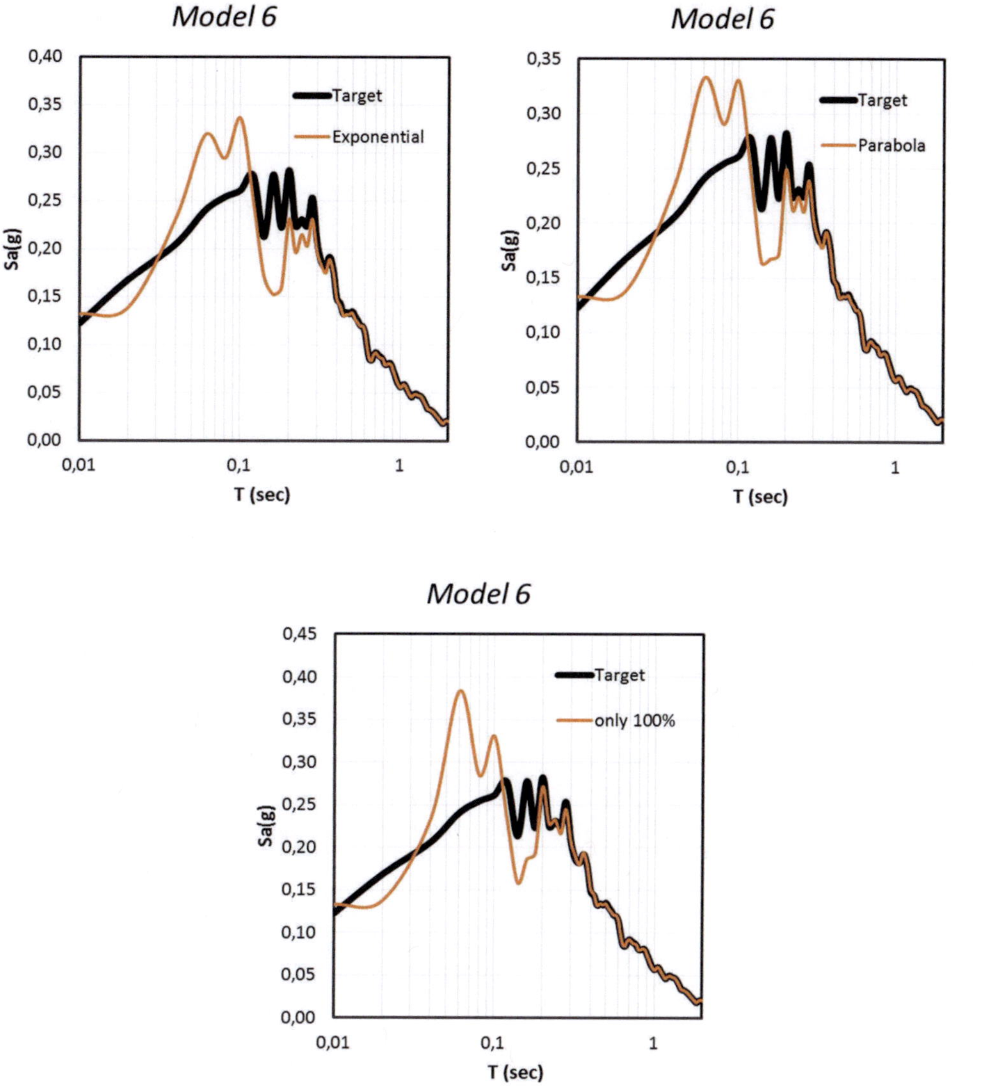

Fig. 6-17 Modifications of the increase in damping in Model 6.

The Rayleigh damping can be given either in terms of mass proportional (α) or stiffness proportional (β) damping or a combination of these two for two frequencies. It is common practice to choose the first frequency and an odd multiple of it in order to calculate the Rayleigh damping parameters. For Model 4 the Rayleigh damping was calculated for three combinations of the natural frequencies of the soil stratum, i.e. 1st and 3rd (f+3f), 1st and 5th (f+5f) and 1st and 7th (f+7f) as well as for mass proportional and stiffness proportional damping only.

Table 37 Values of the Rayleigh damping parameters for a damping ratio of 5%.

Frequencies	Mass proportional damping only	Stiffness proportional damping only	Rayleigh damping	
	α	β	α	β
1st	4.1867	0.002388		
1st and 3rd	-	-	3.1400	0.000597
1st and 5th	-	-	3.4888	0.000398
1st and 7th	-	-	3.6633	0.000298

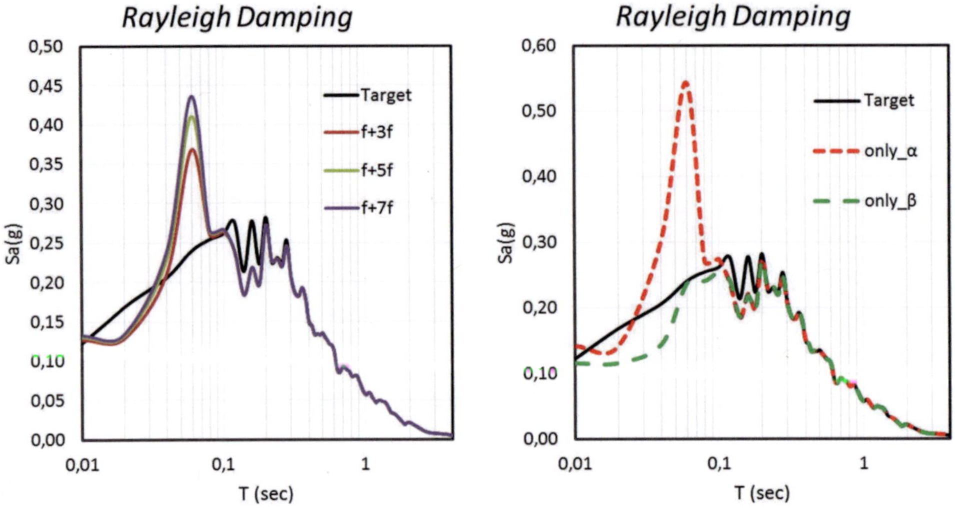

Fig. 6-18 Influence of the calculation of the Rayleigh damping parameters.

For the comparison of the response spectra, the following conclusions are drawn:

a. All the models show a very good agreement at high periods (>0.5 sec) with the exception of Models 6 and 6a.
b. The models with dashpots and infinite elements absorb a lot of energy and thus are non-conservative. Although the infinite elements should serve for meshing reduction, it seems that the boundaries have to be placed at a longer distance from the area of interest in order to avoid such a high damping of the excitation. This problem is mitigated by applying the free field motion at the side boundary nodes.
c. The best agreement is shown by Model 2, which is the massless model. This massless approach of the foundation has obvious restrictions. It cannot be used in explicit procedures and no gravitational load of the soil domain can be applied in order to have real stress conditions in the soil domain. It is mostly used in dam engi-

neering to account for the flexibility of rock, but not for stress failure in rock.

d. Among the remaining models the best results are shown by Models 1a, 4, 5 and 5a. All these models restrict the movement of the soil domain in a way that a shear beam response is achieved and this is the reason why these results are very similar to the one-wave propagation analysis of SHAKE91.
e. The models with increasing damping at the side columns of the soil domain absorb too much energy and they are non-conservative. Another increase of the damping ratios (for example parabolic instead of linear applied here) may give better results.
f. The best results are delivered by the Rayleigh parameters calculated for the 1st and 3rd frequency and even better results by stiffness proportional damping only. This comes in contrast to the results of (Lam, I.P., Law, H., Yang, Ch. 2004), who carried out a similar investigation for a series of weaker soil stratums. In their investigation the best results occur for the Rayleigh parameters calculated for the 1st and 5th natural frequency of each stratum, and the mass proportional damping is more accurate than the stiffness proportional damping. This can be explained due to the different natural frequencies of the investigated systems. The system of this study has a larger natural frequency than the one of (Lam, I.P., Law, H., Yang, Ch. 2004) and the mass proportional damping fails to damp the response of the system in higher frequencies.

This investigation showed which boundaries modelling is more appropriate in order to receive satisfactory results at the free surface of the model in terms of acceleration. These boundaries models, however, are not always applicable for unsymmetrical geometries of the soil domain or for nonlinear analyses, where the soil has a plastic behaviour. It is for the analyst to decide which boundary modification is more appropriate for his problem and which adaptions he must make. This investigation serves only as a guidance.

Chapter 7
Practical application

Summary

The simplified numerical investigation carried out in the chapters 4 and 5 offers a very good overview of the different parameters that affect the dynamic loading of soil and water retaining structures. In this chapter two navigation locks (lock Iffezheim and lock Fankel in Germany) are analysed for a seismic event (for only the horizontal component) and the obtained results are compared with the ones predicted by the analyses of the former chapters. Some deviations are to be expected because of the more complicated geometry of the real structures and the more realistic modelling conditions.

The first lock (Iffezheim) consists of two gravity soil retaining walls which are not monolithically connected to the chamber plate, so a relative rotation of the wall about its base is allowed. The forces acting on the one side wall of the navigation lock are computed with the methology of chapter 5 and are compared with the results of the finite element analysis of this chapter. A half-space is assumed for the soil domain.

The second lock (Fankel) is a monolithical U-chamber. The rotation takes place at the centre of the base of the whole chamber and not at the base of each chamber wall. Because of the monolithic reversed frame and in order to use the tables of chapter 5, the calculation of the springs follows the methology of Bradenberg et al., which splits the springs of the embedded construction in partial springs.

7.1 Navigation lock Iffezheim

The navigation lock of Iffezheim, Germany, is a twin chamber navigation lock with 24 m wide and 270 m long chambers on the river Rhine. The drop height is 12.5 m. The navigation lock Iffezheim consists of three gravity concrete walls which form the chambers. The base plate is made of concrete and connected to the chamber walls by hinges. The gravity walls have a total height of 21.5 m and a width at their base of 12 m. The navigation lock Iffezheim is in the earthquake zone 1 of Germany with a reference acceleration 0,4 m/s^2 (0,04g), which according to (EN 1998-1:2004 Eurocode 8) is a very low seismicity region. The importance factor of the lock is γ=1,2 according to the German code (19702), because its height is bigger than 15.0 m.

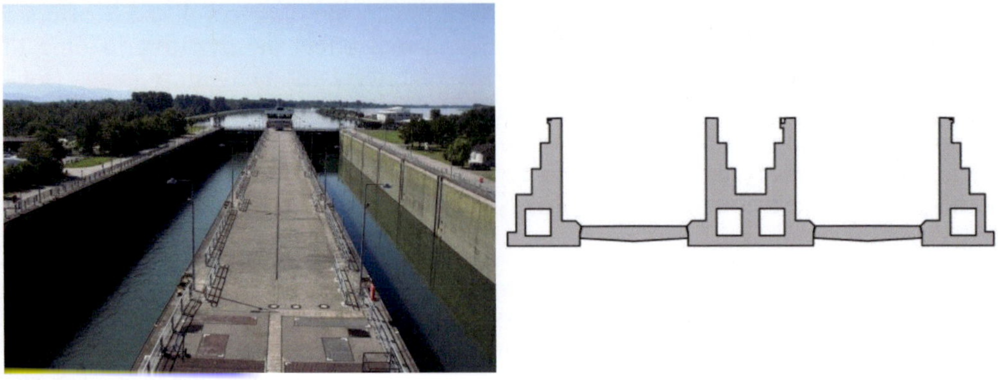

Fig. 7-1 The navigation lock Iffezheim and its cross section.

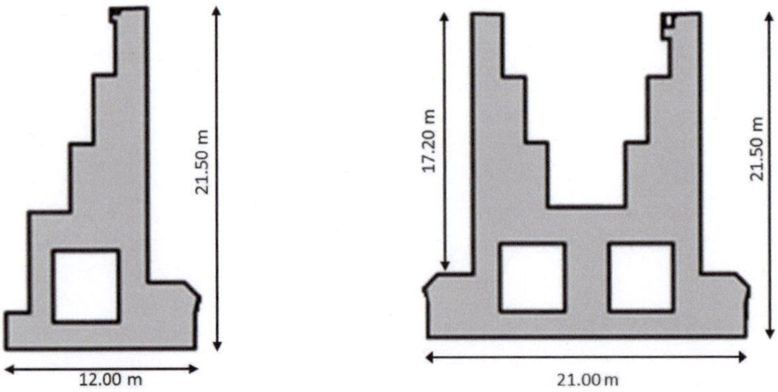

Fig. 7-2 Dimension of the navigation lock monoliths of Iffezheim.

Because of the geological characteristics of the region and the characteristics of the foundation soil the lock is classified to the C-S category of the national annex for Germany (DIN EN 1998-1/NA:2011-01 2011) which is similar to the ground category D of (EN 1998-1:2004 Eurocode 8). Here has to be mentioned that the soil characteristics do not refer to the real soil conditions met along the whole length of the navigation lock and the characteristics of the filling soil material were taken the same as for the static design of the lock. The shear wave velocity of the filling soil material is 170 m/s and for the foundation soil 280 m/s. The soil parameter S is equal to 0.75 according to (DIN EN 1998-1/NA:2011-01 2011). The soil parameter according to (EN 1998-1:2004 Eurocode 8) for the soil category D is 1.8 though. The combination C-S of the German Annex of (EN 1998-1:2004 Eurocode 8) implies a half-space for the underlying soil (or that the bedrock lies very deep from the soil surface). Because of the low seismicity and according to (EN 1998-5:2004 Eurocode 8), for the product $\alpha \times S = 0.04 \times 0.75 = 0.03g$ there is not a corresponding damping factor for the soil. However, a minimum value of 3% was considered. The foundation and the retained soil have the following characteristics:

Table 38 Soil characteristics for navigation lock Iffezheim

NN (m)	Stratum thickness (m)	Density (kg)	E (MPa)	v (-)	C_s (m/s)	φ (deg)	c (kPa)	ψ (deg)
+115	21.5	1900	150	0.33	170	37.5	0.1	3.75
+93.5	43	1900	400	0.33	280	-		-
Halfspace	43	1900	400	0.33	280	-		-

At a first look, the side gravity walls seem able to rotate at their base because they are not monolithically connected to the base plate (hinge connection). But a lateral translational movement seems difficult or very restricted. Soil attaches to both side gravity walls. The concrete is of quality C30/37. The concrete material law used in this analysis it the concrete damaged plasticity (Lee und Fenves 1998) available in Abaqus (ABAQUS 2012).

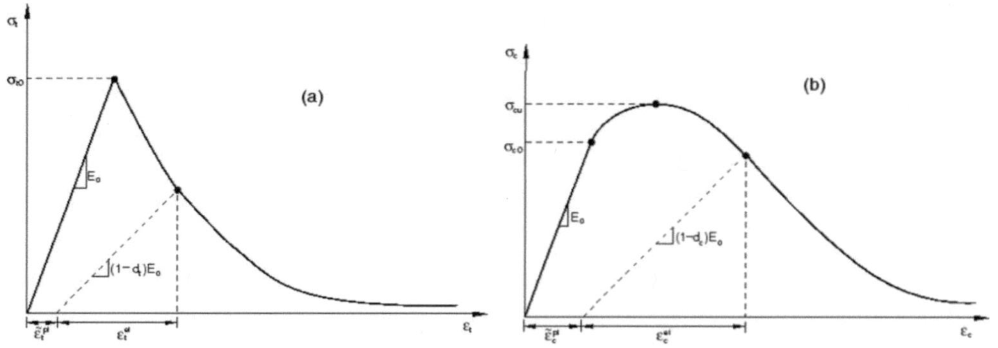

Fig. 7-3 Response of concrete to uniaxial loading in tension (a) and compression (b).

Table 39 Material characteristics of navigation lock Iffezheim

Density (kg)	E (GPa)	ν (-)	σ_{co} (MPa)	σ_{cu} (MPa)	σ_{to} (MPa)	G_f (N/m)	ψ (deg)
2500	33	0.22	15.2	38	3.8	112	36.1

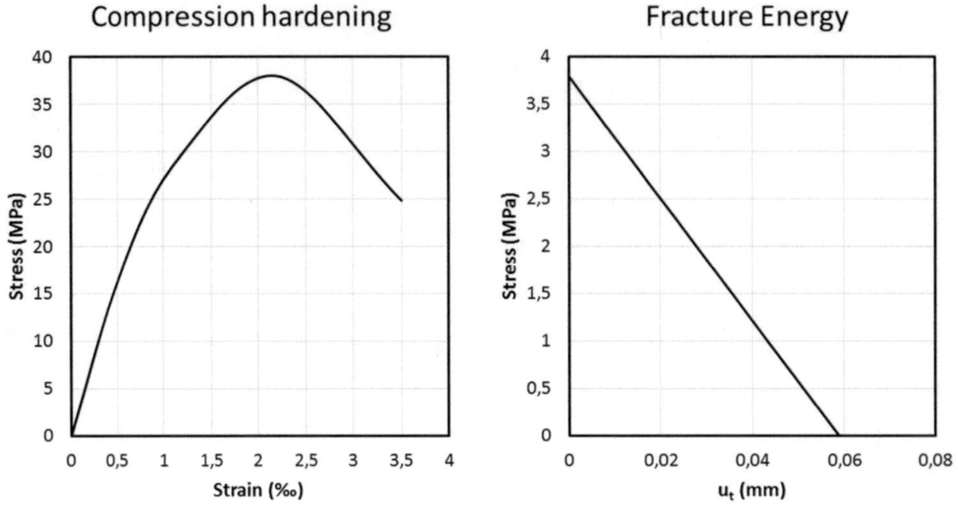

Fig. 7-4 Compression hardening of concrete and fracture energy used in this example.

The boundaries of the numerical model are the same with the Model 3 of the previous chapter. The element size of the soil was chosen 1.0×1.0 m. The damping was given in form of stiffness proportional only damping (β=0.01; the first frequency of the soil layers was calculated with Abaqus and SHAKE91 and equals to 0.94 Hz) The element size for the navigation lock is 0.5×0.5 m and for the water domain 0.5×0.5 m. The target acceleration time history was taken as an artificial time history, which corresponds to the design spectrum at the foundation leven of the lock. This time history was deconvoluted using SHAKE at a depth equal to the base of the numerical model. The deconvoluted 'within' motion was assigned at the model's rigid boundaries following (Dawson, E. M., Mejia, L. H. 2006) and (Kwok et al. 2007).

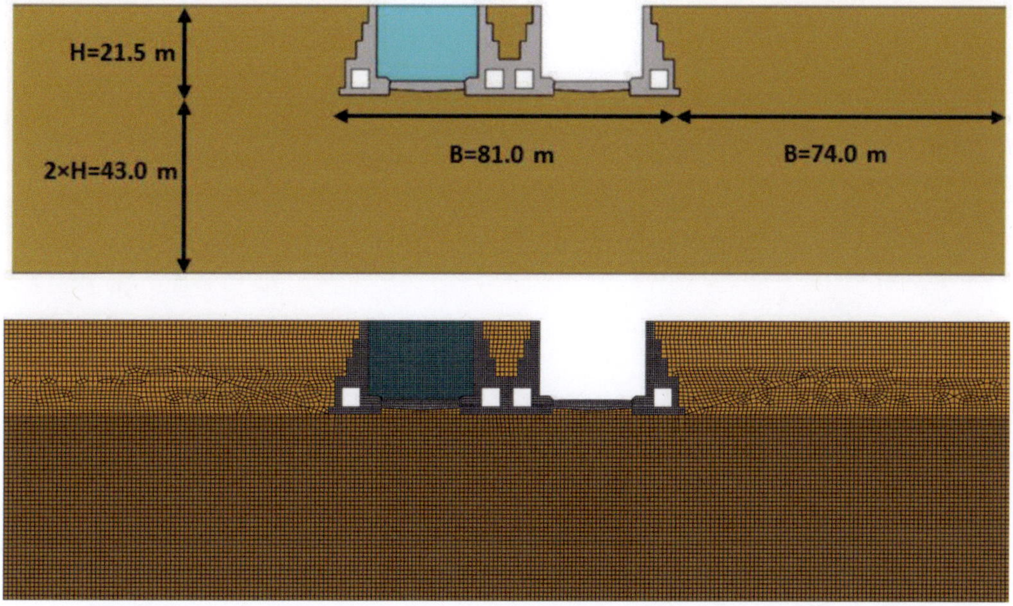

Fig. 7-5 The size and the mesh of the numerical model of lock Iffezheim.

Analysis steps:

- Gravity load is applied to the soil.
- A geostatic step equalizes the strains with the stresses on the soil material so no deformations are observed in the soil material.
- The gravitational load of the navigation lock and the filled soil is applied.
- The hydrostatic load is applied.
- The side boundaries are removed and replaced by the reaction forces.
- The earthquake loading is applied.

The dynamic soil pressures on the right chamber gravity wall can be estimated from the procedure of the fifth chapter. The gravity wall has a base width of 2B=12 m and the navigation lock has a length of 2L=270 m. The ratio L/B equals to 22.5≈20. In order to be able to use the solutions for the impedances for rectangle foundations, as for strip foundations and homogeneous half-space the theoretical values of the translational and rotational spring take infinite values. The area moment of inertia of the foundation-soil contact surface around the axis of the chamber's wall length is: $I_{bx} = (270 \times 12^3)/12 = 38880 m^4$. The factor χ equals:

$$\chi = A_b/4L^2 = (270 \times 12)/(4 \times 135^2) = 0.044 \quad \textbf{(7-1)}$$

The shear modulus of the foundation soil is:

$$G = E/(2(1+v)) = 150375940 \approx 150\,MPa \tag{7-2}$$

And for the fill material:

$$G = E/(2(1+v)) = 56390977\,Pa \approx 56.4\,MPa \tag{7-3}$$

and the static stiffness is given by (Gazetas 1991; Mylonakis et al. 2006):

Horizontal:

$$K_{x,total} = \frac{2.1GL}{2-v}(2+2.50\chi^{0.85}) = \frac{2.1\times150\times10^{6}\times135}{2-0.33}(2+2.50\times0.044^{0.85}) \approx 11\times 5.5^{10}\,\frac{N}{m} \tag{7-4}$$

Rocking:

$$K_{r,total} = \frac{G}{1-v}I_{bx}^{0.75}\left(\frac{L}{B}\right)^{0.25}\left(2.4+0.5\frac{B}{L}\right) = \frac{150\times10^{6}}{1-0.33}38880^{0.75}\left(\frac{135}{6}\right)^{0.25}\left(2.4+0.5\frac{6}{135}\right) \approx 3.28\times10^{12}\,Nm/rad \tag{7-5}$$

And per unit of length:

$$K_x = \frac{5.5\times10^{10}}{270} \approx 205\frac{MN}{m}/m \tag{7-6}$$

$$K_r = \frac{3.28\times10^{12}}{270} \approx 12.1\frac{GNm}{rad}/m \tag{7-7}$$

The translational (horizontal) spring refers to the movement of the wall so as active conditions to can be developed (moving away from the retaining soil that is why the contribution of the retaining soil is not taken into account for the translational spring). Because of the existence of the base plate, it is not sure that the chamber wall is able to move freely in this direction (without slip of the wall, only elastic translation). The plate will also move elastically at the same direction but it will be also restrained by the existence of the other, middle, chamber wall. Due to the type of this construction, where the plate is not monolithically connected to the walls, we cannot use the relations for the embedded structure for the rotational spring but one could use these relations for the translational spring stiffness. Because of the non-monolithical connection and the gap between the plate and the wall additional uncertainties are introduced, whether the wall is able to slide or not. The dimensionless d_x parameter is calculated here but if an unconstrained elastic horizontal translation takes place is doubtful. Moreover, the factor *h/B=21.5/6=3.6* shows also how much more important is the rocking response in comparison to the translational response (NEHRP 2012)

The relative flexibilities are:

$$\mathrm{d}_x = \frac{G}{K_x} = \frac{56 \times 10^6}{205 \times 10^6} = 0.27 \approx 0.3 \tag{7-8}$$

$$\mathrm{d}_\theta = \frac{GH^2}{K_r} = \frac{56.4 \times 10^6 \times 21.5^2}{12.1 \times 10^9} = 2.15 \approx 2.0 \tag{7-9}$$

The bending stiffness of the wall is found by applying a lateral constant load of 1kN/m and back calculation from the flexural displacement of $5.13 \text{x} 10^{-5}$m.

$$d_w = \frac{GH^3}{E_w I_w} = \frac{56.4MPa \times 21.5^3}{33 \times 10^9 \times 15.8} = 1.07 \approx 1.0 \tag{7-10}$$

The soil is homogeneous and has a constant shear modulus distribution. The resultant of the dynamic soil pressures for the statically excited system is estimated to be (for d_x=0 and according to Annex C) $0.444 \times 1900 \times (0.4 \times 1.2 \times 0.75) \times 21.5^2 = 140\ kN$. The Wood solution adopted by (EN 1998-5:2004 Eurocode 8) for rigid walls gives $1900 \times (0.4 \times 1.2 \times 0.75) \times 21.5^2 = 316\ kN$ kN and the dynamic increment between the static soil pressure and the value resulted by the Mononobe-Okabe formula, also adopted by (EN 1998-5:2004 Eurocode 8) gives:

$$dynamic\ increment = K_{a,g+dyn,h} - K_{a,g} = 0.22 - 0.2 = 0.02 \tag{7-11}$$

$$P_{dyn} = 0.02 \times 0.5 \times 1900 \times 9.81 \times 21.5^2 = 86 \text{ kN} \tag{7-12}$$

The dynamic increment according to (Seed H. B., Whitman R. V. 1970) is:

$$\frac{3}{4}\frac{(0.75 \times 0.4 \times 1.2)}{9.81} \times 0.5 \times 1900 \times 9.81 \times 21.5^2 = 119\ kN \qquad \textbf{(7-13)}$$

Table 40 Comparison between the numerical results and the different theories

Method	FEM	Results of this study	Wood	M-O	Seed-Whitman
P_{dyn} (kN)	175	140	316	86	119
Difference (%)	-	-20%	+80%	-50%	-32%

If an amplification factor of about 1.5 for a seismic event (Veletsos, A. S., Younan, A. H. 1997) is considered, the new value is $140 \times 1.5 = 210\ kN$ and the error is +20%. If we consider additionally that the chamber wall is able to move elastically in the horizontal direction (d_x=0.3) and after interpolation the normalized shear factor equals 0.395 the new calculated shear due to the soil pressures is $0.395 \times 1900 \times (0.4 \times 1.2 \times 0.75) \times 21.5^2 = 125\ kN$ and by assuming the same amplification factor the new shear force results 187 kN which is closer to the FE calculation. Of course, for this latter case (considering the horizontal translation of the wall), the AF changes and is not more 1.5. For this type of construction the rotation plays a more important role than the translation and the parameter d_x should not be important. This can be seen by calculating the formula of (Elms, D. G., Richards R. 1979) which shows if the gravity wall will tilt or slide. This wall will tilt for the above calculated forces.

The water pressures are found to be 29 kN. The dimensions of the reservoir are 24.0 m length and 18.0 m depth, so L/H=1.33. The Westergaard's formula gives $0.543 \times 1000 \times (0.4 \times 1.2 \times 0.75) \times 18^2 = 63\ kN$ and the reduction coefficient as suggested by Werner and Sundquist and/or Brahtz and Heilbron equals 0.76 and the total hydrodynamic force reduces to $0.76 \times 63 = 48\ kN$.

Table 41 Comparison between the numerical results and the different theories

Method	FEM	Werner and Sundquist	Housner	EC8-4 (impulsive only)
P_{dyn} (kN)	29	48	55	34
Difference (%)	-	+67%	+92%	+16%

The reduced hydrodynamic pressures can be explained because of the damping of the wall-soil system (see Figure 4.17).

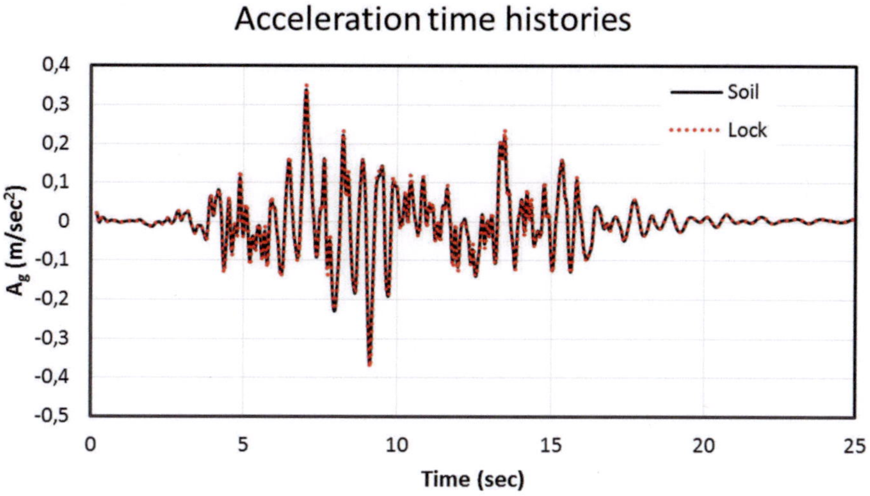

Fig. 7-6 Acceleration time histories of the nodes of the wall and the soil at the wall-soil interface. It can be seen that there is not relative slip between the wall and the soil.

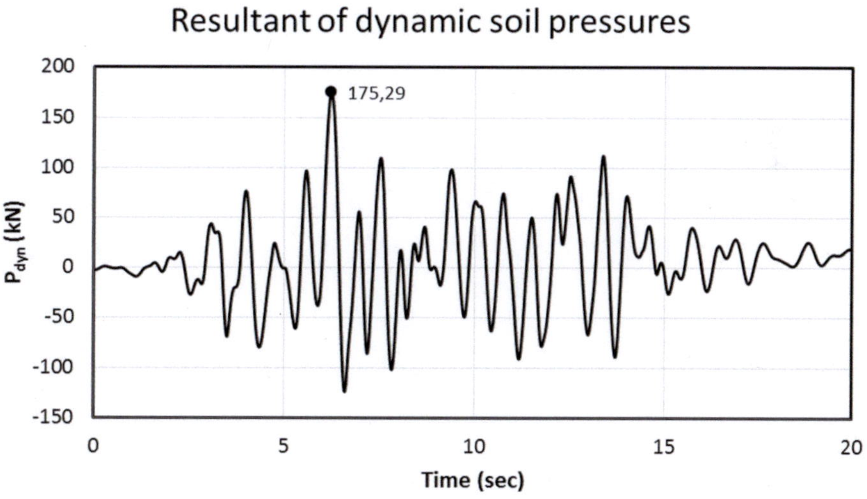

Fig. 7-7 Time history of the dynamic increment of the soil pressures.

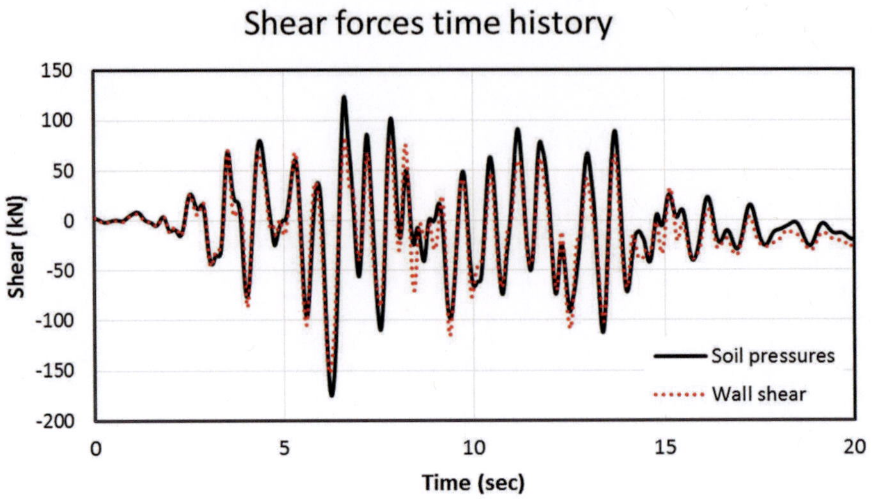

Fig. 7-8 Time history of the shear forces at the wall's base vs the dynamic increment of the soil pressures. The small differences are due to the inertia forces of the wall subjected to this small amplitude seismic excitation.

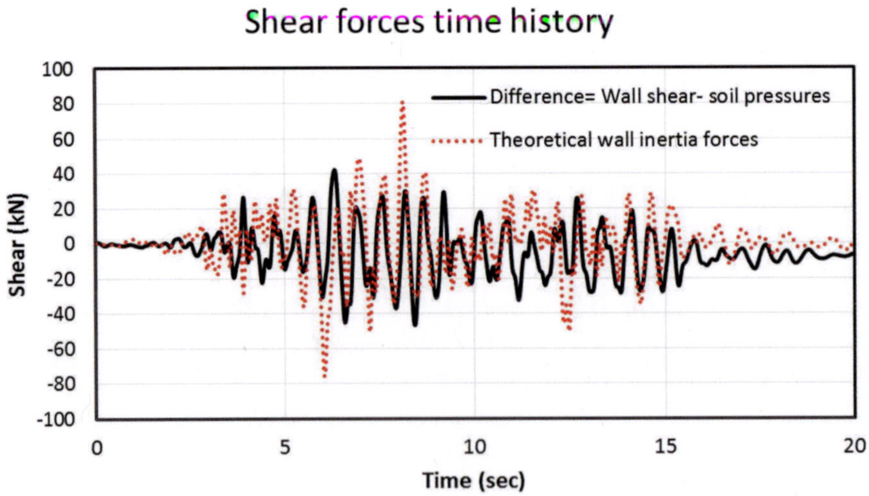

Fig. 7-9 Time history of wall shear forces minus the soil dynamic increment vs the theoretical wall inertia forces (mass × acceleration at wall base).

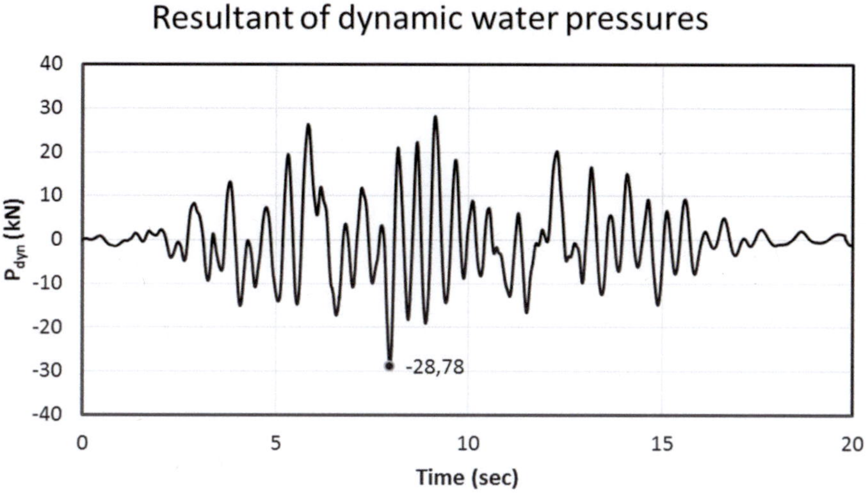

Fig. 7-10 Time history of the hydrodynamic pressures on the wall.

7.2 Navigation lock Fankel

The navigation lock of Fankel, Germany, is a single chamber U-shape navigation lock with 12 m wide and 170 m long chambers on the river Mosel. The drop height is 7 m. The chamber is monolithically constructed and the side walls have a cantilever function in their upper half.

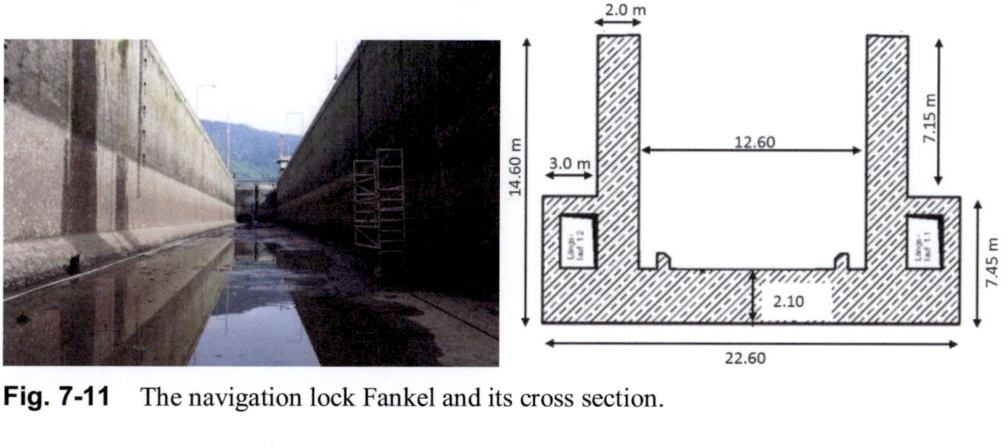

Fig. 7-11 The navigation lock Fankel and its cross section.

The navigation lock Fankel lies also on the seismic zone 1 (0.4 m/s^2) and on the geological category R and the foundation soil category is C (C_s=300 m/sec). The soil parameter S is 1.5 according to (DIN EN 1998-1/NA:2011-01 2011). The height of the lock is less than 15.0 m and the importance factor remains equal to 1.0. The FE model has rigid boundaries which idealize the bedrock and there are no side boundaries for the seismic excitation (Model 3 of the previous chapter). The reaction forces of the static cases are replaced with forces of the same amplitude and the side boundaries are removed for the dynamic analysis.

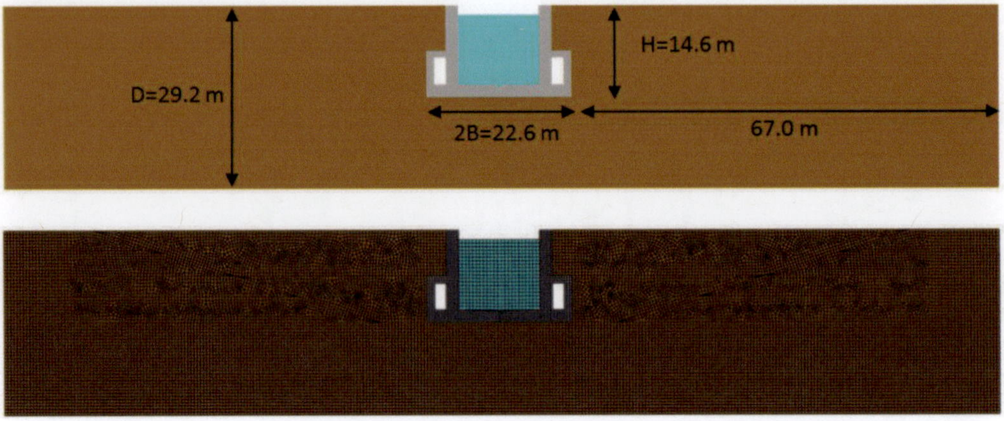

Fig. 7-12 The size and the mesh of the numerical model of Lock Fankel.

Table 42 Soil characteristics for navigation lock Fankel

Stratum thickness (m)	Density (kg)	E (MPa)	v (-)	C_s (m/s)	φ (deg)	ψ (deg)
29.2	2000	478.8	0.33	300	40.0	7.0

Table 43 Material characteristics for navigation lock Fankel

Density (kg)	E (GPa)	v (-)	σ_{co} (MPa)	σ_{co} (MPa)	σ_{to} (MPa)	G_f (N/m)	ψ (deg)
2500	33	0.2	15.2	38	3.8	112	35

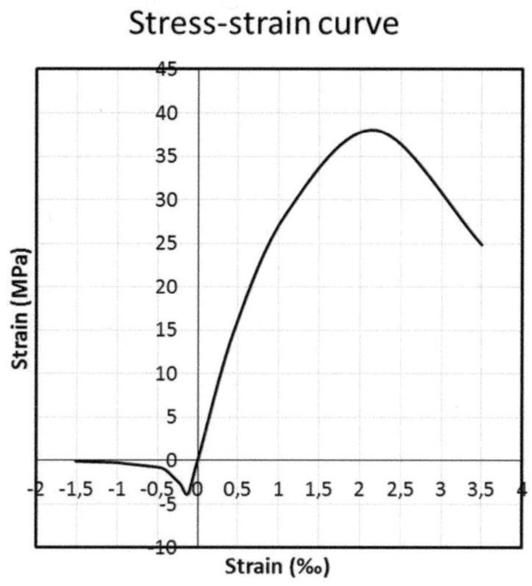

Fig. 7-13 Stress-strain curve for the concrete used in this study.

In order to calculate the dimensionless parameters for the lock Fankel, the formulas suggested in (Brandenberg, S., Mylonakis, G., Stewart, J. 2015) is followed. The procedure followed here is not the same as the one suggested by (Brandenberg, S., Mylonakis, G., Stewart, J. 2015). Here, only use of the separated spring stiffness of the base plate and the walls is made. The translational spring for the base is calculated by:

$$K_x = \chi_x \frac{2.1G}{2-v}\left(1 + 2\frac{B}{D-H}\right) = 0.7\frac{2.1 \times 180\,MPa}{2-0.33}\left(1 + 2\frac{11.3}{29.2-14.6}\right) = 403\frac{MN}{m} \quad \textbf{(7-14)}$$

And for the one wall (the value for χ_x is extrapolated from the diagramms as H/B=1.3):

$$k_x^i = \chi_x \frac{\pi}{\sqrt{(1-v)(2-v)}}\frac{G}{H}\sqrt{1-\left(\frac{2\omega H}{\pi C_s}\right)} = 0.7\frac{\pi}{\sqrt{(1-0.33)(2-0.33)}}\frac{180MPa}{14.6}\sqrt{1-\left(\frac{29.2\omega}{300\pi}\right)} \quad \textbf{(7-15)}$$

The total static spring is found setting ω=0 and equals 25 MN/m/m. The total translational spring equals to:

$$K_{x,total} = 2k_x^i H + K_x = 2 \times 365 + 403 = 1.1\frac{GN}{m}/m \qquad \textbf{(7-16)}$$

The part of the base for the rotational spring is:

$$K_r = \chi_r \frac{\pi G B^2}{2(1-v)}\left(1 + 0.2\frac{B}{D-H}\right) =$$

$$0.7\frac{3.14 \times 180\,MPa \times 11.3^2}{2(1-0.33)}\left(1 + 0.2\frac{11.3}{14.6}\right) \approx 41.7 \times 10^9 \frac{Nm}{rad}/m \qquad \textbf{(7-17)}$$

The total rotational spring can be calculated by taking into account the contribution of the base plate and the contribution of the two walls. The contribution of each wall is calculated by:

$$k_r^i = \chi_r \frac{\pi}{2}\sqrt{\frac{2-v}{1-v}\frac{G}{H}}\sqrt{1-\left(\frac{2\omega H}{\pi C_s}\right)} = 0.8\frac{\pi}{2}\sqrt{\frac{2-0.33}{1-0.33}\frac{180MPa}{14.6}}\sqrt{1-\left(\frac{29.2\omega}{300\pi}\right)}$$

(7-18)

The total static spring for one wall is found setting ω=0 and equals 25 MN/m/m. The total rotational spring equals to:

$$K_{r,total} = k_x^i H^2 + K_r + 2k_r^i \mathrm{H} B^2 = 5459 + 41700 + 61 \times 10^9 = \frac{108\frac{GN}{m}}{m} \qquad \textbf{(7-19)}$$

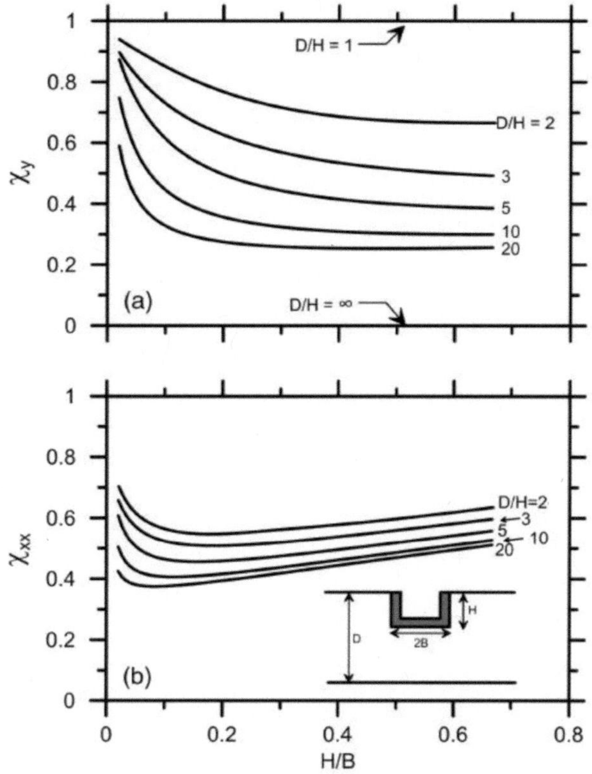

Fig. 7-14 The diagrams for the factors $\chi_y = \chi_x$ and $\chi_{xx} = \chi_r$ from (Brandenberg, S., Mylonakis, G., Stewart, J. 2015).

The relative flexibilities are:

$$d_x = \frac{G}{K_x} = \frac{180\ MPa}{1.1\frac{GN}{m}} = 0.16 \approx 0.1 \tag{7-20}$$

$$d_\theta = \frac{GH^2}{K_r} = \frac{180MPa \times 14.6^2}{108 \times 10^9 \frac{Nm}{rad}} \approx 0.35 \tag{7-21}$$

The bending stiffness of the chamber wall is found as at the former example, by applying a horizontal distributed load of 1 kN. The deflection is now $7.5\text{x}10^{-5}$m:

$$d_w = \frac{GH^3}{E_w I_w} = \frac{180\ \text{MPa} \times 14.6^3}{30 \times 10^9 \times 2.3} \approx 7.4 \tag{7-22}$$

The ratio h/B equas now 1.3 and shows that the translational movement is also important. The soil is homogeneous and has a constant shear modulus distribution. The resultant of the dynamic soil pressures for the statically excited system is estimated to be (for d_x=0 and interpolating between the values of the Annex C) $0.585 \times 2000 \times (0.4 \times 1.5) \times 14.6^2 = 149\ kN$. The translational spring was not considered. This numerical study considered a dimensionless parameter d_x up to the value of 1. Taken this value the calculated normalized factor for the base shear equals 0.48 (with tension forces set to null). The new calculated base shear equals about 122 kN. The dynamic increment of the soil pressures of the finite element model is 106 kN. The Wood solution adopted by (EN 1998-5:2004 Eurocode 8) for rigid walls gives $2000 \times (0.4 \times 1.5) \times 14.6^2 = 256\ kN$ and the dynamic increment between the static soil pressure and the value resulted by the Mononobe-Okabe formula, also adopted by (EN 1998-5:2004 Eurocode 8) gives:

$$dynamic\ increment = K_{a,g+dyn,h} - K_{a,g} = 0.208 - 0.18 = 0.028 \quad \textbf{(7-23)}$$

$$P_{dyn} = 0.028 \times 0.5 \times 2000 \times 9.81 \times 14.6^2 = 58.6\ \mathrm{kN} \quad \textbf{(7-24)}$$

The dynamic increment according to (Seed H. B., Whitman R. V. 1970) is:

$$\frac{0.75 \times (0.4 \times 1.5)}{9.81} \times 0.5 \times 2000 \times 9.81 \times 14.6^2 = 95.9\ kN \quad \textbf{(7-25)}$$

Table 44 Comparison between the numerical results and the different theories

Method	FEM	Results of this study without d_x	Results of this study with d_x	Wood	M-O	Seed-Whitman
P_{dyn} (kN)	106	149	122	256	59	96
Difference (%)	-	+40%	+15%	+142%	-45%	-10%

The very good agreement of the solution of (Seed H. B., Whitman R. V. 1970) for the dynamic increment shall not be assumed as the best solution.

The water reservoir has a depth of 11.2 m and a length of 12.6 m. The L/H ratio equals 1.125. The reduction factor for the finite reservoir length equals 0.7. The next table gives the total hydrodynamic pressure according to the FE calculation and other theories.

Table 45 Comparison between the numerical results and the different theories

Method	FEM	Werner and Sundquist	Housner	EC8-4 (impulsive only)
P_{dyn} (kN)	23	29	33	31
Difference (%)	-	+23%	+40%	+34%

The reduced water pressures can be explained because of the damping of the lock-soil system which results to an overdamped oscillation.

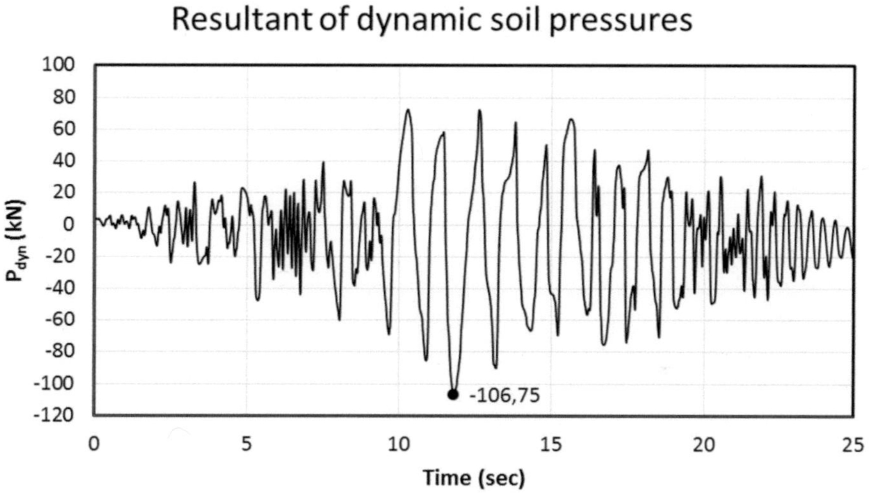

Fig. 7-15 The time history of the dynamic increment of the soil pressures.

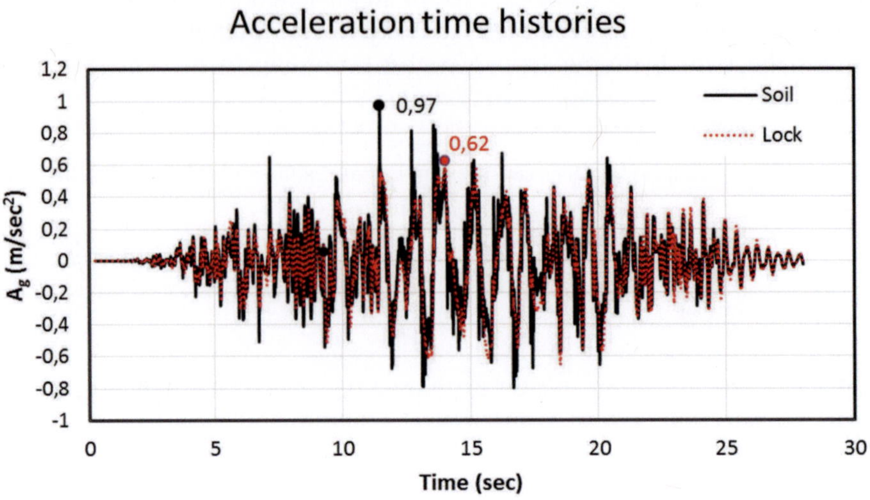

Fig. 7-16 The acceleration time histories of soil and lock nodes at the middle of the soil-wall interface.

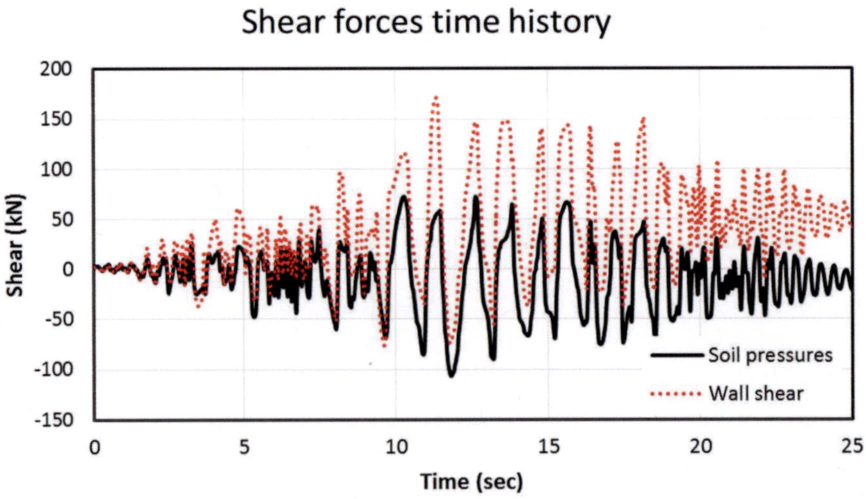

Fig. 7-17 Time history of the dynamic increment of the soil pressures vs the shear at the wall's base. The increase of the shear forces indicates a partially yielding of the soil at the upper half of the lock wall.

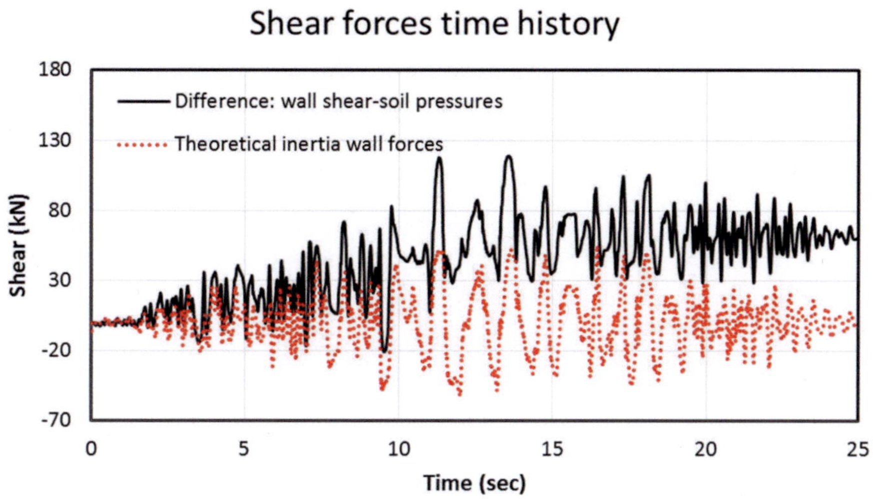

Fig. 7-18 Time history of wall shear forces minus the soil dynamic increment vs the theoretical wall inertia forces (mass × acceleration at wall base).

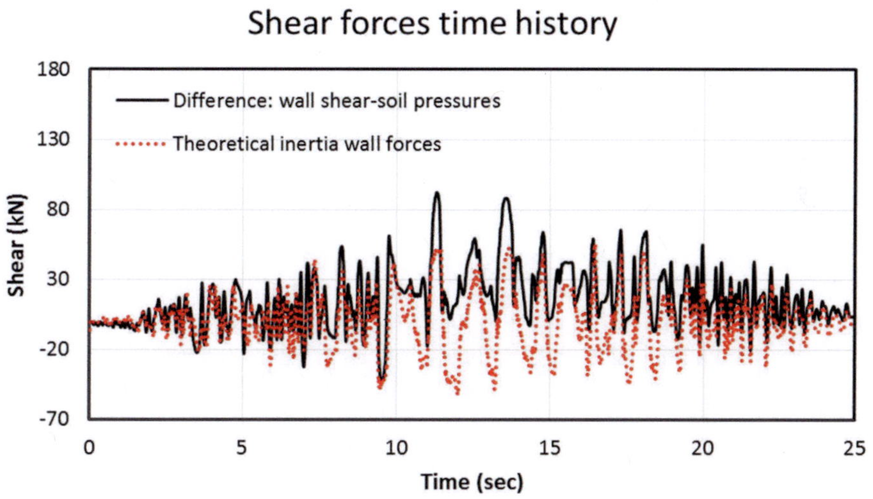

Fig. 7-19 Time history of wall shear forces minus the soil dynamic increment corrected to coincide with the theoretical wall inertia forces at the end of the seismic excitation vs the theoretical wall inertia forces (mass × acceleration at wall base).

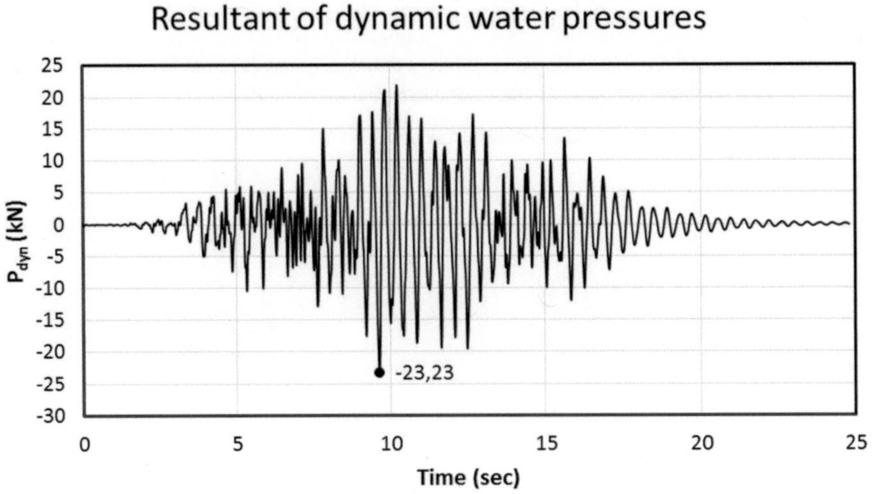

Fig. 7-20 Time history of the hydrodynamic pressures.

7.3 Conclusions

These two practical examples show how this numerical study can be applied at the praxis. However these two examples are academic and some simplifications are assumed. The damping of the lock foundations although exists at the FE analysis, was not considered at the simplified calculation.

Important findings of these two examples are the following:

- The hydrodynamic values are smaller than their theoretical values implying that the damping of the structure-soil system influences the maximum values of the dynamic water pressures.
- For the case of the lock Iffezheim where the soil does not yield, the wall shear is governed by the dynamic soil pressures and not by the inertia forces of the wall, which should also be bigger than the calculated. This fact has to be treated in connection with the very low acceleration applied. For this acceleration no phase effects are obvious, the inertia forces of the wall do not appear amplified, but on the contrary they are less than the theoretical values of the wall mass multiplied with the applied acceleration at the wall-soil interface.
- Similar conclusions are drawn also for the lock Fankel. The partially yielding of the soil at the upper part of the lock wall, deteriorates the explanation of the results but the same motive is also observed here. There are no phase effects present and the slightly bigger values of the calculated base shear in comparison to the theoretical values are due to the partially yielded soil.

As the most research done by other researchers refers to strong seismic motion, these results should not be seen as they opposite other results but should be treated as a further information that complements other results for the case of low seismic motion.

Chapter 8
Conclusions

The numerical investigation carried out here brought new information about the dynamic water and soil pressures considering the water-structure and soil-structure interaction.

The information that resonance cannot occur for the hydrodynamic pressures because the predominant frequency content of earthquakes does not coincide with the natural frequencies of the water domain (except for reservoirs with very big depth) is not more valid, if we consider the soil-structure interaction. Systems of water retaining structures based on flexible foundation, have lower eigenfrequencies and a resonance or amplified values of the hydrodynamic pressures can occur. The possibility of resonance is on the other hand mitigated by the damping of the structure-foundation system and of embedded in soil water retaining structures as locks are. Moreover, the application of the total hydrodynamic force, that is taken at 0.4H in praxis, it is shown to be dependent on the relative flexibilities and depends strongly on the excitation frequency.

As will be discussed also later about the dynamic soil pressures a straightforward comparison of the hydrostatic to the hydrodynamic pressures cannot be done easily, as the latter depend on the frequency content of the excitation and an amplification factor AF is difficult to be assigned. Even when one takes a range of predominant periods of seismic excitation between 0.3-0.6 sec (Rathje und Kottke 2013), it is not easy to come to a firm conclusion as the water-structure system is very sensitive to all the afore mentioned influences. For lock-structures, for which the foundation damping plays an important role on the hydrodynamic pressures, and due to the finite reservoir with common L/H ratios between 0.5 and 3 the hydrodynamic pressures are not expected to become higher than the hydrostatic pressures especially for low seismicity regions.

Referring to the dynamic soil pressures this numerical study as continuation of former studies (Bobet, A., Fernández, G., Jung, C. 2010; Bobet, A., Jung, C. 2008; Wong 1982; Papazafeiropoulos, G., Psarropoulos, P. N. 2010; Psarropoulos et al. 2005; Veletsos, A. S., Younan, A. H. 1992, 1993, 1994a, 1997, 1998b, 1998a; Wu und Liam Finn 1999) tried to bridge the gap between the two extreme cases presented in the current codes; the M-O solution and the Wood solution. This extended parametric study, provided information for three soil profiles and for different wall and base relative flexibilities for semi-infinite and finite wall-soil systems.

The formula of (EN 1998-5:2004 Eurocode 8) for rigid walls is modified to account for the Poisson ratio of the soil, the shear modulus distribution of the soil, the inclination of the wall, the distance of a closed placed wall and the case of resonance as follows (smooth contact):

$$\Delta P_{dyn} = (C_{soil}\ or\ C_{wd}) \times C_v \times C_{P,incl} \times C_{AF} \times a \times S \times \gamma \times H^2 \quad \textbf{(8-1)}$$

where (for rigid wall and foundation):

$$C_{soil} = \begin{cases} 1.00 \ for\ homogeneous\ soil \\ 0.81 \ for\ parabolic\ shear\ modulus\ distribution \\ 0.75 \ for\ linear\ shear\ modulus\ distribution \end{cases}$$

$C_v = \frac{0.81+(6v)^v}{2.05}$, for smooth contact with v the Poisson ratio of the soil

$C_{P,incl} = 0.009(\theta^o) + 0.18$, with θ=90° for vertical wall

C_{wd} is the factor for the influence of a near standing wall (silo-effect):

$$C_{wd} = \begin{cases} 1 - 1.2\ \mathrm{Exp}\left(\frac{-L/H}{1.4}\right),\ when\ L/H < 5 \\ 1,\ when\ L/H \geq 5 \end{cases}, \text{ for homogeneous soil}$$

$C_{wd} = 0.81 - \mathrm{Exp}(-L/H)$, for parabolic shear modulus distribution

$C_{wd} = 0.75 - 0.9\ \mathrm{Exp}\left(\frac{-L/H}{0.95}\right)$, for linear (proportional) shear modulus distribution

α is the peak ground acceleration as ratio of g, S is the soil parameter, γ the unit weight of the soil and H the wall height.

C_{AF} is the factor which takes into account the amplification of the dynamic soil pressures due to the frequency content of the seismic excitation. It must be based on a thorough study of seismic excitations and shear modulus of soil profiles and not on steady state analysis results, as they overestimate it. The analysis of (Veletsos, A. S., Younan, A. H. 1997) gave the average value of 1.2 for rigid walls based on one only seismic excitation for different values of the shear modulus of the soil and for 5% critical damping. For bounded systems (two-walls-soil systems) values for the amplification factor can be found in (Wu und Liam Finn 1999) and they depend on the shear modulus distribution of the soil and the distance between the wall.

In order to account for flexible systems as for example in (DIN 4085:2011-05) is made for the static pressures of the soil due to gravitational load, the aforementioned factor C_{soil} takes the values given in Annex C as a function of the dimensionless parameters d_w, d_θ and d_x.

The retaining walls of lock chambers are calculated in praxis for an earth pressure at rest or for an increased active pressure ($0.25K_a+0.75K_o$ or $0.5K_a+0.5K_o$ or $0.75K_a+0.25K_o$) (DIN 4085:2011-05). A rough approximation is to use the same relations also for the seismic design, i.e. to take the dynamic increment of the M-O formula ($\Delta K=K_{ae}-K_a$) and the dynamic increment for rigid walls and to use the same proportion used for the static design for the earth pressure coefficients also for the dynamic design. This rough approximation should be assumed empirical and has no theoretical background. Otherwise the values presented here can be used for a more precise design.

In order to check until up to which design ground acceleration the static design is adequate also for the seismic case and referring to low seismicity regions the following assumptions are made:

- According to (EN 1998-5:2004 Eurocode 8) the damping for soil for a peak ground acceleration smaller than 0.1g (≈1 m/sec^2) should not be taken bigger than 3% of the critical damping. Furthermore, the elastic moduli reduction is assumed negligible.
- Assuming homogeneous soil, which delivers the biggest soil pressures.
- Assuming null damping for the wall foundation.
- Assuming null damping due to the friction at the wall-soil interface.
- Assuming safety factor equal to 1.5 for the static and 1.0 for the seismic case.
- Assuming rigid wall for the maximum soil pressures for the "static" case.
- For rigid base the soil parameter *S* equals 1 and for flexible soil takes its highest values 1.5 (according to (DIN EN 1998-1/NA:2011-01 2011))
- Assuming a friction angle φ of the soil between 25-45°.
- Assuming an importance factor (IF) of 1.2 for a wall height bigger than 15 m.
- There is no water in the lock and the soil is dry.
- The lock wall is assumed to be rigid enough.
- The amplification Factor AF depends on the damping and on the base flexibility. Here an AF equal to 1.5 is considered.
- Assuming four combinations (CMB) of the static design: A: $100\%K_o$, B: $75\%K_o+25\%K_a$, C: $50\%K_o+50\%K_a$, D:$25\%K_o+75\%K_a$ and $\alpha=\beta=\iota=\delta=0$.
- Assuming only the formula for rigid walls for the seismic case.

Then the check for the horizontal forces is:

$$1.5(0.5\ \text{CMB}\,\gamma H^2) = 1.0[(0.5\ \text{CMB}\,\gamma H^2) + (\alpha S\gamma H^2)AF\ \text{IF}] \rightarrow$$

$$\alpha = \frac{0.25\ \text{CMB}}{S\ \text{AF IF}} \tag{8-2}$$

In this relation the choice of the AF is of great importance. Former studies (Wu und Liam Finn 1999; Veletsos, A. S., Younan, A. H. 1992, 1993, 1994a, 1997) have shown that the AF for seismic events is about 20% of the one calculated for the steady state sinusoidal excitation. Assuming a rigid base has twofold influence; not only the soil parameter S is affected (takes the value 1) but also the AF decrease. It is important to note that the (EN 1998-5:2004 Eurocode 8) considers no AF for the seismic case and gives the “static” value of the soil pressures for a homogeneous soil. Referring to the seismic zonation of Germany and the values suggested by the German Annex of (EN 1998-1:2004 Eurocode 8) the following table can be used as indicator. Attention must be given that the comparison is force-based and not moments-based.

Table 46 Maximum values of the design acceleration (m/s^2) which can be covered by the static design of a retaining wall according to the previous formula considering soil pressures only.

	φ (°)	K_o	K_a	A		B		C		D	
	S=1.0 / AF=1.5										
A-R	25	0.58	0.41	0.79	0.94	0.62	0.74	0.67	0.81	0.73	0.88
	30	0.50	0.33	0.68	0.82	0.51	0.61	0.57	0.68	0.62	0.75
	35	0.43	0.27	0.58	0.70	0.42	0.51	0.47	0.57	0.53	0.63
	40	0.36	0.22	0.49	0.58	0.35	0.42	0.39	0.47	0.44	0.53
	45	0.29	0.17	0.40	0.48	0.27	0.33	0.32	0.38	0.36	0.43
	S=1.25 / AF=1.5										
B-R	25	0.58	0.41	0.63	0.76	0.49	0.59	0.54	0.65	0.58	0.70
	30	0.50	0.33	0.55	0.65	0.41	0.49	0.45	0.54	0.50	0.60
	35	0.43	0.27	0.46	0.56	0.34	0.40	0.38	0.46	0.42	0.51
	40	0.36	0.22	0.39	0.47	0.28	0.33	0.31	0.38	0.35	0.42
	45	0.29	0.17	0.32	0.38	0.22	0.26	0.25	0.30	0.29	0.34
	S=1.5 / AF=1.5										
C-R	25	0.58	0.41	0.52	0.63	0.41	0.49	0.45	0.54	0.49	0.58
	30	0.50	0.33	0.45	0.55	0.34	0.41	0.38	0.45	0.42	0.50
	35	0.43	0.27	0.39	0.46	0.28	0.34	0.32	0.38	0.35	0.42
	40	0.36	0.22	0.32	0.39	0.23	0.28	0.26	0.31	0.29	0.35
	45	0.29	0.17	0.27	0.32	0.18	0.22	0.21	0.25	0.24	0.29
	S=1.0 / AF=1.5										
B-T	25	0.58	0.41	0.79	0.94	0.62	0.74	0.67	0.81	0.73	0.88
	30	0.50	0.33	0.68	0.82	0.51	0.61	0.57	0.68	0.62	0.75
	35	0.43	0.27	0.58	0.70	0.42	0.51	0.47	0.57	0.53	0.63
	40	0.36	0.22	0.49	0.58	0.35	0.42	0.39	0.47	0.44	0.53
	45	0.29	0.17	0.40	0.48	0.27	0.33	0.32	0.38	0.36	0.43
	S=1.25 / AF=1.5										
C-T	25	0.58	0.41	0.63	0.76	0.49	0.59	0.54	0.65	0.58	0.70
	30	0.50	0.33	0.55	0.65	0.41	0.49	0.45	0.54	0.50	0.60
	35	0.43	0.27	0.46	0.56	0.34	0.40	0.38	0.46	0.42	0.51
	40	0.36	0.22	0.39	0.47	0.28	0.33	0.31	0.38	0.35	0.42
	45	0.29	0.17	0.32	0.38	0.22	0.26	0.25	0.30	0.29	0.34
	S=0.75 / AF=1.5										
C-S	25	0.58	0.41	1.05	1.26	0.82	0.99	0.90	1.08	0.97	1.17
	30	0.50	0.33	0.91	1.09	0.68	0.81	0.75	0.90	0.83	1.00
	35	0.43	0.27	0.77	0.93	0.56	0.67	0.63	0.76	0.70	0.84
	40	0.36	0.22	0.65	0.78	0.46	0.55	0.52	0.63	0.59	0.70
	45	0.29	0.17	0.53	0.64	0.36	0.44	0.42	0.50	0.48	0.57

The first values refer to an importance factor (IF) of 1.2 and the second values of an IF 1.0. In order to get the design acceleration without considering the amplification factor (AF) the values of the table have only to be multiplied by the corresponding AF. These accelerations refer only to the soil pressures and do not include the inertia forces of the wall, which have to be considered additionally.

This calculation considers only the dynamic soil pressures. The chamber wall develops further its own inertia forces during a seismic event which have to be added as the state of practice is without considering phase effects (the maximum of the dynamic soil pressures and the maximum of the inertia forces occur at the same time and not at a phase). Assuming that the wall inertia forces follow the design acceleration and not its spectral value, the next formula can be used as indicator:

$$\alpha = \frac{0.25\ \mathrm{CMB}}{S\ \mathrm{AF}\ \mathrm{IF} + (M\ \mathrm{g})} \tag{8-3}$$

Where M the mass of the wall section for which the total shear force has to be calculated and g the acceleration of gravity.

Furthermore, a small numerical investigation has been carried out, which investigates the influence of different boundaries for the analysis of dynamic soil-structure interaction problems and is added to the results of other analyses, which have investigated other important parameters. It has been shown that the sophisticated infinite elements and dashpots influence the soil response as they both damp much the propagating waves. The best results which fit the one dimensional wave propagation and the site response as it is calculated with the SHAKE program are achieved with the shear beam model (without boundaries at all at the sides). When infinite elements and dashpots are going to be used, the free field response of the shear beam model must be given in terms of equivalent forces or stresses at the sides of the numerical model in order satisfactory results to be achieved.

8.1 Further research

This numerical investigation showed the influence of the foundation flexibility and damping on the hydrodynamic pressures of water-retaining structures. However it not concluded to a closed form formula with which the hydrodynamic pressures can be calculated for a specific eigenfrequency for specific values of the dimensionless constants presented here. A more detailed parametric study have to be carried out which should enlighten the influence of the several forms of damping on the hydrodynamic pressures and to show when these increase or decrease. Such an effort was made by Fenves (Chopra, A. K., Fenves, G. 1983, 1984a, 1984b, 1985a), who considered these effects in the total response of a gravity dam. According to author of this thesis, this very precise work is not easily applicable and the range of application may be restricted to the field of dam engineering and cannot be used for the navigation locks.

This investigation has not included some phenomena, which take place during a seismic event and generally in dynamic soil-structure interaction. Some of these are important and other are not expected to influence significantly the conclusions of this study. These points refer to the best knowledge of the author and are presented here succinctly:

8.1.1 Wall-Soil interface

The wall-soil interface was taken into account oversimplified means a smooth or bonded contact. The influence of a no-tension resistance was investigated also simplified means static 1g analyses. Real conditions such as no tension resistance, friction contact etc., shall be taken into consideration. Jung and Bobet (Bobet, A., Jung, C. 2008) investigated on the influence of the friction for a contact interface without tension resistance and showed that the friction can reduce further the dynamic soil pressures.

8.1.2 Liquefaction of the soil

The navigation locks are constructed at sites with saturated soils. A possible liquefaction of the soil during a dynamic effect influences the dynamic soil pressures and the bearing capacity of the underlying soil. These effects can result to loss of the position stability. This study was performed in the fame of elastodynamics and such (non-linear) effects were not taken into consideration.

8.1.3 Consideration of shear strain dependent soil's damping

At the fifth chapter the influence of a shear dependent modulus of elasticity was shown. Apart from the young modulus the damping of soils is also dependent on the shear strains. As the shear strains increase, the soil damping also increases and can lead to combined effects of reduction of the response, shown here separately at the fifth chapter. Psarropoulos et al. (Psarropoulos et al. 2011) used the program QUAD4M, which incorporates the equivalent linear procedure, in order to investigate this effect. The showed, that with an increasing acceleration the soil pressures do not increasing but on the opposite reduce, due to the increasing soil's damping and reducing modulus of elasticity. They pointed though that, the increased damping and the reduced modulus of elasticity, which lead to a reduction of the stratum's eigenfrequency, can more easily lead to an amplified response.

8.1.4 Performance of the navigation locks in the longitudinal direction

The investigation here is based, as the state of practice also is, on the analysis and design of a navigation lock at its transverse direction. As mentioned also before, plane strain conditions apply at the most times for such long structures, especially when they are constructed monolithically without joints between the chamber sections. For big enough lengths and especially for soft soil conditions, which are mostly met at river sites a spatial variation of the ground acceleration is expected. As the case for dams and long bridges are, the structure is subjected not to the same but to spatial differential ground motion. The kinematic interaction (Foundation Input Motion; effects of base slab averaging, deconvolution, embedment wave scattering) is expected to play an important role and it is important to investigate these effects for the longitudinal direction to see in which point a damage (opening of joints for jointed chambers, stresses for monolithical chambers) can occur.

8.1.5 Permanent displacements

Apart from the structural damage, other types of failures such as permanent displacements of the chamber walls or settlements of the base plate can be assumed as damage as they affect the serviceability of the navigation lock and can suspend its operation. New calculation concepts based also on this investigation can be developed further from the existed ones (Elms, D. G., Richards R. 1979; Elms 2000; Zeng 1995; Wong 1982; Nadim F., Whitman R. V. 1983).

8.1.6 Consideration of the Foundation Input Motion

This numerical investigation is based on the bathtub model where the wall and the soil experience the same acceleration. In fact, the foundation input motion (FIM) of the wall or navigation lock's chamber is due to the soil structure interaction (SSI) generally differs from the acceleration at the soil's base. Some steps in this direction have been made recently for U-frame structures considering the kinematic interaction (Brandenberg, S., Mylonakis, G., Stewart, J. 2015). Further research shall include such effects.

8.1.7 Phase effects

For the soil-structure interaction, in order to be able to have a clear influence of the dynamic soil pressures, the soil is assumed massless at the most analyses. For many cases, the flexural rigidity of the wall depends not on the modulus of elasticity of its material but on the section's size (height). For a concrete wall, a rigid behaviour is achieved when the section's height is big enough, which in turn increases the inertia forces of the wall. It is expected that, because of the wave propagation in the wall itself, the maximum of these inertia forces does not occur simultaneously with the maximum of the dynamic soil pressures, but that a phase between these two maxima exists. Steedman and Zeng (Steedman, R. S., Zeng, X. 1990b) have given information for this effect for yielding gravity retaining walls and Sitar et al. (Mikola, R. G., Sitar, N. 2013; Al Atik, L., Sitar, N. 2008) have observed this effect at their experiments tests. A phase at the loading's maxima indicates that the total seismic forces do not need to be added following the absolute maxima rule but other superposition rules such as the SRSS rule can apply. This fact can influence significantly the design or strengthening of retaining walls.

8.1.8 Vertical component of the earthquake

The influence of the vertical component of an earthquake excitation was not investigated and to the author's knowledge is not addressed as a factor due to the physic of the problem in the field of elastodynamics (the vertical component is not expected to have an influence because of the way that the problem is addressed and solved analytically). As the wall and base flexibility increases and the elastic solutions reach the Mononobe-Okabe solution, it is expected that the vertical component should have a small influence on the dynamic soil pressures.

8.1.9 Water pressures of a saturated soil

The problem of the dynamic water pressures of a saturated soil on a wall is treated in the legislations in a "static" manner using the Westergaard solution or the saturated unit weight of the soil. For elastic and flexible systems, where a resonance can also occur, the water pressures depend not only on the permeability of the soil (as the most important factor) but also on the flexibility of the retaining system in means of wall and base flexibility and the frequency content of the excitation. More information about this problem can be found in (Ishibashi, I., Kawamura, M., Matsuzawa, H. 1985), who suggested a correction factor for the water pressures based on the permeability of the soil and predominant period of the excitation. The most recent research in this problem, is according to the author's knowledge the research made by (Theodorakopoulos und Beskos 2003; Theodorakopoulos et al. 2001a, 2001b; Beskos, D. E., Papagiannopoulos, G. A., Triantafyllidis, T. 2015), who treated the soil as a poroelastic medium and showed the influence of the soil's permeability on its natural frequency and the influence of the water pressures as the base rotation increases (in a similar manner the same was shown here numerically for the water pressures of the containing water of the chamber).

8.1.10 Damping of the foundation of the lock

The dynamic soil and water pressures developed during a dynamic event are due to the relative movement of the lock towards the soil or the water. At the chapter 4 was shown that the damping of the foundation (radiation and material damping of the underlying soil) affect strongly the steady state response of the water pressures. For the soil retaining function of a lock, it is also expected to play an important role especially because locks are deep embedded structures. However, it is expected that the reduction of the dynamic soil pressures will be not severe as it happens with the hydrodynamic pressures, as the retaining soil develops a much higher damping than the containing water during a dynamic effect.

References

ABAQUS (2012): Analysis User's Manual 6.11. Dassault Systèmes SimuliaCorp. Providence, USA.

ACI Committee 350.3-06 (2006): Seismic design of liquid-containing concrete structures and commentary. American Concrete Institute. Farmington Hills, MI, U.S.A.

Al Atik, L., Sitar, N. (2008): Experimental and Analytical Study of the Seismic Performance of Retaining Structures. Pacific Earthquake Engineering Research Center (PEER), University of California, Berkeley, 2008-10. In: *PEER-2008/104.*

Al Atik, L., Sitar., N. (2007): Development of Improved Procedures for Seismic Design of Buried and Partially Buried Structures. In: *PEER Report 2007/06.*

Arias, A., Sanchez-Sesma, F. J., Ovando-Shelley, E. (1981): A simplified elastic model for seismic analysis of earth retaining structures with limited displacements. In: *Proceedings: 3rd International Conference on Recent Advances in Geotechnical Earthquake Engineering and Soil Dynamics* 1 (Univ. of Missouri, Rolla, St. Louis), S. 235–240.

Ayre, R. S., Jacobsen, L. S. (1951): Hydrodynamic Experiments with Rigid Cylindical Tanks Subjected to Transient Motions. In: *Bulletin of the Seismological Society of America* 41, S. 313–346.

Basu, U., Chopra, A. K. (2004): Perfectly matched layers for transient elastodynamics of unbounded domains. In: *International Journal for Numerical Methods in Engineering* 59 (8), S. 1039–1074.

Bela, A., Buldgen, L., Philippe, R. (2015): Simplified Analytical Methods to Analyze Lock Gates Submitted to Ship Collisions and Earthquakes. In: *Mathematical Modelling in Civil Engineering* 11 (3), S. 8–22.

Beskos, D. E., Chassiakos, A. P., Theodorakopoulos, D. D. (2001): Dynamic pressures on rigid cantilever walls retaining poroelastic soil media. Part II. Second method of solution. In: *Soil Dynamics and Earthquake Engineering* 21 (4), S. 339–364.

Beskos, D. E., Papagiannopoulos, G. A., Triantafyllidis, T. (2015): Seismic pressures on rigid cantilever walls retaining linear poroelastic soil. An exact solution. In: *Soil Dynamics and Earthquake Engineering* 77, S. 208–219.

Bettess, P., Emson, C., Zienkiewicz, O. C. (1983): A novel boundary infinite element. In: *International Journal for Numerical Methods in Engineering* 19 (3), S. 393–404.

Bishop, R. E. D., Johnson, D. C. (2011): The mechanics of vibration. Paperback re-issue, 1. paperback ed. Cambridge: Cambridge University Press.

Bobet, A., Fernández, G., Jung, C. (2010): Analytical solution for the response of a flexible retaining structure with an elastic backfill. In: *International Journal for Analytical and Numerical Methods in Geomechanics* 34 (13), S. 1387–1408.

Bobet, A., Jung, C. (2008): Seismic Earth Pressures behind Retaining Walls: Effects of Rigid-Body Motions. In: *Geotechnical Earthquake Engineering and Soil Dynamics* 4, S. 1–11.

Bolton M. D., Steedman, R. S. (1982): Centrifugal testing of micro-concrete retaining walls subject to base shaking. Hg. v. Balkema. Proceedings of Conference on Soil dynamics and Earthquake Engineering, 1, 311-329, Balkema. Southhampton.

Bolton M. D., Steedman, R. S. (1985): The behavior of fixed cantilever walls subject to lateral loading. Application of Centrifuge Modeling to Geotechnical Design, Craig (ed.), Balkema. Rotterdam.

Bouaanani, Najib; Miquel, Benjamin (2015): Efficient modal dynamic analysis of flexible beam–fluid systems. In: *Applied Mathematical Modelling* 39 (1), S. 99–116. DOI: 10.1016/j.apm.2014.04.061.

Bouaanani, N., Goulmot, D., Miquel, B. (2014): Efficient seismic analysis of navigation locks. PIANC World Congress San Francisco. USA.

Brahtz, H. A., Heilbron, C. H. (1933): Water pressures on dams during earthquakes. In: *Transactions of the American Society of Civil Engineers* 98, S. 452–460.

Brandenberg, S. J.; Mylonakis, G.; Stewart, J. P. (2015): Closure to Kinematic Framework for Evaluating Seismic Earth Pressures on Retaining Walls. In: *J. Geotech. Geoenviron. Eng.* 141 (7), S. 4015031. DOI: 10.1061/(ASCE)GT.1943-5606.0001312.

Brandenberg, S. J.; Mylonakis, G.; Stewart, J. P. (2017): Approximate solution for seismic earth pressures on rigid walls retaining inhomogeneous elastic soil. In: *Soil Dynamics and Earthquake Engineering* 97, S. 468–477. DOI: 10.1016/j.soildyn.2017.03.028.

Brandenberg, S., Mylonakis, G., Stewart, J. (2015): Kinematic framework for evaluating seismic earth pressures on retaining walls. In: *Journal of Geotechnical and Geoenvironmental Engineering., 10.1061/(ASCE)*.

Bray, J. D., Faraj, F., Rathje, E. M., Russell, S. (2004): Empirical Relationships for Frequency Content Parameters of Earthquake Ground Motions. In: *Earthquake Spectra* 20 (1), S. 119–144. DOI: 10.1193/1.1643356.

Buldgen, L. (2015): Simplified analytical methods for the crashworthiness and the seismic design of lock gates. PhD Thesis. University of Liège, Liège. Structural Engineering Department.

Bustamante, J. I., Flores, A. (1966): Water pressure on dams subjected to earthquakes. In: *Journal of the Engineering Mechanics Division, Proceedings of the ASCE* 92 (EM 5).

Bustamante, J. I., Flores, A., E. Herrera, Rosenblueth I. (1963): Presion hydrodynamic enpresasy depositos. In: *Boletin Sociedad Mexicana de Ingenieria Sismica* 11 (2).

Chakrabatarti, P., Chopra, A. K. (1974): Hydrodynamic Effects in Earthquake Response of Gravity Dams. In: *Journal of the Structural Division, ASCE* 100 (ST6), S. 1211–1224.

Chen, Bang-Fuh; Hung, Tin-Kan (1993): Dynamic Pressure of Water and Sediment on Rigid Dam. In: *J. Eng. Mech.* 119 (7), S. 1411–1433. DOI: 10.1061/(ASCE)0733-9399(1993)119:7(1411).

Chopra, A. K. (1966): Hydrodynamic pressures on dams during earthquakes. Structures and Materials Research, Department of Civil Engineering. Berkley, California.

Chopra, A. K. (1967): Hydrodynamic pressures on dams during earthquakes. In: *Proceedings of the ASCE* (EM6).

Chopra, A. K. (2007): Dynamics of structures. Theory and applications to earthquake engineering. 3rd ed., [Pearson low price ed.]. [New Delhi]: [Pearson Education India] (Prentice-Hall international series in civil engineering and engineering mechanics).

Chopra, A. K., Chakrabarti, P. (1973): Hydrodynamic pressure and response of gravity dams to vertical earthquake component. In: *Earthquake Engineering & Structural Dynamics* 2, S. 143.

Chopra, A. K., Fenves, G. (1983): Effects of reservoir bottom absorption on earthquake response of concrete gravity dams. In: *Earthquake Engineering & Structural Dynamics* 11 (6), S. 809–829.

Chopra, A. K., Fenves, G. (1984a): Earthquake Analysis of Concrete Gravity Dams Including Reservoir Bottom Absorption and Dam-Water-Foundation Rock Interaction. In: *Earthquake Engineering and Structural Dynamics* 12 (5), S. 663–680.

Chopra, A. K., Fenves, G. (1984b): Earthquake Analysis and Response of Concrete Gravity Dams. Hg. v. University of California. Earthquake Engineering Research Center. Berkley (UCB/EERC-84/10).

Chopra, A. K., Fenves, G. (1985a): Effects of Reservoir Bottom Absorption and Dam-Water-Foundation Rock Interaction on Frequency Response Functions for Concrete Gravity Dams. In: *Earthquake Engineering and Structural Dynamics* 13 (1), S. 13–31.

Chopra, A. K., Fenves, G. (1985b): Reservoir Bottom Absorption Effects in Earthquake Response of Concrete Gravity Dams. In: *Journal of Structural Engineering, ASCE* 3 (3), S. 545–562.

Chopra, A. K., Fenves, G. (1985c): Simplified Earthquake Analysis of Concrete Gravity Dams: Separate Hydrodynamic and Foundation Interaction Effects. In: *Journal of Engineering Mechanics* 111 (6), S. 715–735.

Chopra, A. K., Liaw, C.Y. (1973): Earthquake Response of Axisymmetric Tower Structures Surrounded by Water. Hg. v. University of California. Earthquake Engineering Research Center. Berkley (EERC 73-25).

Chwang, A. T. (1977): Hydrodynamic pressures on sloping dam during earthquakes. Part 2. Exact Theory. In: *Journal of Fluid Mechanics* 87, S. 343–348.

Chwang, A. T., Housner, G. W. (1978): Hydrodynamic pressures on sloping dams during earthquakes. Part 1. Momentum Method. In: *Journal of Fluid Mechanics* 87, S. 335–341.

Clough, G. W., Fragaszy, R. F. (1977): A Study of Earth Loadings on Floodway Retaining Structures in the 1971 San Fernando Valley Earthquake. In: *Proceedings of the Sixth World Conference on Earthquake Engineering* 3.

Committee on Seismic Aspects of Dam Design (2012): Position Paper of International Commission on Large Dams. International Commission on Large Dams ICOLD.

Coulomb, C. A. (1776): Essai sur une application des regles de maximis et minimis a quelqesproblemes de stratiquerelatifs a l' architecture. Memoires de mathematique et de physique. In: *Presentes a l' academie royale des sciences* (7), S. 343–382.

Darbre, G. R. (1998): Phenomenological two-parameter model for dynamic dam-reservoir interaction. In: *Journal of Earthquake Engineering* 2 (4), S. 513–524.

Davidson, B. J.; Honey, G. D.; Hopkins, D. C.; Martin, R. J.; Priestley, M. J. N.; Ramsay, G. et al. (1986): Seismic Design of Storage Tanks. Recommendations of a Study Group. In: *New Zealand National Society for Earthquake Engineering*.

Dawson, E. M., Mejia, L. H. (2006): Earthquake deconvolution for FLAC. In: *FLAC and Numerical*.

Di Laora, R. (2015): Discussion on Kinematic Framework for Evaluating Seismic Earth Pressures on Retaining Walls. In: *J. Geotech. Geoenviron. Eng.* 141 (7), S. 4015031. DOI: 10.1061/(ASCE)GT.1943-5606.0001312.

DIN 4085:2011-05: Baugrund - Berechnung des Erddrucks.

DIN EN 1990:2010-12: Eurocode: Grundlagen der Tragwerksplanung.

DIN EN 1992-1: Eurocode 2: Bemessung und Konstruktion von Stahlbeton- und Spannbetontragwerken.

DIN EN 1998-1/NA:2011-01 (2011): National Annex - Nationally determined parameters - Eurocode 8: Design of structures for earthquake resistance - Part 1: General rules, Seismic actions and rules for buildings.

DL 5073-2000: Specifications for seismic design of hydraulic structures.

Dogangün, A. (1995): Earthquake analysis of rectangular water tanks considering soil-structure-liquid interaction using finite element method by comparing with analytical methods. Ph.D. Thesis. Karadeniz Technical University, Trabzon.

Dogangün, A., Livaoglu, R. (2004): Hydrodynamic pressures acting on the walls of rectangular fluid containers. In: *Structural Engineering and Mechanics* 17 (2), S. 203–214.

Dogangün, A., Livaoglu, R. (2007): Effect of foundation embedment on seismic behavior of elevated tanks considering fluid–structure-soil interaction. In: *Soil Dynamics and Earthquake Engineering* 27 (9), S. 855–863.

Dong J.; Duron Z.; Knarr M.; Muto M.; Von Gersdorff N.; Yen J. (2011): The investigation of a concrete gravity dam in a narrow canyon using 3-D nonlinear analysis. 31st Annual USSD Conference. San Diego, California.

Ebeling, R. M., Mosher, R. L., Peters, J. F. (1997): The role of non-linear deformation analyses in the design of a reinforced soil berm at red river u-frame lock no. 1. In: *International Journal for Numerical and Analytical Methods in Geomechanics* 21, S. 753–787.

Eibl, J., Stempniewski, L. (Hg.) (1987a): Dynamic Analysis of Liquid-Filled Tanks Including Plasticity and Fluid Interaction - Earthquake Effects. Unter Mitarbeit von G. De Roeck et al. Proceedings of "Shell and Spatial Structures - Computational Aspects. Heidelberg, 261-269: Springer-Verlag.

Eibl, J., Stempniewski, L. (Hg.) (1987b): Nonlinear Analysis of Liquid Storage Tanks. Unter Mitarbeit von F. H. Wittmann. Transactions of the 9th international conference on structural mechanics in reactor technology. Lausanne. Rotterdam: Balkema (Vol. B).

Eibl, J., Stempniewski, L. (1987c): Über die Beanspruchung von Flüssigkeitsbehältern durch Erdbeben. H.J. Dolling, (ed.): Vortragsband der 4. Jahrestagung der Deutschen Gesellschaft für Erdbebeningenieurwesen und Baudynamik (DGEB). Berlin.

Eibl, J., Stempniewski, L. (1988): Seismic Response of Cylindrical Fluid Tank Systems. Kounadis A.N., Krätzig, W.B. (eds.). Hg. v. Earthquake Engineering & Structural Dynamics. Hellenic Society for Theoretical and Applied Mechanics. Greec-German Seminar, Athens.

Eibl, J., Stempniewski, L. (1989): Water Containers Under Seismic Excitation - Three Dimensional Finite Element Investigation. Proc. of the Int. Conference on Highrise Buildings. Nanjing.

Elms, D. G. (2000): Refinements to the Newmark Sliding Block Model. Proceedings 12th World Conference on Earthquake Engineering, Auckland NZ.

Elms, D. G., Richards R. (1979): Seismic behaviour of gravity retaining walls. In: *Journal of the Geotechnical Engineering Division* 105 (GT4), S. 449–464.

EN 1998-1:2004 Eurocode 8: Design of structures for earthquake resistance – Part 1: General rules, seismic actions and rules for buildings.

EN 1998-4:2006 Eurocode 8: Design of structures for earthquake resistance – Part 4: Silos, tanks and pipelines.

EN 1998-5:2004 Eurocode 8: Design of structures for earthquake resistance – Part 5: Foundations, retaining structures and geotechnical aspects.

Fang, Y. S., Ishibashi, I. (1987): Dynamic earth pressures with different wall movement modes. In: *SOILS AND FOUNDATIONS* 27 (4), S. 11–22.

Fang, Y. S., Sherif, M. A. (1984): Dynamic earth pressures on walls rotating about the top. In: *SOILS AND FOUNDATIONS* 24 (4), S. 109–117.

Fehl, B., Ferhi, A., Truman, K., Petruska, D. (1991): Nonlinear, Incremental Analysis of Mass-Concrete Lock Monolith. In: *Journal of Structural Engineering*.

Gazetas, G. (1983): Analysis of machine foundation vibrations. State of the art. In: *International Journal of Soil Dynamics and Earthquake Engineering* 2 (1), S. 2–42. DOI: 10.1016/0261-7277(83)90025-6.

Gazetas, G. (1991): Foundation Vibrations. In: Hsai-Yang Fang (Hg.): Foundation Engineering Handbook. Boston, MA: Springer US, S. 553–593. Online verfügbar unter http://dx.doi.org/10.1007/978-1-4615-3928-5_15.

Gazetas, G., Travasarou, T. (2004): On the linear seismic response of soils with modulus varying as a power of depth-the Maliakos marine clay. In: *SOILS AND FOUNDATIONS* 44 (5), S. 85–93.

Gazrtas, G., Klonaris, G., Psarropoulos, P. N. (2005): Seismic earth pressures on rigid and flexible retaining walls. In: *Soil Dynamics and Earthquake Engineering* 24 (795–809).

Goto, H.; Toki, K. (1963): Fundamental Studies On Vibration Characteristics And Aseismic Design Of Submerged Bridge Piers. In: *Transactions of the Japan Society of Civil Engineers* 1963 (100), S. 1–8. DOI: 10.2208/jscej1949.1963.100_1.

Govindjee, S., Sagiyama K., Persson P. O. (2014): An efficient time-domain perfectly matched layers formulation for elastodynamics on spherical domains. In: *International Journal for Numerical Methods in Engineering* 100 (419–441).

Graham, E. W., Rodriquez, A. M. (1952): Characteristics of fuel motion which affect airplane dynamics. In: *Journal of Applied Mechanics* 19, S. 381–388.

Hadjian, A. H., Nazarian, H. N. (1979): Earthquake Induced Lateral Soil Pressures on Structures. In: *Journal of Geotechnical Engineering Division, ASCE* 105 (GT9), S. 1049–1066.

Halabian, A. M. (2015): Analysis and Design Issues of Geotechnical Systems: Rigid Walls. Encyclopedia of Earthquake Engineering: Springer Berlin Heidelberg.

Haroun, M. A. (1980): Dynamic analyses of liquid storage tanks. PhD Thesis. California Institute of Technology.

Haroun M. A., Housner G. W. (1981): Earthquake Response of Deformable Liquid Storage Tanks. In: *Journal of Applied Mechanics: ASME DC* 48 (2), S. 411–418.

Haroun M. A., Housner G. W. (1982a): Complications in Free Vibration Analysis of Tanks. In: *Proceedings of the Journal of Engineering Mechanics Division, ASCE* 108 (5), S. 801–818.

Haroun M. A., Housner G. W. (1982b): Dynamic Characteristics of Liquid Storage Tanks. In: *Proceedings of the Journal of Engineering Mechanics Division, ASCE* 108 (5), S. 783–800.

Hettler, A. (Hg.) (2012): EAB. Empfehlungen des Arbeitskreises "Baugruben". 5 ed. Berlin: Ernst & Sohn.

Hoskins, L. M., Jacobsen, L. S. (1934): Water pressure in a tank caused by simulated earthquake. In: *Bulletin of the Seismological Society of America* 24, S. 1–32.

Housner, G. W. (1954): Earthquake pressures on fluid containers. California Institute of Technology.

Housner, G. W. (1957): Dynamic pressures on accelerated fluid containers. In: *Bulletin of the Seismological Society of America* 47 (1), S. 15–35.

Housner, G. W. (1963): The dynamic behavior of water tanks. In: *Bulletin of the Seismological Society of America* 53 (2), S. 381–387.

Ishibashi, I., Kawamura, M., Matsuzawa, H. (1985): Dynamic Soil and Water Pressures of Submerged Soils. In: *Journal of Geotechnical Engineering* 111 (10), S. 1161–1176.

Ishibashi, I., Lee, C. D., Sherif, M. A. (1982): Earth pressure against stiff retaining walls. In: *Journal of Geotechnical Engineering, ASCE* 108 (GT5), S. 679–696.

Jacobsen, L. S. (1949): Impulsive hydrodynamics of fluid inside a cylindrical tank and of fluid surrounding a cylindrical pier. In: *Bulletin of the Seismological Society of America* 39 (3), S. 189–204.

Karman, T. von (1933a): Discussion of water pressures on dams during earthquakes. In: *Transactions of the American Society of Civil Engineers* 98, S. 434–436.

Karman, Th. von (1933b): On pressures on dams during earthquakes - Discussion.

Karnovskiĭ, I. A.; Lebed, Olga I. (2001): Formulas for structural dynamics. Tables, graphs, and solutions. New York: McGraw-Hill.

Kloukinas, P.; Di Scotto Santolo, A.; Penna, A.; Dietz, M.; Evangelista, A.; Simonelli, A. L. et al. (2015): Investigation of seismic response of cantilever retaining walls. Limit analysis vs shaking table testing. In: *Soil Dynamics and Earthquake Engineering* 77, S. 432–445. DOI: 10.1016/j.soildyn.2015.05.018.

Kloukinas, P.; Langousis, M.; Mylonakis, G. (2012): Simple Wave Solution for Seismic Earth Pressures on Nonyielding Walls. In: *J. Geotech. Geoenviron. Eng.* 138 (12), S. 1514–1519. DOI: 10.1061/(ASCE)GT.1943-5606.0000721.

Kotsubo, S. M. (1959): Dynamic water pressure on dams due to irregular Earthquakes. In: *Faculty of Engineering, Kyushu University* 18 (4).

Kotsubo, S. M. (1961): External forces on arch dams during earthquakes. In: *Faculty of Engineering, Kyushu University* 20 (4).

Kotsubo, S. M. (1965a): Dynamic Water Pressures on Arch Dams during Earthquakes. In: *Transactions of the Japan Society of Civil Engineers*, S. 28–37.

Kotsubo, S. M. (1965b): External Forces On Submerged Bridge Piers With Elliptic Cross Sections And Their Vibration During Earthquakes. In: *Transactions of the Japan Society of Civil Engineers* 1965 (120), S. 14–24.

Kramer, Steven Lawrence (2009): Geotechnical earthquake engineering. Upper Saddle River, N.J., London: Prentice Hall; Pearson Education [distributor].

Kuhlemeyer R. L., Lysmer, J. (1969): Finite Dynamic Model for Infinite Media. In: *Engineering Mechanics Division* 4 (7), S. 859–877.

Kuhlemeyer R. L., Lysmer, J. (1973): Finite element method accuracy for wave propagation problems. In: *Journal of the Soil Mechanics and Foundations Division* 99 (5), S. 421–427.

Kwok, Annie O. L.; Stewart, Jonathan P.; Hashash, Youssef M. A.; Matasovic, Neven; Pyke, Robert; Wang, Zhiliang; Yang, Zhaohui (2007): Use of Exact Solutions of Wave Propagation Problems to Guide Implementation of Nonlinear Seismic Ground Response Analysis Procedures. In: *J. Geotech. Geoenviron. Eng.* 133 (11), S. 1385–1398. DOI: 10.1061/(ASCE)1090-0241(2007)133:11(1385).

LAM, Ignatius P.; LAW, Hubert K.; YANG, Chien T. (2007): Modeling of seismic wave scattering on pile groups and caissons. MCEER 07-0017, Buffalo, NY : Multidisciplinary Center for Earthquake Engineering Research, Technical report, 1520-295X.

Lam, I.P., Law, H., Yang, Ch. (2004): Modeling of Seismic Wave Scattering on Pile Group and Caisson. In: *Multidisciplinary Center for Earthquake Engineering Research (MCEER)*.

Lee, Jeeho; Fenves, Gregory L. (1998): Plastic-damage model for cyclic loading of concrete structures. In: *Journal of Engineering Mechanics, ASCE* 124 (8), S. 892–900.

Lee, G. C., Tsai, C. S. (1991): Time-domain analyses of dam-reservoir system. I: exact solution. In: *Journal of Engineering Mechanics, ASCE* 117 (9), S. 1990–2006.

Lee, J., Ortiz, L. A., Scott, R. F. (1983): Dynamic centrifuge testing of a cantilever retaining wall. In: *Earthquake Engineering & Structural Dynamics* 11 (2), S. 251–268.

Leontiev, U. N., Vlasov, V. Z.: Beams, plates, and shells on elastic foundation. Jerusalem: Israel Program for Scientific Translations [translated from Russian].

Li, X. (1999): Dynamic analysis of rigid walls considering flexible foundation. In: *Journal of geotechnical and geoenvironmental engineering* 125 (9), S. 803–806.

Liao, S., Whitman, R. V. (1985): Seismic Design of Gravity Retaining Walls. U. S. Army Engineer Waterways Experiment Station. Vicksburg (MS 39180).

Livaoglu, R. (2008): Investigation of seismic behavior of fluid–rectangular tank–soil/foundation systems in frequency domain. In: *Soil Dynamics and Earthquake Engineering* 28 (2), S. 132–146.

Lysmer, John; Tabatabaie, Mansour; Tajirian, Frederick F.; Vahdani, Shahriar; Ostadan, Farhang: SASSI: A System for Analysis of Soil Structure Interaction. In:

UCB/GT-81/02, University of California, Berkeley, Geotechnical Engineering, 1981-04.

Maltidis, G., Stempniewski, L. (2013): Fluid Structure Interaction: Arch Dam - Reservoir at Seismic Loading. In: *12th International Benchmark Workshop On Numerical Analysis Of Dams.*

Mansur, W. J., Soares Jr., D. (2006): Dynamic analysis of fluid–soil–structure interaction problems by the boundary element method. In: *Journal of Computational Physics* 219 (2), S. 498–512.

19702, 2013: Massivbauwerke im Wasserbau - Tragfähigkeit, Gebrauchstauglichkeit und Dauerhaftigkeit.

Matsuo, H. (1941): Experimental study on the distribution of earth pressures acting on a vertical wall during earthquakes. In: *Journal of the Japanese Society of Civil Engineers* 27 (2).

Matsuo, H., Ohara, S. (1960): Lateral earth pressure and stability of quay walls during earthquakes. 1. Aufl. Proceedings, Earthquake Engineering, Second World Conference. Tokyo.

Matsuo, O., Mononobe, N. (1929): On the determination of earth pressure during earthquakes. In: *Proceeding of the world engineering congress* 9, S. 179–187.

Mikola, R. G., Sitar, N. (2013): Seismic earth pressures on retaining structures in cohesionless soils. Hg. v. University of California. Department of Civil and Environmental Engineering. Berkley (CA13-2170).

Mylonakis, George; Nikolaou, Sissy; Gazetas, George (2006): Footings under seismic loading. Analysis and design issues with emphasis on bridge foundations. In: *Soil Dynamics and Earthquake Engineering* 26 (9), S. 824–853. DOI: 10.1016/j.soildyn.2005.12.005.

Mylonakis G., Kloukinas P., Papantonopoulos C. (2007): An alternative to the Mononobe–Okabe equations for seismic earth pressures. In: *Soil Dynamics and Earthquake Engineering* 27, S. 957–969.

Nadim F., Whitman R. V. (1983): Seismically Induced Movement of Retaining Walls. In: *Journal of Geotechnical Engineering* 109 (7), S. 915–931.

Nakamura, S. (2006): Reexamination of Mononobe-Okabe theory of gravity retaining walls using centrifuge model tests. In: *SOILS AND FOUNDATIONS* 46 (2), S. 135–146.

NEHRP (2012): NIST GCR 12-917-21 Soil-Structure Interaction for Building Structures.

Newmark, N. M., Rosenblueth, E. (1971): Fundamentals of Earthquake Engineering. Prentice-Hall. Englewood Cliffs.

Okabe, S. (1924): General theory on earth pressure and seismic stability of retaining walls and dams. In: *Japan Society of Civil Engineers* 10 (6), S. 1277–1323.

Ostadan, F. (2005): Seismic soil pressure for building walls. An updated approach. In: *Soil Dynamics and Earthquake Engineering* 25 (7-10), S. 785–793.

Pani, P. K.; Bhattacharyya, S. K. (2007): Fluid–structure interaction effects on dynamic pressure of a rectangular lock-gate. In: *Finite Elements in Analysis and Design* 43 (10), S. 739–748. DOI: 10.1016/j.finel.2007.03.003.

Papazafeiropoulos, G.; Tsompanakis, Y.; Psarropoulos, P. N. (2011): Dynamic Interaction of Concrete Dam-Reservoir-Foundation: Analytical and Numerical Solutions. In: M. Papadrakakis, M. Fragiadakis und N. Lagaros (Hg.): Computational Methods in Earthquake Engineering, Bd. 21. Dordrecht: Springer Science+Business Media B.V (Computational Methods in Applied Sciences, 21), S. 455–488.

Papazafeiropoulos, G., Psarropoulos, P. N. (2010): Analytical evaluation of the dynamic distress of rigid fixed-base retaining systems. In: *Soil Dynamics and Earthquake Engineering* 30 (12), S. 1446–1461.

Parikh, V. H., Veletsos, A. S., Younan, A. H. (1995): Dynamic response of a pair of walls retaining a viscoelastic solid. In: *Earthquake Engineering & Structural Dynamics* 24 (12), S. 1567–1589.

Prakash S., Wu Y. (2001): Seismic Displacements of Rigid Retaining Walls. In: *International Conferences on Recent Advances in Geotechnical Earthquake Engineering and Soil Dynamics* 7.05.

Psarropoulos, P. N.; Klonaris, G.; Gazetas, G. (2005): Seismic earth pressures on rigid and flexible retaining walls. In: *Soil Dynamics and Earthquake Engineering* 25 (7-10), S. 795–809. DOI: 10.1016/j.soildyn.2004.11.020.

Psarropoulos, P. N.; Tsompanakis, Y.; Papazafeiropoulos, G. (2011): Effects of soil non-linearity on the seismic response of restrained retaining walls. In: *Structure and Infrastructure Engineering* 7 (12), S. 931–942. DOI: 10.1080/15732470903419677.

Rahman,K. R., Sivakumar, J., Vallabhan,C. V. G. (1988): BEFEC-A Program for Analysis of U-Lock Structures, S. 551–563.

Rashed, A. (1982): Dynamic analyses of fluid-structure systems. PhD Thesis. California Institute of Technology, Pasadena.

Rathje, E. M.; Kottke, A. (2013): Strata. Online verfügbar unter https://nees.org/resources/692.

Schnabel, P. B., Lysmer, J., Seed, H. B.: SHAKE-91: Equivalent Linear Seismic Response Analysis of Horizontally Layered Soil Deposits.

Scott, R. F. (1974): Earthquake-induced pressures on retaining walls. In: *Proceedings: 5th World Conference on Earthquake Engineering* 2, S. 1611–1620.

Seed, H. Bolton; Idriss, I. M.; of Engineering (1970): Soil moduli and damping factors for dynamic response analyses.

Seed H. B., Whitman R. V. (1970): Design of earth retaining structures for dynamic loads. In: *Proceedings of specialty conference on lateral stresses in the ground and design of earth retaining structures* (New York: ASCE), S. 103–147.

Shamoto Y., Tokimatsu, K., Zhang, J. (1998(a)): Evaluation of Earth Pressure under Any Lateral Deformation. In: *SOILS AND FOUNDATIONS* 38 (1), S. 15–33.

Shamoto Y., Tokimatsu, K., Zhang, J. (1998(b)): Seismic earth pressure theory for retaining walls under any lateral displacement. In: *SOILS AND FOUNDATIONS* 38 (2), S. 143–163.

Shamoto, Y., Tokimatsu, K., Zhang, J.-M. (1998(c)): Earth pressures on rigid walls during earthquakes. In: *Geotechnical Special Publication* 75 (2), S. 1057–1070.

SHARE: Seismic Hazard Harmonization in Europe. In: *http://www.share-eu.org/*.

Shivakumar, P., Veletsos, A. S. (1997): Tanks containing liquid or solids. A handbook in computer analysis and design of earthquake resistant structures. Computational Mechanical. Southampton, U.K., S. 725–774.

Sitar, N., Wagner, N. (2015): On Seismic Response of Stiff and Flexible Retaining Structures. 6th International Conference on Earthquake Geotechnical Engineering.

Soares, D.; Mansur, W. J. (2006): Dynamic analysis of fluid–soil–structure interaction problems by the boundary element method. In: *Journal of Computational Physics* 219 (2), S. 498–512. DOI: 10.1016/j.jcp.2006.04.006.

Stadler A. T. (1996): Dynamic centrifuge testing of cantilever retaining walls. PhD Thesis. University of Colorado at Boulder.

Steedman, R. S. (1984): Modeling the behavior of retaining walls in Earthquakes. PhD Thesis. Cambridge University, Cambridge.

Steedman, R. S., Zeng, X. (1990a): The influence of phase on the calculation of pseudo-static earth pressure on a retaining wall. In: *Géotechnique* 40 (1), S. 103–112.

Steedman, R. S., Zeng, X. (1990b): The seismic response of waterfront retaining walls. Design and Performance of Earth Retaining Structures, Conference Proceedings. Cornell University, ASCE Geotechnical Special Publication No. 25. Ithaca, New York.

Steedman, R. S., Zeng, X. (1991): Centrifuge modeling of the effects of earthquakes on free cantilever walls. Proceedings of the International Conference Centrifuge 1991, Boulder, Colorado.

Stempniewski, L. (1990): Flüssigkeitsgefüllte Stahlbetonbehälter unter Erdbebeneinwirkung. PhD Thesis. Universität Karlsruhe, Karlsruhe. Insituts für Massivbau.

Sundquist, K. J., Werner, P. W. (1949): On hydrodynamic earthquake effects. In: *Transactions American Geophysical Union* 30 (5), S. 636–657.

Tajimi, H. (1973): Dynamic earth pressures on basement wall. Proceedings, Earthquake Engineering, Fifth World Congress.

Tang, H. T., Tang, Y., Veletsos, A. S. (1992): Dynamic response of flexibly supported liquid storage tanks. In: *Journal of Structural Engineering* 118, S. 264–283.

Tang, Y., Veletsos, A. S. (1986): Dynamics of vertically excited liquid storage tanks. In: *Journal of Structural Engineering* 112 (6), S. 1228–1246.

Tang, Y., Veletsos, A. S. (1990): Soil-structure interaction effects for laterally excited liquid-storage tanks. In: *Journal of Earthquake Engineering and Structural Dynamics* 19 (4), S. 473–496.

Theodorakopoulos, D. D.; Beskos, D. E. (2003): Dynamic pressures on a pair of rigid walls experiencing base rotation and retaining poroelastic soil. In: *Engineering Structures* 25 (3), S. 359–370. DOI: 10.1016/S0141-0296(02)00171-2.

Theodorakopoulos, D. D.; Chassiakos, A. P.; Beskos, D. E. (2001a): Dynamic pressures on rigid cantilever walls retaining poroelastic soil media. Part I. First method of solution. In: *Soil Dynamics and Earthquake Engineering* 21 (4), S. 315–338. DOI: 10.1016/S0267-7261(01)00009-4.

Theodorakopoulos, D. D.; Chassiakos, A. P.; Beskos, D. E. (2001b): Dynamic pressures on rigid cantilever walls retaining poroelastic soil media. Part II. Second method of solution. In: *Soil Dynamics and Earthquake Engineering* 21 (4), S. 339–364. DOI: 10.1016/S0267-7261(01)00010-0.

U.S. Department of the Interior, Bureau of Reclamation (2006): State-of-practice for the nonlinear analysis of concrete dams at the Bureau of Reclamation. [Denver, Colo.], Springfield, VA: U.S. Dept. of the Interior, Bureau of Reclamation.

US Army Corps of Engineers (1999): Response Spectra and Seismic Analysis for Concrete Hydraulic Structures. Engineer Manual EM 1110-2-6050. Washington DC.

US Army Corps of Engineers (2003): Engineering and Design: Time-History Dynamic Analysis of Concrete Hydraulic Structures. Engineer Manual EM 1110-2-6051. Washington DC.

USSD (Hg.) (2014): Observed Performance of Dams During Earthquakes Volume III. United States Society on Dams. Denver, CO: U.S. Society on Dams.

Veletsos, A. S. (1984): Guidelines for the seismic design of oil and gas pipeline systems. Seismic response and design of liquid storage tanks. In: *Technical Council on Lifeline Earthquake Engineering*, S. 255.

Veletsos, A. S., Yang, J. Y. (1977): Earthquake response of liquid storage tanks. Advances in civil engineering through mechanics. ASCE Proc., 2nd Engineering Mechanics Specially Conf., ASCE, Reston, Va., S. 1–24.

Veletsos, A. S., Younan, A. H. (1992): Dynamic soil pressures on rigid vertical walls. In: *Report No. 52357, Brookhaven National Laboratory, Upton, N.Y.*

Veletsos, A. S., Younan, A. H. (1993): Dynamic modeling and response of soil-wall systems. In: *Journal of geotechnical and geoenvironmental engineering* 120 (12), S. 2155–2179.

Veletsos, A. S., Younan, A. H. (1994a): Dynamic soil pressures on rigid retaining walls. In: *Earthquake Engineering & Structural Dynamics* 23, S. 275–301.

Veletsos, A. S., Younan, A. H. (1994b): Dynamic soil pressures on rigid vertical walls. In: *Earthquake Engineering & Structural Dynamics* 23 (3), S. 275–301.

Veletsos, A. S., Younan, A. H. (1997): Dynamic Response of Cantilever Retaining Walls. In: *Journal of geotechnical and geoenvironmental engineering* 123 (2), S. 161–172. DOI: 10.1061/(ASCE)1090-0241(1997)123:2(161).

Veletsos, A. S., Younan, A. H. (1998a): Dynamics of Solid-Containing Tanks. I: Rigid Tanks. In: *Journal of Structural Engineering* 124 (1), S. 52–61.

Veletsos, A. S., Younan, A. H. (1998b): Dynamics of Solid-Containing Tanks. II: Flexible Tanks. In: *Journal of Structural Engineering* 124 (1), S. 62–70. DOI: 10.1061/(ASCE)0733-9445(1998)124:1(62).

Veletsos, A. S., Younan, A. H. (2000a): Dynamic response of flexible retaining walls. In: *Earthquake Engineering & Structural Dynamics* 29 (12), S. 1815–1844.

Veletsos, A. S., Younan, A. H. (2000b): Dynamic response of flexible retaining walls. In: *Earthquake Engineering & Structural Dynamics* 29 (12), S. 1815–1844.

Vrettos, C. (1988): Oberflächenwellen im Halbraum mit tiefenabhängigem Schubmo- dul (Heft Nr.114).

Vrettos, C.; Beskos, D. E.; Triantafyllidis, T. (2016): Seismic pressures on rigid cantilever walls retaining elastic continuously non-homogeneous soil. An exact solution. In: *Soil Dynamics and Earthquake Engineering* 82, S. 142–153. DOI: 10.1016/j.soildyn.2015.12.006.

Vrettos., C., Feldbusch, A. (2016): Dynamischer Erddruck auf starre und flexible Wände mittels Wellenlösungen: numerische Einzelvergleiche für homogene und geschichtete Bodenprofile. In: *Geotechnik 39*.

Vucetic, M.; Dobry, R. (1991): Effect of Soil Plasticity on Cyclic Response. In: *J. Geotech. Engrg.* 117 (1), S. 89–107. DOI: 10.1061/(ASCE)0733-9410(1991)117:1(89).

Westergaard, H. M. (1933): Water pressures on dams during earthquakes. In: *Transactions of the American Society of Civil Engineers* (98), S. 418–433.

Wieland, M. (2007): Earthquake safety of dams and the importance of emergency planning. 1st National Symposium and Exposition on Dam Safety. Ankara, Turkey.

Wilson, E. L. (2000): Three dimensional static and dynamic analysis of structures. A physical approach with emphasis on earthquake engineering. 3rd ed. Berkeley, Calif.: Computers and Structures.

Wilson, E. L., Khalvati, M. (1983): Finite elements for the dynamic analysis of fluid-solid systems. In: *International Journal for Numerical Methods in Engineering* 19 (11), S. 1657–1668.

Wong, C. P. (1982): Seismic analysis and an improved seismic design procedure for gravity retaining walls. M.Sc. Thesis. Massachusetts Institute of Technology. Dept. of Civil Engineering.

Wood, J. H. (1973): Earthquake induced soil pressures on structures. Doctoral Dissertation. California Institute of Technology, Pasadena.

Wu, G.; Liam Finn, W. D. (1999): Seismic lateral pressures for design of rigid walls. In: *Can. Geotech. J.* 36 (3), S. 509–522. DOI: 10.1139/cgj-36-3-509.

Xu, Chaojin; Spyrakos, C. C. (2001): Seismic analysis of lock–soil–fluid systems by hybrid BEM–FEM. In: *Soil Dynamics and Earthquake Engineering* 21 (3), S. 259–271. DOI: 10.1016/S0267-7261(00)00101-9.

Zangar, C. N. (1952): Hydrodynamic Pressures on Dams due to Horizontal Earthquake Effects.

Zangar, C. N., Haefelri, J. (1952): Electric analog indicates effect of horizontal earthquake shock on dams. In: *Civil Engng* 22, S. 278–279.

Zeng, X. (1990): Modeling behavior of quay walls in earthquakes. PhD Thesis. Cambridge University, Cambridge.

Zeng, X. (1995): Earthquake Induced Displacement of Gravity Retaining Walls. In: *International Conferences on Recent Advances in Geotechnical Earthquake Engineering and Soil Dynamics*.

Softwares

Abaqus 6.11,6.12,6.14, Dassault Systems Simulia Corp., Providende, USA.

Excel, Microsoft Corporation, Redmond, USA.

Mathematica, Wolfram Research, Oxfordshire, United Kingdom.

Seismoartif, Seismosoft, Pavia, Italy.

Seismosignal, Seismosoft, Pavia, Italy.

SHAKE-91, Schnabel, Per B.; Lysmer, John; Seed, H. Bolton; SHAKE-91: Equivalent Linear Seismic Response Analysis of Horizontally Layered Soil Deposits

Annex A

Harmonic vibration with viscous damping (taken by (Chopra 2007)):
The differential equation of an SDOF system to harmonic response is given by:

$$m\ddot{u} + c\dot{u} + ku = p_0 \sin \omega t \tag{A-1}$$

and the initial conditions are:

$$u = u(0) \quad \dot{u} = \dot{u}(0) \tag{A-2}$$

The particular solution of this differential equation is:

$$u_p(t) = C \sin \omega t + D \cos \omega t \tag{A-3}$$

where

$$C = \frac{p_0}{k} \frac{1 - (\omega/\omega_n)^2}{[1 - (\omega/\omega_n)^2]^2 + [2\zeta(\omega/\omega_n)]^2} \tag{A-4}$$

$$D = \frac{p_0}{k} \frac{-2\zeta\, \omega/\omega_n}{[1 - (\omega/\omega_n)^2]^2 + [2\zeta(\omega/\omega_n)]^2} \tag{A-5}$$

and the complementary solution of the differential equation is:

$$u_c(t) = e^{-\zeta \omega_n t}(A \cos \omega_D t + B \sin \omega_D t) \tag{A-6}$$

where

$$\omega_D = \omega_n \sqrt{1-\zeta^2} \quad \textbf{(A-7)}$$

The complete solution of the equation is

$$u(t) = \left\{e^{-\zeta\omega_n t}(A\cos\omega_D t + B\sin\omega_D t)\right\} + \{C\sin\omega t + D\cos\omega t\} \quad \textbf{(A-8)}$$

where the first part of the right side of the equation is the transient response and the second part of the right side the steady state response.

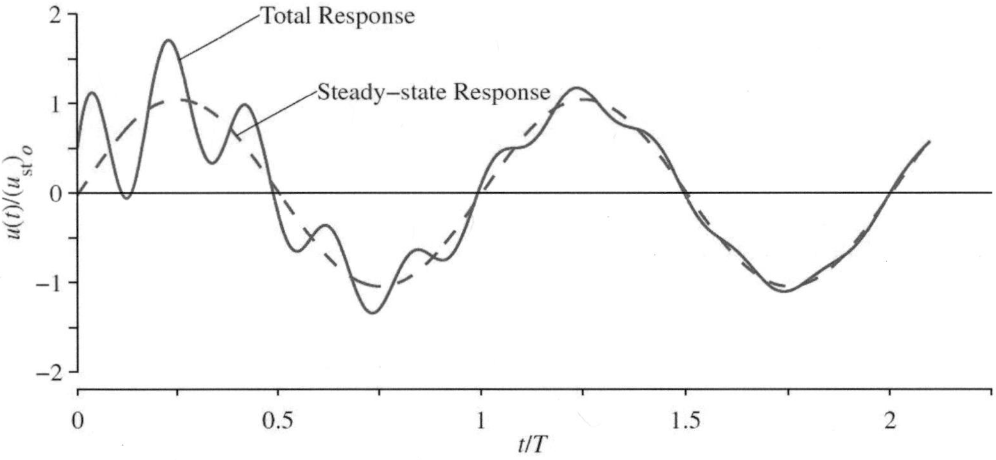

Fig. A-1 Response of damped system to harmonic response ($\omega<\omega_n$).

Dividing Equation A-1 by *m* gives:

$$\ddot{u} + 2\zeta\omega_n\dot{u} + \omega_n^2 u = \frac{p_0}{m}\sin\omega t \quad \textbf{(A-9)}$$

And the particular solution is rewritten:

$$u_p(t) = C \sin \omega t + D \cos \omega t \qquad \textbf{(A-10)}$$

Substituting Equation A-10 in A-9 gives

$$[(\omega_n^2 - \omega^2)C - 2\zeta\omega_n\omega D] \sin \omega t + [2\zeta\omega_n\omega C + (\omega_n^2 - \omega^2)D] \cos \omega t = \frac{p_0}{m} \sin \omega t$$

(A-11)

In order for Equation A-11 to be valid for all t, the coefficients of the sine and cosine terms must be equal. This requirement gives two equations in *C* and *D* which, after dividing by ω_n^2 and using the relation $k = \omega_n^2 m$, become

$$\left[1 - \left(\frac{\omega}{\omega_n}\right)^2\right] C - \left(2\zeta \frac{\omega}{\omega_n}\right) D = \frac{p_0}{k} \qquad \textbf{(A-12)}$$

$$\left(2\zeta \frac{\omega}{\omega_n}\right) C + \left[1 - \left(\frac{\omega}{\omega_n}\right)^2\right] D = 0 \qquad \textbf{(A-13)}$$

Response at resonance (ω=ω$_\mathbf{n}$)

For $\omega = \omega_n$ and zero initial conditions the constants *C*, *D*, *A* and *B* take the values:

$C=0$, $D = -(u_{st})_o/2\,\zeta$, $A = (u_{st})_o/2\,\zeta$ and $B = -(u_{st})_o/2\sqrt{1-\zeta^2}$.

With these solutions for *A*, *B*, *C* and *D*, Equation A-8 becomes

$$u(t) = (u_{st})_0 \frac{1}{2\zeta} \left[e^{-\zeta\omega_n t} \left(\cos \omega_D t + \frac{\zeta}{\sqrt{1-\zeta^2}} \sin \omega_D t \right) - \cos \omega_n t \right] \qquad \textbf{(A-14)}$$

This result is plotted in Figure A-2. Damping lowers each peak and limits the response to the bounded value:

$$u_0 = \frac{(u_{st})_0}{2\zeta} \tag{A-15}$$

For slightly damped systems the sinusoidal term of Equation A-14 is small and $\omega_D \approx \omega_n$; thus:

$$u(t) \simeq (u_{st})_0 \frac{1}{2\zeta}\left(e^{-\zeta\omega_n t} - 1\right) \cos \omega_n t \tag{A-16}$$

The amplitude of the steady state response of a system to harmonic force at $\omega = \omega_n$ and the rate at which a steady state is attained is strongly influenced by damping.

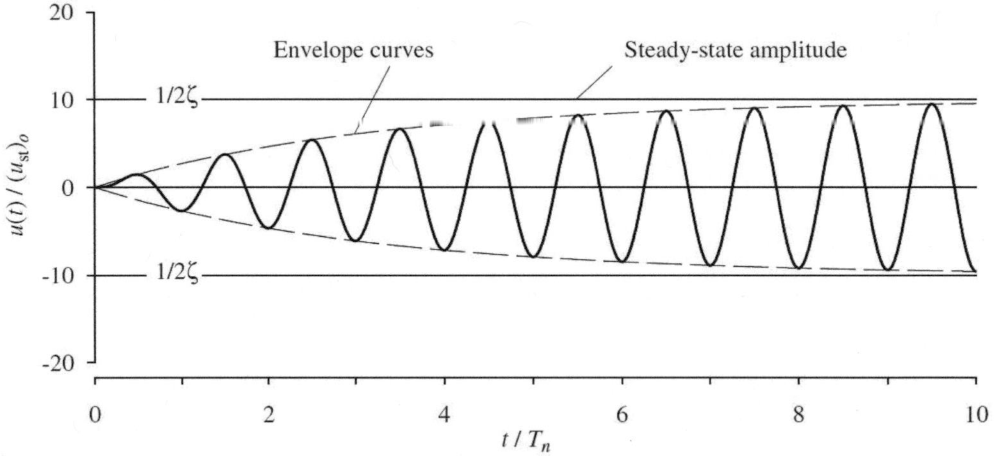

Fig. A-2 Response of damped system to harmonic response ($\omega = \omega_n$).

Maximum deformation

The steady state deformation of the system due to harmonic force, described by Equation A-3 can be rewritten as:

$$u(t) = u_0 \sin(\omega t - \phi) = (u_{st})_0 R_d \sin(\omega t - \phi) \tag{A-17}$$

where

$$u_0 = \sqrt{C^2 + D^2} \quad \textbf{(A-18)}$$

$$\phi = \tan^{-1}(-D/C) \quad \textbf{(A-19)}$$

Substituting for C and D gives

$$R_d = \frac{u_0}{(u_{st})_0} = \frac{1}{\sqrt{[1-(\omega/\omega_n)^2]^2 + [2\zeta(\omega/\omega_n)]^2}} \quad \textbf{(A-20)}$$

$$\phi = \tan^{-1}\frac{2\zeta(\omega/\omega_n)}{1-(\omega/\omega_n)^2} \quad \textbf{(A-21)}$$

Equation A-17 is plotted in Figure A-3 for three values of ω/ω_n and a fixed value of ζ=0.20. The values of R_d and φ computed from Equations A-20 and A-21 are identified. The dashed lines represent the static deformation due to p(t), which varies with time just as the applied force does, except for the constant k. The steady state motion is seen to occur in the forcing period $T=2\pi/\omega$, but with a time lag=$\varphi/2\pi$; φ is called the phase angle or phase lag.

A plot of the amplitude of a response quantity against the excitation frequency is called frequency response curve. Such a plot for the deformation u is given by Figure A-4, where the deformation response factor R_d (or amplification factor AF) is plotted as a function of ω/ω_n) for a few values of ζ.

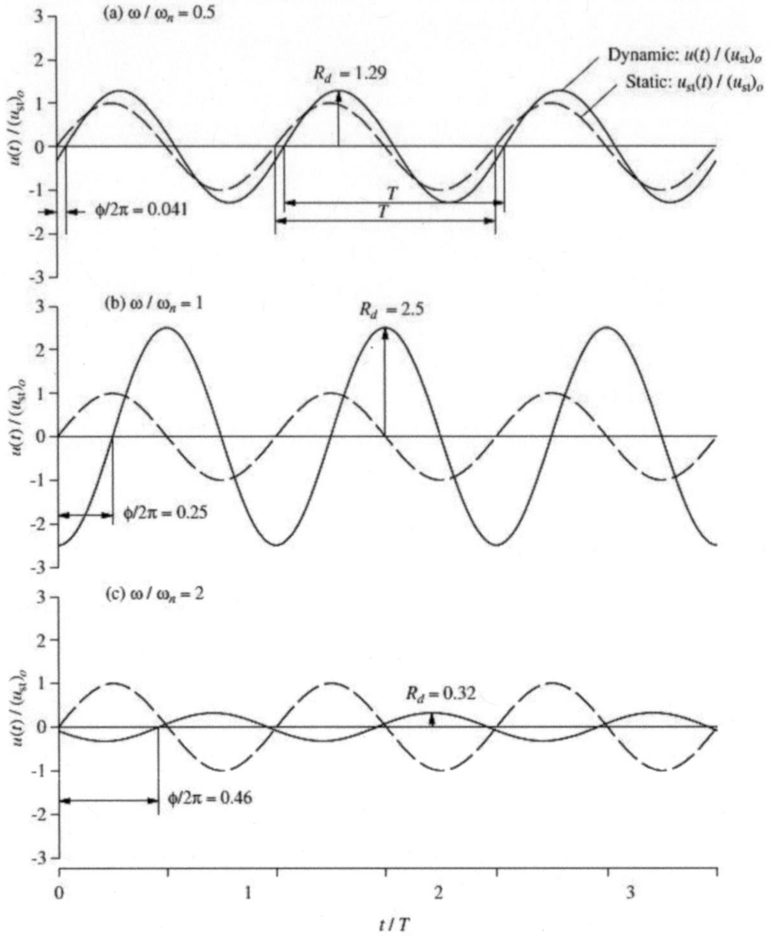

Fig. A-3 Steady state response of damped systems for three values of the frequency ratio.

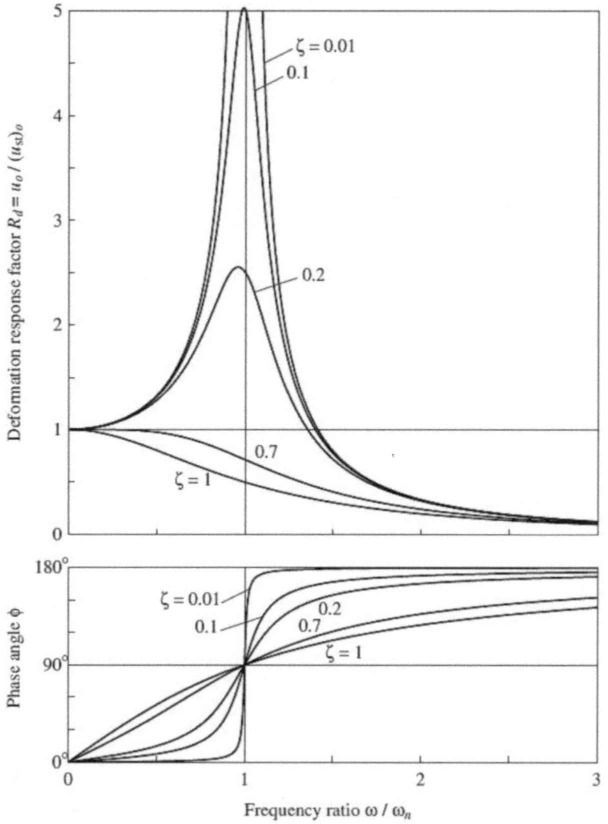

Fig. A-4 Deformation response factor and phase angle for a damped system excited by a harmonic force.

Dynamic Response Factors

The dynamic response factors of displacement, velocity and acceleration are dimensionless factors and define the amplitude of these three response quantities. The steady state displacement of Equation A-17 is:

$$\frac{u(t)}{p_0/k} = R_d \sin(\omega t - \phi) \tag{A-22}$$

Where the deformation response factor R_d is the ratio of the amplitude uo of the dynamic deformation to the static deformation $(u_{st})_0$.

Differentiating Equation A-22 gives an equation for the velocity response:

$$\frac{\dot{u}(t)}{p_0/\sqrt{km}} = R_v \cos(\omega t - \phi) \qquad \text{(A-23)}$$

Where the velocity response factor R_v is related to R_d by

$$R_v = \frac{\omega}{\omega_n} R_d \qquad \text{(A-24)}$$

Differentiating Equation A-23 gives an equation for the acceleration response:

$$\frac{\ddot{u}(t)}{p_0/m} = -R_a \sin(\omega t - \phi) \qquad \text{(A-25)}$$

Where the acceleration response factor R_a is related to R_d by

$$R_a = \left(\frac{\omega}{\omega_n}\right)^2 R_d \qquad \text{(A-26)}$$

Observe from Equation A-25 that R_a is the ratio of the amplitude of the dynamic (vibratory) acceleration to the acceleration due to the force p_o acting on the mass.

The dynamic response factors Rd, Rv and Ra are plotted as functions of ω/ω_n in Figure A-5.

The relation between the three response factors is given by:

$$\frac{R_a}{\omega/\omega_n} = R_v = \frac{\omega}{\omega_n} R_d \qquad \text{(A-27)}$$

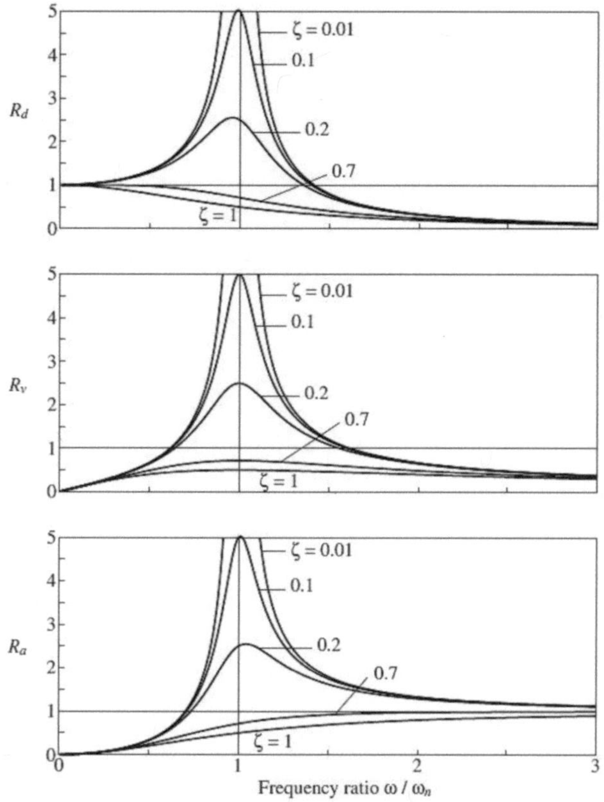

Fig. A-5 Deformation, velocity and acceleration response factors for a damped SDOF system excited by harmonic force.

Resonant frequencies and resonant responses

A resonant frequency is defined as the forcing frequency at which the largest response amplitude occurs. Figure A-5 shows that the peaks in the frequency response curves for displacement, velocity and acceleration occur at slightly different frequencies. These resonant frequencies can be determined by setting to zero the first derivative of Rd, Rv and Ra with respect to ω/ω_n; for $\zeta<1$ they are:

Displacement resonant frequency: $\omega_n\sqrt{1-\zeta^2}$ **(A-28)**

Velocity resonant frequency: ω_n **(A-29)**

Acceleration resonant frequency: $\omega_n/\sqrt{1-\zeta^2}$ **(A-30)**

For an undamped system the three resonant frequencies are identical and equal to the natural frequency ω_n. Intuition might suggest that the resonant frequency for a damped system should be at its natural frequency $\omega_D = \omega_n\sqrt{1-\zeta^2}$, but this does not happen.

The three dynamic response factors at their respective resonant frequencies are:

$$R_d = \frac{1}{2\zeta\sqrt{1-\zeta^2}}$$ **(A-31)**

$$R_v = \frac{1}{2\zeta}$$ **(A-32)**

$$R_a = \frac{1}{2\zeta\sqrt{1-\zeta^2}}$$ **(A-33)**

Annex B

Natural circular frequencies of bounded systems with flexible walls based on flexible base for different ratios of *L/H* (0.5, 1, 2, 3, 5 and 10). These natural frequencies are identiefied as the frequencies, where the maximum shear force due to the water pressures occur. The natural frequency of the unbounded reservoir with depth 10 m is 226 rad/sec. The values written in bold are theoritical values.

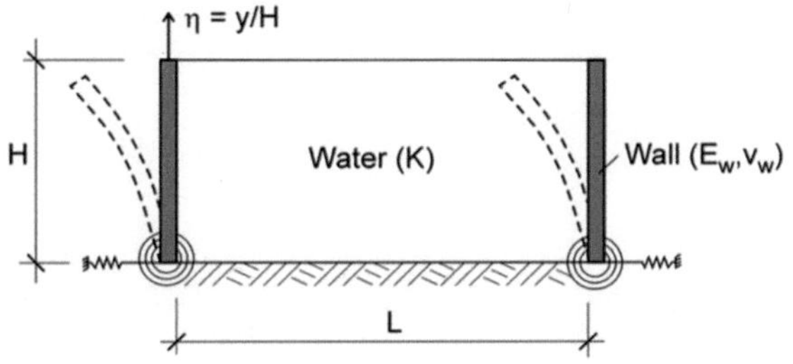

		L/H=0.5																
	dw	0	0.5	1	5	10	25	50	100		0	0.5	1	5	10	25	50	100
	dθ	ω_{11}	ω_{11}	ω_{11}	ω_{11}	ω_{11}	ω_{11}	ω_{11}	ω_{11}		ω_{11}/ω_1	ω_{11}/ω_1	ω_{11}/ω_1	ω_{11}/ω_1	ω_{11}/ω_1	ω_{11}/ω_1	ω_{11}/ω_1	ω_{11}/ω_1
dx=0	0	967.61	917.35	841.95	489.46	357.51	230.59	164.07	116.30		4.28	4.06	3.73	2.17	1.58	1.02	0.73	0.51
	0.5	666.02	625.18	587.48	417.83	326.10	221.80	160.30	115.04		2.95	2.77	2.60	1.85	1.44	0.98	0.71	0.51
	1	514.59	493.23	473.12	370.08	302.22	214.26	157.79	113.78		2.28	2.18	2.09	1.64	1.34	0.95	0.70	0.50
	5	246.93	244.42	241.90	223.05	205.46	170.36	137.67	106.24		1.09	1.08	1.07	0.99	0.91	0.75	0.61	0.47
	10	175.39	175.39	174.13	166.59	159.04	141.44	121.33	97.44		0.78	0.78	0.77	0.74	0.70	0.63	0.54	0.43
	25	111.27	111.27	111.27	108.75	107.5	101.21	92.41	81.09		0.49	0.49	0.49	0.48	0.48	0.45	0.41	0.36
	50	78.58	78.58	78.58	78.58	77.32	74.81	71.04	66.01		0.35	0.35	0.35	0.35	0.34	0.33	0.31	0.29
	100	55.95	55.95	55.95	55.95	55.95	54.69	53.43	50.92		0.25	0.25	0.25	0.25	0.25	0.24	0.24	0.23
dx=0.05	0	816.81	772.83	722.57	466.84	348.72	228.08	162.82	116.3		3.61	3.42	3.20	2.07	1.54	1.01	0.72	0.51
	0.5	603.81	572.4	543.5	402.75	319.81	219.28	160.3	115.04		2.67	2.53	2.40	1.78	1.42	0.97	0.71	0.51
	1	484.43	466.84	450.5	358.77	295.94	211.74	156.53	113.78		2.14	2.07	1.99	1.59	1.31	0.94	0.69	0.50
	5	243.16	240.65	238.13	220.54	204.2	169.1	137.67	104.98		1.08	1.06	1.05	0.98	0.90	0.75	0.61	0.46
	10	174.13	174.13	172.87	165.33	157.79	140.18	120.07	97.44		0.77	0.77	0.76	0.73	0.70	0.62	0.53	0.43
	25	111.27	111.27	111.27	108.75	106.24	101.21	92.41	81.09		0.49	0.49	0.49	0.48	0.47	0.45	0.41	0.36
	50	78.58	78.58	78.58	78.58	77.32	74.81	71.04	66.01		0.35	0.35	0.35	0.35	0.34	0.33	0.31	0.29
	100	55.95	55.95	55.95	55.95	55.95	54.69	53.43	50.92		0.25	0.25	0.25	0.25	0.25	0.24	0.24	0.23
dx=0.1	0	703.72	672.3	640.88	445.48	314.79	225.57	162.82	116.3		3.11	2.97	2.84	1.97	1.39	1.00	0.72	0.51
	0.5	551.04	528.42	507.05	388.93	312.27	218.03	159.04	115.04		2.44	2.34	2.24	1.72	1.38	0.96	0.70	0.51
	1	458.04	442.96	429.14	348.72	290.91	210.49	156.53	113.78		2.03	1.96	1.90	1.54	1.29	0.93	0.69	0.50
	5	239.39	236.88	235.62	218.03	201.69	167.84	136.41	104.98		1.06	1.05	1.04	0.96	0.89	0.74	0.60	0.46
	10	172.87	172.87	171.62	165.33	157.79	140.18	120.07	97.44		0.76	0.76	0.76	0.73	0.70	0.62	0.53	0.43
	25	111.27	111.27	110.01	108.75	106.24	101.21	92.41	81.09		0.49	0.49	0.49	0.48	0.47	0.45	0.41	0.36
	50	78.58	78.58	78.58	78.58	77.32	74.81	71.04	66.01		0.35	0.35	0.35	0.35	0.34	0.33	0.31	0.29
	100	55.95	55.95	55.95	55.95	54.69	54.69	53.43	50.92		0.25	0.25	0.25	0.25	0.24	0.24	0.24	0.23
dx=0.5	0	387.67	381.39	375.11	327.35	283.37	209.23	156.53	113.78		1.72	1.69	1.66	1.45	1.25	0.93	0.69	0.50
	0.5	356.26	349.97	344.95	304.73	267.04	201.69	152.76	112.52		1.58	1.55	1.53	1.35	1.18	0.89	0.68	0.50
	1	328.61	323.58	319.81	285.88	253.21	195.41	150.24	111.27		1.45	1.43	1.42	1.26	1.12	0.86	0.66	0.49
	5	216.77	214.26	213	200.43	187.87	160.3	132.64	103.72		0.96	0.95	0.94	0.89	0.83	0.71	0.59	0.46
	10	164.07	164.07	162.82	156.53	150.24	135.16	117.55	96.18		0.73	0.73	0.72	0.69	0.66	0.60	0.52	0.43
	25	108.75	108.75	108.75	106.24	104.98	98.7	91.15	79.84		0.48	0.48	0.48	0.47	0.46	0.44	0.40	0.35
	50	78.58	78.58	77.32	77.32	76.06	73.55	71.04	64.75		0.35	0.35	0.34	0.34	0.34	0.33	0.31	0.29
	100	55.95	55.95	55.95	54.69	54.69	54.69	52.18	50.92		0.25	0.25	0.25	0.24	0.24	0.24	0.23	0.23

L/H=0.5																	
	dw	0	0.5	1	5	10	25	50	100	0	0.5	1	5	10	25	50	100
	dθ	ω_{11}	ω_{11}	ω_{11}	ω_{11}	ω_{11}	ω_{11}	ω_{11}	ω_{11}	ω_{11}/ω_1	ω_{11}/ω_1	ω_{11}/ω_1	ω_{11}/ω_1	ω_{11}/ω_1	ω_{11}/ω_1	ω_{11}/ω_1	ω_{11}/ω_1
dx=1	0	282.12	279.6	277.09	258.24	235.62	190.38	147.73	110.01	1.25	1.24	1.23	1.14	1.04	0.84	0.65	0.49
	0.5	269.55	267.04	264.52	246.93	226.82	184.10	145.21	110.01	1.19	1.18	1.17	1.09	1.00	0.81	0.64	0.49
	1	258.24	255.73	253.21	236.88	218.03	180.33	143.96	108.75	1.14	1.13	1.12	1.05	0.96	0.80	0.64	0.48
	5	194.15	192.89	191.64	182.84	174.13	151.50	127.61	101.21	0.86	0.85	0.85	0.81	0.77	0.67	0.56	0.45
	10	154.01	154.01	152.76	147.73	142.7	130.13	113.78	93.67	0.68	0.68	0.68	0.65	0.63	0.58	0.50	0.41
	25	106.24	104.98	104.98	103.72	101.21	96.18	89.89	78.58	0.47	0.46	0.46	0.46	0.45	0.43	0.40	0.35
	50	77.32	77.32	77.32	76.06	74.81	73.55	69.78	64.75	0.34	0.34	0.34	0.34	0.33	0.33	0.31	0.29
	100	54.69	54.69	54.69	54.69	54.69	53.43	52.18	49.66	0.24	0.24	0.24	0.24	0.24	0.24	0.23	0.22
dx=2.5	0	181.58	181.58	180.33	175.39	169.10	151.50	128.87	102.47	0.80	0.80	0.80	0.78	0.75	0.67	0.57	0.45
	0.5	177.81	177.81	176.65	171.62	165.33	148.99	127.61	101.21	0.79	0.79	0.78	0.76	0.73	0.66	0.56	0.45
	1	175.39	174.13	174.13	167.84	162.82	146.47	126.35	101.21	0.78	0.77	0.77	0.74	0.72	0.65	0.56	0.45
	5	151.50	151.50	150.24	146.47	142.70	130.13	115.04	94.92	0.67	0.67	0.66	0.65	0.63	0.58	0.51	0.42
	10	131.38	131.38	130.13	127.61	125.10	116.3	104.98	88.64	0.58	0.58	0.58	0.56	0.55	0.51	0.46	0.39
	25	97.44	97.44	97.44	96.18	94.92	91.15	84.87	76.06	0.43	0.43	0.43	0.43	0.42	0.40	0.38	0.34
	50	73.55	73.55	73.55	73.55	72.29	71.04	67.26	62.23	0.33	0.33	0.33	0.33	0.32	0.31	0.30	0.28
	100	54.69	53.43	53.43	53.43	53.43	52.18	50.92	49.66	0.24	0.24	0.24	0.24	0.24	0.23	0.23	0.22
dx=5	0	128.87	128.87	128.87	126.35	123.84	117.55	107.5	91.15	0.57	0.57	0.57	0.56	0.55	0.52	0.48	0.40
	0.5	127.61	127.61	127.61	125.1	123.84	116.30	106.24	91.15	0.56	0.56	0.56	0.55	0.55	0.51	0.47	0.40
	1	126.35	126.35	126.35	123.84	122.58	115.04	104.98	89.89	0.56	0.56	0.56	0.55	0.54	0.51	0.46	0.40
	5	117.55	117.55	117.55	115.04	113.78	107.50	98.7	86.12	0.52	0.52	0.52	0.51	0.50	0.48	0.44	0.38
	10	107.50	107.50	107.50	106.24	103.72	99.95	92.41	81.09	0.48	0.48	0.48	0.47	0.46	0.44	0.41	0.36
	25	87.38	87.38	87.38	86.12	84.87	82.35	78.58	71.04	0.39	0.39	0.39	0.38	0.38	0.36	0.35	0.31
	50	69.78	69.78	69.78	68.52	68.52	67.26	64.75	59.72	0.31	0.31	0.31	0.30	0.30	0.30	0.29	0.26
	100	52.18	52.18	52.18	52.18	52.18	50.92	49.66	48.40	0.23	0.23	0.23	0.23	0.23	0.23	0.22	0.21
dx=10	0	91.15	91.15	91.15	91.15	89.89	87.38	83.61	76.06	0.40	0.40	0.40	0.40	0.40	0.39	0.37	0.34
	0.5	91.15	91.15	91.15	89.89	89.89	87.38	82.35	76.06	0.40	0.40	0.40	0.40	0.40	0.39	0.36	0.34
	1	91.15	91.15	91.15	89.89	88.64	86.12	82.35	74.81	0.40	0.40	0.40	0.40	0.39	0.38	0.36	0.33
	5	87.38	87.38	87.38	86.12	86.12	83.61	79.84	72.29	0.39	0.39	0.39	0.38	0.38	0.37	0.35	0.32
	10	83.61	83.61	83.61	82.35	81.09	79.84	76.06	69.78	0.37	0.37	0.37	0.36	0.36	0.35	0.34	0.31
	25	73.55	73.55	73.55	72.29	72.29	71.04	67.26	63.49	0.33	0.33	0.33	0.32	0.32	0.31	0.30	0.28
	50	62.23	62.23	62.23	60.98	60.98	59.72	58.46	54.69	0.28	0.28	0.28	0.27	0.27	0.26	0.26	0.24
	100	49.66	48.4	48.4	48.4	48.4	48.4	47.15	45.89	0.22	0.21	0.21	0.21	0.21	0.21	0.21	0.20

		L/H=1																
	dw	0	0.5	1	5	10	25	50	100		0	0.5	1	5	10	25	50	100
	dθ	ω_{11}	ω_{11}	ω_{11}	ω_{11}	ω_{11}	ω_{11}	ω_{11}	ω_{11}		ω_{11}/ω_1	ω_{11}/ω_1	ω_{11}/ω_1	ω_{11}/ω_1	ω_{11}/ω_1	ω_{11}/ω_1	ω_{11}/ω_1	ω_{11}/ω_1
dx=0	0	525.9	515.85	504.54	393.96	304.73	202.95	145.21	103.72		2.33	2.28	2.23	1.74	1.35	0.90	0.64	0.46
	0.5	458.04	445.48	431.65	343.69	279.6	194.15	142.7	102.47		2.03	1.97	1.91	1.52	1.24	0.86	0.63	0.45
	1	392.7	382.65	372.59	308.5	258.24	187.87	138.93	101.21		1.74	1.69	1.65	1.37	1.14	0.83	0.61	0.45
	5	209.23	206.72	205.46	191.64	177.81	148.99	121.33	93.67		0.93	0.91	0.91	0.85	0.79	0.66	0.54	0.41
	10	150.24	150.24	148.99	143.96	137.67	122.58	106.24	86.12		0.66	0.66	0.66	0.64	0.61	0.54	0.47	0.38
	25	96.18	96.18	96.18	94.92	92.41	87.38	81.09	71.04		0.43	0.43	0.43	0.42	0.41	0.39	0.36	0.31
	50	68.52	68.52	68.52	67.26	67.26	64.75	62.23	57.21		0.30	0.30	0.30	0.30	0.30	0.29	0.28	0.25
	100	48.40	48.40	48.40	48.40	48.40	47.15	45.89	44.63		0.21	0.21	0.21	0.21	0.21	0.21	0.20	0.20
dx=0.05	0	484.43	474.38	464.33	371.34	294.68	200.43	143.96	103.72		2.14	2.10	2.05	1.64	1.30	0.89	0.64	0.46
	0.5	424.12	412.81	402.75	328.61	272.06	192.89	141.44	102.47		1.88	1.83	1.78	1.45	1.20	0.85	0.63	0.45
	1	370.08	361.28	352.49	297.19	253.21	185.35	138.93	101.21		1.64	1.60	1.56	1.32	1.12	0.82	0.61	0.45
	5	205.46	202.95	201.69	189.12	175.39	147.73	120.07	93.67		0.91	0.90	0.89	0.84	0.78	0.65	0.53	0.41
	10	148.99	148.99	147.73	142.7	136.41	122.58	104.98	86.12		0.66	0.66	0.65	0.63	0.60	0.54	0.46	0.38
	25	96.18	96.18	96.18	93.67	92.41	87.38	81.09	71.04		0.43	0.43	0.43	0.41	0.41	0.39	0.36	0.31
	50	68.52	68.52	68.52	67.26	67.26	64.75	62.23	57.21		0.30	0.30	0.30	0.30	0.30	0.29	0.28	0.25
	100	48.40	48.40	48.40	48.40	48.40	47.15	45.89	44.63		0.21	0.21	0.21	0.21	0.21	0.21	0.20	0.20
dx=0.1	0	447.99	439.19	429.14	352.49	285.88	197.92	143.96	102.47		1.98	1.94	1.90	1.56	1.26	0.88	0.64	0.45
	0.5	395.21	386.42	377.62	316.04	264.52	190.38	140.18	102.47		1.75	1.71	1.67	1.40	1.17	0.84	0.62	0.45
	1	349.97	342.43	336.15	287.14	246.93	182.84	137.67	101.21		1.55	1.52	1.49	1.27	1.09	0.81	0.61	0.45
	5	201.69	200.43	199.18	186.61	172.87	146.47	120.07	93.67		0.89	0.89	0.88	0.83	0.76	0.65	0.53	0.41
	10	147.73	147.73	146.47	141.44	135.16	121.33	104.98	86.12		0.65	0.65	0.65	0.63	0.60	0.54	0.46	0.38
	25	96.18	96.18	94.92	93.67	92.41	87.38	81.09	71.04		0.43	0.43	0.42	0.41	0.41	0.39	0.36	0.31
	50	68.52	68.52	68.52	67.26	67.26	64.75	62.23	57.21		0.30	0.30	0.30	0.30	0.30	0.29	0.28	0.25
	100	48.40	48.40	48.40	48.40	48.40	47.15	45.89	44.63		0.21	0.21	0.21	0.21	0.21	0.21	0.20	0.20
dx=0.5	0	289.65	287.14	283.37	258.24	230.59	177.81	136.41	99.95		1.28	1.27	1.25	1.14	1.02	0.79	0.60	0.44
	0.5	272.06	269.55	265.78	243.16	219.28	172.87	133.9	98.70		1.20	1.19	1.18	1.08	0.97	0.76	0.59	0.44
	1	255.73	253.21	250.7	230.59	209.23	167.84	131.38	98.70		1.13	1.12	1.11	1.02	0.93	0.74	0.58	0.44
	5	179.07	177.81	177.81	169.10	159.04	137.67	115.04	91.15		0.79	0.79	0.79	0.75	0.70	0.61	0.51	0.40
	10	138.93	138.93	137.67	133.90	128.87	116.30	102.47	83.61		0.61	0.61	0.61	0.59	0.57	0.51	0.45	0.37
	25	93.67	93.67	92.41	91.15	89.89	84.87	78.58	69.78		0.41	0.41	0.41	0.40	0.40	0.38	0.35	0.31
	50	67.26	67.26	67.26	67.26	66.01	64.75	60.98	57.21		0.30	0.30	0.30	0.30	0.29	0.29	0.27	0.25
	100	48.40	48.40	48.40	48.40	47.15	47.15	45.89	43.38		0.21	0.21	0.21	0.21	0.21	0.21	0.20	0.19

		L/H=1															
	dw	0	0.5	1	5	10	25	50	100	0	0.5	1	5	10	25	50	100
	dθ	ω_{11}	ω_{11}	ω_{11}	ω_{11}	ω_{11}	ω_{11}	ω_{11}	ω_{11}	ω_{11}/ω_1	ω_{11}/ω_1	ω_{11}/ω_1	ω_{11}/ω_1	ω_{11}/ω_1	ω_{11}/ω_1	ω_{11}/ω_1	ω_{11}/ω_1
dx=1	0	218.03	216.77	215.51	204.2	190.38	159.04	127.61	97.44	0.96	0.96	0.95	0.90	0.84	0.70	0.56	0.43
	0.5	210.49	209.23	207.97	196.66	184.1	155.27	125.1	96.18	0.93	0.93	0.92	0.87	0.81	0.69	0.55	0.43
	1	202.95	201.69	200.43	189.12	177.81	151.5	123.84	94.92	0.90	0.89	0.89	0.84	0.79	0.67	0.55	0.42
	5	159.04	157.79	157.79	151.5	145.21	128.87	110.01	88.64	0.70	0.70	0.70	0.67	0.64	0.57	0.49	0.39
	10	128.87	128.87	128.87	125.1	121.33	111.27	98.7	82.35	0.57	0.57	0.57	0.55	0.54	0.49	0.44	0.36
	25	89.89	89.89	89.89	88.64	87.38	83.61	77.32	68.52	0.40	0.40	0.40	0.39	0.39	0.37	0.34	0.30
	50	66.01	66.01	66.01	66.01	64.75	63.49	60.98	55.95	0.29	0.29	0.29	0.29	0.29	0.28	0.27	0.25
	100	47.15	47.15	47.15	47.15	47.15	47.15	45.89	43.38	0.21	0.21	0.21	0.21	0.21	0.21	0.20	0.19
dx=2.5	0	143.96	142.7	142.7	138.93	135.16	123.84	108.75	88.64	0.64	0.63	0.63	0.61	0.60	0.55	0.48	0.39
	0.5	141.44	140.18	140.18	136.41	132.64	121.33	107.5	87.38	0.63	0.62	0.62	0.60	0.59	0.54	0.48	0.39
	1	138.93	138.93	137.67	135.16	130.13	120.07	106.24	87.38	0.61	0.61	0.61	0.60	0.58	0.53	0.47	0.39
	5	122.58	122.58	122.58	120.07	116.3	108.75	97.44	82.35	0.54	0.54	0.54	0.53	0.51	0.48	0.43	0.36
	10	107.5	107.5	107.5	106.24	103.72	97.44	88.64	76.06	0.48	0.48	0.48	0.47	0.46	0.43	0.39	0.34
	25	82.35	82.35	82.35	81.09	79.84	77.32	72.29	66.01	0.36	0.36	0.36	0.36	0.35	0.34	0.32	0.29
	50	63.49	63.49	63.49	62.23	62.23	60.98	58.46	54.69	0.28	0.28	0.28	0.28	0.28	0.27	0.26	0.24
	100	45.89	45.89	45.89	45.89	45.89	45.89	44.63	42.12	0.20	0.20	0.20	0.20	0.20	0.20	0.20	0.19
dx=5	0	102.47	102.47	102.47	101.21	99.95	94.92	88.64	77.32	0.45	0.45	0.45	0.45	0.44	0.42	0.39	0.34
	0.5	102.47	101.21	101.21	99.95	98.7	93.67	87.38	76.06	0.45	0.45	0.45	0.44	0.44	0.41	0.39	0.34
	1	101.21	101.21	101.21	99.95	97.44	93.67	87.38	76.06	0.45	0.45	0.45	0.44	0.43	0.41	0.39	0.34
	5	94.92	94.92	93.67	93.67	91.15	87.38	82.35	72.29	0.42	0.42	0.41	0.41	0.40	0.39	0.36	0.32
	10	87.38	87.38	87.38	86.12	84.87	82.35	77.32	68.52	0.39	0.39	0.39	0.38	0.38	0.36	0.34	0.30
	25	72.29	72.29	72.29	72.29	71.04	68.52	66.01	60.98	0.32	0.32	0.32	0.32	0.31	0.30	0.29	0.27
	50	58.46	58.46	58.46	58.46	57.21	55.95	54.69	50.92	0.26	0.26	0.26	0.26	0.25	0.25	0.24	0.23
	100	44.63	44.63	44.63	44.63	44.63	43.38	43.38	40.86	0.20	0.20	0.20	0.20	0.20	0.19	0.19	0.18
dx=10	0	73.55	73.55	73.55	72.29	72.29	69.78	67.26	62.23	0.33	0.33	0.33	0.32	0.32	0.31	0.30	0.28
	0.5	72.29	72.29	72.29	72.29	72.29	69.78	67.26	62.23	0.32	0.32	0.32	0.32	0.32	0.31	0.30	0.28
	1	72.29	72.29	72.29	72.29	71.04	69.78	67.26	62.23	0.32	0.32	0.32	0.32	0.31	0.31	0.30	0.28
	5	69.78	69.78	69.78	69.78	68.52	67.26	64.75	59.72	0.31	0.31	0.31	0.31	0.30	0.30	0.29	0.26
	10	67.26	67.26	67.26	67.26	66.01	64.75	62.23	58.46	0.30	0.30	0.30	0.30	0.29	0.29	0.28	0.26
	25	59.72	59.72	59.72	59.72	59.72	58.46	55.95	53.43	0.26	0.26	0.26	0.26	0.26	0.26	0.25	0.24
	50	50.92	50.92	50.92	50.92	50.92	49.66	48.4	47.15	0.23	0.23	0.23	0.23	0.23	0.22	0.21	0.21
	100	40.86	40.86	40.86	40.86	40.86	40.86	39.6	38.35	0.18	0.18	0.18	0.18	0.18	0.18	0.18	0.17

L/H=2																		
	dw	0	0.5	1	5	10	25	50	100		0	0.5	1	5	10	25	50	100
	dθ	ω_{11}	ω_{11}	ω_{11}	ω_{11}	ω_{11}	ω_{11}	ω_{11}	ω_{11}		ω_{11}/ω_1	ω_{11}/ω_1	ω_{11}/ω_1	ω_{11}/ω_1	ω_{11}/ω_1	ω_{11}/ω_1	ω_{11}/ω_1	ω_{11}/ω_1
dx=0	0	332.38	329.87	327.35	297.19	256.98	182.84	133.9	96.18		1.47	1.46	1.45	1.32	1.14	0.81	0.59	0.43
	0.5	313.53	309.76	305.99	274.58	239.39	176.65	131.38	94.92		1.39	1.37	1.35	1.21	1.06	0.78	0.58	0.42
	1	292.17	288.4	284.63	254.47	224.31	170.36	128.87	93.67		1.29	1.28	1.26	1.13	0.99	0.75	0.57	0.41
	5	182.84	181.58	180.33	170.36	159.04	135.16	111.27	87.38		0.81	0.80	0.80	0.75	0.70	0.60	0.49	0.39
	10	135.16	135.16	133.9	130.13	123.84	112.52	97.44	79.84		0.60	0.60	0.59	0.58	0.55	0.50	0.43	0.35
	25	87.38	87.38	87.38	86.12	84.87	79.84	74.81	66.01		0.39	0.39	0.39	0.38	0.38	0.35	0.33	0.29
	50	62.23	62.23	62.23	62.23	60.98	59.72	57.21	53.43		0.28	0.28	0.28	0.28	0.27	0.26	0.25	0.24
	100	44.63	44.63	44.63	44.63	43.38	43.38	42.12	40.86		0.20	0.20	0.20	0.20	0.19	0.19	0.19	0.18
dx=0.05	0	319.81	316.04	313.53	284.63	248.19	180.33	132.64	96.18		1.42	1.40	1.39	1.26	1.10	0.80	0.59	0.43
	0.5	299.71	297.19	293.42	264.52	231.85	174.13	130.13	94.92		1.33	1.32	1.30	1.17	1.03	0.77	0.58	0.42
	1	279.60	277.09	273.32	246.93	218.03	167.84	127.61	93.67		1.24	1.23	1.21	1.09	0.96	0.74	0.56	0.41
	5	180.33	179.07	177.81	167.84	156.53	133.90	111.27	86.12		0.80	0.79	0.79	0.74	0.69	0.59	0.49	0.38
	10	133.90	133.90	132.64	128.87	123.84	111.27	97.44	79.84		0.59	0.59	0.59	0.57	0.55	0.49	0.43	0.35
	25	87.38	87.38	87.38	86.12	83.61	79.84	73.55	66.01		0.39	0.39	0.39	0.38	0.37	0.35	0.33	0.29
	50	62.23	62.23	62.23	62.23	60.98	59.72	57.21	52.18		0.28	0.28	0.28	0.28	0.27	0.26	0.25	0.23
	100	44.63	44.63	44.63	44.63	43.38	43.38	42.12	40.86		0.20	0.20	0.20	0.20	0.19	0.19	0.19	0.18
dx=0.1	0	305.99	303.48	300.96	273.32	240.65	177.81	131.38	96.18		1.35	1.34	1.33	1.21	1.06	0.79	0.58	0.43
	0.5	288.4	284.63	280.86	254.47	225.57	171.62	123.87	94.92		1.28	1.26	1.24	1.13	1.00	0.76	0.57	0.42
	1	269.55	265.78	263.27	238.13	213	165.33	125.35	93.67		1.19	1.18	1.16	1.05	0.94	0.73	0.56	0.41
	5	176.65	175.39	174.13	165.33	155.27	132.64	110.01	86.12		0.78	0.78	0.77	0.73	0.69	0.59	0.49	0.38
	10	132.64	132.64	131.38	127.61	122.58	110.01	96.18	79.84		0.59	0.59	0.58	0.56	0.54	0.49	0.43	0.35
	25	87.38	87.38	86.12	84.87	83.61	79.84	73.55	64.75		0.39	0.39	0.38	0.38	0.37	0.35	0.33	0.29
	50	62.23	62.23	62.23	62.23	60.98	59.72	57.21	52.18		0.28	0.28	0.28	0.28	0.27	0.26	0.25	0.23
	100	44.63	44.63	44.63	44.63	43.38	43.38	42.12	40.86		0.20	0.20	0.20	0.20	0.19	0.19	0.19	0.18
dx=0.5	0	229.34	226.82	225.57	211.74	194.15	157.79	122.84	92.41		1.01	1.00	1.00	0.94	0.86	0.70	0.55	0.41
	0.5	219.28	216.77	215.51	201.69	186.61	154.01	121.33	91.15		0.97	0.96	0.95	0.89	0.83	0.68	0.54	0.40
	1	209.23	207.97	205.46	192.89	179.07	148.99	118.81	91.15		0.93	0.92	0.91	0.85	0.79	0.66	0.53	0.40
	5	156.53	155.27	154.01	147.73	141.44	123.84	104.98	83.61		0.69	0.69	0.68	0.65	0.63	0.55	0.46	0.37
	10	123.84	122.58	122.58	118.81	115.04	104.98	93.67	77.32		0.55	0.54	0.54	0.53	0.51	0.46	0.41	0.34
	25	84.87	84.87	83.61	82.35	81.09	77.32	72.29	64.75		0.38	0.38	0.37	0.36	0.36	0.34	0.32	0.29
	50	60.98	60.98	60.98	60.98	59.72	58.46	55.95	52.18		0.27	0.27	0.27	0.27	0.26	0.26	0.25	0.23
	100	43.38	43.38	43.38	43.38	43.38	43.38	42.12	39.6		0.19	0.19	0.19	0.19	0.19	0.19	0.19	0.18

L/H=2																	
	dw	0	0.5	1	5	10	25	50	100	0	0.5	1	5	10	25	50	100
	dθ	ω_{11}	ω_{11}	ω_{11}	ω_{11}	ω_{11}	ω_{11}	ω_{11}	ω_{11}	ω_{11}/ω_1	ω_{11}/ω_1	ω_{11}/ω_1	ω_{11}/ω_1	ω_{11}/ω_1	ω_{11}/ω_1	ω_{11}/ω_1	ω_{11}/ω_1
dx=1	0	180.33	179.07	177.81	170.36	161.56	140.18	115.04	88.64	0.80	0.79	0.79	0.75	0.71	0.62	0.51	0.39
	0.5	175.39	174.13	172.87	165.33	157.79	136.41	113.78	88.64	0.78	0.77	0.76	0.73	0.70	0.60	0.50	0.39
	1	169.1	169.1	167.84	161.56	152.76	133.9	111.27	87.38	0.75	0.75	0.74	0.71	0.68	0.59	0.49	0.39
	5	137.67	137.67	136.41	132.64	127.61	115.04	99.95	81.09	0.61	0.61	0.60	0.59	0.56	0.51	0.44	0.36
	10	113.78	113.78	113.78	111.27	107.5	99.95	88.64	74.81	0.50	0.50	0.50	0.49	0.48	0.44	0.39	0.33
	25	81.09	81.09	81.09	79.84	78.58	74.81	71.04	63.49	0.36	0.36	0.36	0.35	0.35	0.33	0.31	0.28
	50	59.72	59.72	59.72	59.72	58.46	57.21	54.69	50.92	0.26	0.26	0.26	0.26	0.26	0.25	0.24	0.23
	100	43.38	43.38	43.38	43.38	43.38	42.12	42.12	39.6	0.19	0.19	0.19	0.19	0.19	0.19	0.19	0.18
dx=2.5	0	122.58	121.33	121.33	118.81	116.3	107.5	96.18	79.84	0.54	0.54	0.54	0.53	0.51	0.48	0.43	0.35
	0.5	120.07	120.07	120.07	117.55	113.78	106.24	94.92	78.58	0.53	0.53	0.53	0.52	0.50	0.47	0.42	0.35
	1	118.81	118.81	117.55	115.04	112.52	104.98	93.67	78.58	0.53	0.53	0.52	0.51	0.50	0.46	0.41	0.35
	5	106.24	106.24	106.24	103.72	101.21	94.92	86.12	73.55	0.47	0.47	0.47	0.46	0.45	0.42	0.38	0.33
	10	94.92	94.92	94.92	92.41	91.15	86.12	79.84	69.78	0.42	0.42	0.42	0.41	0.40	0.38	0.35	0.31
	25	73.55	73.55	73.55	72.29	72.29	69.78	66.01	59.72	0.33	0.33	0.33	0.32	0.32	0.31	0.29	0.26
	50	57.21	57.21	57.21	55.95	55.95	54.69	53.43	49.66	0.25	0.25	0.25	0.25	0.25	0.24	0.24	0.22
	100	42.12	42.12	42.12	42.12	42.12	40.86	40.86	38.35	0.19	0.19	0.19	0.19	0.19	0.18	0.18	0.17
dx=5	0	88.64	88.64	88.64	87.38	86.12	82.35	77.32	68.52	0.39	0.39	0.39	0.39	0.38	0.36	0.34	0.30
	0.5	87.38	87.38	87.38	86.12	84.87	82.35	76.06	68.52	0.39	0.39	0.39	0.38	0.38	0.36	0.34	0.30
	1	87.38	87.38	87.38	86.12	84.87	81.09	76.06	67.26	0.39	0.39	0.39	0.38	0.38	0.36	0.34	0.30
	5	82.35	82.35	81.09	81.09	79.84	77.32	72.29	64.75	0.36	0.36	0.36	0.36	0.35	0.34	0.32	0.29
	10	76.06	76.06	76.06	76.06	74.81	72.29	68.52	62.23	0.34	0.34	0.34	0.34	0.33	0.32	0.30	0.28
	25	64.75	64.75	64.75	63.49	63.49	60.98	58.46	54.69	0.29	0.29	0.29	0.28	0.28	0.27	0.26	0.24
	50	52.18	52.18	52.18	52.18	52.18	50.92	49.66	47.15	0.23	0.23	0.23	0.23	0.23	0.23	0.22	0.21
	100	40.86	40.86	40.86	40.86	39.60	39.60	38.35	37.09	0.18	0.18	0.18	0.18	0.18	0.18	0.17	0.16
dx=10	0	63.49	63.49	63.49	62.23	62.23	60.98	58.46	54.69	0.28	0.28	0.28	0.28	0.28	0.27	0.26	0.24
	0.5	63.49	63.49	63.49	62.23	62.23	60.98	58.46	54.69	0.28	0.28	0.28	0.28	0.28	0.27	0.26	0.24
	1	62.23	62.23	62.23	62.23	62.23	60.98	58.46	54.69	0.28	0.28	0.28	0.28	0.28	0.27	0.26	0.24
	5	60.98	60.98	60.98	59.72	59.72	58.46	57.21	53.43	0.27	0.27	0.27	0.26	0.26	0.26	0.25	0.24
	10	58.46	58.46	58.46	58.46	57.21	55.95	54.69	50.92	0.26	0.26	0.26	0.26	0.25	0.25	0.24	0.23
	25	52.18	52.18	52.18	52.18	52.18	50.92	49.66	47.15	0.23	0.23	0.23	0.23	0.23	0.23	0.22	0.21
	50	45.89	45.89	45.89	45.89	45.89	44.63	43.38	42.12	0.20	0.20	0.20	0.20	0.20	0.20	0.19	0.19
	100	37.09	37.09	37.09	37.09	37.09	37.09	35.83	34.57	0.16	0.16	0.16	0.16	0.16	0.16	0.16	0.15

L/H=3																		
	dw	0	0.5	1	5	10	25	50	100		0	0.5	1	5	10	25	50	100
	dθ	ω_{11}	ω_{11}	ω_{11}	ω_{11}	ω_{11}	ω_{11}	ω_{11}	ω_{11}		ω_{11}/ω_1	ω_{11}/ω_1	ω_{11}/ω_1	ω_{11}/ω_1	ω_{11}/ω_1	ω_{11}/ω_1	ω_{11}/ω_1	ω_{11}/ω_1
dx=0	0	282.12	280.86	279.6	263.27	236.88	177.81	131.38	94.92		1.25	1.24	1.24	1.16	1.05	0.79	0.58	0.42
	0.5	272.06	269.55	268.29	249.44	224.31	170.36	128.87	93.67		1.20	1.19	1.19	1.10	0.99	0.75	0.57	0.41
	1	259.50	256.98	254.47	235.62	211.74	164.07	125.10	92.41		1.15	1.14	1.13	1.04	0.94	0.73	0.55	0.41
	5	175.39	174.13	172.87	164.07	154.01	131.38	108.75	84.87		0.78	0.77	0.76	0.73	0.68	0.58	0.48	0.38
	10	131.38	131.38	130.13	126.35	121.33	108.75	94.92	78.58		0.58	0.58	0.58	0.56	0.54	0.48	0.42	0.35
	25	86.12	86.12	84.87	83.61	82.35	78.58	72.29	64.75		0.38	0.38	0.38	0.37	0.36	0.35	0.32	0.29
	50	60.98	60.98	60.98	60.98	59.72	58.46	55.95	52.18		0.27	0.27	0.27	0.27	0.26	0.26	0.25	0.23
	100	43.38	43.38	43.38	43.38	43.38	42.12	40.86	39.60		0.19	0.19	0.19	0.19	0.19	0.19	0.18	0.18
dx=0.05	0	274.58	273.32	272.06	254.47	230.59	174.13	130.13	93.67		1.21	1.21	1.20	1.13	1.02	0.77	0.58	0.41
	0.5	263.27	262.01	259.5	241.9	218.03	167.84	127.61	93.67		1.16	1.16	1.15	1.07	0.96	0.74	0.56	0.41
	1	250.70	249.44	246.93	228.08	206.72	162.82	125.1	92.41		1.11	1.10	1.09	1.01	0.91	0.72	0.55	0.41
	5	172.87	171.62	170.36	161.56	151.5	130.13	108.75	84.87		0.76	0.76	0.75	0.71	0.67	0.58	0.48	0.38
	10	130.13	130.13	128.87	125.10	120.07	108.75	94.92	78.58		0.58	0.58	0.57	0.55	0.53	0.48	0.42	0.35
	25	84.87	84.87	84.87	83.61	82.35	78.58	72.29	64.75		0.38	0.38	0.38	0.37	0.36	0.35	0.32	0.29
	50	60.98	60.98	60.98	60.98	59.72	58.46	55.95	52.18		0.27	0.27	0.27	0.27	0.26	0.26	0.25	0.23
	100	43.38	43.38	43.38	43.38	43.38	42.12	40.86	39.60		0.19	0.19	0.19	0.19	0.19	0.19	0.18	0.18
dx=0.1	0	267.04	265.78	263.27	246.93	224.31	171.62	128.87	93.67		1.18	1.18	1.16	1.09	0.99	0.76	0.57	0.41
	0.5	255.73	253.21	251.96	234.36	213.00	165.33	126.35	92.41		1.13	1.12	1.11	1.04	0.94	0.73	0.56	0.41
	1	243.16	241.90	239.39	221.80	201.69	160.30	123.84	92.41		1.08	1.07	1.06	0.98	0.89	0.71	0.55	0.41
	5	170.36	169.10	167.84	159.04	150.24	128.87	107.50	84.87		0.75	0.75	0.74	0.70	0.66	0.57	0.48	0.38
	10	128.87	128.87	127.61	123.84	118.81	107.50	94.92	77.32		0.57	0.57	0.56	0.55	0.53	0.48	0.42	0.34
	25	84.87	84.87	84.87	83.61	82.35	78.58	72.29	63.49		0.38	0.38	0.38	0.37	0.36	0.35	0.32	0.28
	50	60.98	60.98	60.98	60.98	59.72	58.46	55.95	52.18		0.27	0.27	0.27	0.27	0.26	0.26	0.25	0.23
	100	43.38	43.38	43.38	43.38	43.38	42.12	40.86	39.60		0.19	0.19	0.19	0.19	0.19	0.19	0.18	0.18
dx=0.5	0	211.74	210.49	209.23	197.92	184.1	152.76	121.33	91.15		0.94	0.93	0.93	0.88	0.81	0.68	0.54	0.40
	0.5	204.20	202.95	201.69	190.38	177.81	147.73	118.81	89.89		0.90	0.90	0.89	0.84	0.79	0.65	0.53	0.40
	1	196.66	195.41	194.15	182.84	171.62	143.96	116.3	88.64		0.87	0.86	0.86	0.81	0.76	0.64	0.51	0.39
	5	150.24	148.99	148.99	142.7	136.41	120.07	102.47	82.35		0.66	0.66	0.66	0.63	0.60	0.53	0.45	0.36
	10	120.07	120.07	118.81	116.3	112.52	102.47	91.15	76.06		0.53	0.53	0.53	0.51	0.50	0.45	0.40	0.34
	25	82.35	82.35	82.35	81.09	79.84	76.06	71.04	63.49		0.36	0.36	0.36	0.36	0.35	0.34	0.31	0.28
	50	59.72	59.72	59.72	59.72	58.46	57.21	54.69	50.92		0.26	0.26	0.26	0.26	0.26	0.25	0.24	0.23
	100	43.38	43.38	43.38	43.38	42.12	42.12	40.86	39.6		0.19	0.19	0.19	0.19	0.19	0.19	0.18	0.18

		L/H=3																
	dw	0	0.5	1	5	10	25	50	100		0	0.5	1	5	10	25	50	100
	dθ	ω_{11}	ω_{11}	ω_{11}	ω_{11}	ω_{11}	ω_{11}	ω_{11}	ω_{11}		ω_{11}/ω_1	ω_{11}/ω_1	ω_{11}/ω_1	ω_{11}/ω_1	ω_{11}/ω_1	ω_{11}/ω_1	ω_{11}/ω_1	ω_{11}/ω_1
dx=1	0	170.36	170.36	169.1	162.82	155.27	135.16	112.52	87.38		0.75	0.75	0.75	0.72	0.69	0.60	0.50	0.39
	0.5	166.59	165.33	164.07	157.79	150.24	131.38	110.01	86.12		0.74	0.73	0.73	0.70	0.66	0.58	0.49	0.38
	1	161.56	160.3	160.3	154.01	146.47	128.87	108.75	84.87		0.71	0.71	0.71	0.68	0.65	0.57	0.48	0.38
	5	132.64	132.64	131.38	127.61	123.84	111.27	97.44	79.84		0.59	0.59	0.58	0.56	0.55	0.49	0.43	0.35
	10	111.27	110.01	110.01	107.5	104.98	97.44	87.38	73.55		0.49	0.49	0.49	0.48	0.46	0.43	0.39	0.33
	25	79.84	79.84	78.58	78.58	77.32	73.55	68.52	62.23		0.35	0.35	0.35	0.35	0.34	0.33	0.30	0.28
	50	58.46	58.46	58.46	58.46	57.21	55.95	54.69	50.92		0.26	0.26	0.26	0.26	0.25	0.25	0.24	0.23
	100	42.12	42.12	42.12	42.12	42.12	42.12	40.86	39.6		0.19	0.19	0.19	0.19	0.19	0.19	0.18	0.18
dx=2.5	0	117.55	117.55	116.3	115.04	111.27	103.72	93.67	77.32		0.52	0.52	0.51	0.51	0.49	0.46	0.41	0.34
	0.5	116.30	115.04	115.04	112.52	110.01	102.47	92.41	77.32		0.51	0.51	0.51	0.50	0.49	0.45	0.41	0.34
	1	113.78	113.78	113.78	111.27	108.75	101.21	91.15	76.06		0.50	0.50	0.50	0.49	0.48	0.45	0.40	0.34
	5	102.47	102.47	102.47	99.95	98.7	92.41	84.87	72.29		0.45	0.45	0.45	0.44	0.44	0.41	0.38	0.32
	10	92.41	91.15	91.15	89.89	88.64	83.61	77.32	68.52		0.41	0.40	0.40	0.40	0.39	0.37	0.34	0.30
	25	72.29	72.29	71.04	71.04	69.78	67.26	63.49	58.46		0.32	0.32	0.31	0.31	0.31	0.30	0.28	0.26
	50	55.95	55.95	55.95	54.69	54.69	53.43	52.18	48.4		0.25	0.25	0.25	0.24	0.24	0.24	0.23	0.21
	100	40.86	40.86	40.86	40.86	40.86	40.86	39.60	38.35		0.18	0.18	0.18	0.18	0.18	0.18	0.18	0.17
dx=5	0	84.87	84.87	84.87	84.87	83.61	79.84	74.81	67.26		0.38	0.38	0.38	0.38	0.37	0.35	0.33	0.30
	0.5	84.87	84.87	84.87	83.61	82.35	79.84	74.81	66.01		0.38	0.38	0.38	0.37	0.36	0.35	0.33	0.29
	1	83.61	83.61	83.61	83.61	82.35	78.58	73.55	66.01		0.37	0.37	0.37	0.37	0.36	0.35	0.33	0.29
	5	79.84	78.58	78.58	78.58	77.32	74.81	69.78	63.49		0.35	0.35	0.35	0.35	0.34	0.33	0.31	0.28
	10	73.55	73.55	73.55	73.55	72.29	69.78	66.01	59.72		0.33	0.33	0.33	0.33	0.32	0.31	0.29	0.26
	25	62.23	62.23	62.23	62.23	60.98	59.72	57.21	53.43		0.28	0.28	0.28	0.28	0.27	0.26	0.25	0.24
	50	50.92	50.92	50.92	50.92	50.92	49.66	48.40	45.89		0.23	0.23	0.23	0.23	0.23	0.22	0.21	0.20
	100	39.60	39.60	39.60	39.60	39.60	38.35	38.35	37.09		0.18	0.18	0.18	0.18	0.18	0.17	0.17	0.16
dx=10	0	60.98	60.98	60.98	60.98	60.98	59.72	57.21	53.43		0.27	0.27	0.27	0.27	0.27	0.26	0.25	0.24
	0.5	60.98	60.98	60.98	60.98	59.72	58.46	57.21	53.43		0.27	0.27	0.27	0.27	0.26	0.26	0.25	0.24
	1	60.98	60.98	60.98	60.98	59.72	58.46	57.21	53.43		0.27	0.27	0.27	0.27	0.26	0.26	0.25	0.24
	5	58.46	58.46	58.46	58.46	58.46	57.21	54.69	52.18		0.26	0.26	0.26	0.26	0.26	0.25	0.24	0.23
	10	57.21	57.21	57.21	55.95	55.95	54.69	53.43	49.66		0.25	0.25	0.25	0.25	0.25	0.24	0.24	0.22
	25	50.92	50.92	50.92	50.92	50.92	49.66	48.40	45.89		0.23	0.23	0.23	0.23	0.23	0.22	0.21	0.20
	50	44.63	44.63	44.63	44.63	44.63	43.38	42.12	40.86		0.20	0.20	0.20	0.20	0.20	0.19	0.19	0.18
	100	35.83	35.83	35.83	35.83	35.83	35.83	34.57	34.57		0.16	0.16	0.16	0.16	0.16	0.16	0.15	0.15

L/H=5										
	dw	0	25	50	100		0	25	50	100
	dθ	ω_{11}	ω_{11}	ω_{11}	ω_{11}		ω_{11}/ω_1	ω_{11}/ω_1	ω_{11}/ω_1	ω_{11}/ω_1
dx=0	0	253.21	174.13	130.13	93.67		1.12	0.77	0.58	0.41
	25	84.87	78.58	72.29	63.49		0.38	0.35	0.32	0.28
	50	60.98	58.46	55.95	52.18		0.27	0.26	0.25	0.23
	100	43.38	42.12	40.86	39.60		0.19	0.19	0.18	0.18
dx=2.5	0	116.30	102.47	92.41	77.32		0.51	0.45	0.41	0.34
	25	71.04	67.26	63.49	58.46		0.31	0.30	0.28	0.26
	50	54.69	53.43	50.92	48.40		0.24	0.24	0.23	0.21
	100	40.86	40.86	39.60	38.35		0.18	0.18	0.18	0.17
dx=5	0	84.87	78.58	74.81	66.01		0.38	0.35	0.33	0.29
	25	62.23	59.72	57.21	53.43		0.28	0.26	0.25	0.24
	50	50.92	49.66	48.40	45.89		0.23	0.22	0.21	0.20
	100	39.60	38.35	38.35	37.09		0.18	0.17	0.17	0.16
dx=10	0	60.98	58.46	57.21	53.43		0.27	0.26	0.25	0.24
	25	50.92	49.66	48.40	45.89		0.23	0.22	0.21	0.20
	50	44.63	43.38	42.12	40.86		0.20	0.19	0.19	0.18
	100	35.83	35.83	34.57	33.32		0.16	0.16	0.15	0.15

L/H=10										
	dw	0	25	50	100		0	25	50	100
	dθ	ω_{11}	ω_{11}	ω_{11}	ω_{11}		ω_{11}/ω_1	ω_{11}/ω_1	ω_{11}/ω_1	ω_{11}/ω_1
dx=0	0	239.39	174.13	130.13	93.67		1.06	0.77	0.58	0.41
	25	84.87	78.58	72.29	63.49		0.38	0.35	0.32	0.28
	50	60.98	58.46	55.95	52.18		0.27	0.26	0.25	0.23
	100	43.38	42.12	40.86	39.6		0.19	0.19	0.18	0.18
dx=2.5	0	116.30	102.47	92.41	77.32		0.51	0.45	0.41	0.34
	25	71.04	67.26	63.49	58.46		0.31	0.30	0.28	0.26
	50	54.69	53.43	50.92	48.40		0.24	0.24	0.23	0.21
	100	40.86	40.86	39.60	38.35		0.18	0.18	0.18	0.17
dx=5	0	84.87	78.58	74.81	66.01		0.38	0.35	0.33	0.29
	25	62.23	59.72	57.21	53.43		0.28	0.26	0.25	0.24
	50	50.92	49.66	48.40	44.63		0.23	0.22	0.21	0.20
	100	39.60	38.35	38.35	37.09		0.18	0.17	0.17	0.16
dx=10	0	60.98	58.46	57.21	53.43		0.27	0.26	0.25	0.24
	25	50.92	49.66	48.40	45.89		0.23	0.22	0.21	0.20
	50	44.63	43.38	42.12	40.86		0.20	0.19	0.19	0.18
	100	35.83	35.83	34.57	33.32		0.16	0.16	0.15	0.15

Natural circular frequencies of U-section systems with flexible walls based on flexible base for different ratios of *L/H* (0.5, 1, 2, 3, 5 and 10). These natural frequencies are identiefied as the frequencies, where the maximum shear force due to the water pressures occur. The natural frequency of the unbounded reservoir with depth 10 m is 226 rad/sec. The values written in bold are theoritical values.

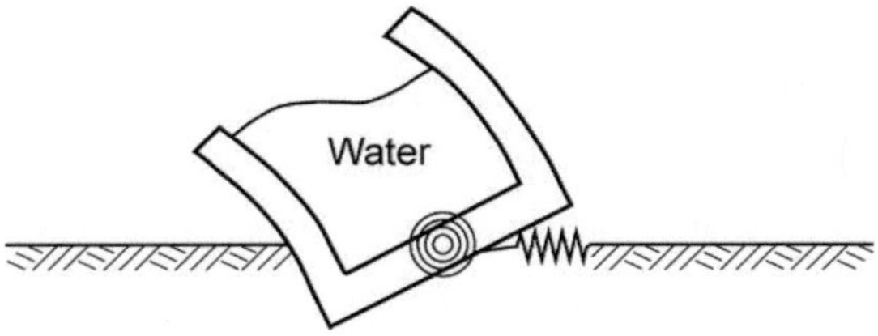

L/H=0.5																	
	dw	0	0.5	1	5	10	25	50	100	0	0.5	1	5	10	25	50	100
	dθ	ω_{11}	ω_{11}	ω_{11}	ω_{11}	ω_{11}	ω_{11}	ω_{11}	ω_{11}	ω_{11}/ω_1	ω_{11}/ω_1	ω_{11}/ω_1	ω_{11}/ω_1	ω_{11}/ω_1	ω_{11}/ω_1	ω_{11}/ω_1	ω_{11}/ω_1
dx=0	0	967.61	917.35	841.95	489.46	357.51	230.59	164.07	116.30	4.28	4.06	3.73	2.17	1.58	1.02	0.73	0.51
	0.5	498.26	479.41	461.81	366.31	300.96	213.00	157.79	113.78	2.20	2.12	2.04	1.62	1.33	0.94	0.70	0.50
	1	366.31	358.77	351.23	303.48	263.27	199.18	151.50	111.27	1.62	1.59	1.55	1.34	1.16	0.88	0.67	0.49
	5	169.10	169.10	167.84	161.56	155.27	138.93	120.07	97.44	0.75	0.75	0.74	0.71	0.69	0.61	0.53	0.43
	10	120.07	120.07	120.07	117.55	115.04	107.50	98.70	84.87	0.53	0.53	0.53	0.52	0.51	0.48	0.44	0.38
	25	76.06	76.06	76.06	76.06	74.81	73.55	69.78	64.75	0.34	0.34	0.34	0.34	0.33	0.33	0.31	0.29
	50	54.69	54.69	54.69	54.69	53.43	53.43	52.18	49.66	0.24	0.24	0.24	0.24	0.24	0.24	0.23	0.22
	100	38.35	38.35	38.35	38.35	38.35	38.35	38.35	37.09	0.17	0.17	0.17	0.17	0.17	0.17	0.17	0.16
dx=0.05	0	697.43	666.02	634.6	444.22	339.92	225.57	162.82	116.3	3.09	2.95	2.81	1.97	1.50	1.00	0.72	0.51
	0.5	445.48	431.65	419.09	344.95	289.65	209.23	156.53	113.78	1.97	1.91	1.85	1.53	1.28	0.93	0.69	0.50
	1	344.95	338.66	332.38	290.91	255.73	195.41	150.24	111.27	1.53	1.50	1.47	1.29	1.13	0.86	0.66	0.49
	5	166.59	166.59	165.33	160.3	152.76	137.67	118.81	96.18	0.74	0.74	0.73	0.71	0.68	0.61	0.53	0.43
	10	120.07	118.81	118.81	116.3	113.78	107.5	97.44	84.87	0.53	0.53	0.53	0.51	0.50	0.48	0.43	0.38
	25	76.06	76.06	76.06	76.06	74.81	72.29	69.78	64.75	0.34	0.34	0.34	0.34	0.33	0.32	0.31	0.29
	50	54.69	54.69	54.69	54.69	53.43	53.43	52.18	49.66	0.24	0.24	0.24	0.24	0.24	0.24	0.23	0.22
	100	38.35	38.35	38.35	38.35	38.35	38.35	38.35	37.09	0.17	0.17	0.17	0.17	0.17	0.17	0.17	0.16
dx=0.1	0	557.32	539.73	522.13	405.27	323.58	221.8	160.3	115.04	2.47	2.39	2.31	1.79	1.43	0.98	0.71	0.51
	0.5	402.75	393.96	385.16	326.1	278.35	205.46	154.01	112.52	1.78	1.74	1.70	1.44	1.23	0.91	0.68	0.50
	1	324.84	319.81	314.79	279.6	248.19	192.89	148.99	110.01	1.44	1.42	1.39	1.24	1.10	0.85	0.66	0.49
	5	165.33	164.07	164.07	157.79	151.5	136.41	117.55	96.18	0.73	0.73	0.73	0.70	0.67	0.60	0.52	0.43
	10	118.81	118.81	118.81	116.3	113.78	106.24	97.44	84.87	0.53	0.53	0.53	0.51	0.50	0.47	0.43	0.38
	25	76.06	76.06	76.06	74.81	74.81	72.29	69.78	64.75	0.34	0.34	0.34	0.33	0.33	0.32	0.31	0.29
	50	54.69	54.69	54.69	53.43	53.43	53.43	52.18	49.66	0.24	0.24	0.24	0.24	0.24	0.24	0.23	0.22
	100	38.35	38.35	38.35	38.35	38.35	38.35	38.35	37.09	0.17	0.17	0.17	0.17	0.17	0.17	0.17	0.16
dx=0.5	0	277.09	274.58	272.06	254.47	234.36	189.12	147.73	110.01	1.23	1.21	1.20	1.13	1.04	0.84	0.65	0.49
	0.5	251.96	250.7	248.19	233.11	216.77	179.07	142.7	108.75	1.11	1.11	1.10	1.03	0.96	0.79	0.63	0.48
	1	231.85	229.34	228.08	215.51	201.69	170.36	138.93	106.24	1.03	1.01	1.01	0.95	0.89	0.75	0.61	0.47
	5	148.99	148.99	148.99	143.96	140.18	127.61	112.52	93.67	0.66	0.66	0.66	0.64	0.62	0.56	0.50	0.41
	10	112.52	112.52	112.52	110.01	108.75	102.47	93.67	82.35	0.50	0.50	0.50	0.49	0.48	0.45	0.41	0.36
	25	74.81	74.81	74.81	73.55	73.55	71.04	68.52	63.49	0.33	0.33	0.33	0.33	0.33	0.31	0.30	0.28
	50	53.43	53.43	53.43	53.43	53.43	52.18	50.92	49.66	0.24	0.24	0.24	0.24	0.24	0.23	0.23	0.22
	100	38.35	38.35	38.35	38.35	38.35	38.35	37.09	37.09	0.17	0.17	0.17	0.17	0.17	0.17	0.16	0.16

		L/H=0.5																
	dw	0	0.5	1	5	10	25	50	100		0	0.5	1	5	10	25	50	100
	dθ	ω_{11}	ω_{11}	ω_{11}	ω_{11}	ω_{11}	ω_{11}	ω_{11}	ω_{11}		ω_{11}/ω_1	ω_{11}/ω_1	ω_{11}/ω_1	ω_{11}/ω_1	ω_{11}/ω_1	ω_{11}/ω_1	ω_{11}/ω_1	ω_{11}/ω_1
dx=1	0	197.92	197.92	196.66	190.38	181.58	160.3	133.9	104.98		0.88	0.88	0.87	0.84	0.80	0.71	0.59	0.46
	0.5	189.12	187.87	187.87	181.58	174.13	154.01	130.13	103.72		0.84	0.83	0.83	0.80	0.77	0.68	0.58	0.46
	1	180.33	179.07	179.07	172.87	166.59	148.99	127.61	101.21		0.80	0.79	0.79	0.76	0.74	0.66	0.56	0.45
	5	133.9	133.9	133.9	130.13	127.61	118.81	106.24	89.89		0.59	0.59	0.59	0.58	0.56	0.53	0.47	0.40
	10	106.24	106.24	106.24	103.72	102.47	97.44	89.89	79.84		0.47	0.47	0.47	0.46	0.45	0.43	0.40	0.35
	25	72.29	72.29	72.29	72.29	71.04	69.78	67.26	62.23		0.32	0.32	0.32	0.32	0.31	0.31	0.30	0.28
	50	53.43	53.43	53.43	52.18	52.18	52.18	50.92	48.40		0.24	0.24	0.24	0.23	0.23	0.23	0.23	0.21
	100	38.35	38.35	38.35	38.35	38.35	38.35	37.09	37.09		0.17	0.17	0.17	0.17	0.17	0.17	0.16	0.16
dx=2.5	0	126.35	126.35	126.35	123.84	122.58	116.3	106.24	91.15		0.56	0.56	0.56	0.55	0.54	0.51	0.47	0.40
	0.5	123.84	123.84	123.84	121.33	120.07	113.78	103.72	89.89		0.55	0.55	0.55	0.54	0.53	0.50	0.46	0.40
	1	121.33	121.33	121.33	120.07	117.55	111.27	102.47	88.64		0.54	0.54	0.54	0.53	0.52	0.49	0.45	0.39
	5	104.98	104.98	104.98	103.72	102.47	97.44	91.15	81.09		0.46	0.46	0.46	0.46	0.45	0.43	0.40	0.36
	10	91.15	89.89	89.89	89.89	88.64	84.87	81.09	73.55		0.40	0.40	0.40	0.40	0.39	0.38	0.36	0.33
	25	67.26	67.26	67.26	67.26	66.01	64.75	63.49	59.72		0.30	0.30	0.30	0.30	0.29	0.29	0.28	0.26
	50	50.92	50.92	50.92	50.92	50.92	49.66	48.4	47.15		0.23	0.23	0.23	0.23	0.23	0.22	0.21	0.21
	100	37.09	37.09	37.09	37.09	37.09	37.09	37.09	35.83		0.16	0.16	0.16	0.16	0.16	0.16	0.16	0.16
dx=5	0	89.89	89.89	89.89	88.64	88.64	86.12	82.35	74.81		0.40	0.40	0.40	0.39	0.39	0.38	0.36	0.33
	0.5	88.64	88.64	88.64	88.64	87.38	84.87	81.09	74.81		0.39	0.39	0.39	0.39	0.39	0.38	0.36	0.33
	1	87.38	87.38	87.38	87.38	86.12	83.61	79.84	73.55		0.39	0.39	0.39	0.39	0.38	0.37	0.35	0.33
	5	81.09	81.09	81.09	81.09	79.84	78.58	74.81	69.78		0.36	0.36	0.36	0.36	0.35	0.35	0.33	0.31
	10	74.81	74.81	74.81	73.55	73.55	72.29	68.52	64.75		0.33	0.33	0.33	0.33	0.33	0.32	0.30	0.29
	25	60.98	60.98	59.72	59.72	59.72	58.46	57.21	54.69		0.27	0.27	0.26	0.26	0.26	0.26	0.25	0.24
	50	48.40	48.40	48.40	47.15	47.15	47.15	45.89	44.63		0.21	0.21	0.21	0.21	0.21	0.21	0.20	0.20
	100	35.83	35.83	35.83	35.83	35.83	35.83	35.83	34.57		0.16	0.16	0.16	0.16	0.16	0.16	0.16	0.15
dx=10	0	63.49	63.49	63.49	63.49	63.49	62.23	60.98	58.46		0.28	0.28	0.28	0.28	0.28	0.28	0.27	0.26
	0.5	63.49	63.49	63.49	63.49	62.23	62.23	60.98	57.21		0.28	0.28	0.28	0.28	0.28	0.28	0.27	0.25
	1	63.49	62.23	62.23	62.23	62.23	60.98	59.72	57.21		0.28	0.28	0.28	0.28	0.28	0.27	0.26	0.25
	5	60.98	60.98	60.98	59.72	59.72	59.72	57.21	55.95		0.27	0.27	0.27	0.26	0.26	0.26	0.25	0.25
	10	57.21	57.21	57.21	57.21	57.21	55.95	54.69	53.43		0.25	0.25	0.25	0.25	0.25	0.25	0.24	0.24
	25	50.92	50.92	50.92	50.92	49.66	49.66	48.4	47.15		0.23	0.23	0.23	0.23	0.22	0.22	0.21	0.21
	50	43.38	43.38	43.38	42.12	42.12	42.12	42.12	40.86		0.19	0.19	0.19	0.19	0.19	0.19	0.19	0.18
	100	34.57	34.57	34.57	34.57	34.57	33.32	33.32	33.32		0.15	0.15	0.15	0.15	0.15	0.15	0.15	0.15

		L/H=1															
	dw	0	0.5	1	5	10	25	50	100	0	0.5	1	5	10	25	50	100
	dθ	ω_{11}	ω_{11}	ω_{11}	ω_{11}	ω_{11}	ω_{11}	ω_{11}	ω_{11}	ω_{11}/ω_1	ω_{11}/ω_1	ω_{11}/ω_1	ω_{11}/ω_1	ω_{11}/ω_1	ω_{11}/ω_1	ω_{11}/ω_1	ω_{11}/ω_1
dx=0	0	525.9	515.85	504.54	393.96	304.73	202.95	145.21	103.72	2.33	2.28	2.23	1.74	1.35	0.90	0.64	0.46
	0.5	338.66	333.64	327.35	285.88	248.19	184.1	138.93	101.21	1.50	1.48	1.45	1.26	1.10	0.81	0.61	0.45
	1	259.5	256.98	254.47	233.11	211.74	169.1	131.38	98.7	1.15	1.14	1.13	1.03	0.94	0.75	0.58	0.44
	5	123.84	123.84	123.84	121.33	117.55	110.01	98.7	82.35	0.55	0.55	0.55	0.54	0.52	0.49	0.44	0.36
	10	88.64	88.64	88.64	87.38	86.12	82.35	77.32	69.78	0.39	0.39	0.39	0.39	0.38	0.36	0.34	0.31
	25	55.95	55.95	55.95	55.95	55.95	54.69	53.43	50.92	0.25	0.25	0.25	0.25	0.25	0.24	0.24	0.23
	50	39.60	39.60	39.60	39.60	39.60	39.60	38.35	37.09	0.18	0.18	0.18	0.18	0.18	0.18	0.17	0.16
	100	28.29	28.29	28.29	28.29	28.29	28.29	28.29	27.03	0.13	0.13	0.13	0.13	0.13	0.13	0.13	0.12
dx=0.05	0	445.48	437.94	427.88	352.49	285.88	197.92	143.96	102.47	1.97	1.94	1.89	1.56	1.26	0.88	0.64	0.45
	0.5	311.02	305.99	302.22	268.29	236.88	180.33	136.41	99.95	1.38	1.35	1.34	1.19	1.05	0.80	0.60	0.44
	1	246.93	244.42	241.9	224.31	204.20	165.33	130.13	97.44	1.09	1.08	1.07	0.99	0.90	0.73	0.58	0.43
	5	122.58	122.58	121.33	120.07	116.30	108.75	97.44	82.35	0.54	0.54	0.54	0.53	0.51	0.48	0.43	0.36
	10	87.38	87.38	87.38	86.12	86.12	82.35	77.32	69.78	0.39	0.39	0.39	0.38	0.38	0.36	0.34	0.31
	25	55.95	55.95	55.95	55.95	55.95	54.69	53.43	50.92	0.25	0.25	0.25	0.25	0.25	0.24	0.24	0.23
	50	39.60	39.60	39.60	39.60	39.60	39.60	38.35	37.09	0.18	0.18	0.18	0.18	0.18	0.18	0.17	0.16
	100	28.29	28.29	28.29	28.29	28.29	28.29	28.29	27.03	0.13	0.13	0.13	0.13	0.13	0.13	0.13	0.12
dx=0.1	0	386.42	378.88	372.59	318.56	268.29	192.89	141.44	102.47	1.71	1.68	1.65	1.41	1.19	0.85	0.63	0.45
	0.5	288.4	284.63	280.86	254.47	226.82	175.39	135.16	99.95	1.28	1.26	1.24	1.13	1.00	0.78	0.60	0.44
	1	235.62	233.11	230.59	215.51	197.92	162.82	128.87	97.44	1.04	1.03	1.02	0.95	0.88	0.72	0.57	0.43
	5	121.33	121.33	120.07	117.55	115.04	107.50	97.44	82.35	0.54	0.54	0.53	0.52	0.51	0.48	0.43	0.36
	10	87.38	87.38	87.38	86.12	84.87	82.35	77.32	69.78	0.39	0.39	0.39	0.38	0.38	0.36	0.34	0.31
	25	55.95	55.95	55.95	55.95	54.69	54.69	53.43	50.92	0.25	0.25	0.25	0.25	0.24	0.24	0.24	0.23
	50	39.60	39.60	39.60	39.60	39.60	39.60	38.35	37.09	0.18	0.18	0.18	0.18	0.18	0.18	0.17	0.16
	100	28.29	28.29	28.29	28.29	28.29	28.29	28.29	27.03	0.13	0.13	0.13	0.13	0.13	0.13	0.13	0.12
dx=0.5	0	213.00	211.74	210.49	200.43	187.87	157.79	127.61	97.44	0.94	0.94	0.93	0.89	0.83	0.70	0.56	0.43
	0.5	192.89	191.64	190.38	181.58	172.87	148.99	122.58	94.92	0.85	0.85	0.84	0.80	0.76	0.66	0.54	0.42
	1	175.39	174.13	174.13	166.59	159.04	140.18	117.55	92.41	0.78	0.77	0.77	0.74	0.70	0.62	0.52	0.41
	5	111.27	111.27	111.27	108.75	106.24	99.95	91.15	78.58	0.49	0.49	0.49	0.48	0.47	0.44	0.40	0.35
	10	83.61	83.61	83.61	82.35	81.09	78.58	74.81	67.26	0.37	0.37	0.37	0.36	0.36	0.35	0.33	0.30
	25	54.69	54.69	54.69	54.69	54.69	53.43	52.18	49.66	0.24	0.24	0.24	0.24	0.24	0.24	0.23	0.22
	50	39.60	39.60	39.60	39.60	39.60	38.35	38.35	37.09	0.18	0.18	0.18	0.18	0.18	0.17	0.17	0.16
	100	28.29	28.29	28.29	28.29	28.29	28.29	27.03	27.03	0.13	0.13	0.13	0.13	0.13	0.13	0.12	0.12

L/H=1																		
	dw	0	0.5	1	5	10	25	50	100		0	0.5	1	5	10	25	50	100
	dθ	ω_{11}	ω_{11}	ω_{11}	ω_{11}	ω_{11}	ω_{11}	ω_{11}	ω_{11}		ω_{11}/ω_1	ω_{11}/ω_1	ω_{11}/ω_1	ω_{11}/ω_1	ω_{11}/ω_1	ω_{11}/ω_1	ω_{11}/ω_1	ω_{11}/ω_1
dx=1	0	155.27	155.27	154.01	150.24	145.21	131.38	113.78	91.15		0.69	0.69	0.68	0.66	0.64	0.58	0.50	0.40
	0.5	146.47	146.47	146.47	142.7	137.67	125.1	110.01	88.64		0.65	0.65	0.65	0.63	0.61	0.55	0.49	0.39
	1	138.93	138.93	138.93	135.16	131.38	120.07	106.24	87.38		0.61	0.61	0.61	0.60	0.58	0.53	0.47	0.39
	5	101.21	101.21	101.21	99.95	97.44	92.41	86.12	74.81		0.45	0.45	0.45	0.44	0.43	0.41	0.38	0.33
	10	78.58	78.58	78.58	78.58	77.32	74.81	71.04	64.75		0.35	0.35	0.35	0.35	0.34	0.33	0.31	0.29
	25	53.43	53.43	53.43	53.43	53.43	52.18	50.92	48.40		0.24	0.24	0.24	0.24	0.24	0.23	0.23	0.21
	50	38.35	38.35	38.35	38.35	38.35	38.35	38.35	37.09		0.17	0.17	0.17	0.17	0.17	0.17	0.17	0.16
	100	28.29	28.29	28.29	28.29	28.29	27.03	27.03	27.03		0.13	0.13	0.13	0.13	0.13	0.12	0.12	0.12
dx=2.5	0	99.95	99.95	99.95	98.7	97.44	93.67	87.38	76.06		0.44	0.44	0.44	0.44	0.43	0.41	0.39	0.34
	0.5	97.44	97.44	97.44	96.18	94.92	91.15	84.87	74.81		0.43	0.43	0.43	0.43	0.42	0.40	0.38	0.33
	1	94.92	94.92	94.92	93.67	92.41	88.64	83.61	73.55		0.42	0.42	0.42	0.41	0.41	0.39	0.37	0.33
	5	81.09	81.09	81.09	79.84	79.84	77.32	73.55	66.01		0.36	0.36	0.36	0.35	0.35	0.34	0.33	0.29
	10	68.52	68.52	68.52	68.52	67.26	66.01	63.49	59.72		0.30	0.30	0.30	0.30	0.30	0.29	0.28	0.26
	25	50.92	49.66	49.66	49.66	49.66	49.66	48.4	45.89		0.23	0.22	0.22	0.22	0.22	0.22	0.21	0.20
	50	37.09	37.09	37.09	37.09	37.09	37.09	37.09	35.83		0.16	0.16	0.16	0.16	0.16	0.16	0.16	0.16
	100	27.03	27.03	27.03	27.03	27.03	27.03	27.03	27.03		0.12	0.12	0.12	0.12	0.12	0.12	0.12	0.12
dx=5	0	71.04	71.04	71.04	71.04	69.78	68.52	66.01	60.98		0.31	0.31	0.31	0.31	0.31	0.30	0.29	0.27
	0.5	69.78	69.78	69.78	69.78	69.78	67.26	66.01	60.98		0.31	0.31	0.31	0.31	0.31	0.30	0.29	0.27
	1	69.78	69.78	69.78	68.52	68.52	67.26	64.75	60.98		0.31	0.31	0.31	0.30	0.30	0.30	0.29	0.27
	5	63.49	63.49	63.49	63.49	62.23	60.98	59.72	55.95		0.28	0.28	0.28	0.28	0.28	0.27	0.26	0.25
	10	57.21	57.21	57.21	57.21	57.21	55.95	54.69	52.18		0.25	0.25	0.25	0.25	0.25	0.25	0.24	0.23
	25	45.89	45.89	45.89	45.89	45.89	44.63	44.63	42.12		0.20	0.20	0.20	0.20	0.20	0.20	0.20	0.19
	50	35.83	35.83	35.83	35.83	35.83	35.83	34.57	34.57		0.16	0.16	0.16	0.16	0.16	0.16	0.15	0.15
	100	27.03	27.03	27.03	27.03	27.03	27.03	25.77	25.77		0.12	0.12	0.12	0.12	0.12	0.12	0.11	0.11
dx=10	0	50.92	50.92	50.92	49.66	49.66	49.66	48.4	47.15		0.23	0.23	0.23	0.22	0.22	0.22	0.21	0.21
	0.5	49.66	49.66	49.66	49.66	49.66	49.66	48.4	47.15		0.22	0.22	0.22	0.22	0.22	0.22	0.21	0.21
	1	49.66	49.66	49.66	49.66	49.66	48.4	48.4	45.89		0.22	0.22	0.22	0.22	0.22	0.21	0.21	0.20
	5	47.15	47.15	47.15	47.15	47.15	47.15	45.89	44.63		0.21	0.21	0.21	0.21	0.21	0.21	0.20	0.20
	10	44.63	44.63	44.63	44.63	44.63	44.63	43.38	42.12		0.20	0.20	0.20	0.20	0.20	0.20	0.19	0.19
	25	38.35	38.35	38.35	38.35	38.35	38.35	38.35	37.09		0.17	0.17	0.17	0.17	0.17	0.17	0.17	0.16
	50	32.06	32.06	32.06	32.06	32.06	32.06	32.06	30.8		0.14	0.14	0.14	0.14	0.14	0.14	0.14	0.14
	100	25.77	25.77	25.77	25.77	24.52	24.52	24.52	24.52		0.11	0.11	0.11	0.11	0.11	0.11	0.11	0.11

	dw	0	0.5	1	5	10	25	50	100	0	0.5	1	5	10	25	50	100
L/H=2	dθ	ω_{11}	ω_{11}	ω_{11}	ω_{11}	ω_{11}	ω_{11}	ω_{11}	ω_{11}	ω_{11}/ω_1	ω_{11}/ω_1	ω_{11}/ω_1	ω_{11}/ω_1	ω_{11}/ω_1	ω_{11}/ω_1	ω_{11}/ω_1	ω_{11}/ω_1
dx=0	0	332.38	329.87	327.35	297.19	256.98	182.84	133.9	96.18	1.47	1.46	1.45	1.32	1.14	0.81	0.59	0.43
	0.5	186.61	185.35	184.1	177.81	170.36	147.73	120.07	91.15	0.83	0.82	0.81	0.79	0.75	0.65	0.53	0.40
	1	140.18	140.18	140.18	137.67	133.9	123.84	107.5	86.12	0.62	0.62	0.62	0.61	0.59	0.55	0.48	0.38
	5	66.01	66.01	66.01	66.01	66.01	64.75	62.23	58.46	0.29	0.29	0.29	0.29	0.29	0.29	0.28	0.26
	10	47.15	47.15	47.15	47.15	47.15	47.15	45.89	44.63	0.21	0.21	0.21	0.21	0.21	0.21	0.20	0.20
	25	29.55	29.55	29.55	29.55	29.55	29.55	29.55	29.55	0.13	0.13	0.13	0.13	0.13	0.13	0.13	0.13
	50	20.74	20.74	20.74	20.74	20.74	20.74	20.74	20.74	0.09	0.09	0.09	0.09	0.09	0.09	0.09	0.09
	100	14.46	14.46	14.46	14.46	14.46	14.46	14.46	14.46	0.06	0.06	0.06	0.06	0.06	0.06	0.06	0.06
dx=0.05	0	305.99	303.48	299.71	273.32	240.65	177.81	131.38	96.18	1.35	1.34	1.33	1.21	1.06	0.79	0.58	0.43
	0.5	180.33	180.33	179.07	172.87	165.33	143.96	118.81	91.15	0.80	0.80	0.79	0.76	0.73	0.64	0.53	0.40
	1	137.67	137.67	137.67	135.16	131.38	121.33	106.24	86.12	0.61	0.61	0.61	0.60	0.58	0.54	0.47	0.38
	5	66.01	66.01	66.01	66.01	66.01	64.75	62.23	58.46	0.29	0.29	0.29	0.29	0.29	0.29	0.28	0.26
	10	47.15	47.15	47.15	47.15	47.15	47.15	45.89	44.63	0.21	0.21	0.21	0.21	0.21	0.21	0.20	0.20
	25	29.55	29.55	29.55	29.55	29.55	29.55	29.55	29.55	0.13	0.13	0.13	0.13	0.13	0.13	0.13	0.13
	50	20.74	20.74	20.74	20.74	20.74	20.74	20.74	20.74	0.09	0.09	0.09	0.09	0.09	0.09	0.09	0.09
	100	14.46	14.46	14.46	14.46	14.46	14.46	14.46	14.46	0.06	0.06	0.06	0.06	0.06	0.06	0.06	0.06
dx=0.1	0	280.86	278.35	275.83	251.96	225.57	171.62	130.13	94.92	1.24	1.23	1.22	1.11	1.00	0.76	0.58	0.42
	0.5	175.39	174.13	174.13	167.84	161.56	141.44	116.3	89.89	0.78	0.77	0.77	0.74	0.71	0.63	0.51	0.40
	1	136.41	135.16	135.16	132.64	128.87	120.07	104.98	84.87	0.60	0.60	0.60	0.59	0.57	0.53	0.46	0.38
	5	66.01	66.01	66.01	66.01	64.75	63.49	62.23	58.46	0.29	0.29	0.29	0.29	0.29	0.28	0.28	0.26
	10	47.15	47.15	47.15	47.15	47.15	45.89	45.89	44.63	0.21	0.21	0.21	0.21	0.21	0.20	0.20	0.20
	25	29.55	29.55	29.55	29.55	29.55	29.55	29.55	29.55	0.13	0.13	0.13	0.13	0.13	0.13	0.13	0.13
	50	20.74	20.74	20.74	20.74	20.74	20.74	20.74	20.74	0.09	0.09	0.09	0.09	0.09	0.09	0.09	0.09
	100	14.46	14.46	14.46	14.46	14.46	14.46	14.46	14.46	0.06	0.06	0.06	0.06	0.06	0.06	0.06	0.06
dx=0.5	0	175.39	174.13	174.13	166.59	159.04	138.93	115.04	88.64	0.78	0.77	0.77	0.74	0.70	0.61	0.51	0.39
	0.5	141.44	140.18	140.18	136.41	132.64	121.33	104.98	84.87	0.63	0.62	0.62	0.60	0.59	0.54	0.46	0.38
	1	118.81	118.81	118.81	116.3	113.78	107.5	96.18	81.09	0.53	0.53	0.53	0.51	0.50	0.48	0.43	0.36
	5	63.49	63.49	63.49	63.49	63.49	62.23	60.98	57.21	0.28	0.28	0.28	0.28	0.28	0.28	0.27	0.25
	10	45.89	45.89	45.89	45.89	45.89	45.89	44.63	43.38	0.20	0.20	0.20	0.20	0.20	0.20	0.20	0.19
	25	29.55	29.55	29.55	29.55	29.55	29.55	29.55	29.55	0.13	0.13	0.13	0.13	0.13	0.13	0.13	0.13
	50	20.74	20.74	20.74	20.74	20.74	20.74	20.74	20.74	0.09	0.09	0.09	0.09	0.09	0.09	0.09	0.09
	100	14.46	14.46	14.46	14.46	14.46	14.46	14.46	14.46	0.06	0.06	0.06	0.06	0.06	0.06	0.06	0.06

	L/H=2																	
	dw	0	0.5	1	5	10	25	50	100		0	0.5	1	5	10	25	50	100
	dθ	ω_{11}	ω_{11}	ω_{11}	ω_{11}	ω_{11}	ω_{11}	ω_{11}	ω_{11}		ω_{11}/ω_1	ω_{11}/ω_1	ω_{11}/ω_1	ω_{11}/ω_1	ω_{11}/ω_1	ω_{11}/ω_1	ω_{11}/ω_1	ω_{11}/ω_1
dx=1	0	130.13	130.13	130.13	126.35	123.84	113.78	99.95	82.35		0.58	0.58	0.58	0.56	0.55	0.50	0.44	0.36
	0.5	115.04	115.04	115.04	112.52	110.01	103.72	93.67	78.58		0.51	0.51	0.51	0.50	0.49	0.46	0.41	0.35
	1	103.72	103.72	102.47	101.21	99.95	94.92	87.38	74.81		0.46	0.46	0.45	0.45	0.44	0.42	0.39	0.33
	5	62.23	62.23	62.23	60.98	60.98	59.72	58.46	54.69		0.28	0.28	0.28	0.27	0.27	0.26	0.26	0.24
	10	45.89	45.89	45.89	45.89	45.89	44.63	44.63	43.38		0.20	0.20	0.20	0.20	0.20	0.20	0.20	0.19
	25	29.55	29.55	29.55	29.55	29.55	29.55	29.55	28.29		0.13	0.13	0.13	0.13	0.13	0.13	0.13	0.13
	50	20.74	20.74	20.74	20.74	20.74	20.74	20.74	20.74		0.09	0.09	0.09	0.09	0.09	0.09	0.09	0.09
	100	14.46	14.46	14.46	14.46	14.46	14.46	14.46	14.46		0.06	0.06	0.06	0.06	0.06	0.06	0.06	0.06
dx=2.5	0	84.87	84.87	84.87	83.61	82.35	79.84	74.81	67.26		0.38	0.38	0.38	0.37	0.36	0.35	0.33	0.30
	0.5	81.09	81.09	81.09	79.84	78.58	76.06	72.29	66.01		0.36	0.36	0.36	0.35	0.35	0.34	0.32	0.29
	1	77.32	77.32	76.06	76.06	74.81	73.55	69.78	63.49		0.34	0.34	0.34	0.34	0.33	0.33	0.31	0.28
	5	55.95	55.95	55.95	54.69	54.69	54.69	53.43	50.92		0.25	0.25	0.25	0.24	0.24	0.24	0.24	0.23
	10	43.38	43.38	43.38	43.38	43.38	42.12	42.12	40.86		0.19	0.19	0.19	0.19	0.19	0.19	0.19	0.18
	25	28.29	28.29	28.29	28.29	28.29	28.29	28.29	28.29		0.13	0.13	0.13	0.13	0.13	0.13	0.13	0.13
	50	20.74	20.74	20.74	20.74	20.74	20.74	20.74	20.74		0.09	0.09	0.09	0.09	0.09	0.09	0.09	0.09
	100	14.46	14.46	14.46	14.46	14.46	14.46	14.46	14.46		0.06	0.06	0.06	0.06	0.06	0.06	0.06	0.06
dx=5	0	60.98	60.98	60.98	59.72	59.72	58.46	57.21	53.43		0.27	0.27	0.27	0.26	0.26	0.26	0.25	0.24
	0.5	59.72	59.72	58.46	58.46	58.46	57.21	55.95	52.18		0.26	0.26	0.26	0.26	0.26	0.25	0.25	0.23
	1	57.21	57.21	57.21	57.21	57.21	55.95	54.69	52.18		0.25	0.25	0.25	0.25	0.25	0.25	0.24	0.23
	5	47.15	47.15	47.15	47.15	47.15	47.15	45.89	44.63		0.21	0.21	0.21	0.21	0.21	0.21	0.20	0.20
	10	39.60	39.60	39.60	39.60	39.60	38.35	38.35	37.09		0.18	0.18	0.18	0.18	0.18	0.17	0.17	0.16
	25	28.29	28.29	28.29	28.29	28.29	28.29	27.03	27.03		0.13	0.13	0.13	0.13	0.13	0.13	0.12	0.12
	50	20.74	20.74	20.74	20.74	20.74	20.74	20.74	20.74		0.09	0.09	0.09	0.09	0.09	0.09	0.09	0.09
	100	14.46	14.46	14.46	14.46	14.46	14.46	14.46	14.46		0.06	0.06	0.06	0.06	0.06	0.06	0.06	0.06
dx=10	0	43.38	43.38	43.38	43.38	43.38	42.12	42.12	40.86		0.19	0.19	0.19	0.19	0.19	0.19	0.19	0.18
	0.5	42.12	42.12	42.12	42.12	42.12	42.12	40.86	39.6		0.19	0.19	0.19	0.19	0.19	0.19	0.18	0.18
	1	42.12	42.12	42.12	42.12	42.12	40.86	40.86	39.6		0.19	0.19	0.19	0.19	0.19	0.18	0.18	0.18
	5	38.35	38.35	38.35	38.35	38.35	37.09	37.09	35.83		0.17	0.17	0.17	0.17	0.17	0.16	0.16	0.16
	10	33.32	33.32	33.32	33.32	33.32	33.32	33.32	32.06		0.15	0.15	0.15	0.15	0.15	0.15	0.15	0.14
	25	25.77	25.77	25.77	25.77	25.77	25.77	25.77	25.77		0.11	0.11	0.11	0.11	0.11	0.11	0.11	0.11
	50	19.49	19.49	19.49	19.49	19.49	19.49	19.49	19.49		0.09	0.09	0.09	0.09	0.09	0.09	0.09	0.09
	100	14.46	14.46	14.46	14.46	14.46	14.46	14.46	14.46		0.06	0.06	0.06	0.06	0.06	0.06	0.06	0.06

	L/H=3																
	dw	0	0.5	1	5	10	25	50	100	0	0.5	1	5	10	25	50	100
	dθ	ω_{11}	ω_{11}	ω_{11}	ω_{11}	ω_{11}	ω_{11}	ω_{11}	ω_{11}	ω_{11}/ω_1	ω_{11}/ω_1	ω_{11}/ω_1	ω_{11}/ω_1	ω_{11}/ω_1	ω_{11}/ω_1	ω_{11}/ω_1	ω_{11}/ω_1
dx=0	0	283.37	280.86	279.6	263.27	236.88	177.81	131.38	94.92	1.25	1.24	1.24	1.16	1.05	0.79	0.58	0.42
	0.5	118.81	118.81	117.55	116.3	115.04	110.01	99.95	83.61	0.53	0.53	0.52	0.51	0.51	0.49	0.44	0.37
	1	86.12	86.12	86.12	86.12	86.12	83.61	79.84	72.29	0.38	0.38	0.38	0.38	0.38	0.37	0.35	0.32
	5	39.60	39.60	39.60	39.60	39.60	39.60	39.60	38.35	0.18	0.18	0.18	0.18	0.18	0.18	0.18	0.17
	10	28.29	28.29	28.29	28.29	28.29	28.29	28.29	28.29	0.13	0.13	0.13	0.13	0.13	0.13	0.13	0.13
	25	18.23	18.23	18.23	18.23	18.23	18.23	18.23	18.23	0.08	0.08	0.08	0.08	0.08	0.08	0.08	0.08
	50	13.20	13.20	13.20	13.20	13.20	13.20	13.20	13.20	0.06	0.06	0.06	0.06	0.06	0.06	0.06	0.06
	100	9.43	9.43	9.43	9.43	9.43	9.43	9.43	9.43	0.04	0.04	0.04	0.04	0.04	0.04	0.04	0.04
dx=0.05	0	267.04	264.52	263.27	246.93	224.31	171.62	128.87	93.67	1.18	1.17	1.16	1.09	0.99	0.76	0.57	0.41
	0.5	117.55	117.55	116.3	116.3	113.78	108.75	99.95	83.61	0.52	0.52	0.51	0.51	0.50	0.48	0.44	0.37
	1	86.12	86.12	86.12	86.12	84.87	83.61	79.84	72.29	0.38	0.38	0.38	0.38	0.38	0.37	0.35	0.32
	5	39.60	39.60	39.60	39.60	39.60	39.60	39.60	38.35	0.18	0.18	0.18	0.18	0.18	0.18	0.18	0.17
	10	28.29	28.29	28.29	28.29	28.29	28.29	28.29	28.29	0.13	0.13	0.13	0.13	0.13	0.13	0.13	0.13
	25	18.23	18.23	18.23	18.23	18.23	18.23	18.23	18.23	0.08	0.08	0.08	0.08	0.08	0.08	0.08	0.08
	50	13.20	13.20	13.20	13.20	13.20	13.20	13.20	13.20	0.06	0.06	0.06	0.06	0.06	0.06	0.06	0.06
	100	9.43	9.43	9.43	9.43	9.43	9.43	9.43	9.43	0.04	0.04	0.04	0.04	0.04	0.04	0.04	0.04
dx=0.1	0	250.70	248.19	246.93	231.85	211.74	166.59	126.35	93.67	1.11	1.10	1.09	1.03	0.94	0.74	0.56	0.41
	0.5	116.30	116.30	115.04	115.04	112.52	107.50	98.70	82.35	0.51	0.51	0.51	0.51	0.50	0.48	0.44	0.36
	1	86.12	86.12	86.12	84.87	84.87	82.35	78.58	72.29	0.38	0.38	0.38	0.38	0.38	0.36	0.35	0.32
	5	39.60	39.60	39.60	39.60	39.60	39.60	39.60	38.35	0.18	0.18	0.18	0.18	0.18	0.18	0.18	0.17
	10	28.29	28.29	28.29	28.29	28.29	28.29	28.29	28.29	0.13	0.13	0.13	0.13	0.13	0.13	0.13	0.13
	25	18.23	18.23	18.23	18.23	18.23	18.23	18.23	18.23	0.08	0.08	0.08	0.08	0.08	0.08	0.08	0.08
	50	13.20	13.20	13.20	13.20	13.20	13.20	13.20	13.20	0.06	0.06	0.06	0.06	0.06	0.06	0.06	0.06
	100	9.43	9.43	9.43	9.43	9.43	9.43	9.43	9.43	0.04	0.04	0.04	0.04	0.04	0.04	0.04	0.04
dx=0.5	0	165.33	164.07	164.07	157.79	151.5	132.64	111.27	87.38	0.73	0.73	0.73	0.70	0.67	0.59	0.49	0.39
	0.5	106.24	106.24	106.24	104.98	102.47	98.7	91.15	77.32	0.47	0.47	0.47	0.46	0.45	0.44	0.40	0.34
	1	82.35	82.35	82.35	81.09	81.09	78.58	76.06	68.52	0.36	0.36	0.36	0.36	0.36	0.35	0.34	0.30
	5	39.60	39.60	39.60	39.60	39.60	39.60	38.35	38.35	0.18	0.18	0.18	0.18	0.18	0.18	0.17	0.17
	10	28.29	28.29	28.29	28.29	28.29	28.29	28.29	28.29	0.13	0.13	0.13	0.13	0.13	0.13	0.13	0.13
	25	18.23	18.23	18.23	18.23	18.23	18.23	18.23	18.23	0.08	0.08	0.08	0.08	0.08	0.08	0.08	0.08
	50	13.20	13.20	13.20	13.20	13.20	13.20	13.20	13.20	0.06	0.06	0.06	0.06	0.06	0.06	0.06	0.06
	100	9.43	9.43	9.43	9.43	9.43	9.43	9.43	9.43	0.04	0.04	0.04	0.04	0.04	0.04	0.04	0.04

	L/H=3																
	dw	0	0.5	1	5	10	25	50	100	0	0.5	1	5	10	25	50	100
	dθ	ω_{11}	ω_{11}	ω_{11}	ω_{11}	ω_{11}	ω_{11}	ω_{11}	ω_{11}	ω_{11}/ω_1	ω_{11}/ω_1	ω_{11}/ω_1	ω_{11}/ω_1	ω_{11}/ω_1	ω_{11}/ω_1	ω_{11}/ω_1	ω_{11}/ω_1
dx=1	0	81.09	81.09	81.09	79.84	78.58	76.06	72.29	66.01	0.36	0.36	0.36	0.35	0.35	0.34	0.32	0.29
	0.5	72.29	72.29	72.29	72.29	71.04	69.78	66.01	60.98	0.32	0.32	0.32	0.32	0.31	0.31	0.29	0.27
	1	64.75	64.75	64.75	64.75	63.49	62.23	60.98	57.21	0.29	0.29	0.29	0.29	0.28	0.28	0.27	0.25
	5	37.09	37.09	37.09	37.09	37.09	37.09	37.09	35.83	0.16	0.16	0.16	0.16	0.16	0.16	0.16	0.16
	10	27.03	27.03	27.03	27.03	27.03	27.03	27.03	27.03	0.12	0.12	0.12	0.12	0.12	0.12	0.12	0.12
	25	18.23	18.23	18.23	18.23	18.23	18.23	18.23	16.97	0.08	0.08	0.08	0.08	0.08	0.08	0.08	0.08
	50	11.94	11.94	11.94	11.94	11.94	11.94	11.94	11.94	0.05	0.05	0.05	0.05	0.05	0.05	0.05	0.05
	100	9.43	9.43	9.43	9.43	9.43	9.43	9.43	9.43	0.04	0.04	0.04	0.04	0.04	0.04	0.04	0.04
dx=2.5	0	81.09	81.09	81.09	79.84	78.58	76.06	72.29	66.01	0.36	0.36	0.36	0.35	0.35	0.34	0.32	0.29
	0.5	72.29	72.29	72.29	72.29	71.04	69.78	66.01	60.98	0.32	0.32	0.32	0.32	0.31	0.31	0.29	0.27
	1	64.75	64.75	64.75	64.75	63.49	62.23	60.98	57.21	0.29	0.29	0.29	0.29	0.28	0.28	0.27	0.25
	5	37.09	37.09	37.09	37.09	37.09	37.09	37.09	35.83	0.16	0.16	0.16	0.16	0.16	0.16	0.16	0.16
	10	27.03	27.03	27.03	27.03	27.03	27.03	27.03	27.03	0.12	0.12	0.12	0.12	0.12	0.12	0.12	0.12
	25	18.23	18.23	18.23	18.23	18.23	18.23	18.23	16.97	0.08	0.08	0.08	0.08	0.08	0.08	0.08	0.08
	50	11.94	11.94	11.94	11.94	11.94	11.94	11.94	11.94	0.05	0.05	0.05	0.05	0.05	0.05	0.05	0.05
	100	9.43	9.43	9.43	9.43	9.43	9.43	9.43	9.43	0.04	0.04	0.04	0.04	0.04	0.04	0.04	0.04
dx=5	0	58.46	57.21	57.21	57.21	57.21	55.95	54.69	52.18	0.26	0.25	0.25	0.25	0.25	0.25	0.24	0.23
	0.5	54.69	54.69	54.69	54.69	54.69	53.43	52.18	49.66	0.24	0.24	0.24	0.24	0.24	0.24	0.23	0.22
	1	52.18	52.18	52.18	50.92	50.92	50.92	49.66	47.15	0.23	0.23	0.23	0.23	0.23	0.23	0.22	0.21
	5	35.83	35.83	35.83	34.57	34.57	34.57	34.57	34.57	0.16	0.16	0.16	0.15	0.15	0.15	0.15	0.15
	10	27.03	27.03	27.03	27.03	27.03	27.03	27.03	25.77	0.12	0.12	0.12	0.12	0.12	0.12	0.12	0.11
	25	16.97	16.97	16.97	16.97	16.97	16.97	16.97	16.97	0.08	0.08	0.08	0.08	0.08	0.08	0.08	0.08
	50	11.94	11.94	11.94	11.94	11.94	11.94	11.94	11.94	0.05	0.05	0.05	0.05	0.05	0.05	0.05	0.05
	100	9.43	9.43	9.43	9.43	9.43	9.43	9.43	9.43	0.04	0.04	0.04	0.04	0.04	0.04	0.04	0.04
dx=10	0	40.86	40.86	40.86	40.86	40.86	40.86	39.60	38.35	0.18	0.18	0.18	0.18	0.18	0.18	0.18	0.17
	0.5	39.60	39.60	39.60	39.60	39.60	39.60	38.35	38.35	0.18	0.18	0.18	0.18	0.18	0.18	0.17	0.17
	1	38.35	38.35	38.35	38.35	38.35	38.35	38.35	37.09	0.17	0.17	0.17	0.17	0.17	0.17	0.17	0.16
	5	30.80	30.80	30.80	30.80	30.80	30.80	30.80	30.80	0.14	0.14	0.14	0.14	0.14	0.14	0.14	0.14
	10	24.52	24.52	24.52	24.52	24.52	24.52	24.52	24.52	0.11	0.11	0.11	0.11	0.11	0.11	0.11	0.11
	25	16.97	16.97	16.97	16.97	16.97	16.97	16.97	16.97	0.08	0.08	0.08	0.08	0.08	0.08	0.08	0.08
	50	11.94	11.94	11.94	11.94	11.94	11.94	63.49	11.94	0.05	0.05	0.05	0.05	0.05	0.05	0.28	0.05
	100	9.43	9.43	9.43	9.43	9.43	8.17	8.17	8.17	0.04	0.04	0.04	0.04	0.04	0.04	0.04	0.04

		L/H=5								
	dw	0	25	50	100		0	25	50	100
	dθ	ω_{11}	ω_{11}	ω_{11}	ω_{11}		ω_{11}/ω_1	ω_{11}/ω_1	ω_{11}/ω_1	ω_{11}/ω_1
dx=0	0	225.57	169.10	128.87	93.67		1.00	0.75	0.57	0.41
	25	8.17	8.17	8.17	8.17		0.04	0.04	0.04	0.04
	50	5.65	5.65	5.65	5.65		0.03	0.03	0.03	0.03
	100	4.40	4.40	4.40	4.40		0.02	0.02	0.02	0.02
dx=2.5	0	77.32	73.55	69.78	63.49		0.34	0.33	0.31	0.28
	25	8.17	8.17	8.17	8.17		0.04	0.04	0.04	0.04
	50	5.65	5.65	5.65	5.65		0.03	0.03	0.03	0.03
	100	4.40	4.40	4.40	4.40		0.02	0.02	0.02	0.02
dx=5	0	54.69	53.43	52.18	49.66		0.24	0.24	0.23	0.22
	25	8.17	8.17	8.17	8.17		0.04	0.04	0.04	0.04
	50	5.65	5.65	5.65	5.65		0.03	0.03	0.03	0.03
	100	4.40	4.40	4.40	4.40		0.02	0.02	0.02	0.02
dx=10	0	39.60	38.35	38.35	37.09		0.18	0.17	0.17	0.16
	25	8.17	8.17	8.17	8.17		0.04	0.04	0.04	0.04
	50	5.65	5.65	5.65	5.65		0.03	0.03	0.03	0.03
	100	4.40	4.40	4.40	4.40		0.02	0.02	0.02	0.02

		L/H=10								
	dw	0	25	50	100		0	25	50	100
	dθ	ω_{11}	ω_{11}	ω_{11}	ω_{11}		ω_{11}/ω_1	ω_{11}/ω_1	ω_{11}/ω_1	ω_{11}/ω_1
dx=0	0	135.16	130.13	116.3	91.15		0.60	0.58	0.51	0.40
	25	3.14	3.14	3.14	3.14		0.01	0.01	0.01	0.01
	50	1.88	1.88	1.88	1.88		0.01	0.01	0.01	0.01
	100	0.63	0.63	0.63	0.63		0.00	0.00	0.00	0.00
dx=2.5	0	71.04	68.52	66.01	60.98		0.31	0.30	0.29	0.27
	25	3.14	3.14	3.14	3.14		0.01	0.01	0.01	0.01
	50	1.88	1.88	1.88	1.88		0.01	0.01	0.01	0.01
	100	0.63	0.63	0.63	0.63		0.00	0.00	0.00	0.00
dx=5	0	50.92	49.66	48.4	47.15		0.23	0.22	0.21	0.21
	25	3.14	3.14	3.14	3.14		0.01	0.01	0.01	0.01
	50	1.88	1.88	1.88	1.88		0.01	0.01	0.01	0.01
	100	0.63	0.63	0.63	0.63		0.00	0.00	0.00	0.00
dx=10	0	35.83	35.83	35.83	34.57		0.16	0.16	0.16	0.15
	25	3.14	3.14	3.14	3.14		0.01	0.01	0.01	0.01
	50	1.88	1.88	1.88	1.88		0.01	0.01	0.01	0.01
	100	0.63	0.63	0.63	0.63		0.00	0.00	0.00	0.00

Annex C

Normalized base shear forces and effective heights for the three soil profiles investigated here and for **bonded** contact. For the effective heights (points of application of the resultant force) because of the bonded contact the paradox is observed that the height for increasing d_x parameter also increases and for some cases lies beyond the wall's height. This paradox is mitigated when the tension forces are set to null. By setting the tension forces to null though the results are not the same as with a contact, which allows separation between the wall and the soil. However a good approximation is achieved. The results arise from the **time domain** analysis.

Homogeneous soil								
$P/\rho AgH^2$	d_θ	d_w=0	d_w=1	d_w=5	d_w=10	d_w=20	d_w=30	d_w=40
dx=0	0	1.00	0.89	0.67	0.57	0.48	0.44	0.40
	0.5	0.74	0.69	0.58	0.52	0.45	0.41	0.38
	1	0.60	0.57	0.51	0.47	0.42	0.39	0.37
	2	0.45	0.44	0.42	0.40	0.38	0.35	0.34
	3	0.38	0.38	0.37	0.36	0.34	0.33	0.31
	4	0.34	0.34	0.33	0.33	0.32	0.30	0.29
	5	0.31	0.31	0.31	0.30	0.29	0.28	0.28
dx=0.1	0	0.67	0.60	0.47	0.41	0.35	0.32	0.29
	0.5	0.51	0.48	0.41	0.37	0.33	0.30	0.28
	1	0.42	0.41	0.37	0.34	0.31	0.28	0.27
	2	0.33	0.32	0.31	0.29	0.28	0.26	0.25
	3	0.28	0.28	0.27	0.26	0.25	0.24	0.23
	4	0.25	0.25	0.25	0.24	0.23	0.23	0.22
	5	0.23	0.23	0.23	0.22	0.22	0.21	0.21
dx=0.5	0	0.35	0.32	0.26	0.22	0.19	0.18	0.17
	0.5	0.28	0.26	0.23	0.21	0.18	0.17	0.16
	1	0.23	0.23	0.20	0.19	0.17	0.16	0.15
	2	0.19	0.18	0.17	0.17	0.16	0.15	0.14
	3	0.16	0.16	0.15	0.15	0.15	0.14	0.13
	4	0.14	0.14	0.14	0.14	0.14	0.13	0.13
	5	0.13	0.13	0.13	0.13	0.13	0.12	0.12
dx=1	0	0.26	0.24	0.19	0.17	0.15	0.14	0.13
	0.5	0.21	0.20	0.17	0.16	0.14	0.13	0.12
	1	0.18	0.17	0.16	0.14	0.13	0.12	0.12
	2	0.14	0.14	0.13	0.13	0.12	0.11	0.11
	3	0.12	0.12	0.12	0.12	0.11	0.11	0.10
	4	0.11	0.11	0.11	0.11	0.10	0.10	0.10
	5	0.10	0.10	0.10	0.10	0.10	0.09	0.09

Parabolic soil profile								
$P/\rho AgH^2$	d_θ	d_w=0	d_w=1	d_w=5	d_w=10	d_w=20	d_w=30	d_w=40
dx=0	0	0.88	0.79	0.62	0.52	0.43	0.39	0.36
	0.5	0.65	0.61	0.52	0.46	0.39	0.36	0.34
	1	0.52	0.5	0.45	0.4	0.36	0.34	0.32
	2	0.38	0.37	0.36	0.34	0.32	0.3	0.29
	3	0.31	0.31	0.30	0.29	0.28	0.27	0.26
	4	0.26	0.26	0.26	0.26	0.25	0.25	0.24
	5	0.23	0.23	0.23	0.23	0.23	0.23	0.22
dx=0.1	0	0.52	0.48	0.38	0.32	0.27	0.24	0.23
	0.5	0.39	0.37	0.32	0.28	0.25	0.23	0.21
	1	0.32	0.31	0.28	0.25	0.23	0.21	0.20
	2	0.24	0.23	0.23	0.21	0.20	0.19	0.19
	3	0.19	0.19	0.19	0.19	0.18	0.17	0.17
	4	0.17	0.17	0.16	0.16	0.16	0.16	0.16
	5	0.15	0.15	0.15	0.15	0.15	0.15	0.15
dx=0.5	0	0.27	0.25	0.20	0.17	0.14	0.13	0.12
	0.5	0.21	0.20	0.17	0.15	0.13	0.12	0.12
	1	0.17	0.17	0.15	0.14	0.13	0.12	0.11
	2	0.13	0.13	0.12	0.12	0.11	0.11	0.10
	3	0.11	0.11	0.10	0.10	0.10	0.10	0.09
	4	0.09	0.09	0.09	0.09	0.09	0.09	0.09
	5	0.08	0.08	0.08	0.08	0.08	0.08	0.08
dx=1	0	0.21	0.19	0.16	0.14	0.11	0.10	0.10
	0.5	0.16	0.16	0.13	0.12	0.11	0.10	0.09
	1	0.14	0.13	0.12	0.11	0.10	0.09	0.09
	2	0.10	0.10	0.10	0.09	0.09	0.08	0.08
	3	0.08	0.08	0.08	0.08	0.08	0.08	0.07
	4	0.07	0.07	0.07	0.07	0.07	0.07	0.07
	5	0.06	0.06	0.06	0.06	0.07	0.07	0.06

Linear soil profile								
$P/\rho AgH^2$	d_θ	d_w=0	d_w=1	d_w=5	d_w=10	d_w=20	d_w=30	d_w=40
dx=0	0	0.71	0.65	0.51	0.42	0.35	0.31	0.28
	0.5	0.52	0.49	0.42	0.37	0.31	0.28	0.26
	1	0.41	0.4	0.35	0.32	0.28	0.26	0.25
	2	0.29	0.28	0.27	0.26	0.24	0.23	0.22
	3	0.22	0.22	0.22	0.21	0.21	0.20	0.20
	4	0.18	0.18	0.18	0.18	0.18	0.18	0.18
	5	0.15	0.15	0.16	0.16	0.16	0.16	0.16
dx=0.1	0	0.42	0.38	0.31	0.26	0.22	0.19	0.17
	0.5	0.32	0.3	0.26	0.22	0.20	0.18	0.16
	1	0.26	0.25	0.22	0.20	0.18	0.16	0.15
	2	0.18	0.18	0.17	0.16	0.16	0.14	0.14
	3	0.14	0.14	0.14	0.14	0.14	0.13	0.13
	4	0.12	0.12	0.12	0.12	0.12	0.12	0.11
	5	0.10	0.10	0.10	0.10	0.11	0.11	0.10
dx=0.5	0	0.22	0.2	0.16	0.14	0.11	0.10	0.10
	0.5	0.17	0.16	0.14	0.12	0.10	0.10	0.09
	1	0.14	0.13	0.12	0.11	0.09	0.09	0.09
	2	0.10	0.10	0.09	0.09	0.08	0.08	0.08
	3	0.08	0.08	0.08	0.07	0.07	0.07	0.07
	4	0.06	0.06	0.06	0.06	0.06	0.06	0.06
	5	0.05	0.05	0.05	0.06	0.06	0.06	0.06
dx=1	0	0.18	0.16	0.13	0.11	0.09	0.08	0.08
	0.5	0.14	0.13	0.11	0.10	0.08	0.07	0.07
	1	0.11	0.11	0.09	0.09	0.08	0.07	0.07
	2	0.08	0.08	0.07	0.07	0.07	0.06	0.06
	3	0.06	0.06	0.06	0.06	0.06	0.05	0.06
	4	0.05	0.05	0.05	0.05	0.05	0.05	0.05
	5	0.04	0.04	0.04	0.04	0.05	0.04	0.05

Homogeneous soil								
h/H	d_θ	d_w=0	d_w=1	d_w=5	d_w=10	d_w=20	d_w=30	d_w=40
dx=0	0	0.59	0.55	0.44	0.37	0.31	0.27	0.25
	0.5	0.55	0.51	0.42	0.36	0.30	0.27	0.25
	1	0.51	0.48	0.40	0.35	0.29	0.27	0.25
	2	0.45	0.43	0.37	0.33	0.28	0.26	0.24
	3	0.40	0.39	0.34	0.31	0.27	0.25	0.24
	4	0.36	0.35	0.32	0.29	0.26	0.24	0.23
	5	0.33	0.32	0.30	0.28	0.25	0.24	0.23
dx=0.1	0	0.75	0.70	0.56	0.47	0.38	0.34	0.31
	0.5	0.70	0.65	0.53	0.45	0.37	0.33	0.31
	1	0.65	0.62	0.51	0.44	0.37	0.33	0.30
	2	0.58	0.55	0.47	0.41	0.35	0.32	0.30
	3	0.52	0.50	0.44	0.39	0.34	0.31	0.29
	4	0.47	0.45	0.41	0.37	0.33	0.30	0.28
	5	0.43	0.42	0.38	0.35	0.32	0.29	0.28
dx=0.5	0	1.21	1.12	0.88	0.73	0.59	0.52	0.47
	0.5	1.13	1.05	0.85	0.71	0.58	0.51	0.47
	1	1.06	0.99	0.81	0.69	0.57	0.50	0.46
	2	0.94	0.89	0.75	0.65	0.55	0.49	0.45
	3	0.84	0.81	0.70	0.62	0.53	0.48	0.44
	4	0.77	0.74	0.65	0.59	0.51	0.47	0.44
	5	0.70	0.68	0.61	0.56	0.50	0.46	0.43
dx=1	0	1.54	1.42	1.11	0.92	0.74	0.65	0.59
	0.5	1.44	1.33	1.07	0.90	0.73	0.64	0.58
	1	1.35	1.26	1.03	0.87	0.72	0.63	0.58
	2	1.20	1.13	0.95	0.83	0.69	0.62	0.57
	3	1.08	1.03	0.89	0.79	0.67	0.60	0.56
	4	0.98	0.95	0.84	0.75	0.65	0.59	0.55
	5	0.90	0.87	0.79	0.71	0.63	0.57	0.54

Parabolic soil profile								
h/H	d_θ	d_w=0	d_w=1	d_w=5	d_w=10	d_w=20	d_w=30	d_w=40
dx=0	0	0.51	0.48	0.41	0.36	0.30	0.26	0.51
	0.5	0.49	0.47	0.40	0.35	0.29	0.26	0.49
	1	0.48	0.46	0.39	0.34	0.29	0.26	0.48
	2	0.45	0.43	0.38	0.33	0.28	0.25	0.45
	3	0.43	0.41	0.36	0.32	0.28	0.25	0.43
	4	0.41	0.39	0.34	0.31	0.27	0.25	0.41
	5	0.38	0.37	0.33	0.30	0.26	0.24	0.38
dx=0.1	0	0.72	0.68	0.57	0.49	0.40	0.35	0.72
	0.5	0.70	0.66	0.56	0.48	0.39	0.35	0.70
	1	0.68	0.65	0.55	0.47	0.39	0.34	0.68
	2	0.64	0.62	0.53	0.45	0.38	0.34	0.64
	3	0.61	0.58	0.50	0.44	0.37	0.33	0.61
	4	0.58	0.56	0.48	0.43	0.36	0.33	0.58
	5	0.55	0.53	0.46	0.42	0.36	0.32	0.55
dx=0.5	0	1.21	1.15	0.95	0.81	0.65	0.57	1.21
	0.5	1.18	1.12	0.94	0.80	0.65	0.56	1.18
	1	1.16	1.10	0.92	0.79	0.64	0.56	1.16
	2	1.10	1.05	0.89	0.77	0.63	0.55	1.10
	3	1.05	1.01	0.87	0.75	0.62	0.55	1.05
	4	1.01	0.97	0.84	0.74	0.61	0.54	1.01
	5	0.97	0.93	0.81	0.72	0.60	0.53	0.97
dx=1	0	1.50	1.42	1.18	1.00	0.79	0.69	1.50
	0.5	1.47	1.39	1.16	0.99	0.79	0.69	1.47
	1	1.44	1.36	1.14	0.98	0.79	0.69	1.44
	2	1.38	1.31	1.11	0.96	0.78	0.68	1.38
	3	1.32	1.26	1.07	0.94	0.77	0.68	1.32
	4	1.28	1.22	1.05	0.92	0.76	0.67	1.28
	5	1.23	1.18	1.03	0.90	0.75	0.66	1.23

Linear soil profile								
h/H	d_θ	d_w=0	d_w=1	d_w=5	d_w=10	d_w=20	d_w=30	d_w=40
dx=0	0	0.50	0.48	0.42	0.37	0.31	0.27	0.25
	0.5	0.50	0.48	0.42	0.37	0.31	0.27	0.25
	1	0.50	0.48	0.42	0.37	0.31	0.27	0.25
	2	0.50	0.48	0.42	0.37	0.31	0.27	0.25
	3	0.50	0.48	0.42	0.37	0.31	0.27	0.25
	4	0.50	0.48	0.42	0.37	0.31	0.27	0.25
	5	0.50	0.48	0.42	0.37	0.31	0.27	0.25
dx=0.1	0	0.72	0.70	0.60	0.54	0.43	0.39	0.35
	0.5	0.72	0.70	0.60	0.54	0.43	0.39	0.35
	1	0.73	0.70	0.60	0.54	0.43	0.39	0.35
	2	0.73	0.71	0.60	0.54	0.43	0.39	0.35
	3	0.73	0.71	0.62	0.54	0.44	0.39	0.36
	4	0.74	0.72	0.63	0.55	0.44	0.39	0.36
	5	0.74	0.72	0.63	0.55	0.44	0.39	0.36
dx=0.5	0	1.24	1.19	1.03	0.90	0.76	0.64	0.57
	0.5	1.26	1.20	1.04	0.90	0.76	0.64	0.58
	1	1.27	1.22	1.05	0.91	0.77	0.65	0.58
	2	1.30	1.24	1.07	0.92	0.77	0.65	0.58
	3	1.32	1.27	1.09	0.94	0.78	0.66	0.59
	4	1.35	1.29	1.10	0.95	0.79	0.66	0.59
	5	1.38	1.32	1.12	0.97	0.80	0.67	0.60
dx=1	0	1.50	1.45	1.29	1.11	0.89	0.82	0.69
	0.5	1.52	1.47	1.30	1.12	0.90	0.82	0.70
	1	1.54	1.49	1.31	1.13	0.91	0.83	0.70
	2	1.60	1.53	1.34	1.15	0.92	0.84	0.71
	3	1.64	1.57	1.37	1.17	0.93	0.85	0.72
	4	1.69	1.61	1.40	1.19	0.95	0.86	0.73
	5	1.75	1.66	1.43	1.21	0.96	0.87	0.74

Normalized base shear forces and effective heights for <u>tension forces</u> developed at the wall <u>set to null</u>. The results arise from the time domain analysis and for **<u>bonded</u>** contact.

Homogeneous soil								
$P/\rho A g H^2$	d_θ	$d_w=0$	$d_w=1$	$d_w=5$	$d_w=10$	$d_w=20$	$d_w=30$	$d_w=40$
dx=0	0	1.00	0.89	0.68	0.60	0.53	0.48	0.45
	0.5	0.74	0.69	0.59	0.55	0.50	0.46	0.43
	1	0.6	0.57	0.53	0.51	0.47	0.44	0.41
	2	0.46	0.46	0.46	0.45	0.43	0.40	0.39
	3	0.41	0.41	0.41	0.41	0.39	0.38	0.36
	4	0.38	0.38	0.38	0.38	0.37	0.35	0.34
	5	0.36	0.36	0.36	0.36	0.35	0.34	0.32
dx=0.1	0	0.79	0.70	0.54	0.48	0.42	0.39	0.36
	0.5	0.60	0.56	0.47	0.44	0.40	0.37	0.35
	1	0.49	0.47	0.43	0.41	0.38	0.36	0.34
	2	0.39	0.38	0.38	0.37	0.35	0.33	0.32
	3	0.34	0.34	0.34	0.34	0.33	0.31	0.30
	4	0.32	0.32	0.32	0.32	0.31	0.30	0.28
	5	0.30	0.30	0.30	0.30	0.29	0.28	0.27
dx=0.5	0	0.65	0.57	0.44	0.39	0.35	0.32	0.30
	0.5	0.50	0.47	0.39	0.37	0.33	0.31	0.29
	1	0.42	0.40	0.36	0.34	0.32	0.30	0.28
	2	0.33	0.33	0.32	0.31	0.29	0.28	0.26
	3	0.30	0.30	0.30	0.29	0.28	0.26	0.25
	4	0.28	0.28	0.28	0.27	0.26	0.25	0.24
	5	0.26	0.26	0.26	0.26	0.25	0.24	0.23
dx=1	0	0.62	0.54	0.41	0.37	0.33	0.30	0.28
	0.5	0.48	0.45	0.37	0.35	0.31	0.29	0.27
	1	0.40	0.38	0.35	0.33	0.30	0.28	0.26
	2	0.32	0.32	0.31	0.30	0.28	0.26	0.25
	3	0.29	0.28	0.28	0.28	0.27	0.25	0.24
	4	0.27	0.27	0.27	0.26	0.25	0.24	0.23
	5	0.25	0.25	0.25	0.25	0.24	0.23	0.22

Parabolic soil profile								
$P/\rho AgH^2$	d_θ	$d_w=0$	$d_w=1$	$d_w=5$	$d_w=10$	$d_w=20$	$d_w=30$	$d_w=40$
dx=0	0	0.88	0.79	0.62	0.52	0.43	0.39	0.37
	0.5	0.65	0.61	0.52	0.46	0.40	0.37	0.35
	1	0.52	0.50	0.45	0.41	0.37	0.34	0.33
	2	0.38	0.37	0.36	0.34	0.32	0.31	0.30
	3	0.31	0.31	0.30	0.29	0.29	0.28	0.27
	4	0.26	0.26	0.26	0.26	0.26	0.26	0.25
	5	0.23	0.23	0.23	0.24	0.24	0.24	0.24
dx=0.1	0	0.66	0.60	0.46	0.38	0.31	0.28	0.27
	0.5	0.50	0.46	0.39	0.34	0.29	0.27	0.25
	1	0.40	0.38	0.34	0.30	0.27	0.25	0.24
	2	0.30	0.29	0.27	0.25	0.24	0.23	0.22
	3	0.24	0.24	0.23	0.22	0.21	0.21	0.20
	4	0.20	0.20	0.20	0.20	0.20	0.19	0.19
	5	0.18	0.18	0.18	0.18	0.18	0.18	0.18
dx=0.5	0	0.56	0.50	0.38	0.32	0.26	0.23	0.21
	0.5	0.43	0.40	0.33	0.28	0.24	0.22	0.20
	1	0.36	0.34	0.29	0.26	0.22	0.21	0.20
	2	0.27	0.26	0.24	0.22	0.20	0.19	0.18
	3	0.22	0.21	0.20	0.19	0.18	0.17	0.17
	4	0.18	0.18	0.18	0.17	0.17	0.16	0.16
	5	0.16	0.16	0.16	0.16	0.15	0.15	0.15
dx=1	0	0.54	0.49	0.37	0.30	0.25	0.22	0.20
	0.5	0.42	0.39	0.32	0.27	0.23	0.21	0.20
	1	0.35	0.33	0.28	0.25	0.21	0.20	0.19
	2	0.26	0.25	0.23	0.21	0.19	0.18	0.17
	3	0.21	0.21	0.20	0.18	0.17	0.17	0.16
	4	0.18	0.18	0.17	0.16	0.16	0.16	0.15
	5	0.16	0.16	0.15	0.15	0.15	0.15	0.14

Linear soil profile								
$P/\rho AgH^2$	d_θ	d_w=0	d_w=1	d_w=5	d_w=10	d_w=20	d_w=30	d_w=40
dx=0	0	0.71	0.65	0.51	0.42	0.35	0.31	0.28
	0.5	0.52	0.49	0.42	0.37	0.31	0.28	0.26
	1	0.41	0.40	0.35	0.32	0.28	0.26	0.25
	2	0.29	0.28	0.27	0.26	0.24	0.23	0.22
	3	0.22	0.22	0.22	0.21	0.21	0.20	0.20
	4	0.18	0.18	0.18	0.18	0.18	0.18	0.18
	5	0.15	0.15	0.16	0.16	0.16	0.16	0.16
dx=0.1	0	0.54	0.50	0.38	0.31	0.26	0.22	0.20
	0.5	0.42	0.39	0.32	0.27	0.23	0.21	0.19
	1	0.34	0.32	0.28	0.24	0.21	0.19	0.18
	2	0.24	0.23	0.22	0.20	0.18	0.17	0.16
	3	0.19	0.19	0.18	0.17	0.16	0.15	0.15
	4	0.16	0.15	0.15	0.15	0.14	0.14	0.13
	5	0.13	0.13	0.13	0.13	0.13	0.13	0.12
dx=0.5	0	0.47	0.43	0.33	0.27	0.21	0.18	0.17
	0.5	0.37	0.34	0.28	0.24	0.19	0.17	0.16
	1	0.30	0.29	0.24	0.21	0.18	0.16	0.15
	2	0.22	0.21	0.19	0.18	0.15	0.14	0.14
	3	0.18	0.17	0.16	0.15	0.14	0.13	0.12
	4	0.15	0.14	0.14	0.13	0.12	0.12	0.11
	5	0.12	0.12	0.12	0.12	0.11	0.11	0.11
dx=1	0	0.46	0.42	0.32	0.26	0.20	0.17	0.16
	0.5	0.36	0.34	0.27	0.23	0.19	0.16	0.15
	1	0.30	0.28	0.23	0.20	0.17	0.15	0.14
	2	0.22	0.21	0.19	0.17	0.15	0.14	0.13
	3	0.17	0.17	0.16	0.15	0.13	0.12	0.12
	4	0.14	0.14	0.13	0.13	0.12	0.11	0.11
	5	0.12	0.12	0.12	0.11	0.11	0.10	0.10

Homogeneous soil								
h/H	d_θ	d_w=0	d_w=1	d_w=5	d_w=10	d_w=20	d_w=30	d_w=40
dx=0	0	0.59	0.55	0.44	0.40	0.36	0.34	0.33
	0.5	0.55	0.51	0.43	0.40	0.36	0.34	0.33
	1	0.51	0.48	0.42	0.39	0.36	0.34	0.33
	2	0.46	0.45	0.41	0.39	0.36	0.34	0.33
	3	0.44	0.43	0.41	0.39	0.36	0.35	0.33
	4	0.43	0.42	0.40	0.39	0.36	0.35	0.33
	5	0.42	0.41	0.40	0.39	0.36	0.35	0.33
dx=0.1	0	0.65	0.61	0.50	0.45	0.41	0.39	0.37
	0.5	0.61	0.57	0.49	0.45	0.41	0.39	0.37
	1	0.57	0.54	0.48	0.44	0.41	0.39	0.37
	2	0.52	0.5	0.47	0.44	0.41	0.39	0.37
	3	0.49	0.48	0.46	0.44	0.41	0.39	0.38
	4	0.48	0.47	0.45	0.43	0.41	0.39	0.38
	5	0.47	0.46	0.45	0.43	0.41	0.39	0.38
dx=0.5	0	0.68	0.64	0.54	0.49	0.45	0.43	0.41
	0.5	0.64	0.61	0.53	0.49	0.45	0.43	0.41
	1	0.61	0.58	0.52	0.48	0.45	0.43	0.41
	2	0.56	0.54	0.50	0.48	0.45	0.43	0.41
	3	0.53	0.52	0.50	0.47	0.45	0.43	0.41
	4	0.52	0.51	0.49	0.47	0.45	0.43	0.42
	5	0.50	0.50	0.48	0.47	0.45	0.43	0.42
dx=1	0	0.69	0.65	0.55	0.50	0.46	0.44	0.42
	0.5	0.65	0.62	0.54	0.50	0.46	0.44	0.42
	1	0.62	0.59	0.53	0.49	0.46	0.44	0.43
	2	0.57	0.55	0.51	0.49	0.46	0.44	0.43
	3	0.54	0.53	0.50	0.48	0.46	0.44	0.43
	4	0.53	0.52	0.50	0.48	0.46	0.44	0.43
	5	0.51	0.51	0.49	0.48	0.46	0.44	0.43

Parabolic soil profile								
h/H	d_θ	d_w=0	d_w=1	d_w=5	d_w=10	d_w=20	d_w=30	d_w=40
dx=0	0	0.51	0.48	0.41	0.36	0.30	0.27	0.51
	0.5	0.49	0.47	0.40	0.35	0.30	0.27	0.49
	1	0.48	0.46	0.39	0.34	0.29	0.27	0.48
	2	0.45	0.43	0.38	0.33	0.29	0.27	0.45
	3	0.43	0.41	0.36	0.32	0.29	0.27	0.43
	4	0.41	0.39	0.35	0.32	0.29	0.27	0.41
	5	0.38	0.37	0.34	0.31	0.28	0.27	0.38
dx=0.1	0	0.58	0.55	0.48	0.42	0.35	0.32	0.58
	0.5	0.56	0.54	0.47	0.41	0.35	0.32	0.56
	1	0.55	0.53	0.46	0.40	0.34	0.32	0.55
	2	0.52	0.50	0.44	0.39	0.34	0.32	0.52
	3	0.50	0.48	0.42	0.38	0.34	0.31	0.50
	4	0.48	0.46	0.41	0.37	0.33	0.31	0.48
	5	0.45	0.44	0.40	0.36	0.33	0.31	0.45
dx=0.5	0	0.61	0.58	0.52	0.46	0.38	0.35	0.61
	0.5	0.59	0.57	0.51	0.45	0.38	0.35	0.59
	1	0.58	0.56	0.50	0.44	0.38	0.35	0.58
	2	0.56	0.54	0.48	0.43	0.38	0.35	0.56
	3	0.54	0.52	0.47	0.42	0.37	0.35	0.54
	4	0.52	0.50	0.45	0.41	0.37	0.35	0.52
	5	0.50	0.49	0.44	0.40	0.36	0.34	0.50
dx=1	0	0.61	0.59	0.52	0.47	0.39	0.36	0.61
	0.5	0.60	0.58	0.52	0.46	0.39	0.36	0.60
	1	0.59	0.57	0.51	0.45	0.39	0.36	0.59
	2	0.57	0.55	0.49	0.44	0.38	0.36	0.57
	3	0.55	0.53	0.48	0.43	0.38	0.35	0.55
	4	0.53	0.51	0.46	0.42	0.38	0.35	0.53
	5	0.51	0.50	0.45	0.41	0.38	0.35	0.51

Linear soil profile								
h/H	d_θ	d_w=0	d_w=1	d_w=5	d_w=10	d_w=20	d_w=30	d_w=40
dx=0	0	0.50	0.48	0.42	0.37	0.31	0.27	0.25
	0.5	0.50	0.48	0.42	0.37	0.31	0.27	0.25
	1	0.50	0.48	0.42	0.37	0.31	0.27	0.25
	2	0.50	0.48	0.42	0.37	0.31	0.27	0.25
	3	0.50	0.48	0.42	0.37	0.31	0.27	0.25
	4	0.50	0.48	0.42	0.37	0.31	0.27	0.25
	5	0.50	0.48	0.42	0.37	0.31	0.27	0.25
dx=0.1	0	0.57	0.55	0.49	0.44	0.37	0.33	0.31
	0.5	0.56	0.55	0.49	0.44	0.37	0.34	0.31
	1	0.56	0.55	0.49	0.44	0.37	0.34	0.31
	2	0.56	0.55	0.49	0.45	0.37	0.34	0.31
	3	0.56	0.55	0.50	0.45	0.37	0.34	0.31
	4	0.57	0.55	0.50	0.45	0.37	0.34	0.31
	5	0.57	0.55	0.50	0.45	0.37	0.34	0.31
dx=0.5	0	0.59	0.58	0.53	0.48	0.42	0.37	0.34
	0.5	0.59	0.58	0.53	0.48	0.42	0.37	0.34
	1	0.59	0.58	0.53	0.48	0.42	0.37	0.34
	2	0.60	0.58	0.53	0.48	0.42	0.37	0.34
	3	0.60	0.58	0.53	0.49	0.42	0.37	0.34
	4	0.60	0.59	0.54	0.49	0.42	0.37	0.34
	5	0.60	0.59	0.54	0.49	0.42	0.38	0.35
dx=1	0	0.60	0.59	0.54	0.49	0.43	0.38	0.35
	0.5	0.60	0.59	0.54	0.49	0.43	0.38	0.35
	1	0.60	0.59	0.54	0.49	0.43	0.39	0.35
	2	0.60	0.59	0.54	0.49	0.43	0.39	0.35
	3	0.60	0.59	0.54	0.50	0.43	0.39	0.35
	4	0.60	0.59	0.54	0.50	0.43	0.39	0.35
	5	0.60	0.59	0.55	0.50	0.43	0.39	0.35

Amplification factors for different soil profiles for the one-wall system with bonded contact (time domain analysis).

Homogeneous soil							
	d_w=0	d_w=1	d_w=5	d_w=10	d_w=20	d_w=30	d_w=40
d_θ=0	3.53	4.06	5.40	6.17	6.86	7.21	7.42
d_θ=0.5	4.65	5.01	5.93	6.49	7.02	7.30	7.49
d_θ=1	5.45	5.70	6.32	6.73	7.15	7.39	7.54
d_θ=2	6.40	6.49	6.81	7.04	7.33	7.50	7.64
d_θ=3	6.83	6.87	7.06	7.23	7.44	7.59	7.70
d_θ=4	7.03	7.06	7.20	7.33	7.51	7.65	7.75
d_θ=5	7.14	7.14	7.26	7.38	7.56	7.68	7.78

Parabolic soil							
	d_w=0	d_w=1	d_w=5	d_w=10	d_w=20	d_w=30	d_w=40
d_θ=0	3.30	3.67	4.74	5.53	6.31	6.68	6.90
d_θ=0.5	4.15	4.43	5.25	5.85	6.46	6.76	6.95
d_θ=1	4.79	5.01	5.62	6.08	6.57	6.83	6.99
d_θ=2	5.60	5.72	6.09	6.37	6.72	6.91	7.05
d_θ=3	5.98	6.07	6.31	6.52	6.78	6.95	7.06
d_θ=4	6.13	6.20	6.39	6.56	6.80	6.95	7.06
d_θ=5	6.14	6.20	6.38	6.55	6.78	6.93	7.05

Linear soil							
	d_w=0	d_w=1	d_w=5	d_w=10	d_w=20	d_w=30	d_w=40
d_θ=0	3.18	3.49	4.42	5.13	5.84	6.17	6.35
d_θ=0.5	3.90	4.13	4.84	5.38	5.94	6.21	6.36
d_θ=1	4.41	4.60	5.14	5.56	6.00	6.23	6.36
d_θ=2	5.02	5.12	5.45	5.72	6.03	6.21	6.34
d_θ=3	5.21	5.29	5.51	5.72	5.99	6.16	6.29
d_θ=4	5.18	5.22	5.43	5.62	5.89	6.08	6.21
d_θ=5	4.96	5.02	5.24	5.45	5.76	5.97	6.12

Normalized base shear forces and amplification factors (AF) for the three soil profiles investigated here and for non-bonded contact (separation is allowed). The results are almost identical with the results of the **smooth** contact and arise from a **frequency domain** (steady state analysis).

Homogeneous soil								
$P/\rho A g H^2$	d_θ	d_w=0	d_w=1	d_w=5	d_w=10	d_w=20	d_w=30	d_w=40
dx=0	0	0.970	0.897	0.729	0.638	0.552	0.506	0.474
	0.5	0.769	0.736	0.642	0.582	0.519	0.482	0.456
	1	0.650	0.634	0.578	0.538	0.491	0.461	0.439
	2	0.517	0.515	0.493	0.474	0.446	0.426	0.409
	3	0.445	0.448	0.439	0.429	0.412	0.398	0.385
	4	0.401	0.404	0.401	0.396	0.385	0.375	0.365
	5	0.370	0.374	0.373	0.371	0.363	0.355	0.348
dx=0.1	0	0.694	0.646	0.533	0.471	0.413	0.381	0.359
	0.5	0.563	0.540	0.476	0.434	0.390	0.364	0.346
	1	0.483	0.471	0.432	0.404	0.371	0.350	0.334
	2	0.391	0.389	0.373	0.360	0.340	0.326	0.314
	3	0.340	0.341	0.335	0.328	0.316	0.306	0.297
	4	0.307	0.310	0.308	0.304	0.297	0.290	0.283
	5	0.285	0.288	0.288	0.286	0.281	0.276	0.270
dx=0.5	0	0.382	0.358	0.301	0.270	0.239	0.222	0.211
	0.5	0.318	0.307	0.273	0.251	0.228	0.214	0.205
	1	0.277	0.272	0.251	0.236	0.218	0.207	0.199
	2	0.229	0.228	0.220	0.213	0.202	0.194	0.188
	3	0.201	0.202	0.199	0.196	0.189	0.184	0.179
	4	0.183	0.185	0.184	0.183	0.179	0.175	0.172
	5	0.171	0.173	0.173	0.173	0.171	0.168	0.165
dx=1	0	0.286	0.269	0.228	0.205	0.182	0.170	0.162
	0.5	0.241	0.232	0.207	0.191	0.174	0.164	0.157
	1	0.211	0.207	0.191	0.180	0.167	0.159	0.153
	2	0.176	0.175	0.169	0.163	0.155	0.150	0.145
	3	0.155	0.156	0.153	0.151	0.146	0.142	0.139
	4	0.142	0.143	0.142	0.141	0.138	0.136	0.133
	5	0.132	0.133	0.134	0.134	0.132	0.130	0.128

Parabolic soil profile								
$P/\rho AgH^2$	d_θ	d_w=0	d_w=1	d_w=5	d_w=10	d_w=20	d_w=30	d_w=40
dx=0	0	0.806	0.763	0.640	0.561	0.483	0.441	0.413
	0.5	0.646	0.627	0.555	0.504	0.448	0.416	0.394
	1	0.548	0.537	0.494	0.460	0.419	0.394	0.376
	2	0.429	0.427	0.411	0.395	0.374	0.359	0.347
	3	0.362	0.362	0.357	0.351	0.340	0.331	0.323
	4	0.316	0.320	0.319	0.318	0.313	0.308	0.303
	5	0.285	0.289	0.291	0.292	0.292	0.289	0.286
dx=0.1	0	0.531	0.506	0.430	0.381	0.332	0.305	0.288
	0.5	0.437	0.424	0.379	0.346	0.310	0.290	0.276
	1	0.375	0.369	0.340	0.319	0.292	0.276	0.265
	2	0.299	0.299	0.287	0.278	0.264	0.254	0.246
	3	0.255	0.256	0.252	0.248	0.242	0.236	0.231
	4	0.225	0.227	0.227	0.227	0.224	0.221	0.218
	5	0.204	0.206	0.208	0.210	0.210	0.209	0.207
dx=0.5	0	0.287	0.274	0.236	0.211	0.186	0.172	0.163
	0.5	0.241	0.234	0.211	0.194	0.175	0.165	0.157
	1	0.209	0.206	0.191	0.180	0.166	0.158	0.152
	2	0.170	0.170	0.164	0.159	0.151	0.146	0.142
	3	0.146	0.147	0.145	0.143	0.140	0.137	0.135
	4	0.130	0.132	0.132	0.132	0.131	0.129	0.128
	5	0.119	0.120	0.122	0.123	0.123	0.123	0.122
dx=1	0	0.222	0.213	0.184	0.165	0.146	0.135	0.128
	0.5	0.188	0.183	0.165	0.152	0.138	0.129	0.124
	1	0.164	0.162	0.150	0.141	0.131	0.124	0.120
	2	0.134	0.134	0.129	0.125	0.120	0.116	0.113
	3	0.116	0.116	0.115	0.113	0.111	0.108	0.107
	4	0.103	0.104	0.104	0.104	0.104	0.103	0.101
	5	0.094	0.095	0.096	0.097	0.098	0.097	0.097

Linear soil profile								
$P/\rho A g H^2$	d_θ	d_w=0	d_w=1	d_w=5	d_w=10	d_w=20	d_w=30	d_w=40
dx=0	0	0.749	0.713	0.602	0.528	0.452	0.411	0.385
	0.5	0.600	0.583	0.518	0.470	0.416	0.385	0.364
	1	0.504	0.496	0.456	0.424	0.386	0.363	0.346
	2	0.386	0.387	0.372	0.358	0.339	0.326	0.316
	3	0.318	0.321	0.317	0.312	0.304	0.297	0.291
	4	0.272	0.277	0.278	0.278	0.277	0.274	0.271
	5	0.241	0.245	0.249	0.252	0.255	0.255	0.254
dx=0.1	0	0.469	0.451	0.386	0.342	0.296	0.271	0.255
	0.5	0.387	0.377	0.337	0.308	0.275	0.256	0.243
	1	0.331	0.325	0.301	0.281	0.257	0.243	0.232
	2	0.258	0.258	0.249	0.240	0.229	0.220	0.214
	3	0.215	0.217	0.214	0.211	0.207	0.203	0.199
	4	0.186	0.188	0.189	0.190	0.189	0.188	0.186
	5	0.165	0.168	0.171	0.173	0.175	0.176	0.176
dx=0.5	0	0.258	0.248	0.214	0.191	0.167	0.154	0.146
	0.5	0.216	0.211	0.189	0.174	0.156	0.146	0.139
	1	0.186	0.184	0.170	0.160	0.147	0.139	0.134
	2	0.148	0.148	0.143	0.138	0.132	0.128	0.124
	3	0.124	0.125	0.124	0.123	0.120	0.118	0.116
	4	0.108	0.110	0.111	0.111	0.111	0.110	0.110
	5	0.097	0.098	0.100	0.102	0.103	0.104	0.104
dx=1	0	0.207	0.198	0.172	0.153	0.134	0.124	0.118
	0.5	0.173	0.169	0.152	0.140	0.126	0.118	0.113
	1	0.150	0.148	0.137	0.129	0.119	0.113	0.108
	2	0.120	0.120	0.116	0.112	0.107	0.103	0.101
	3	0.101	0.102	0.101	0.100	0.098	0.096	0.095
	4	0.088	0.089	0.090	0.090	0.090	0.090	0.089
	5	0.079	0.080	0.081	0.083	0.084	0.085	0.085

Homogeneous soil								
AF	d_θ	d_w=0	d_w=1	d_w=5	d_w=10	d_w=20	d_w=30	d_w=40
dx=0	0	3.59	4.17	6.15	7.71	9.17	9.80	10.16
	0.5	4.94	5.45	7.13	8.35	9.48	9.99	10.28
	1	6.20	6.61	7.95	8.87	9.74	10.15	10.40
	2	8.13	8.31	9.06	9.58	10.11	10.39	10.56
	3	9.25	9.29	9.70	10.00	10.34	10.55	10.69
	4	9.83	9.82	10.06	10.25	10.50	10.66	10.78
	5	10.16	10.09	10.25	10.41	10.60	10.74	10.84
dx=0.1	0	4.23	4.87	7.01	8.54	9.83	10.35	10.63
	0.5	5.63	6.19	7.93	9.09	10.07	10.48	10.71
	1	6.90	7.32	8.66	9.52	10.26	10.59	10.78
	2	8.70	8.91	9.62	10.09	10.52	10.75	10.90
	3	9.70	9.75	10.14	10.41	10.70	10.87	10.98
	4	10.20	10.18	10.41	10.59	10.81	10.94	11.04
	5	10.45	10.39	10.56	10.69	10.87	10.99	11.08
dx=0.5	0	5.21	5.94	8.18	9.56	10.58	10.94	11.12
	0.5	6.65	7.23	8.96	9.96	10.72	11.01	11.17
	1	7.87	8.29	9.54	10.27	10.84	11.08	11.20
	2	9.45	9.63	10.27	10.66	11.00	11.16	11.27
	3	10.28	10.31	10.64	10.87	11.10	11.22	11.30
	4	10.67	10.62	10.83	10.97	11.14	11.26	11.32
	5	10.82	10.77	10.91	11.03	11.18	11.27	11.34
dx=1	0	5.58	6.33	8.55	9.87	10.79	11.09	11.25
	0.5	7.02	7.60	9.28	10.23	10.91	11.16	11.28
	1	8.20	8.60	9.82	10.49	11.01	11.21	11.32
	2	9.68	9.86	10.47	10.82	11.14	11.27	11.37
	3	10.42	10.47	10.79	11.00	11.21	11.32	11.39
	4	10.78	10.75	10.95	11.09	11.25	11.34	11.41
	5	10.94	10.88	11.02	11.12	11.27	11.35	11.42

Parabolic soil profile								
AF	d_θ	d_w=0	d_w=1	d_w=5	d_w=10	d_w=20	d_w=30	d_w=40
dx=0	0	3.37	3.77	5.54	7.56	10.00	10.81	11.06
	0.5	4.50	4.91	6.70	8.53	10.38	10.94	11.12
	1	5.65	6.06	7.75	9.28	10.65	11.03	11.14
	2	7.74	8.07	9.31	10.22	10.91	11.10	11.14
	3	9.18	9.38	10.12	10.61	10.98	11.08	11.11
	4	9.94	9.99	10.43	10.72	10.95	11.03	11.06
	5	10.19	10.19	10.48	10.68	10.85	10.94	10.98
dx=0.1	0	3.96	4.42	6.44	8.59	10.75	11.31	11.42
	0.5	5.14	5.61	7.58	9.41	10.99	11.36	11.43
	1	6.32	6.77	8.56	10.01	11.14	11.38	11.41
	2	8.36	8.68	9.88	10.70	11.25	11.36	11.36
	3	9.65	9.80	10.51	10.94	11.23	11.29	11.29
	4	10.25	10.30	10.70	10.95	11.14	11.18	11.20
	5	10.41	10.41	10.67	10.85	11.02	11.09	11.12
dx=0.5	0	4.64	5.18	7.43	9.57	11.33	11.64	11.66
	0.5	5.87	6.39	8.48	10.21	11.43	11.64	11.62
	1	7.06	7.52	9.33	10.65	11.50	11.62	11.58
	2	8.98	9.26	10.40	11.09	11.50	11.54	11.50
	3	10.07	10.20	10.83	11.21	11.42	11.43	11.41
	4	10.54	10.56	10.93	11.15	11.29	11.31	11.31
	5	10.61	10.60	10.85	11.02	11.14	11.19	11.21
dx=1	0	4.86	5.41	7.71	9.83	11.45	11.70	11.70
	0.5	6.09	6.63	8.74	10.40	11.54	11.69	11.65
	1	7.27	7.74	9.53	10.79	11.57	11.65	11.62
	2	9.11	9.41	10.52	11.19	11.54	11.57	11.52
	3	10.16	10.30	10.91	11.27	11.45	11.46	11.44
	4	10.59	10.63	10.99	11.20	11.33	11.33	11.33
	5	10.64	10.66	10.90	11.06	11.18	11.22	11.24

Linear soil profile								
AF	d_θ	d_w=0	d_w=1	d_w=5	d_w=10	d_w=20	d_w=30	d_w=40
dx=0	0	3.25	3.59	5.12	7.02	9.53	10.37	10.57
	0.5	4.21	4.55	6.13	7.90	9.88	10.44	10.56
	1	5.17	5.51	7.05	8.60	10.10	10.48	10.55
	2	6.93	7.20	8.44	9.45	10.26	10.45	10.46
	3	8.16	8.32	9.16	9.77	10.23	10.34	10.35
	4	8.78	8.85	9.39	9.78	10.10	10.20	10.24
	5	8.92	8.95	9.34	9.64	9.92	10.03	10.10
dx=0.1	0	3.75	4.11	5.82	7.85	10.16	10.75	10.83
	0.5	4.70	5.08	6.79	8.61	10.37	10.76	10.79
	1	5.65	6.02	7.65	9.18	10.47	10.74	10.73
	2	7.35	7.63	8.86	9.82	10.52	10.64	10.62
	3	8.45	8.63	9.44	10.00	10.42	10.48	10.49
	4	8.97	9.06	9.59	9.95	10.24	10.32	10.34
	5	9.07	9.10	9.48	9.76	10.03	10.15	10.20
dx=0.5	0	4.18	4.60	6.46	8.53	10.57	10.97	10.97
	0.5	5.15	5.56	7.38	9.16	10.68	10.95	10.92
	1	6.10	6.49	8.16	9.61	10.73	10.89	10.85
	2	7.69	7.99	9.21	10.09	10.68	10.75	10.71
	3	8.71	8.88	9.66	10.19	10.54	10.59	10.56
	4	9.15	9.23	9.73	10.09	10.35	10.40	10.42
	5	9.20	9.23	9.60	9.88	10.13	10.23	10.27
dx=1	0	4.31	4.73	6.63	8.70	10.65	11.02	11.00
	0.5	5.29	5.69	7.53	9.29	10.75	10.98	10.94
	1	6.21	6.61	8.29	9.71	10.77	10.92	10.87
	2	7.80	8.08	9.29	10.15	10.71	10.77	10.73
	3	8.77	8.94	9.71	10.24	10.56	10.61	10.58
	4	9.21	9.28	9.78	10.13	10.37	10.43	10.44
	5	9.23	9.27	9.64	9.92	10.16	10.25	10.29

Annex D

Normalized total soil pressure on rigid walls for different values of the non-homogeneity constant α (0.125, 0.250) and for the non-homogeneity parameter Ξo (0.7, 0.8, 0.9), for different ratios of L/H (1.0, 1.5, 2.0, 3.0, 5.0, 10.0).

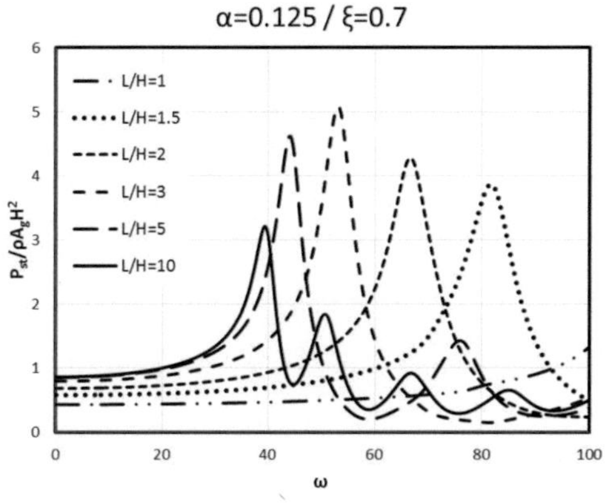

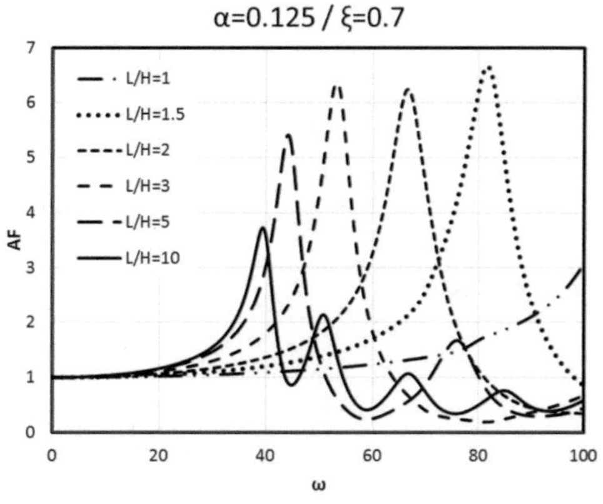

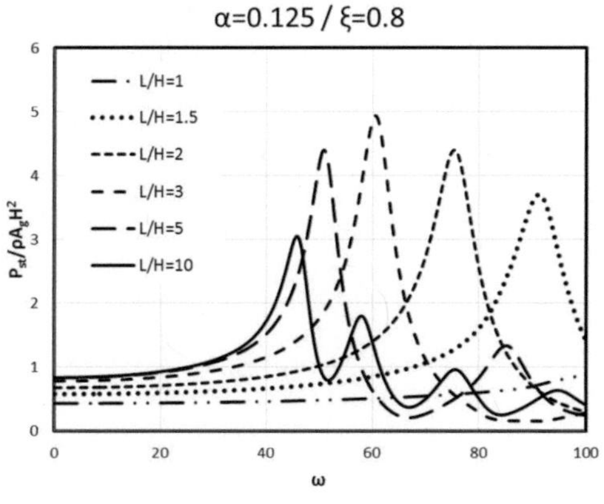

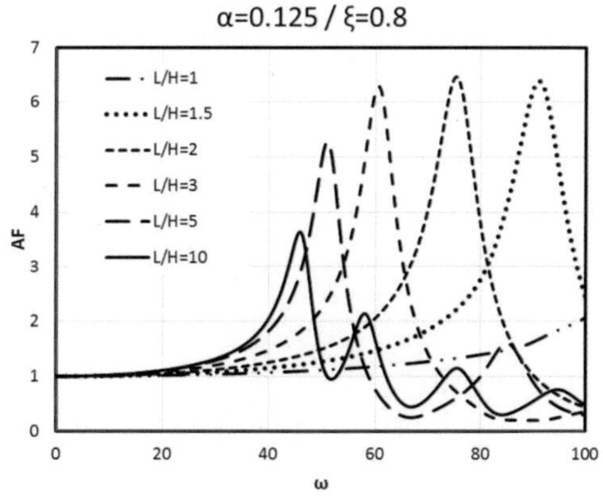

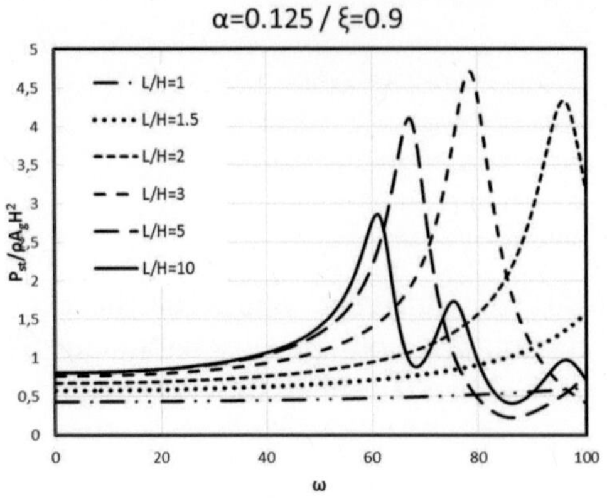
α=0.125 / ξ=0.9
L/H=1
L/H=1.5
L/H=2
L/H=3
L/H=5
L/H=10
$P_{st}/\rho A_g H^2$
ω

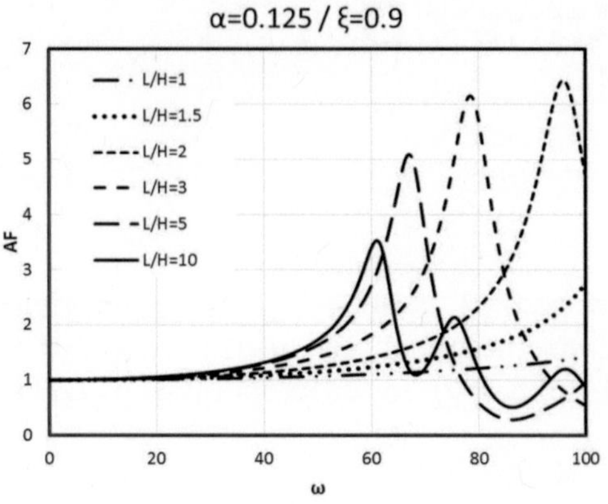
α=0.125 / ξ=0.9
L/H=1
L/H=1.5
L/H=2
L/H=3
L/H=5
L/H=10
AF
ω

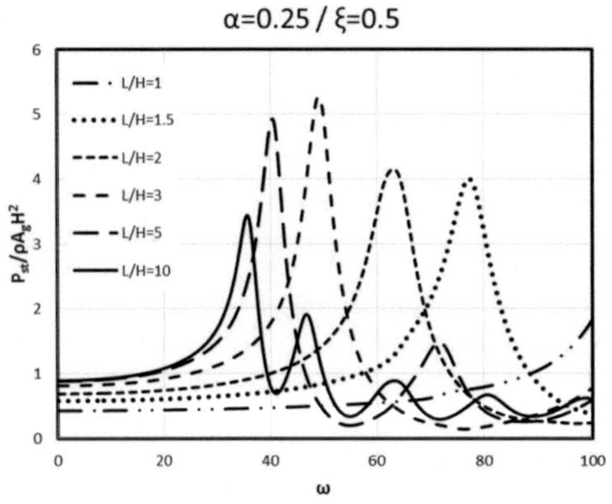
α=0.25 / ξ=0.5
L/H=1
L/H=1.5
L/H=2
L/H=3
L/H=5
L/H=10
$P_{st}/\rho A_g H^2$
ω

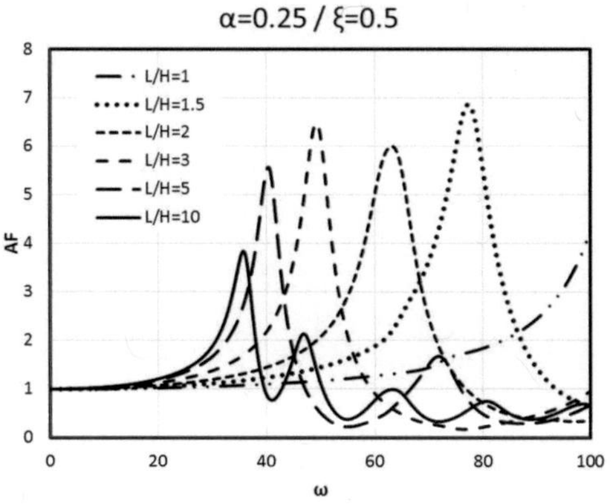
α=0.25 / ξ=0.5
L/H=1
L/H=1.5
L/H=2
L/H=3
L/H=5
L/H=10
AF
ω

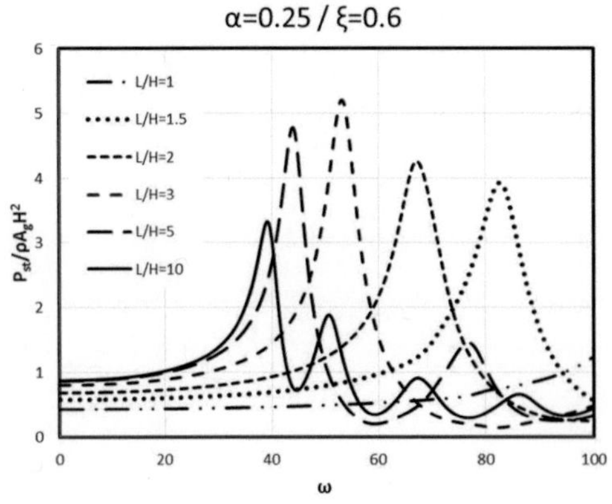
α=0.25 / ξ=0.6
L/H=1
L/H=1.5
L/H=2
L/H=3
L/H=5
L/H=10
$P_{st}/\rho A_g H^2$
ω

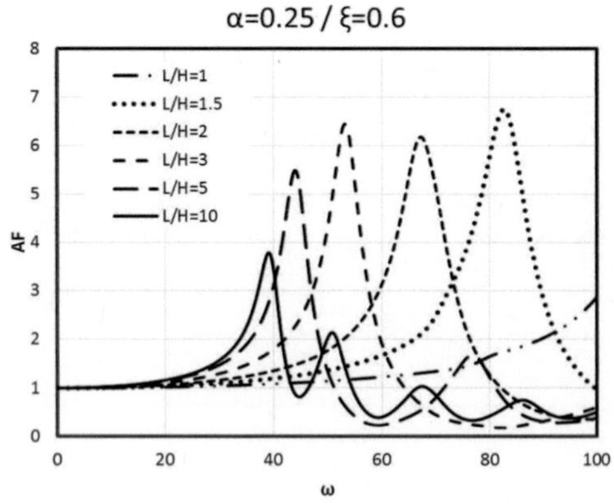
α=0.25 / ξ=0.6
L/H=1
L/H=1.5
L/H=2
L/H=3
L/H=5
L/H=10
AF
ω

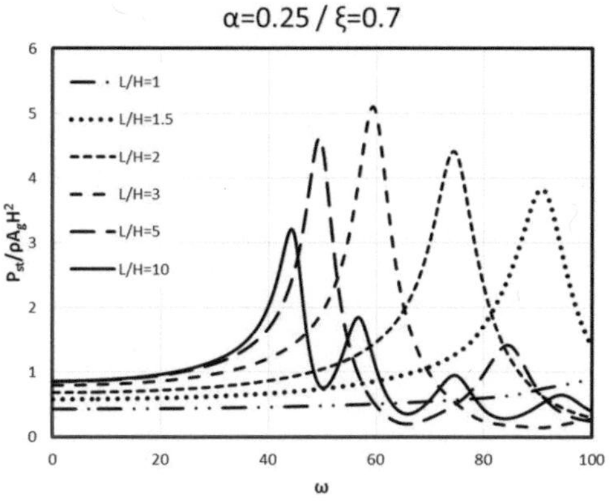
α=0.25 / ξ=0.7
L/H=1
L/H=1.5
L/H=2
L/H=3
L/H=5
L/H=10
$P_{st}/\rho A_g H^2$
ω

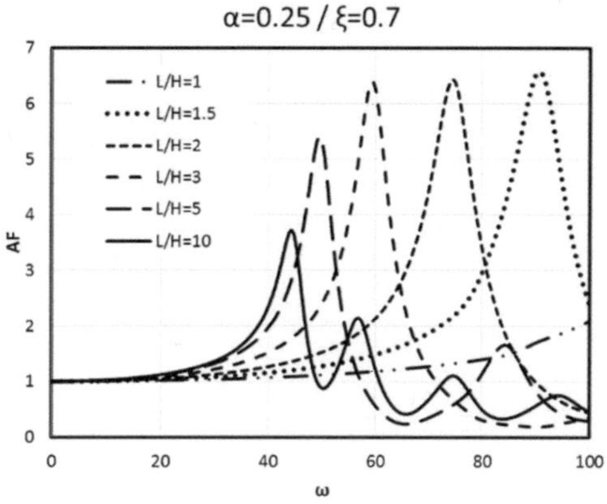
α=0.25 / ξ=0.7
L/H=1
L/H=1.5
L/H=2
L/H=3
L/H=5
L/H=10
AF
ω

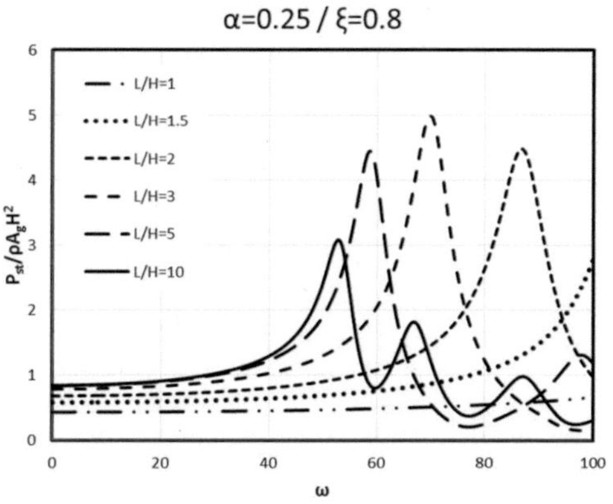
α=0.25 / ξ=0.8
L/H=1
L/H=1.5
L/H=2
L/H=3
L/H=5
L/H=10
$P_{st}/\rho A_g H^2$
ω

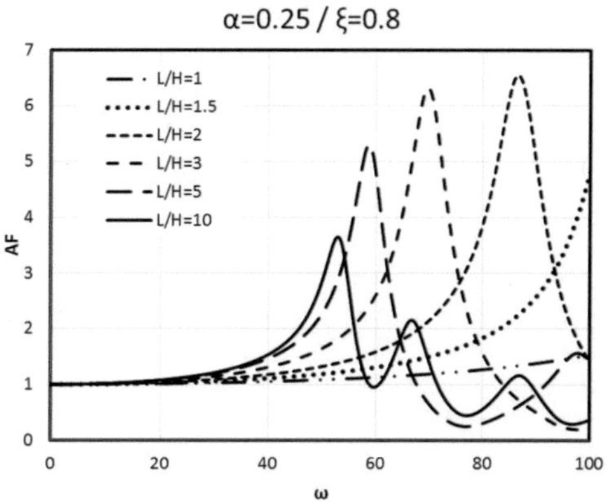
α=0.25 / ξ=0.8
L/H=1
L/H=1.5
L/H=2
L/H=3
L/H=5
L/H=10
AF
ω

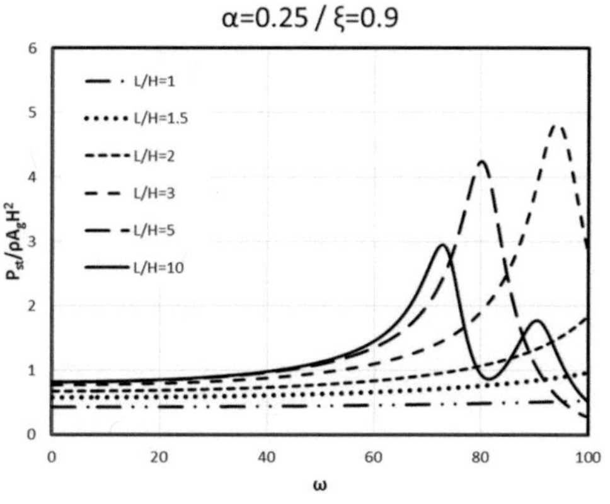
α=0.25 / ξ=0.9
L/H=1
L/H=1.5
L/H=2
L/H=3
L/H=5
L/H=10
$P_{st}/\rho A_g H^2$
ω

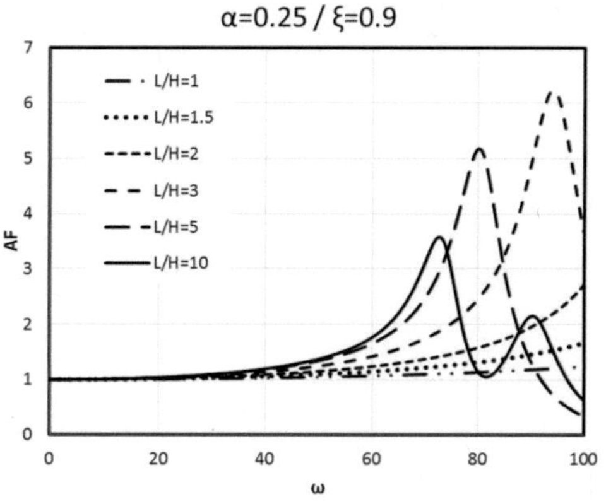
α=0.25 / ξ=0.9
L/H=1
L/H=1.5
L/H=2
L/H=3
L/H=5
L/H=10
AF
ω

Annex E

Natural circular frequencies of bounded systems with flexible walls based on flexible base for different ratios of L/H (1.0, 1.5, 2.0, 3.0, 5.0, 10.0) and for the three investigated soil profiles: homogeneous, parabolic and linear. The contact between the walls and the soil is smooth contact. These natural frequencies are identified as the frequencies, where the maximum shear force due to the soil pressures. The natural frequencies of the unbounded soil stratum for constant, parabolic and linear shear modulus distribution are 19.63 rad/s, 22.59 rad/s and 22.38 rad/sec respectively.

Homogeneous soil – smooth contact															
L/H=1	dw	0	1	5	10	20	30	40		0	1	5	10	20	30	40
	dθ	ω_{11}	ω_{11}	ω_{11}	ω_{11}	ω_{11}	ω_{11}	ω_{11}		ω_{11}/ω_1	ω_{11}/ω_1	ω_{11}/ω_1	ω_{11}/ω_1	ω_{11}/ω_1	ω_{11}/ω_1	ω_{11}/ω_1
dx=0	0	66.27	51.02	29.08	23.96	20.32	18.80	17.91		3.38	2.60	1.48	1.22	1.04	0.96	0.91
	0.5	38.52	33.96	26.14	22.61	19.74	18.44	17.74		1.96	1.73	1.33	1.15	1.01	0.94	0.90
	1	31.12	28.80	24.19	21.54	19.17	18.09	17.57		1.59	1.47	1.23	1.10	0.98	0.92	0.90
	2	24.67	23.96	21.54	20.12	18.44	17.57	17.07		1.26	1.22	1.10	1.03	0.94	0.90	0.87
	3	21.96	21.33	20.12	18.99	17.91	17.23	16.90		1.12	1.09	1.03	0.97	0.91	0.88	0.86
	4	20.12	19.93	18.99	18.27	17.40	16.90	16.58		1.03	1.02	0.97	0.93	0.89	0.86	0.84
	5	19.17	18.99	18.27	17.74	17.07	16.58	16.42		0.98	0.97	0.93	0.90	0.87	0.84	0.84
dx=0.1	0	51.51	39.66	27.44	22.83	19.55	18.27	17.57		2.62	2.02	1.40	1.16	1.00	0.93	0.90
	0.5	35.30	31.73	24.91	21.75	19.17	17.91	17.23		1.80	1.62	1.27	1.11	0.98	0.91	0.88
	1	29.37	27.44	23.27	20.92	18.80	17.74	17.07		1.50	1.40	1.19	1.07	0.96	0.90	0.87
	2	23.96	23.05	21.12	19.55	18.09	17.23	16.74		1.22	1.17	1.08	1.00	0.92	0.88	0.85
	3	21.33	20.92	19.74	18.62	17.57	16.90	16.58		1.09	1.07	1.01	0.95	0.90	0.86	0.84
	4	19.93	19.55	18.80	18.09	17.23	16.74	16.42		1.02	1.00	0.96	0.92	0.88	0.85	0.84
	5	18.80	18.62	18.09	17.57	16.90	16.42	16.10		0.96	0.95	0.92	0.90	0.86	0.84	0.82
dx=0.5	0	39.66	34.29	25.39	21.75	18.80	17.57	16.90		2.02	1.75	1.29	1.11	0.96	0.90	0.86
	0.5	31.43	28.80	23.50	20.72	18.44	17.40	16.74		1.60	1.47	1.20	1.06	0.94	0.89	0.85
	1	27.18	25.64	22.17	20.12	18.09	17.23	16.58		1.38	1.31	1.13	1.03	0.92	0.88	0.84
	2	22.83	22.17	20.32	18.99	17.57	16.90	16.42		1.16	1.13	1.04	0.97	0.90	0.86	0.84
	3	20.72	20.32	19.17	18.27	17.23	16.58	16.26		1.06	1.04	0.98	0.93	0.88	0.84	0.83
	4	19.55	19.17	18.27	17.57	16.74	16.26	15.95		1.00	0.98	0.93	0.90	0.85	0.83	0.81
	5	18.62	18.27	17.74	17.23	16.58	16.10	15.79		0.95	0.93	0.90	0.88	0.84	0.82	0.80
dx=1	0	37.05	32.67	24.67	21.33	18.62	17.40	16.74		1.89	1.66	1.26	1.09	0.95	0.89	0.85
	0.5	30.23	27.98	23.05	20.52	18.27	17.23	16.58		1.54	1.43	1.17	1.05	0.93	0.88	0.84
	1	26.40	25.15	21.75	19.74	17.91	17.07	16.42		1.35	1.28	1.11	1.01	0.91	0.87	0.84
	2	22.61	21.96	20.12	18.80	17.40	16.74	16.26		1.15	1.12	1.03	0.96	0.89	0.85	0.83
	3	20.52	20.12	18.99	18.09	17.07	16.42	16.10		1.05	1.03	0.97	0.92	0.87	0.84	0.82
	4	19.36	18.99	18.27	17.57	16.74	16.26	15.95		0.99	0.97	0.93	0.90	0.85	0.83	0.81
	5	18.44	18.27	17.57	17.07	16.42	15.95	15.79		0.94	0.93	0.90	0.87	0.84	0.81	0.80

Homogeneous soil – smooth contact																
L/H=2	dw	0	1	5	10	20	30	40		0	1	5	10	20	30	40
	dθ	ω_{11}	ω_{11}	ω_{11}	ω_{11}	ω_{11}	ω_{11}	ω_{11}		ω_{11}/ω_1	ω_{11}/ω_1	ω_{11}/ω_1	ω_{11}/ω_1	ω_{11}/ω_1	ω_{11}/ω_1	ω_{11}/ω_1
dx=0	0	40.49	33.90	25.76	22.92	20.81	19.98	19.48		2.06	1.73	1.31	1.17	1.06	1.02	0.99
	0.5	30.01	27.95	23.87	22.01	20.50	19.78	19.38		1.53	1.42	1.22	1.12	1.04	1.01	0.99
	1	26.03	24.99	22.69	21.35	20.19	19.58	19.19		1.33	1.27	1.16	1.09	1.03	1.00	0.98
	2	22.69	22.23	21.13	20.39	19.68	19.29	18.99		1.16	1.13	1.08	1.04	1.00	0.98	0.97
	3	21.13	20.92	20.29	19.88	19.29	18.99	18.80		1.08	1.07	1.03	1.01	0.98	0.97	0.96
	4	20.19	20.09	19.68	19.38	18.99	18.80	18.61		1.03	1.02	1.00	0.99	0.97	0.96	0.95
	5	19.68	19.58	19.29	19.09	18.80	18.61	18.42		1.00	1.00	0.98	0.97	0.96	0.95	0.94
dx=0.1	0	34.24	30.47	24.36	22.01	20.29	19.58	19.19		1.74	1.55	1.24	1.12	1.03	1.00	0.98
	0.5	27.80	26.16	22.92	21.35	19.98	19.38	18.99		1.42	1.33	1.17	1.09	1.02	0.99	0.97
	1	24.86	23.99	22.01	20.81	19.78	19.19	18.90		1.27	1.22	1.12	1.06	1.01	0.98	0.96
	2	22.01	21.68	20.71	20.09	19.38	18.99	18.71		1.12	1.10	1.06	1.02	0.99	0.97	0.95
	3	20.71	20.50	19.98	19.58	18.99	18.71	18.52		1.06	1.04	1.02	1.00	0.97	0.95	0.94
	4	19.98	19.78	19.48	19.19	18.80	18.61	18.42		1.02	1.01	0.99	0.98	0.96	0.95	0.94
	5	19.38	19.38	19.09	18.90	18.61	18.42	18.33		0.99	0.99	0.97	0.96	0.95	0.94	0.93
dx=0.5	0	29.40	27.11	22.81	21.03	19.58	18.99	18.71		1.50	1.38	1.16	1.07	1.00	0.97	0.95
	0.5	25.50	24.36	21.79	20.50	19.38	18.90	18.61		1.30	1.24	1.11	1.04	0.99	0.96	0.95
	1	23.51	22.81	21.13	20.09	19.19	18.80	18.52		1.20	1.16	1.08	1.02	0.98	0.96	0.94
	2	21.35	21.03	20.19	19.58	18.90	18.61	18.33		1.09	1.07	1.03	1.00	0.96	0.95	0.93
	3	20.9	20.09	19.58	19.09	18.71	18.42	18.24		1.03	1.02	1.00	0.97	0.95	0.94	0.93
	4	19.58	19.48	19.09	18.80	18.52	18.24	18.15		1.00	0.99	0.97	0.96	0.94	0.93	0.92
	5	19.09	19.09	18.80	18.61	18.33	18.15	18.05		0.97	0.97	0.96	0.95	0.93	0.92	0.92
dx=1	0	28.23	26.16	22.35	20.71	19.38	18.90	18.52		1.44	1.33	1.14	1.06	0.99	0.96	0.94
	0.5	24.86	23.75	21.46	20.29	19.19	18.71	18.42		1.27	1.21	1.09	1.03	0.98	0.95	0.94
	1	23.04	22.35	20.81	19.88	19.09	18.61	18.33		1.17	1.14	1.06	1.01	0.97	0.95	0.93
	2	21.13	20.81	19.98	19.38	18.80	18.42	18.24		1.08	1.06	1.02	0.99	0.96	0.94	0.93
	3	20.09	19.88	19.38	18.99	18.52	18.33	18.15		1.02	1.01	0.99	0.97	0.94	0.93	0.92
	4	19.48	19.38	18.99	18.71	18.42	18.15	18.05		0.99	0.99	0.97	0.95	0.94	0.92	0.92
	5	18.99	18.99	18.71	18.52	18.24	18.05	17.96		0.97	0.97	0.95	0.94	0.93	0.92	0.92

Homogeneous soil – smooth contact																
L/H=3	dw	0	1	5	10	20	30	40		0	1	5	10	20	30	40
	dθ	ω_{11}	ω_{11}	ω_{11}	ω_{11}	ω_{11}	ω_{11}	ω_{11}		ω_{11}/ω_1	ω_{11}/ω_1	ω_{11}/ω_1	ω_{11}/ω_1	ω_{11}/ω_1	ω_{11}/ω_1	ω_{11}/ω_1
dx=0	0	29.50	27.81	23.64	21.80	20.55	19.95	19.66		1.50	1.42	1.20	1.11	1.05	1.02	1.00
	0.5	26.02	24.90	22.62	21.32	20.24	19.80	19.51		1.33	1.27	1.15	1.09	1.03	1.01	0.99
	1	23.82	23.29	21.80	20.85	20.10	19.66	19.37		1.21	1.19	1.11	1.06	1.02	1.00	0.99
	2	21.64	21.48	20.70	20.24	19.66	19.37	19.22		1.10	1.09	1.05	1.03	1.00	0.99	0.98
	3	20.70	20.55	20.10	19.80	19.51	19.22	19.08		1.05	1.05	1.02	1.01	0.99	0.98	0.97
	4	20.10	19.95	19.80	19.51	19.22	19.08	18.94		1.02	1.02	1.01	0.99	0.98	0.97	0.97
	5	19.66	19.66	19.51	19.37	19.08	18.94	18.94		1.00	1.00	0.99	0.99	0.97	0.97	0.97
dx=0.1	0	28.02	26.22	22.78	21.32	20.10	19.66	19.37		1.43	1.34	1.16	1.09	1.02	1.00	0.99
	0.5	24.90	23.99	21.96	20.85	19.95	19.51	19.22		1.27	1.22	1.12	1.06	1.02	0.99	0.98
	1	23.12	22.62	21.32	20.55	19.80	19.37	19.22		1.18	1.15	1.09	1.05	1.01	0.99	0.98
	2	21.32	21.01	20.39	19.95	19.51	19.22	19.08		1.09	1.07	1.04	1.02	0.99	0.98	0.97
	3	20.39	20.24	19.95	19.66	19.22	19.08	18.94		1.04	1.03	1.02	1.00	0.98	0.97	0.97
	4	19.95	19.80	19.51	19.37	19.08	18.94	18.80		1.02	1.01	0.99	0.99	0.97	0.97	0.96
	5	19.51	19.51	19.37	19.22	18.94	18.80	18.80		0.99	0.99	0.99	0.98	0.97	0.96	0.96
dx=0.5	0	25.64	24.53	21.96	20.70	19.66	19.22	19.08		1.31	1.25	1.12	1.05	1.00	0.98	0.97
	0.5	23.47	22.78	21.16	20.39	19.51	19.22	18.94		1.20	1.16	1.08	1.04	0.99	0.98	0.97
	1	22.12	21.80	20.70	20.10	19.37	19.08	18.94		1.13	1.11	1.05	1.02	0.99	0.97	0.97
	2	20.85	20.55	20.10	19.66	19.22	18.94	18.80		1.06	1.05	1.02	1.00	0.98	0.97	0.96
	3	20.10	19.95	19.66	19.37	19.08	18.80	18.67		1.02	1.02	1.00	0.99	0.97	0.96	0.95
	4	19.66	19.51	19.37	19.08	18.94	18.80	18.67		1.00	0.99	0.99	0.97	0.97	0.96	0.95
	5	19.37	19.22	19.08	18.94	18.80	18.67	18.53		0.99	0.98	0.97	0.97	0.96	0.95	0.94
dx=1	0	25.08	23.99	21.64	20.39	19.51	19.08	18.94		1.28	1.22	1.10	1.04	0.99	0.97	0.97
	0.5	23.12	22.45	21.01	20.10	19.37	19.08	18.80		1.18	1.14	1.07	1.02	0.99	0.97	0.96
	1	21.96	21.48	20.55	19.95	19.22	18.94	18.80		1.12	1.09	1.05	1.02	0.98	0.97	0.96
	2	20.70	20.39	19.95	19.51	19.08	18.94	18.67		1.05	1.04	1.02	0.99	0.97	0.97	0.95
	3	19.95	19.80	19.51	19.22	18.94	18.80	18.67		1.02	1.01	0.99	0.98	0.97	0.96	0.95
	4	19.51	19.51	19.22	19.08	18.80	18.67	18.53		0.99	0.99	0.98	0.97	0.96	0.95	0.94
	5	19.22	19.22	19.08	18.94	18.67	18.67	18.53		0.98	0.98	0.97	0.97	0.95	0.95	0.94

Homogeneous soil – smooth contact																
L/H=5	dw	0	1	5	10	20	30	40		0	1	5	10	20	30	40
	dθ	ω_{11}	ω_{11}	ω_{11}	ω_{11}	ω_{11}	ω_{11}	ω_{11}		ω_{11}/ω_1	ω_{11}/ω_1	ω_{11}/ω_1	ω_{11}/ω_1	ω_{11}/ω_1	ω_{11}/ω_1	ω_{11}/ω_1
dx=0	0	23.82	23.29	21.80	20.85	20.10	19.80	19.66		1.21	1.19	1.11	1.06	1.02	1.01	1.00
	0.5	22.62	22.28	21.16	20.55	19.95	19.66	19.51		1.15	1.14	1.08	1.05	1.02	1.00	0.99
	1	21.80	21.48	20.85	20.39	19.80	19.66	19.51		1.11	1.09	1.06	1.04	1.01	1.00	0.99
	2	20.85	20.70	20.24	19.95	19.66	19.51	19.37		1.06	1.05	1.03	1.02	1.00	0.99	0.99
	3	20.24	20.10	19.95	19.80	19.51	19.37	19.22		1.03	1.02	1.02	1.01	0.99	0.99	0.98
	4	19.95	19.80	19.66	19.51	19.37	19.22	19.22		1.02	1.01	1.00	0.99	0.99	0.98	0.98
	5	19.66	19.66	19.51	19.37	19.37	19.22	19.08		1.00	1.00	0.99	0.99	0.99	0.98	0.97
dx=0.1	0	14.76	14.34	13.14	12.52	12.04	11.81	11.70		0.75	0.73	0.67	0.64	0.61	0.60	0.60
	0.5	13.93	13.53	12.76	12.40	11.93	11.70	11.58		0.71	0.69	0.65	0.63	0.61	0.60	0.59
	1	13.27	13.01	12.52	12.16	11.81	11.70	11.58		0.68	0.66	0.64	0.62	0.60	0.60	0.59
	2	12.52	12.40	12.16	11.93	11.70	11.58	11.47		0.64	0.63	0.62	0.61	0.60	0.59	0.58
	3	12.16	12.04	11.93	11.81	11.58	11.47	11.47		0.62	0.61	0.61	0.60	0.59	0.58	0.58
	4	11.93	11.81	11.70	11.70	11.47	11.47	11.36		0.61	0.60	0.60	0.60	0.58	0.58	0.58
	5	11.70	11.70	11.58	11.58	11.47	11.36	11.36		0.60	0.60	0.59	0.59	0.58	0.58	0.58
dx=0.5	0	14.20	13.79	12.76	12.28	11.81	11.58	11.47		0.72	0.70	0.65	0.63	0.60	0.59	0.58
	0.5	13.40	13.14	12.52	12.04	11.70	11.58	11.47		0.68	0.67	0.64	0.61	0.60	0.59	0.58
	1	12.89	12.76	12.28	11.93	11.70	11.47	11.47		0.66	0.65	0.63	0.61	0.60	0.58	0.58
	2	12.40	12.28	11.93	11.81	11.58	11.47	11.36		0.63	0.63	0.61	0.60	0.59	0.58	0.58
	3	12.04	11.93	11.81	11.70	11.47	11.36	11.36		0.61	0.61	0.60	0.60	0.58	0.58	0.58
	4	11.81	11.70	11.58	11.58	11.47	11.36	11.25		0.60	0.60	0.59	0.59	0.58	0.58	0.57
	5	11.58	11.58	11.58	11.47	11.36	11.36	11.25		0.59	0.59	0.59	0.58	0.58	0.58	0.57
dx=1	0	13.93	13.53	12.64	12.16	11.70	11.58	11.47		0.71	0.69	0.64	0.62	0.60	0.59	0.58
	0.5	13.27	13.01	12.40	12.04	11.70	11.47	11.36		0.68	0.66	0.63	0.61	0.60	0.58	0.58
	1	12.76	12.64	12.16	11.93	11.58	11.47	11.36		0.65	0.64	0.62	0.61	0.59	0.58	0.58
	2	12.28	12.16	11.93	11.70	11.47	11.36	11.36		0.63	0.62	0.61	0.60	0.58	0.58	0.58
	3	11.93	11.93	11.70	11.58	11.47	11.36	11.25		0.61	0.61	0.60	0.59	0.58	0.58	0.57
	4	11.70	11.70	11.58	11.47	11.36	11.36	11.25		0.60	0.60	0.59	0.58	0.58	0.58	0.57
	5	11.58	11.58	11.47	11.47	11.36	11.25	11.25		0.59	0.59	0.58	0.58	0.58	0.57	0.57

Homogeneous soil – smooth contact																
L/H=10	dw	0	1	5	10	20	30	40		0	1	5	10	20	30	40
	dθ	ω_{11}	ω_{11}	ω_{11}	ω_{11}	ω_{11}	ω_{11}	ω_{11}		ω_{11}/ω_1	ω_{11}/ω_1	ω_{11}/ω_1	ω_{11}/ω_1	ω_{11}/ω_1	ω_{11}/ω_1	ω_{11}/ω_1
dx=0	0	20.55	20.55	20.24	20.10	19.80	19.66	19.66		1.05	1.05	1.03	1.02	1.01	1.00	1.00
	0.5	20.39	20.39	20.24	19.95	19.80	19.66	19.66		1.04	1.04	1.03	1.02	1.01	1.00	1.00
	1	20.24	20.24	20.10	19.95	19.80	19.66	19.51		1.03	1.03	1.02	1.02	1.01	1.00	0.99
	2	20.10	20.10	19.95	19.80	19.66	19.51	19.51		1.02	1.02	1.02	1.01	1.00	0.99	0.99
	3	19.95	19.80	19.80	19.66	19.51	19.51	19.51		1.02	1.01	1.01	1.00	0.99	0.99	0.99
	4	19.80	19.66	19.66	19.66	19.51	19.51	19.37		1.01	1.00	1.00	1.00	0.99	0.99	0.99
	5	19.66	19.66	19.51	19.51	19.51	19.37	19.37		1.00	1.00	0.99	0.99	0.99	0.99	0.99
dx=0.1	0	20.55	20.39	20.24	19.95	19.80	19.66	19.51		1.05	1.04	1.03	1.02	1.01	1.00	0.99
	0.5	20.39	20.39	20.10	19.95	19.66	19.66	19.51		1.04	1.04	1.02	1.02	1.00	1.00	0.99
	1	20.24	20.24	19.95	19.80	19.66	19.51	19.51		1.03	1.03	1.02	1.01	1.00	0.99	0.99
	2	19.95	19.95	19.80	19.66	19.51	19.51	19.51		1.02	1.02	1.01	1.00	0.99	0.99	0.99
	3	19.80	19.80	19.66	19.66	19.51	19.51	19.37		1.01	1.01	1.00	1.00	0.99	0.99	0.99
	4	19.66	19.66	19.66	19.51	19.51	19.37	19.37		1.00	1.00	1.00	0.99	0.99	0.99	0.99
	5	19.66	19.51	19.51	19.51	19.37	19.37	19.37		1.00	0.99	0.99	0.99	0.99	0.99	0.99
dx=0.5	0	20.39	20.39	20.10	19.95	19.66	19.51	19.37		1.04	1.04	1.02	1.02	1.00	0.99	0.99
	0.5	20.24	20.24	19.95	19.80	19.66	19.51	19.37		1.03	1.03	1.02	1.01	1.00	0.99	0.99
	1	20.10	20.10	19.95	19.80	19.51	19.51	19.37		1.02	1.02	1.02	1.01	0.99	0.99	0.99
	2	19.95	19.95	19.80	19.66	19.51	19.37	19.37		1.02	1.02	1.01	1.00	0.99	0.99	0.99
	3	19.80	19.66	19.66	19.51	19.37	19.37	19.37		1.01	1.00	1.00	0.99	0.99	0.99	0.99
	4	19.66	19.66	19.51	19.51	19.37	19.37	19.37		1.00	1.00	0.99	0.99	0.99	0.99	0.99
	5	19.51	19.51	19.51	19.37	19.37	19.37	19.22		0.99	0.99	0.99	0.99	0.99	0.99	0.98
dx=1	0	20.39	20.39	20.10	19.80	19.66	19.51	19.37		1.04	1.04	1.02	1.01	1.00	0.99	0.99
	0.5	20.24	20.24	19.95	19.80	19.51	19.51	19.37		1.03	1.03	1.02	1.01	0.99	0.99	0.99
	1	20.10	20.10	19.80	19.66	19.51	19.37	19.37		1.02	1.02	1.01	1.00	0.99	0.99	0.99
	2	19.95	19.80	19.66	19.66	19.51	19.37	19.37		1.02	1.01	1.00	1.00	0.99	0.99	0.99
	3	19.80	19.66	19.66	19.51	19.37	19.37	19.22		1.01	1.00	1.00	0.99	0.99	0.99	0.98
	4	19.66	19.66	19.51	19.51	19.37	19.37	19.22		1.00	1.00	0.99	0.99	0.99	0.99	0.98
	5	19.51	19.51	19.51	19.37	19.37	19.22	19.22		0.99	0.99	0.99	0.99	0.99	0.98	0.98

Parabolic soil – smooth contact																
L/H=1	dw	0	1	5	10	20	30	40		0	1	5	10	20	30	40
	dθ	ω_{11}	ω_{11}	ω_{11}	ω_{11}	ω_{11}	ω_{11}	ω_{11}		ω_{11}/ω_1	ω_{11}/ω_1	ω_{11}/ω_1	ω_{11}/ω_1	ω_{11}/ω_1	ω_{11}/ω_1	ω_{11}/ω_1
dx=0	0	51.72	40.23	30.39	25.64	21.96	20.55	19.66		2.29	1.78	1.35	1.14	0.97	0.91	0.87
	0.5	36.82	33.95	27.61	24.35	21.48	20.24	19.51		1.63	1.50	1.22	1.08	0.95	0.90	0.86
	1	31.53	29.72	25.83	23.29	21.01	19.95	19.22		1.40	1.32	1.14	1.03	0.93	0.88	0.85
	2	26.22	25.45	23.47	21.96	20.39	19.51	18.94		1.16	1.13	1.04	0.97	0.90	0.86	0.84
	3	23.82	23.29	21.96	21.01	19.80	19.08	18.67		1.05	1.03	0.97	0.93	0.88	0.84	0.83
	4	22.28	21.96	21.01	20.24	19.37	18.80	18.53		0.99	0.97	0.93	0.90	0.86	0.83	0.82
	5	21.32	21.01	20.39	19.80	19.08	18.67	18.26		0.94	0.93	0.90	0.88	0.84	0.83	0.81
dx=0.1	0	42.06	37.93	28.86	24.71	21.48	20.10	19.37		1.86	1.68	1.28	1.09	0.95	0.89	0.86
	0.5	34.45	32.00	26.61	23.64	21.01	19.80	19.08		1.53	1.42	1.18	1.05	0.93	0.88	0.84
	1	30.17	28.65	25.08	22.78	20.55	19.66	18.94		1.34	1.27	1.11	1.01	0.91	0.87	0.84
	2	25.64	24.90	22.95	21.48	19.95	19.22	18.67		1.14	1.10	1.02	0.95	0.88	0.85	0.83
	3	23.47	22.95	21.64	20.70	19.51	18.94	18.53		1.04	1.02	0.96	0.92	0.86	0.84	0.82
	4	21.96	21.80	20.85	20.10	19.22	18.67	18.26		0.97	0.96	0.92	0.89	0.85	0.83	0.81
	5	21.01	20.85	20.24	19.66	18.80	18.39	18.12		0.93	0.92	0.90	0.87	0.83	0.81	0.80
dx=0.5	0	38.49	34.97	27.40	23.82	20.85	19.66	18.94		1.70	1.55	1.21	1.05	0.92	0.87	0.84
	0.5	32.24	30.39	25.45	22.78	20.55	19.37	18.80		1.43	1.35	1.13	1.01	0.91	0.86	0.83
	1	28.86	27.61	24.17	22.12	20.10	19.22	18.67		1.28	1.22	1.07	0.98	0.89	0.85	0.83
	2	25.08	24.35	22.45	21.16	19.66	18.94	18.39		1.11	1.08	0.99	0.94	0.87	0.84	0.81
	3	22.95	22.62	21.32	20.39	19.22	18.67	18.26		1.02	1.00	0.94	0.90	0.85	0.83	0.81
	4	21.80	21.48	20.55	19.80	18.94	18.39	17.99		0.96	0.95	0.91	0.88	0.84	0.81	0.80
	5	20.85	20.70	19.95	19.37	18.67	18.26	17.86		0.92	0.92	0.88	0.86	0.83	0.81	0.79
dx=1	0	37.37	33.95	27.00	23.47	20.70	19.51	18.80		1.65	1.50	1.20	1.04	0.92	0.86	0.83
	0.5	31.77	29.72	25.27	22.62	20.39	19.22	18.67		1.41	1.32	1.12	1.00	0.90	0.85	0.83
	1	28.43	27.20	23.99	21.96	20.10	19.08	18.53		1.26	1.20	1.06	0.97	0.89	0.84	0.82
	2	24.90	24.17	22.28	21.01	19.51	18.80	18.26		1.10	1.07	0.99	0.93	0.86	0.83	0.81
	3	22.95	22.45	21.32	20.24	19.22	18.53	18.12		1.02	0.99	0.94	0.90	0.85	0.82	0.80
	4	21.64	21.32	20.55	19.80	18.80	18.39	17.99		0.96	0.94	0.91	0.88	0.83	0.81	0.80
	5	20.85	20.55	19.95	19.37	18.67	18.12	17.86		0.92	0.91	0.88	0.86	0.83	0.80	0.79

Parabolic soil – smooth contact																
L/H=2	dw	0	1	5	10	20	30	40		0	1	5	10	20	30	40
	dθ	ω_{11}	ω_{11}	ω_{11}	ω_{11}	ω_{11}	ω_{11}	ω_{11}		ω_{11}/ω_1	ω_{11}/ω_1	ω_{11}/ω_1	ω_{11}/ω_1	ω_{11}/ω_1	ω_{11}/ω_1	ω_{11}/ω_1
dx=0	0	24.67	22.83	18.27	15.95	14.20	13.53	13.14		1.09	1.01	0.81	0.71	0.63	0.60	0.58
	0.5	20.72	19.55	16.74	15.19	13.93	13.40	13.01		0.92	0.87	0.74	0.67	0.62	0.59	0.58
	1	18.27	17.57	15.79	14.76	13.66	13.27	12.89		0.81	0.78	0.70	0.65	0.60	0.59	0.57
	2	15.95	15.64	14.76	14.06	13.40	13.01	12.76		0.71	0.69	0.65	0.62	0.59	0.58	0.56
	3	14.76	14.62	14.06	13.53	13.14	12.76	12.64		0.65	0.65	0.62	0.60	0.58	0.56	0.56
	4	14.06	13.93	13.53	13.27	12.89	12.64	12.52		0.62	0.62	0.60	0.59	0.57	0.56	0.55
	5	13.53	13.53	13.27	13.01	12.76	12.52	12.40		0.60	0.60	0.59	0.58	0.56	0.55	0.55
dx=0.1	0	22.61	20.92	17.23	15.34	13.93	13.27	12.89		1.00	0.93	0.76	0.68	0.62	0.59	0.57
	0.5	19.36	18.44	16.10	14.76	13.66	13.14	12.76		0.86	0.82	0.71	0.65	0.60	0.58	0.56
	1	17.57	16.90	15.34	14.34	13.40	13.01	12.76		0.78	0.75	0.68	0.63	0.59	0.58	0.56
	2	15.49	15.19	14.48	13.79	13.14	12.76	12.64		0.69	0.67	0.64	0.61	0.58	0.56	0.56
	3	14.48	14.34	13.79	13.40	12.89	12.64	12.52		0.64	0.63	0.61	0.59	0.57	0.56	0.55
	4	13.93	13.79	13.40	13.14	12.76	12.52	12.40		0.62	0.61	0.59	0.58	0.56	0.55	0.55
	5	13.53	13.40	13.14	12.89	12.64	12.40	12.28		0.60	0.59	0.58	0.57	0.56	0.55	0.54
dx=0.5	0	20.52	19.17	16.26	14.76	13.53	13.01	12.64		0.91	0.85	0.72	0.65	0.60	0.58	0.56
	0.5	18.27	17.40	15.49	14.34	13.40	12.89	12.64		0.81	0.77	0.69	0.63	0.59	0.57	0.56
	1	16.74	16.26	14.90	14.06	13.14	12.76	12.52		0.74	0.72	0.66	0.62	0.58	0.56	0.55
	2	15.19	14.90	14.06	13.53	12.89	12.64	12.40		0.67	0.66	0.62	0.60	0.57	0.56	0.55
	3	14.34	14.06	13.66	13.27	12.76	12.52	12.28		0.63	0.62	0.60	0.59	0.56	0.55	0.54
	4	13.79	13.66	13.27	13.01	12.64	12.40	12.28		0.61	0.60	0.59	0.58	0.56	0.55	0.54
	5	13.40	13.27	13.01	12.76	12.52	12.28	12.16		0.59	0.59	0.58	0.56	0.55	0.54	0.54
dx=1	0	19.93	18.80	16.10	14.62	13.40	12.89	12.64		0.88	0.83	0.71	0.65	0.59	0.57	0.56
	0.5	17.91	17.07	15.34	14.20	13.27	12.76	12.52		0.79	0.76	0.68	0.63	0.59	0.56	0.55
	1	16.58	16.10	14.76	13.93	13.14	12.76	12.52		0.73	0.71	0.65	0.62	0.58	0.56	0.55
	2	15.05	14.76	14.06	13.53	12.89	12.52	12.40		0.67	0.65	0.62	0.60	0.57	0.55	0.55
	3	14.20	14.06	13.53	13.14	12.76	12.40	12.28		0.63	0.62	0.60	0.58	0.56	0.55	0.54
	4	13.66	13.53	13.27	12.89	12.52	12.40	12.28		0.60	0.60	0.59	0.57	0.55	0.55	0.54
	5	13.40	13.27	13.01	12.76	12.40	12.28	12.16		0.59	0.59	0.58	0.56	0.55	0.54	0.54

Parabolic soil – smooth contact																
L/H=3	dw	0	1	5	10	20	30	40		0	1	5	10	20	30	40
	dθ	ω_{11}	ω_{11}	ω_{11}	ω_{11}	ω_{11}	ω_{11}	ω_{11}		ω_{11}/ω_1	ω_{11}/ω_1	ω_{11}/ω_1	ω_{11}/ω_1	ω_{11}/ω_1	ω_{11}/ω_1	ω_{11}/ω_1
dx=0	0	28.80	27.98	25.64	23.96	22.61	21.96	21.75		1.27	1.24	1.14	1.06	1.00	0.97	0.96
	0.5	26.91	26.40	24.67	23.50	22.39	21.96	21.54		1.19	1.17	1.09	1.04	0.99	0.97	0.95
	1	25.64	25.15	23.96	23.05	22.17	21.75	21.54		1.14	1.11	1.06	1.02	0.98	0.96	0.95
	2	23.96	23.73	23.05	22.39	21.96	21.54	21.33		1.06	1.05	1.02	0.99	0.97	0.95	0.94
	3	23.05	22.83	22.39	22.17	21.54	21.33	21.12		1.02	1.01	0.99	0.98	0.95	0.94	0.94
	4	22.39	22.39	22.17	21.75	21.54	21.33	21.12		0.99	0.99	0.98	0.96	0.95	0.94	0.94
	5	22.17	21.96	21.75	21.54	21.33	21.12	20.92		0.98	0.97	0.96	0.95	0.94	0.94	0.93
dx=0.1	0	27.98	27.18	24.91	23.50	22.39	21.75	21.54		1.24	1.20	1.10	1.04	0.99	0.96	0.95
	0.5	26.40	25.64	24.19	23.05	22.17	21.75	21.33		1.17	1.14	1.07	1.02	0.98	0.96	0.94
	1	25.15	24.67	23.50	22.83	21.96	21.54	21.33		1.11	1.09	1.04	1.01	0.97	0.95	0.94
	2	23.73	23.50	22.83	22.17	21.75	21.33	21.12		1.05	1.04	1.01	0.98	0.96	0.94	0.94
	3	22.83	22.83	22.39	21.96	21.54	21.33	21.12		1.01	1.01	0.99	0.97	0.95	0.94	0.94
	4	22.39	22.17	21.96	21.75	21.33	21.12	20.92		0.99	0.98	0.97	0.96	0.94	0.94	0.93
	5	21.96	21.96	21.75	21.54	21.12	21.12	20.92		0.97	0.97	0.96	0.95	0.94	0.94	0.93
dx=0.5	0	26.91	26.14	24.19	23.05	21.96	21.54	21.33		1.19	1.16	1.07	1.02	0.97	0.95	0.94
	0.5	25.64	25.15	23.73	22.83	21.96	21.33	21.12		1.14	1.11	1.05	1.01	0.97	0.94	0.94
	1	24.67	24.19	23.27	22.39	21.75	21.33	21.12		1.09	1.07	1.03	0.99	0.96	0.94	0.94
	2	23.27	23.27	22.61	21.96	21.54	21.12	20.92		1.03	1.03	1.00	0.97	0.95	0.94	0.93
	3	22.61	22.61	22.17	21.75	21.33	21.12	20.92		1.00	1.00	0.98	0.96	0.94	0.94	0.93
	4	22.17	22.17	21.75	21.54	21.12	20.92	20.92		0.98	0.98	0.96	0.95	0.94	0.93	0.93
	5	21.96	21.75	21.54	21.33	21.12	20.92	20.72		0.97	0.96	0.95	0.94	0.94	0.93	0.92
dx=1	0	26.65	25.89	24.19	23.05	21.96	21.54	21.12		1.18	1.15	1.07	1.02	0.97	0.95	0.94
	0.5	25.39	24.91	23.50	22.61	21.75	21.33	21.12		1.12	1.10	1.04	1.00	0.96	0.94	0.94
	1	24.43	24.19	23.05	22.39	21.75	21.33	21.12		1.08	1.07	1.02	0.99	0.96	0.94	0.94
	2	23.27	23.05	22.39	21.96	21.54	21.12	20.92		1.03	1.02	0.99	0.97	0.95	0.94	0.93
	3	22.61	22.39	21.96	21.75	21.33	21.12	20.92		1.00	0.99	0.97	0.96	0.94	0.94	0.93
	4	22.17	22.17	21.75	21.54	21.12	20.92	20.92		0.98	0.98	0.96	0.95	0.94	0.93	0.93
	5	21.96	21.75	21.54	21.33	21.12	20.92	20.72		0.97	0.96	0.95	0.94	0.94	0.93	0.92

Parabolic soil – smooth contact																
L/H=5	dw	0	1	5	10	20	30	40		0	1	5	10	20	30	40
	dθ	ω_{11}	ω_{11}	ω_{11}	ω_{11}	ω_{11}	ω_{11}	ω_{11}		ω_{11}/ω_1	ω_{11}/ω_1	ω_{11}/ω_1	ω_{11}/ω_1	ω_{11}/ω_1	ω_{11}/ω_1	ω_{11}/ω_1
dx=0	0	26.02	25.83	25.08	24.35	23.47	23.12	22.78		1.15	1.14	1.11	1.08	1.04	1.02	1.01
	0.5	24.53	25.27	24.53	23.99	23.29	22.95	22.78		1.09	1.12	1.09	1.06	1.03	1.02	1.01
	1	25.08	24.90	24.17	23.82	23.12	22.95	22.78		1.11	1.10	1.07	1.05	1.02	1.02	1.01
	2	24.17	24.17	23.64	23.47	22.95	22.78	22.62		1.07	1.07	1.05	1.04	1.02	1.01	1.00
	3	23.82	22.78	23.29	23.12	22.78	22.62	22.45		1.05	1.01	1.03	1.02	1.01	1.00	0.99
	4	23.47	23.29	23.12	22.95	22.62	22.62	22.45		1.04	1.03	1.02	1.02	1.00	1.00	0.99
	5	23.12	23.12	22.95	22.78	22.62	22.45	22.45		1.02	1.02	1.02	1.01	1.00	0.99	0.99
dx=0.1	0	25.83	25.64	23.82	23.99	23.29	22.95	22.62		1.14	1.14	1.05	1.06	1.03	1.02	1.00
	0.5	25.27	25.08	24.35	23.82	22.28	22.78	22.62		1.12	1.11	1.08	1.05	0.99	1.01	1.00
	1	24.90	24.71	23.99	23.64	23.12	22.78	22.62		1.10	1.09	1.06	1.05	1.02	1.01	1.00
	2	24.17	23.99	23.64	23.29	22.78	22.62	22.45		1.07	1.06	1.05	1.03	1.01	1.00	0.99
	3	23.64	23.47	22.45	22.95	22.78	21.64	22.45		1.05	1.04	0.99	1.02	1.01	0.96	0.99
	4	23.29	23.29	22.95	22.78	22.62	22.45	22.45		1.03	1.03	1.02	1.01	1.00	0.99	0.99
	5	23.12	22.95	22.78	22.78	21.64	22.45	22.28		1.02	1.02	1.01	1.01	0.96	0.99	0.99
dx=0.5	0	25.45	24.17	24.53	23.82	23.12	22.78	22.62		1.13	1.07	1.09	1.05	1.02	1.01	1.00
	0.5	25.08	24.71	23.12	22.62	22.95	22.62	22.45		1.11	1.09	1.02	1.00	1.02	1.00	0.99
	1	24.53	24.35	23.82	23.47	22.95	22.62	22.45		1.09	1.08	1.05	1.04	1.02	1.00	0.99
	2	23.99	23.82	23.47	23.12	22.78	22.45	22.45		1.06	1.05	1.04	1.02	1.01	0.99	0.99
	3	23.47	23.47	23.12	22.95	22.62	22.45	22.28		1.04	1.04	1.02	1.02	1.00	0.99	0.99
	4	23.29	23.12	22.95	22.78	22.45	22.45	22.28		1.03	1.02	1.02	1.01	0.99	0.99	0.99
	5	22.12	22.95	22.78	21.80	22.45	22.28	22.28		0.98	1.02	1.01	0.96	0.99	0.99	0.99
dx=1	0	25.45	25.08	24.35	22.78	22.95	22.62	22.45		1.13	1.11	1.08	1.01	1.02	1.00	0.99
	0.5	24.90	24.71	23.99	23.47	22.95	22.62	22.45		1.10	1.09	1.06	1.04	1.02	1.00	0.99
	1	24.53	24.35	23.82	23.29	22.78	22.62	22.45		1.09	1.08	1.05	1.03	1.01	1.00	0.99
	2	23.82	23.82	23.47	23.12	22.62	21.64	22.28		1.05	1.05	1.04	1.02	1.00	0.96	0.99
	3	23.47	23.47	23.12	22.95	21.64	21.48	21.48		1.04	1.04	1.02	1.02	0.96	0.95	0.95
	4	22.28	22.28	22.12	21.80	21.64	21.48	21.32		0.99	0.99	0.98	0.96	0.96	0.95	0.94
	5	22.12	22.12	21.96	21.80	21.48	21.48	21.32		0.98	0.98	0.97	0.96	0.95	0.95	0.94

Parabolic soil – smooth contact																
L/H=10	dw	0	1	5	10	20	30	40		0	1	5	10	20	30	40
	dθ	ω_{11}	ω_{11}	ω_{11}	ω_{11}	ω_{11}	ω_{11}	ω_{11}		ω_{11}/ω_1	ω_{11}/ω_1	ω_{11}/ω_1	ω_{11}/ω_1	ω_{11}/ω_1	ω_{11}/ω_1	ω_{11}/ω_1
dx=0	0	23.99	23.99	23.82	23.82	23.47	23.29	23.12		1.06	1.06	1.05	1.05	1.04	1.03	1.02
	0.5	23.99	23.99	23.82	23.64	23.47	23.29	23.12		1.06	1.06	1.05	1.05	1.04	1.03	1.02
	1	22.95	23.82	22.78	23.64	22.45	23.12	23.12		1.02	1.05	1.01	1.05	0.99	1.02	1.02
	2	23.82	23.82	23.64	23.47	23.29	23.12	22.95		1.05	1.05	1.05	1.04	1.03	1.02	1.02
	3	23.64	23.64	23.47	23.29	23.12	22.95	22.95		1.05	1.05	1.04	1.03	1.02	1.02	1.02
	4	23.47	23.47	22.45	23.29	23.12	22.95	22.95		1.04	1.04	0.99	1.03	1.02	1.02	1.02
	5	23.29	23.29	23.29	23.12	22.95	22.95	22.78		1.03	1.03	1.03	1.02	1.02	1.02	1.01
dx=0.1	0	23.99	23.99	23.82	23.82	23.47	23.12	23.12		1.06	1.06	1.05	1.05	1.04	1.02	1.02
	0.5	23.99	23.82	23.82	23.64	23.29	23.12	22.95		1.06	1.05	1.05	1.05	1.03	1.02	1.02
	1	23.82	23.82	23.82	23.64	23.29	23.12	22.95		1.05	1.05	1.05	1.05	1.03	1.02	1.02
	2	23.82	23.64	23.64	23.47	23.12	22.95	22.95		1.05	1.05	1.05	1.04	1.02	1.02	1.02
	3	23.64	23.47	23.47	23.29	23.12	22.95	22.78		1.05	1.04	1.04	1.03	1.02	1.02	1.01
	4	23.47	23.47	23.29	23.12	22.95	22.95	22.78		1.04	1.04	1.03	1.02	1.02	1.02	1.01
	5	23.29	23.29	23.12	23.12	22.95	21.96	22.78		1.03	1.03	1.02	1.02	1.02	0.97	1.01
dx=0.5	0	23.99	23.99	23.82	23.64	23.29	23.12	22.95		1.06	1.06	1.05	1.05	1.03	1.02	1.02
	0.5	23.82	23.82	23.82	23.64	23.29	23.12	22.95		1.05	1.05	1.05	1.05	1.03	1.02	1.02
	1	23.82	23.82	23.64	23.47	23.12	22.95	22.78		1.05	1.05	1.05	1.04	1.02	1.02	1.01
	2	23.64	23.64	23.47	23.29	23.12	22.95	22.78		1.05	1.05	1.04	1.03	1.02	1.02	1.01
	3	23.64	23.47	23.29	23.12	22.95	22.78	22.78		1.05	1.04	1.03	1.02	1.02	1.01	1.01
	4	23.47	23.29	23.29	23.12	22.95	22.78	22.78		1.04	1.03	1.03	1.02	1.02	1.01	1.01
	5	23.29	23.29	23.12	22.95	22.78	22.78	22.62		1.03	1.03	1.02	1.02	1.01	1.01	1.00
dx=1	0	23.99	23.82	23.82	23.64	23.29	23.12	22.95		1.06	1.05	1.05	1.05	1.03	1.02	1.02
	0.5	23.82	23.82	23.64	23.47	23.29	22.95	22.78		1.05	1.05	1.05	1.04	1.03	1.02	1.01
	1	22.78	22.78	22.62	22.45	23.12	22.95	22.78		1.01	1.01	1.00	0.99	1.02	1.02	1.01
	2	23.64	23.64	23.47	23.29	23.12	22.95	22.78		1.05	1.05	1.04	1.03	1.02	1.02	1.01
	3	23.47	22.62	23.29	23.12	22.95	22.78	22.78		1.04	1.00	1.03	1.02	1.02	1.01	1.01
	4	23.47	23.29	23.12	23.12	22.95	22.78	22.78		1.04	1.03	1.02	1.02	1.02	1.01	1.01
	5	23.29	23.29	23.12	22.95	22.78	22.78	22.62		1.03	1.03	1.02	1.02	1.01	1.01	1.00

Linear soil – smooth contact																
L/H=1	dw	0	1	5	10	20	30	40		0	1	5	10	20	30	40
	dθ	ω_{11}	ω_{11}	ω_{11}	ω_{11}	ω_{11}	ω_{11}	ω_{11}		ω_{11}/ω_1	ω_{11}/ω_1	ω_{11}/ω_1	ω_{11}/ω_1	ω_{11}/ω_1	ω_{11}/ω_1	ω_{11}/ω_1
dx=0	0	40.83	37.93	29.72	25.45	21.96	20.55	19.66		1.82	1.69	1.33	1.14	0.98	0.92	0.88
	0.5	34.97	32.72	27.40	24.17	21.48	20.24	19.51		1.56	1.46	1.22	1.08	0.96	0.90	0.87
	1	30.84	29.29	25.64	23.29	21.01	19.95	19.37		1.38	1.31	1.15	1.04	0.94	0.89	0.87
	2	26.22	25.45	23.47	21.96	20.39	19.51	19.08		1.17	1.14	1.05	0.98	0.91	0.87	0.85
	3	23.99	23.47	22.28	21.16	19.95	19.22	18.80		1.07	1.05	1.00	0.95	0.89	0.86	0.84
	4	22.62	22.28	21.32	20.55	19.51	18.94	18.67		1.01	1.00	0.95	0.92	0.87	0.85	0.83
	5	21.64	21.48	20.70	20.10	19.22	18.80	18.53		0.97	0.96	0.92	0.90	0.86	0.84	0.83
dx=0.1	0	39.35	36.01	28.43	24.71	21.48	20.10	19.37		1.76	1.61	1.27	1.10	0.96	0.90	0.87
	0.5	33.45	31.53	26.41	23.64	21.01	19.95	19.22		1.49	1.41	1.18	1.06	0.94	0.89	0.86
	1	29.72	28.43	25.08	22.78	20.70	19.66	19.08		1.33	1.27	1.12	1.02	0.92	0.88	0.85
	2	25.83	25.08	23.12	21.80	20.24	19.37	18.80		1.15	1.12	1.03	0.97	0.90	0.87	0.84
	3	23.64	23.29	21.96	21.01	19.80	19.08	18.67		1.06	1.04	0.98	0.94	0.88	0.85	0.83
	4	22.45	22.12	21.16	20.39	19.37	18.80	18.53		1.00	0.99	0.95	0.91	0.87	0.84	0.83
	5	21.48	21.32	20.55	19.95	19.22	18.67	18.39		0.96	0.95	0.92	0.89	0.86	0.83	0.82
dx=0.5	0	37.37	34.20	27.61	23.99	21.16	19.80	19.08		1.67	1.53	1.23	1.07	0.95	0.88	0.85
	0.5	32.24	30.17	25.83	23.12	20.70	19.66	18.94		1.44	1.35	1.15	1.03	0.92	0.88	0.85
	1	29.07	27.81	24.53	22.45	20.39	19.51	18.80		1.30	1.24	1.10	1.00	0.91	0.87	0.84
	2	25.45	24.71	22.95	21.48	19.95	19.08	18.67		1.14	1.10	1.03	0.96	0.89	0.85	0.83
	3	23.47	23.12	21.80	20.85	19.51	18.94	18.53		1.05	1.03	0.97	0.93	0.87	0.85	0.83
	4	22.28	21.96	21.01	20.24	19.22	18.67	18.39		1.00	0.98	0.94	0.90	0.86	0.83	0.82
	5	21.48	21.16	20.55	19.95	19.08	18.53	18.26		0.96	0.95	0.92	0.89	0.85	0.83	0.82
dx=1	0	36.82	33.70	27.20	23.82	21.01	19.80	19.08		1.65	1.51	1.22	1.06	0.94	0.88	0.85
	0.5	31.77	29.94	25.64	22.95	20.70	19.51	18.94		1.42	1.34	1.15	1.03	0.92	0.87	0.85
	1	28.65	27.61	24.35	22.28	20.39	19.37	18.80		1.28	1.23	1.09	1.00	0.91	0.87	0.84
	2	25.27	24.71	22.78	21.32	19.95	19.08	18.67		1.13	1.10	1.02	0.95	0.89	0.85	0.83
	3	23.47	22.95	21.80	20.70	19.51	18.80	18.39		1.05	1.03	0.97	0.92	0.87	0.84	0.82
	4	22.28	21.96	21.01	20.24	19.22	18.67	18.26		1.00	0.98	0.94	0.90	0.86	0.83	0.82
	5	21.32	21.16	20.39	19.80	19.08	18.53	18.26		0.95	0.95	0.91	0.88	0.85	0.83	0.82

Linear soil – smooth contact																
L/H=2	dw	0	1	5	10	20	30	40		0	1	5	10	20	30	40
	dθ	ω_{11}	ω_{11}	ω_{11}	ω_{11}	ω_{11}	ω_{11}	ω_{11}		ω_{11}/ω_1	ω_{11}/ω_1	ω_{11}/ω_1	ω_{11}/ω_1	ω_{11}/ω_1	ω_{11}/ω_1	ω_{11}/ω_1
dx=0	0	24.53	23.99	21.96	20.24	18.53	17.72	17.08		1.10	1.07	0.98	0.90	0.83	0.79	0.76
	0.5	23.12	22.62	21.01	19.66	18.26	17.46	16.96		1.03	1.01	0.94	0.88	0.82	0.78	0.76
	1	21.96	21.64	20.24	19.22	17.99	17.34	16.96		0.98	0.97	0.90	0.86	0.80	0.77	0.76
	2	20.39	20.10	19.22	18.53	17.59	17.08	16.71		0.91	0.90	0.86	0.83	0.79	0.76	0.75
	3	19.37	19.22	18.53	17.99	17.34	16.83	16.58		0.87	0.86	0.83	0.80	0.77	0.75	0.74
	4	18.67	18.53	17.99	17.59	17.08	16.71	16.46		0.83	0.83	0.80	0.79	0.76	0.75	0.74
	5	18.12	17.99	17.59	17.34	16.83	16.58	16.34		0.81	0.80	0.79	0.77	0.75	0.74	0.73
dx=0.1	0	23.82	23.12	21.32	19.80	18.26	17.46	16.96		1.06	1.03	0.95	0.88	0.82	0.78	0.76
	0.5	22.45	21.96	20.55	19.37	17.99	17.34	16.83		1.00	0.98	0.92	0.87	0.80	0.77	0.75
	1	21.48	21.16	19.95	18.94	17.86	17.21	16.71		0.96	0.95	0.89	0.85	0.80	0.77	0.75
	2	20.10	19.80	18.94	18.26	17.46	16.96	16.58		0.90	0.88	0.85	0.82	0.78	0.76	0.74
	3	19.08	18.94	18.39	17.86	17.21	16.71	16.46		0.85	0.85	0.82	0.80	0.77	0.75	0.74
	4	18.53	18.39	17.86	17.46	16.96	16.58	16.34		0.83	0.82	0.80	0.78	0.76	0.74	0.73
	5	17.99	17.86	17.59	17.21	16.83	16.46	16.22		0.80	0.80	0.79	0.77	0.75	0.74	0.72
dx=0.5	0	22.78	22.28	20.55	19.37	17.86	17.21	16.71		1.02	1.00	0.92	0.87	0.80	0.77	0.75
	0.5	21.64	21.32	19.95	18.94	17.72	17.08	16.71		0.97	0.95	0.89	0.85	0.79	0.76	0.75
	1	20.85	20.55	19.51	18.53	17.59	16.96	16.58		0.93	0.92	0.87	0.83	0.79	0.76	0.74
	2	19.66	19.51	18.67	17.99	17.21	16.71	16.46		0.88	0.87	0.83	0.80	0.77	0.75	0.74
	3	18.80	18.67	18.12	17.59	16.96	16.58	16.34		0.84	0.83	0.81	0.79	0.76	0.74	0.73
	4	18.26	18.12	17.72	17.34	16.83	16.46	16.22		0.82	0.81	0.79	0.77	0.75	0.74	0.72
	5	17.86	17.72	17.46	17.08	16.71	16.34	16.10		0.80	0.79	0.78	0.76	0.75	0.73	0.72
dx=1	0	22.45	21.96	20.39	19.22	17.86	17.08	16.71		1.00	0.98	0.91	0.86	0.80	0.76	0.75
	0.5	21.48	21.01	19.80	18.80	17.59	16.96	16.58		0.96	0.94	0.88	0.84	0.79	0.76	0.74
	1	20.70	20.39	19.37	18.53	17.46	16.83	16.58		0.92	0.91	0.87	0.83	0.78	0.75	0.74
	2	19.51	19.37	18.67	17.99	17.21	16.71	16.46		0.87	0.87	0.83	0.80	0.77	0.75	0.74
	3	18.80	18.67	18.12	17.59	16.96	16.58	16.34		0.84	0.83	0.81	0.79	0.76	0.74	0.73
	4	18.26	18.12	17.72	17.34	16.83	16.46	16.22		0.82	0.81	0.79	0.77	0.75	0.74	0.72
	5	17.86	17.72	17.34	17.08	16.58	16.34	16.10		0.80	0.79	0.77	0.76	0.74	0.73	0.72

Linear soil – smooth contact																
L/H=3	dw	0	1	5	10	20	30	40		0	1	5	10	20	30	40
	dθ	ω_{11}	ω_{11}	ω_{11}	ω_{11}	ω_{11}	ω_{11}	ω_{11}		ω_{11}/ω_1	ω_{11}/ω_1	ω_{11}/ω_1	ω_{11}/ω_1	ω_{11}/ω_1	ω_{11}/ω_1	ω_{11}/ω_1
dx=0	0	27.44	26.91	25.15	23.73	22.39	21.96	21.54		1.23	1.20	1.12	1.06	1.00	0.98	0.96
	0.5	26.14	25.64	24.43	23.27	22.17	21.75	21.33		1.17	1.15	1.09	1.04	0.99	0.97	0.95
	1	25.15	24.67	23.73	22.83	21.96	21.54	21.33		1.12	1.10	1.06	1.02	0.98	0.96	0.95
	2	23.73	23.50	23.05	22.39	21.75	21.33	21.12		1.06	1.05	1.03	1.00	0.97	0.95	0.94
	3	23.05	22.83	22.39	21.96	21.54	21.33	21.12		1.03	1.02	1.00	0.98	0.96	0.95	0.94
	4	22.61	22.39	22.17	21.75	21.33	21.12	21.12		1.01	1.00	0.99	0.97	0.95	0.94	0.94
	5	22.17	22.17	21.96	21.54	21.33	21.12	20.92		0.99	0.99	0.98	0.96	0.95	0.94	0.93
dx=0.1	0	26.65	26.14	24.67	23.50	22.17	21.75	21.33		1.19	1.17	1.10	1.05	0.99	0.97	0.95
	0.5	25.64	25.15	23.96	23.05	21.96	21.54	21.33		1.15	1.12	1.07	1.03	0.98	0.96	0.95
	1	24.67	24.43	23.50	22.61	21.96	21.54	21.12		1.10	1.09	1.05	1.01	0.98	0.96	0.94
	2	23.50	23.50	22.83	22.17	21.75	21.33	21.12		1.05	1.05	1.02	0.99	0.97	0.95	0.94
	3	22.83	22.83	22.39	21.96	21.54	21.12	21.12		1.02	1.02	1.00	0.98	0.96	0.94	0.94
	4	22.39	22.39	21.96	21.75	21.33	21.12	20.92		1.00	1.00	0.98	0.97	0.95	0.94	0.93
	5	22.17	22.17	21.75	21.54	21.33	21.12	20.92		0.99	0.99	0.97	0.96	0.95	0.94	0.93
dx=0.5	0	26.14	25.64	24.19	23.05	21.96	21.54	21.12		1.17	1.15	1.08	1.03	0.98	0.96	0.94
	0.5	25.15	24.91	23.73	22.83	21.96	21.33	21.12		1.12	1.11	1.06	1.02	0.98	0.95	0.94
	1	24.43	24.19	23.27	22.61	21.75	21.33	21.12		1.09	1.08	1.04	1.01	0.97	0.95	0.94
	2	23.50	23.27	22.61	22.17	21.54	21.12	21.12		1.05	1.04	1.01	0.99	0.96	0.94	0.94
	3	22.83	22.61	22.17	21.75	21.33	21.12	20.92		1.02	1.01	0.99	0.97	0.95	0.94	0.93
	4	22.39	22.39	21.96	21.54	21.33	21.12	20.92		1.00	1.00	0.98	0.96	0.95	0.94	0.93
	5	22.17	21.96	21.75	21.54	21.12	20.92	20.92		0.99	0.98	0.97	0.96	0.94	0.93	0.93
dx=1	0	26.14	25.64	24.19	23.05	21.96	21.54	21.12		1.17	1.15	1.08	1.03	0.98	0.96	0.94
	0.5	25.15	24.67	23.50	22.83	21.75	21.33	21.12		1.12	1.10	1.05	1.02	0.97	0.95	0.94
	1	24.43	24.19	23.27	22.39	21.75	21.33	21.12		1.09	1.08	1.04	1.00	0.97	0.95	0.94
	2	23.50	23.27	22.61	22.17	21.54	21.12	20.92		1.05	1.04	1.01	0.99	0.96	0.94	0.93
	3	22.83	22.61	22.17	21.75	21.33	21.12	20.92		1.02	1.01	0.99	0.97	0.95	0.94	0.93
	4	22.39	22.17	21.96	21.54	21.33	21.12	20.92		1.00	0.99	0.98	0.96	0.95	0.94	0.93
	5	22.17	21.96	21.75	21.54	21.12	20.92	20.92		0.99	0.98	0.97	0.96	0.94	0.93	0.93

Linear soil – smooth contact															
L/H=5	dw	0	1	5	10	20	30	40		0	1	5	10	20	30	40
	dθ	ω_{11}	ω_{11}	ω_{11}	ω_{11}	ω_{11}	ω_{11}	ω_{11}		ω_{11}/ω_1	ω_{11}/ω_1	ω_{11}/ω_1	ω_{11}/ω_1	ω_{11}/ω_1	ω_{11}/ω_1	ω_{11}/ω_1
dx=0	0	24.35	24.17	23.82	23.12	22.45	22.12	21.80		1.09	1.08	1.06	1.03	1.00	0.99	0.97
	0.5	23.99	23.99	23.47	22.95	22.28	21.96	21.80		1.07	1.07	1.05	1.03	1.00	0.98	0.97
	1	23.82	23.64	23.12	22.78	22.28	21.96	21.80		1.06	1.06	1.03	1.02	1.00	0.98	0.97
	2	23.12	23.12	22.78	22.45	21.96	21.80	21.64		1.03	1.03	1.02	1.00	0.98	0.97	0.97
	3	22.78	22.78	22.45	22.28	21.96	21.64	21.64		1.02	1.02	1.00	1.00	0.98	0.97	0.97
	4	22.62	22.45	22.28	22.12	21.80	21.64	21.48		1.01	1.00	1.00	0.99	0.97	0.97	0.96
	5	22.28	22.28	22.12	21.96	21.64	21.64	21.48		1.00	1.00	0.99	0.98	0.97	0.97	0.96
dx=0.1	0	24.17	23.99	23.47	22.95	22.28	21.96	21.80		1.08	1.07	1.05	1.03	1.00	0.98	0.97
	0.5	23.82	23.82	23.29	22.78	22.28	21.96	21.64		1.06	1.06	1.04	1.02	1.00	0.98	0.97
	1	23.64	23.47	23.12	22.62	22.12	21.80	21.64		1.06	1.05	1.03	1.01	0.99	0.97	0.97
	2	23.12	22.95	22.62	22.28	21.96	21.64	21.64		1.03	1.03	1.01	1.00	0.98	0.97	0.97
	3	22.78	22.62	22.45	22.12	21.80	21.64	21.48		1.02	1.01	1.00	0.99	0.97	0.97	0.96
	4	22.45	22.45	22.28	21.96	21.80	21.64	21.48		1.00	1.00	1.00	0.98	0.97	0.97	0.96
	5	22.28	22.28	22.12	21.96	21.64	21.48	21.48		1.00	1.00	0.99	0.98	0.97	0.96	0.96
dx=0.5	0	23.99	23.82	23.29	22.78	22.28	21.80	21.64		1.07	1.06	1.04	1.02	1.00	0.97	0.97
	0.5	23.82	23.64	23.12	22.62	22.12	21.80	21.64		1.06	1.06	1.03	1.01	0.99	0.97	0.97
	1	23.47	23.29	22.95	22.45	21.96	21.80	21.48		1.05	1.04	1.03	1.00	0.98	0.97	0.96
	2	22.95	22.95	22.62	22.28	21.80	21.64	21.48		1.03	1.03	1.01	1.00	0.97	0.97	0.96
	3	22.78	22.62	22.28	22.12	21.80	21.64	21.48		1.02	1.01	1.00	0.99	0.97	0.97	0.96
	4	22.45	22.45	22.12	21.96	21.64	21.48	21.48		1.00	1.00	0.99	0.98	0.97	0.96	0.96
	5	22.28	22.28	22.12	21.96	21.64	21.48	21.32		1.00	1.00	0.99	0.98	0.97	0.96	0.95
dx=1	0	23.99	23.82	23.29	22.78	22.12	21.80	21.64		1.07	1.06	1.04	1.02	0.99	0.97	0.97
	0.5	23.64	23.64	23.12	22.62	22.12	21.80	21.64		1.06	1.06	1.03	1.01	0.99	0.97	0.97
	1	23.47	23.29	22.95	22.45	21.96	21.64	21.48		1.05	1.04	1.03	1.00	0.98	0.97	0.96
	2	22.95	22.95	22.62	22.28	21.80	21.64	21.48		1.03	1.03	1.01	1.00	0.97	0.97	0.96
	3	22.62	22.62	22.28	22.12	21.80	21.48	21.48		1.01	1.01	1.00	0.99	0.97	0.96	0.96
	4	22.45	22.45	22.12	21.96	21.64	21.48	21.48		1.00	1.00	0.99	0.98	0.97	0.96	0.96
	5	22.28	22.28	22.12	21.96	21.64	21.48	21.32		1.00	1.00	0.99	0.98	0.97	0.96	0.95

Linear soil – smooth contact																
L/H=10	dw	0	1	5	10	20	30	40		0	1	5	10	20	30	40
	dθ	ω_{11}	ω_{11}	ω_{11}	ω_{11}	ω_{11}	ω_{11}	ω_{11}		ω_{11}/ω_1	ω_{11}/ω_1	ω_{11}/ω_1	ω_{11}/ω_1	ω_{11}/ω_1	ω_{11}/ω_1	ω_{11}/ω_1
dx=0	0	22.78	22.78	22.78	22.78	22.45	22.28	22.12		1.02	1.02	1.02	1.02	1.00	1.00	0.99
	0.5	22.78	22.78	22.78	22.62	22.45	22.28	22.12		1.02	1.02	1.02	1.01	1.00	1.00	0.99
	1	22.78	22.78	22.78	22.62	22.45	22.12	22.12		1.02	1.02	1.02	1.01	1.00	0.99	0.99
	2	22.78	22.78	22.62	22.45	22.28	22.12	21.96		1.02	1.02	1.01	1.00	1.00	0.99	0.98
	3	22.62	22.62	22.45	22.45	22.12	21.96	21.96		1.01	1.01	1.00	1.00	0.99	0.98	0.98
	4	22.62	22.62	22.45	22.28	22.12	21.96	21.80		1.01	1.01	1.00	1.00	0.99	0.98	0.97
	5	22.45	22.45	22.28	22.28	22.12	21.96	21.80		1.00	1.00	1.00	1.00	0.99	0.98	0.97
dx=0.1	0	22.78	22.78	22.78	22.62	22.45	22.28	21.96		1.02	1.02	1.02	1.01	1.00	1.00	0.98
	0.5	22.78	22.78	22.78	22.62	22.45	22.12	21.96		1.02	1.02	1.02	1.01	1.00	0.99	0.98
	1	22.78	22.78	22.62	22.62	22.28	22.12	21.96		1.02	1.02	1.01	1.01	1.00	0.99	0.98
	2	22.78	22.62	22.62	22.45	22.28	21.96	21.96		1.02	1.01	1.01	1.00	1.00	0.98	0.98
	3	22.62	22.62	22.45	22.28	22.12	21.96	21.96		1.01	1.01	1.00	1.00	0.99	0.98	0.98
	4	22.62	22.62	22.45	22.28	22.12	21.96	21.80		1.01	1.01	1.00	1.00	0.99	0.98	0.97
	5	22.45	22.45	22.45	22.28	22.12	21.96	21.80		1.00	1.00	1.00	1.00	0.99	0.98	0.97
dx=0.5	0	22.78	22.78	22.78	22.62	22.45	22.12	21.96		1.02	1.02	1.02	1.01	1.00	0.99	0.98
	0.5	22.78	22.78	22.78	22.62	22.28	22.12	21.96		1.02	1.02	1.02	1.01	1.00	0.99	0.98
	1	22.78	22.78	22.62	22.62	22.28	22.12	21.96		1.02	1.02	1.01	1.01	1.00	0.99	0.98
	2	22.78	22.62	22.62	22.45	22.12	21.96	21.80		1.02	1.01	1.01	1.00	0.99	0.98	0.97
	3	22.62	22.62	22.45	22.28	22.12	21.96	21.80		1.01	1.01	1.00	1.00	0.99	0.98	0.97
	4	22.62	22.45	22.45	22.28	21.96	21.96	21.80		1.01	1.00	1.00	1.00	0.98	0.98	0.97
	5	22.45	22.45	22.28	22.28	21.96	21.80	21.80		1.00	1.00	1.00	1.00	0.98	0.97	0.97
dx=1	0	22.78	22.78	22.78	22.62	22.28	22.12	21.96		1.02	1.02	1.02	1.01	1.00	0.99	0.98
	0.5	22.78	22.78	22.78	22.62	22.28	22.12	21.96		1.02	1.02	1.02	1.01	1.00	0.99	0.98
	1	22.78	22.78	22.62	22.45	22.28	21.96	21.96		1.02	1.02	1.01	1.00	1.00	0.98	0.98
	2	22.62	22.62	22.62	22.45	22.12	21.96	21.80		1.01	1.01	1.01	1.00	0.99	0.98	0.97
	3	22.62	22.62	22.45	22.28	22.12	21.96	21.80		1.01	1.01	1.00	1.00	0.99	0.98	0.97
	4	22.45	22.45	22.45	22.28	21.96	21.96	21.80		1.00	1.00	1.00	1.00	0.98	0.98	0.97
	5	22.45	22.45	22.28	22.28	21.96	21.80	21.80		1.00	1.00	1.00	1.00	0.98	0.97	0.97

Annex F

Natural circular frequencies of bounded systems with flexible walls based on flexible base for different ratios of *L/H* (1.0, 1.5, 2.0, 3.0, 5.0, 10.0) and for the three investigated soil profiles: homogeneous, parabolic and linear. The contact between the walls and the soil is bonded contact. These natural frequencies are identified as the frequencies, where the maximum shear force due to the soil pressures. The natural frequencies of the unbounded soil stratum for constant, parabolic and linear shear modulus distribution are 19.63 rad/s, 22.59 rad/s and 22.38 rad/sec respectively.

Homogeneous soil – bonded contact																
L/H=1	dw	0	1	5	10	20	30	40		0	1	5	10	20	30	40
	dθ	ω_{11}	ω_{11}	ω_{11}	ω_{11}	ω_{11}	ω_{11}	ω_{11}		ω_{11}/ω_1	ω_{11}/ω_1	ω_{11}/ω_1	ω_{11}/ω_1	ω_{11}/ω_1	ω_{11}/ω_1	ω_{11}/ω_1
dx=0	0	70.73	52.41	34.69	29.86	26.44	25.03	24.22		3.60	2.67	1.77	1.52	1.35	1.28	1.23
	0.5	42.75	38.52	31.47	28.25	25.83	24.83	24.02		2.18	1.96	1.60	1.44	1.32	1.27	1.22
	1	35.10	33.28	29.46	27.24	25.43	24.42	23.82		1.79	1.70	1.50	1.39	1.30	1.24	1.21
	2	29.26	28.65	27.04	25.83	24.63	23.82	23.42		1.49	1.46	1.38	1.32	1.26	1.21	1.19
	3	26.84	26.44	25.63	24.83	24.02	23.42	23.02		1.37	1.35	1.31	1.27	1.22	1.19	1.17
	4	25.43	25.23	24.63	24.22	23.62	23.22	22.81		1.30	1.29	1.26	1.23	1.20	1.18	1.16
	5	24.42	24.42	24.02	23.62	23.22	22.81	22.61		1.24	1.24	1.22	1.20	1.18	1.16	1.15
dx=0.1	0	54.63	44.76	32.48	28.65	25.83	24.63	23.82		2.78	2.28	1.66	1.46	1.32	1.26	1.21
	0.5	38.72	35.70	30.06	27.44	25.23	24.22	23.62		1.97	1.82	1.53	1.40	1.29	1.23	1.20
	1	33.08	31.67	28.45	26.44	24.83	24.02	23.42		1.69	1.61	1.45	1.35	1.27	1.22	1.19
	2	28.45	27.85	26.24	25.23	24.22	23.62	23.02		1.45	1.42	1.34	1.29	1.23	1.20	1.17
	3	26.24	26.04	25.23	24.42	23.62	23.22	22.81		1.34	1.33	1.29	1.24	1.20	1.18	1.16
	4	25.03	24.83	24.42	23.82	23.22	22.81	22.61		1.28	1.27	1.24	1.21	1.18	1.16	1.15
	5	24.22	24.02	23.82	23.42	23.02	22.61	22.41		1.23	1.22	1.21	1.19	1.17	1.15	1.14
dx=0.5	0	43.55	38.32	30.26	27.24	24.83	23.82	23.22		2.22	1.95	1.54	1.39	1.27	1.21	1.18
	0.5	35.10	32.88	28.45	26.44	24.63	23.62	23.02		1.79	1.68	1.45	1.35	1.26	1.20	1.17
	1	31.07	29.86	27.24	25.63	24.22	23.42	23.02		1.58	1.52	1.39	1.31	1.23	1.19	1.17
	2	27.44	27.04	25.63	24.63	23.62	23.02	22.61		1.40	1.38	1.31	1.26	1.20	1.17	1.15
	3	25.63	25.43	24.63	24.02	23.22	22.81	22.41		1.31	1.30	1.26	1.22	1.18	1.16	1.14
	4	24.63	24.42	24.02	23.42	23.02	22.61	22.21		1.26	1.24	1.22	1.19	1.17	1.15	1.13
	5	23.82	23.82	23.42	23.02	22.61	22.41	22.21		1.21	1.21	1.19	1.17	1.15	1.14	1.13
dx=1	0	41.14	36.71	29.66	26.84	24.63	23.62	23.02		2.10	1.87	1.51	1.37	1.26	1.20	1.17
	0.5	34.09	32.08	28.05	26.04	24.22	23.42	23.02		1.74	1.63	1.43	1.33	1.23	1.19	1.17
	1	30.67	29.46	26.84	25.43	24.02	23.22	22.81		1.56	1.50	1.37	1.30	1.22	1.18	1.16
	2	27.24	26.64	25.43	24.42	23.42	23.02	22.61		1.39	1.36	1.30	1.24	1.19	1.17	1.15
	3	25.43	25.23	24.42	23.82	23.22	22.61	22.41		1.30	1.29	1.24	1.21	1.18	1.15	1.14
	4	24.42	24.22	23.82	23.42	22.81	22.41	22.21		1.24	1.23	1.21	1.19	1.16	1.14	1.13
	5	23.82	23.62	23.22	23.02	22.61	22.21	22.01		1.21	1.20	1.18	1.17	1.15	1.13	1.12

Homogeneous soil – bonded contact																
L/H=2	dw	0	1	5	10	20	30	40		0	1	5	10	20	30	40
	dθ	ω_{11}	ω_{11}	ω_{11}	ω_{11}	ω_{11}	ω_{11}	ω_{11}		ω_{11}/ω_1	ω_{11}/ω_1	ω_{11}/ω_1	ω_{11}/ω_1	ω_{11}/ω_1	ω_{11}/ω_1	ω_{11}/ω_1
dx=0	0	39.32	34.29	27.24	24.83	23.22	22.41	22.01		2.00	1.75	1.39	1.27	1.18	1.14	1.12
	0.5	30.87	29.06	25.63	24.02	22.81	22.21	21.81		1.57	1.48	1.31	1.22	1.16	1.13	1.11
	1	27.44	26.44	24.63	23.62	22.61	22.01	21.81		1.40	1.35	1.26	1.20	1.15	1.12	1.11
	2	24.42	24.22	23.42	22.81	22.21	21.81	21.61		1.24	1.23	1.19	1.16	1.13	1.11	1.10
	3	23.22	23.02	22.61	22.21	21.81	21.61	21.40		1.18	1.17	1.15	1.13	1.11	1.10	1.09
	4	22.61	22.41	22.21	22.01	21.61	21.40	21.20		1.15	1.14	1.13	1.12	1.10	1.09	1.08
	5	22.01	22.01	21.81	21.61	21.40	21.20	21.20		1.12	1.12	1.11	1.10	1.09	1.08	1.08
dx=0.1	0	35.10	31.47	26.04	24.22	22.81	22.21	21.81		1.79	1.60	1.33	1.23	1.16	1.13	1.11
	0.5	29.06	27.65	24.83	23.62	22.41	22.01	21.61		1.48	1.41	1.27	1.20	1.14	1.12	1.10
	1	26.44	25.63	24.02	23.22	22.21	21.81	21.61		1.35	1.31	1.22	1.18	1.13	1.11	1.10
	2	24.02	23.82	23.02	22.61	22.01	21.61	21.40		1.22	1.21	1.17	1.15	1.12	1.10	1.09
	3	23.02	22.81	22.41	22.01	21.61	21.40	21.20		1.17	1.16	1.14	1.12	1.10	1.09	1.08
	4	22.41	22.21	22.01	21.81	21.40	21.20	21.20		1.14	1.13	1.12	1.11	1.09	1.08	1.08
	5	22.01	21.81	21.81	21.61	21.40	21.20	21.00		1.12	1.11	1.11	1.10	1.09	1.08	1.07
dx=0.5	0	31.27	28.85	25.03	23.42	22.21	21.81	21.40		1.59	1.47	1.28	1.19	1.13	1.11	1.09
	0.5	27.44	26.24	24.02	23.02	22.21	21.61	21.40		1.40	1.34	1.22	1.17	1.13	1.10	1.09
	1	25.43	24.83	23.42	22.61	22.01	21.61	21.40		1.30	1.27	1.19	1.15	1.12	1.10	1.09
	2	23.62	23.42	22.61	22.21	21.61	21.40	21.20		1.20	1.19	1.15	1.13	1.10	1.09	1.08
	3	22.61	22.61	22.21	21.81	21.40	21.20	21.00		1.15	1.15	1.13	1.11	1.09	1.08	1.07
	4	22.21	22.01	21.81	21.61	21.20	21.20	21.00		1.13	1.12	1.11	1.10	1.08	1.08	1.07
	5	21.81	21.81	21.61	21.40	21.20	21.00	20.80		1.11	1.11	1.10	1.09	1.08	1.07	1.06
dx=1	0	30.26	28.05	24.63	23.22	22.21	21.61	21.40		1.54	1.43	1.26	1.18	1.13	1.10	1.09
	0.5	26.84	25.83	23.82	22.81	22.01	21.61	21.40		1.37	1.32	1.21	1.16	1.12	1.10	1.09
	1	25.23	24.63	23.22	22.61	21.81	21.40	21.20		1.29	1.26	1.18	1.15	1.11	1.09	1.08
	2	23.42	23.22	22.61	22.01	21.61	21.20	21.20		1.19	1.18	1.15	1.12	1.10	1.08	1.08
	3	22.61	22.41	22.01	21.81	21.40	21.20	21.00		1.15	1.14	1.12	1.11	1.09	1.08	1.07
	4	22.01	22.01	21.81	21.61	21.20	21.00	21.00		1.12	1.12	1.11	1.10	1.08	1.07	1.07
	5	21.81	21.61	21.40	21.40	21.00	21.00	20.80		1.11	1.10	1.09	1.09	1.07	1.07	1.06

Homogeneous soil – bonded contact																
L/H=3	dw	0	1	5	10	20	30	40		0	1	5	10	20	30	40
	dθ	ω_{11}	ω_{11}	ω_{11}	ω_{11}	ω_{11}	ω_{11}	ω_{11}		ω_{11}/ω_1	ω_{11}/ω_1	ω_{11}/ω_1	ω_{11}/ω_1	ω_{11}/ω_1	ω_{11}/ω_1	ω_{11}/ω_1
dx=0	0	30.06	27.85	24.42	23.02	21.81	21.40	21.20		1.53	1.42	1.24	1.17	1.11	1.09	1.08
	0.5	26.24	25.23	23.42	22.41	21.61	21.40	21.00		1.34	1.29	1.19	1.14	1.10	1.09	1.07
	1	24.42	24.02	22.81	22.21	21.61	21.20	21.00		1.24	1.22	1.16	1.13	1.10	1.08	1.07
	2	22.81	22.61	22.01	21.61	21.20	21.00	20.80		1.16	1.15	1.12	1.10	1.08	1.07	1.06
	3	22.01	21.81	21.61	21.40	21.00	21.00	20.80		1.12	1.11	1.10	1.09	1.07	1.07	1.06
	4	21.61	21.40	21.20	21.20	21.00	20.80	20.60		1.10	1.09	1.08	1.08	1.07	1.06	1.05
	5	21.20	21.20	21.00	21.00	20.80	20.60	20.60		1.08	1.08	1.07	1.07	1.06	1.05	1.05
dx=0.1	0	28.25	26.64	23.62	22.61	21.61	21.20	21.00		1.44	1.36	1.20	1.15	1.10	1.08	1.07
	0.5	25.43	24.63	23.02	22.21	21.40	21.20	21.00		1.30	1.26	1.17	1.13	1.09	1.08	1.07
	1	23.82	23.42	22.41	21.81	21.40	21.00	20.80		1.21	1.19	1.14	1.11	1.09	1.07	1.06
	2	22.41	22.41	21.81	21.40	21.20	21.00	20.80		1.14	1.14	1.11	1.09	1.08	1.07	1.06
	3	21.81	21.81	21.40	21.20	21.00	20.80	20.60		1.11	1.11	1.09	1.08	1.07	1.06	1.05
	4	21.40	21.40	21.20	21.00	20.80	20.80	20.60		1.09	1.09	1.08	1.07	1.06	1.06	1.05
	5	21.20	21.00	21.00	20.80	20.80	20.60	20.60		1.08	1.07	1.07	1.06	1.06	1.05	1.05
dx=0.5	0	30.06	27.85	24.42	23.02	21.81	21.40	21.20		1.53	1.42	1.24	1.17	1.11	1.09	1.08
	0.5	26.24	25.23	23.42	22.41	21.61	21.40	21.00		1.34	1.29	1.19	1.14	1.10	1.09	1.07
	1	24.42	24.02	22.81	22.21	21.61	21.20	21.00		1.24	1.22	1.16	1.13	1.10	1.08	1.07
	2	22.81	22.61	22.01	21.61	21.20	21.00	20.80		1.16	1.15	1.12	1.10	1.08	1.07	1.06
	3	22.01	21.81	21.61	21.40	21.00	20.80	20.80		1.12	1.11	1.10	1.09	1.07	1.06	1.06
	4	21.61	21.40	21.20	21.20	21.00	20.80	20.60		1.10	1.09	1.08	1.08	1.07	1.06	1.05
	5	21.20	21.20	21.00	21.00	20.80	20.60	20.60		1.08	1.08	1.07	1.07	1.06	1.05	1.05
dx=1	0	26.04	24.83	22.81	22.01	21.20	21.00	20.80		1.33	1.27	1.16	1.12	1.08	1.07	1.06
	0.5	24.22	23.62	22.41	21.81	21.20	20.80	20.80		1.23	1.20	1.14	1.11	1.08	1.06	1.06
	1	23.22	22.81	22.01	21.61	21.00	20.80	20.60		1.18	1.16	1.12	1.10	1.07	1.06	1.05
	2	22.01	22.01	21.61	21.20	21.00	20.80	20.60		1.12	1.12	1.10	1.08	1.07	1.06	1.05
	3	21.61	21.40	21.20	21.00	20.80	20.60	20.60		1.10	1.09	1.08	1.07	1.06	1.05	1.05
	4	21.20	21.20	21.00	20.80	20.60	20.60	20.40		1.08	1.08	1.07	1.06	1.05	1.05	1.04
	5	21.00	21.00	20.80	20.80	20.60	20.40	20.40		1.07	1.07	1.06	1.06	1.05	1.04	1.04

Homogeneous soil – bonded contact																
L/H=5	dw	0	1	5	10	20	30	40		0	1	5	10	20	30	40
	dθ	ω_{11}	ω_{11}	ω_{11}	ω_{11}	ω_{11}	ω_{11}	ω_{11}		ω_{11}/ω_1	ω_{11}/ω_1	ω_{11}/ω_1	ω_{11}/ω_1	ω_{11}/ω_1	ω_{11}/ω_1	ω_{11}/ω_1
dx=0	0	23.82	23.22	22.01	21.40	20.80	20.60	20.40		1.21	1.18	1.12	1.09	1.06	1.05	1.04
	0.5	22.81	22.41	21.61	21.20	20.80	20.60	20.40		1.16	1.14	1.10	1.08	1.06	1.05	1.04
	1	22.01	21.81	21.40	21.00	20.60	20.60	20.40		1.12	1.11	1.09	1.07	1.05	1.05	1.04
	2	21.20	21.20	21.00	20.80	20.60	20.40	20.40		1.08	1.08	1.07	1.06	1.05	1.04	1.04
	3	21.00	20.80	20.80	20.60	20.40	20.40	20.20		1.07	1.06	1.06	1.05	1.04	1.04	1.03
	4	20.60	20.60	20.60	20.40	20.40	20.20	20.20		1.05	1.05	1.05	1.04	1.04	1.03	1.03
	5	20.60	20.60	20.40	20.40	20.20	20.20	20.20		1.05	1.05	1.04	1.04	1.03	1.03	1.03
dx=0.1	0	23.42	22.81	21.81	21.20	20.80	20.60	20.40		1.19	1.16	1.11	1.08	1.06	1.05	1.04
	0.5	22.41	22.01	21.40	21.00	20.60	20.40	20.40		1.14	1.12	1.09	1.07	1.05	1.04	1.04
	1	21.81	21.61	21.20	20.80	20.60	20.40	20.40		1.11	1.10	1.08	1.06	1.05	1.04	1.04
	2	21.20	21.00	20.80	20.60	20.40	20.40	20.20		1.08	1.07	1.06	1.05	1.04	1.04	1.03
	3	20.80	20.80	20.60	20.60	20.40	20.20	20.20		1.06	1.06	1.05	1.05	1.04	1.03	1.03
	4	20.60	20.60	20.40	20.40	20.40	20.20	20.20		1.05	1.05	1.04	1.04	1.04	1.03	1.03
	5	20.40	20.40	20.40	20.40	20.20	20.20	20.20		1.04	1.04	1.04	1.04	1.03	1.03	1.03
dx=0.5	0	22.81	22.41	21.40	21.00	20.60	20.40	20.40		1.16	1.14	1.09	1.07	1.05	1.04	1.04
	0.5	22.01	21.81	21.20	20.80	20.60	20.40	20.20		1.12	1.11	1.08	1.06	1.05	1.04	1.03
	1	21.61	21.40	21.00	20.80	20.40	20.40	20.20		1.10	1.09	1.07	1.06	1.04	1.04	1.03
	2	21.00	21.00	20.80	20.60	20.40	20.20	20.20		1.07	1.07	1.06	1.05	1.04	1.03	1.03
	3	20.80	20.60	20.60	20.40	20.40	20.20	20.20		1.06	1.05	1.05	1.04	1.04	1.03	1.03
	4	20.60	20.60	20.40	20.40	20.20	20.20	20.20		1.05	1.05	1.04	1.04	1.03	1.03	1.03
	5	20.40	20.40	20.40	20.20	20.20	20.20	20.00		1.04	1.04	1.04	1.03	1.03	1.03	1.02
dx=1	0	22.61	22.21	21.40	21.00	20.60	20.40	20.20		1.15	1.13	1.09	1.07	1.05	1.04	1.03
	0.5	22.01	21.61	21.20	20.80	20.40	20.40	20.20		1.12	1.10	1.08	1.06	1.04	1.04	1.03
	1	21.40	21.40	21.00	20.60	20.40	20.40	20.20		1.09	1.09	1.07	1.05	1.04	1.04	1.03
	2	21.00	20.80	20.60	20.60	20.40	20.20	20.20		1.07	1.06	1.05	1.05	1.04	1.03	1.03
	3	20.60	20.60	20.60	20.40	20.20	20.20	20.20		1.05	1.05	1.05	1.04	1.03	1.03	1.03
	4	20.60	20.40	20.40	20.40	20.20	20.20	20.20		1.05	1.04	1.04	1.04	1.03	1.03	1.03
	5	20.40	20.40	20.40	20.20	20.20	20.20	20.00		1.04	1.04	1.04	1.03	1.03	1.03	1.02

Homogeneous soil – bonded contact																
L/H=10	dw	0	1	5	10	20	30	40		0	1	5	10	20	30	40
	dθ	ω_{11}	ω_{11}	ω_{11}	ω_{11}	ω_{11}	ω_{11}	ω_{11}		ω_{11}/ω_1	ω_{11}/ω_1	ω_{11}/ω_1	ω_{11}/ω_1	ω_{11}/ω_1	ω_{11}/ω_1	ω_{11}/ω_1
dx=0	0	20.60	20.60	20.40	20.20	20.00	20.00	20.00		1.05	1.05	1.04	1.03	1.02	1.02	1.02
	0.5	20.40	20.40	20.20	20.20	20.00	20.00	20.00		1.04	1.04	1.03	1.03	1.02	1.02	1.02
	1	20.40	20.40	20.20	20.20	20.00	20.00	20.00		1.04	1.04	1.03	1.03	1.02	1.02	1.02
	2	20.20	20.20	20.20	20.00	20.00	20.00	20.00		1.03	1.03	1.03	1.02	1.02	1.02	1.02
	3	20.00	20.00	20.00	20.00	20.00	20.00	19.79		1.02	1.02	1.02	1.02	1.02	1.02	1.01
	4	20.00	20.00	20.00	20.00	20.00	19.79	19.79		1.02	1.02	1.02	1.02	1.02	1.01	1.01
	5	20.00	20.00	20.00	20.00	19.79	19.79	19.79		1.02	1.02	1.02	1.02	1.01	1.01	1.01
dx=0.1	0	20.60	20.40	20.20	20.20	20.00	20.00	20.00		1.05	1.04	1.03	1.03	1.02	1.02	1.02
	0.5	20.40	20.40	20.20	20.20	20.00	20.00	20.00		1.04	1.04	1.03	1.03	1.02	1.02	1.02
	1	20.40	20.20	20.20	20.00	20.00	20.00	20.00		1.04	1.03	1.03	1.02	1.02	1.02	1.02
	2	20.20	20.20	20.00	20.00	20.00	20.00	19.79		1.03	1.03	1.02	1.02	1.02	1.02	1.01
	3	20.00	20.00	20.00	20.00	20.00	19.79	19.79		1.02	1.02	1.02	1.02	1.02	1.01	1.01
	4	20.00	20.00	20.00	20.00	20.00	19.79	19.79		1.02	1.02	1.02	1.02	1.02	1.01	1.01
	5	20.00	20.00	20.00	20.00	19.79	19.79	19.79		1.02	1.02	1.02	1.02	1.01	1.01	1.01
dx=0.5	0	20.40	20.40	20.20	20.20	20.00	20.00	20.00		1.04	1.04	1.03	1.03	1.02	1.02	1.02
	0.5	20.40	20.40	20.20	20.00	20.00	20.00	19.79		1.04	1.04	1.03	1.02	1.02	1.02	1.01
	1	20.20	20.20	20.20	20.00	20.00	20.00	19.79		1.03	1.03	1.03	1.02	1.02	1.02	1.01
	2	20.20	20.20	20.00	20.00	20.00	19.79	19.79		1.03	1.03	1.02	1.02	1.02	1.01	1.01
	3	20.00	20.00	20.00	20.00	20.00	19.79	19.79		1.02	1.02	1.02	1.02	1.02	1.01	1.01
	4	20.00	20.00	20.00	20.00	19.79	19.79	19.79		1.02	1.02	1.02	1.02	1.01	1.01	1.01
	5	20.00	20.00	20.00	19.79	19.79	19.79	19.79		1.02	1.02	1.02	1.01	1.01	1.01	1.01
dx=1	0	20.40	20.40	20.20	20.20	20.00	20.00	20.00		1.04	1.04	1.03	1.03	1.02	1.02	1.02
	0.5	20.40	20.40	20.20	20.00	20.00	20.00	19.79		1.04	1.04	1.03	1.02	1.02	1.02	1.01
	1	20.20	20.20	20.20	20.00	20.00	20.00	19.79		1.03	1.03	1.03	1.02	1.02	1.02	1.01
	2	20.20	20.20	20.00	20.00	20.00	19.79	19.79		1.03	1.03	1.02	1.02	1.02	1.01	1.01
	3	20.00	20.00	20.00	20.00	20.00	19.79	19.79		1.02	1.02	1.02	1.02	1.02	1.01	1.01
	4	20.00	20.00	20.00	20.00	19.79	19.79	19.79		1.02	1.02	1.02	1.02	1.01	1.01	1.01
	5	20.00	20.00	20.00	19.79	19.79	19.79	19.79		1.02	1.02	1.02	1.01	1.01	1.01	1.01

Parabolic soil – bonded contact																
L/H=1	dw	0	1	5	10	20	30	40		0	1	5	10	20	30	40
	dθ	ω_{11}	ω_{11}	ω_{11}	ω_{11}	ω_{11}	ω_{11}	ω_{11}		ω_{11}/ω_1	ω_{11}/ω_1	ω_{11}/ω_1	ω_{11}/ω_1	ω_{11}/ω_1	ω_{11}/ω_1	ω_{11}/ω_1
dx=0	0	42.14	38.72	31.87	28.85	26.44	25.43	25.03		1.87	1.71	1.41	1.28	1.17	1.13	1.11
	0.5	36.30	34.49	30.26	28.05	26.24	25.43	24.83		1.61	1.53	1.34	1.24	1.16	1.13	1.10
	1	33.08	31.87	29.26	27.44	25.83	25.23	24.63		1.47	1.41	1.30	1.22	1.14	1.12	1.09
	2	29.66	29.26	27.65	26.64	25.43	24.83	24.42		1.31	1.30	1.22	1.18	1.13	1.10	1.08
	3	28.05	27.65	26.64	26.04	25.03	24.63	24.42		1.24	1.22	1.18	1.15	1.11	1.09	1.08
	4	26.84	26.64	26.04	25.43	24.83	24.42	24.22		1.19	1.18	1.15	1.13	1.10	1.08	1.07
	5	26.24	26.04	25.63	25.23	24.63	24.22	24.02		1.16	1.15	1.14	1.12	1.09	1.07	1.06
dx=0.1	0	46.77	42.34	33.49	29.86	27.04	26.04	25.43		2.07	1.88	1.48	1.32	1.20	1.15	1.13
	0.5	38.72	36.30	31.47	28.85	26.64	25.83	25.23		1.71	1.61	1.39	1.28	1.18	1.14	1.12
	1	34.49	33.08	29.86	28.05	26.24	25.43	25.03		1.53	1.47	1.32	1.24	1.16	1.13	1.11
	2	30.26	29.66	28.05	27.04	25.83	25.23	24.83		1.34	1.31	1.24	1.20	1.14	1.12	1.10
	3	28.25	28.05	27.04	26.24	25.43	24.83	24.63		1.25	1.24	1.20	1.16	1.13	1.10	1.09
	4	27.04	26.84	26.24	25.63	25.03	24.63	24.42		1.20	1.19	1.16	1.14	1.11	1.09	1.08
	5	26.44	26.24	25.63	25.23	24.83	24.42	24.22		1.17	1.16	1.14	1.12	1.10	1.08	1.07
dx=0.5	0	43.15	39.32	32.28	29.06	26.64	25.63	25.03		1.91	1.74	1.43	1.29	1.18	1.14	1.11
	0.5	36.71	34.89	30.47	28.25	26.24	25.43	24.83		1.63	1.55	1.35	1.25	1.16	1.13	1.10
	1	33.28	32.08	29.26	27.65	26.04	25.23	24.83		1.47	1.42	1.30	1.22	1.15	1.12	1.10
	2	29.86	29.26	27.85	26.64	25.63	25.03	24.63		1.32	1.30	1.23	1.18	1.14	1.11	1.09
	3	28.05	27.65	26.84	26.04	25.23	24.63	24.42		1.24	1.22	1.19	1.15	1.12	1.09	1.08
	4	27.04	26.64	26.04	25.63	24.83	24.42	24.22		1.20	1.18	1.15	1.14	1.10	1.08	1.07
	5	26.24	26.04	25.63	25.23	24.63	24.42	24.02		1.16	1.15	1.14	1.12	1.09	1.08	1.06
dx=1	0	42.14	38.72	31.87	28.85	26.44	25.43	25.03		1.87	1.71	1.41	1.28	1.17	1.13	1.11
	0.5	36.30	34.49	30.26	28.05	26.24	25.43	24.83		1.61	1.53	1.34	1.24	1.16	1.13	1.10
	1	33.08	31.87	29.26	27.44	25.83	25.23	24.63		1.47	1.41	1.30	1.22	1.14	1.12	1.09
	2	29.66	29.26	27.65	26.64	25.43	24.83	24.42		1.31	1.30	1.22	1.18	1.13	1.10	1.08
	3	28.05	27.65	26.64	26.04	25.03	24.63	24.42		1.24	1.22	1.18	1.15	1.11	1.09	1.08
	4	26.84	26.64	26.04	25.43	24.83	24.42	24.22		1.19	1.18	1.15	1.13	1.10	1.08	1.07
	5	26.24	26.04	25.63	25.23	24.63	24.22	24.02		1.16	1.15	1.14	1.12	1.09	1.07	1.06

Parabolic soil – bonded contact																
L/H=2	dw	0	1	5	10	20	30	40		0	1	5	10	20	30	40
	dθ	ω_{11}	ω_{11}	ω_{11}	ω_{11}	ω_{11}	ω_{11}	ω_{11}		ω_{11}/ω_1	ω_{11}/ω_1	ω_{11}/ω_1	ω_{11}/ω_1	ω_{11}/ω_1	ω_{11}/ω_1	ω_{11}/ω_1
dx=0	0	42.55	37.91	28.85	28.45	26.84	26.24	25.83		1.88	1.68	1.28	1.26	1.19	1.16	1.14
	0.5	34.29	32.48	29.26	27.85	26.44	26.04	25.63		1.52	1.44	1.30	1.23	1.17	1.15	1.14
	1	30.87	30.06	28.25	27.24	26.24	25.83	25.43		1.37	1.33	1.25	1.21	1.16	1.14	1.13
	2	28.25	27.85	27.04	26.44	25.83	25.63	25.23		1.25	1.23	1.20	1.17	1.14	1.14	1.12
	3	27.04	26.84	26.44	26.04	25.63	25.43	25.23		1.20	1.19	1.17	1.15	1.14	1.13	1.12
	4	26.24	26.24	25.83	25.63	25.43	25.23	25.03		1.16	1.16	1.14	1.14	1.13	1.12	1.11
	5	25.83	25.83	25.63	25.43	25.23	25.03	25.03		1.14	1.14	1.14	1.13	1.12	1.11	1.11
dx=0.1	0	38.52	35.10	29.86	27.85	26.44	25.83	25.43		1.71	1.55	1.32	1.23	1.17	1.14	1.13
	0.5	32.68	31.27	28.65	27.24	26.24	25.63	25.43		1.45	1.38	1.27	1.21	1.16	1.14	1.13
	1	30.06	29.46	27.85	26.84	26.04	25.63	25.23		1.33	1.30	1.23	1.19	1.15	1.14	1.12
	2	27.85	27.65	26.84	26.24	25.63	25.43	25.23		1.23	1.22	1.19	1.16	1.14	1.13	1.12
	3	26.84	26.64	26.24	25.83	25.43	25.23	25.03		1.19	1.18	1.16	1.14	1.13	1.12	1.11
	4	26.24	26.04	25.83	25.63	25.23	25.03	24.83		1.16	1.15	1.14	1.14	1.12	1.11	1.10
	5	25.83	25.63	25.43	25.23	25.03	25.03	24.83		1.14	1.14	1.13	1.12	1.11	1.11	1.10
dx=0.5	0	35.50	33.08	29.06	27.44	26.04	25.63	25.23		1.57	1.47	1.29	1.22	1.15	1.14	1.12
	0.5	31.47	30.26	28.05	26.84	25.83	25.43	25.23		1.39	1.34	1.24	1.19	1.14	1.13	1.12
	1	29.46	28.85	27.44	26.44	25.83	25.43	25.23		1.30	1.28	1.22	1.17	1.14	1.13	1.12
	2	27.44	27.24	26.44	26.04	25.43	25.23	25.03		1.22	1.21	1.17	1.15	1.13	1.12	1.11
	3	26.64	26.44	26.04	25.63	25.23	25.03	24.83		1.18	1.17	1.15	1.14	1.12	1.11	1.10
	4	26.04	25.83	25.63	25.43	25.03	25.03	24.83		1.15	1.14	1.14	1.13	1.11	1.11	1.10
	5	25.63	25.63	25.43	25.23	25.03	24.83	24.83		1.14	1.14	1.13	1.12	1.11	1.10	1.10
dx=1	0	34.89	32.68	28.85	27.24	26.04	25.63	25.23		1.55	1.45	1.28	1.21	1.15	1.14	1.12
	0.5	31.27	30.06	26.24	26.84	25.83	25.43	25.23		1.38	1.33	1.16	1.19	1.14	1.13	1.12
	1	29.26	28.65	27.24	26.44	25.63	25.23	25.03		1.30	1.27	1.21	1.17	1.14	1.12	1.11
	2	27.44	27.24	26.44	26.04	25.43	25.23	25.03		1.22	1.21	1.17	1.15	1.13	1.12	1.11
	3	26.44	26.44	26.04	25.63	25.23	25.03	24.83		1.17	1.17	1.15	1.14	1.12	1.11	1.10
	4	26.04	25.83	25.63	25.43	25.03	24.83	24.83		1.15	1.14	1.14	1.13	1.11	1.10	1.10
	5	25.63	25.63	25.43	25.23	25.03	24.83	24.63		1.14	1.14	1.13	1.12	1.11	1.10	1.09

Parabolic soil – bonded contact																
L/H=3	dw	0	1	5	10	20	30	40		0	1	5	10	20	30	40
	dθ	ω_{11}	ω_{11}	ω_{11}	ω_{11}	ω_{11}	ω_{11}	ω_{11}		ω_{11}/ω_1	ω_{11}/ω_1	ω_{11}/ω_1	ω_{11}/ω_1	ω_{11}/ω_1	ω_{11}/ω_1	ω_{11}/ω_1
dx=0	0	29.06	28.25	26.24	25.23	24.22	23.82	23.62		1.29	1.25	1.16	1.12	1.07	1.05	1.05
	0.5	27.24	26.84	25.63	24.83	24.02	23.82	23.62		1.21	1.19	1.13	1.10	1.06	1.05	1.05
	1	26.24	25.83	25.03	24.42	24.02	23.62	23.42		1.16	1.14	1.11	1.08	1.06	1.05	1.04
	2	25.03	24.83	24.42	24.02	23.62	23.42	23.42		1.11	1.10	1.08	1.06	1.05	1.04	1.04
	3	24.42	24.42	24.02	23.82	23.62	23.42	23.22		1.08	1.08	1.06	1.05	1.05	1.04	1.03
	4	24.02	24.02	23.82	23.62	23.42	23.22	23.22		1.06	1.06	1.05	1.05	1.04	1.03	1.03
	5	23.82	23.82	23.62	23.42	23.42	23.22	23.22		1.05	1.05	1.05	1.04	1.04	1.03	1.03
dx=0.1	0	28.25	27.44	25.83	24.83	24.02	23.62	23.42		1.25	1.21	1.14	1.10	1.06	1.05	1.04
	0.5	26.84	26.44	25.23	24.63	23.82	23.62	23.42		1.19	1.17	1.12	1.09	1.05	1.05	1.04
	1	25.83	25.63	24.83	24.22	23.82	23.62	23.42		1.14	1.13	1.10	1.07	1.05	1.05	1.04
	2	24.83	24.83	24.22	24.02	23.62	23.42	23.22		1.10	1.10	1.07	1.06	1.05	1.04	1.03
	3	24.42	24.22	24.02	23.82	23.42	23.42	23.22		1.08	1.07	1.06	1.05	1.04	1.04	1.03
	4	24.02	23.82	23.82	23.62	23.42	23.22	23.22		1.06	1.05	1.05	1.05	1.04	1.03	1.03
	5	23.82	23.62	23.62	23.42	23.22	23.22	23.02		1.05	1.05	1.05	1.04	1.03	1.03	1.02
dx=0.5	0	30.47	29.26	27.04	26.04	25.23	25.03	24.83		1.35	1.30	1.20	1.15	1.12	1.11	1.10
	0.5	28.45	27.85	26.44	25.83	25.23	24.83	24.63		1.26	1.23	1.17	1.14	1.12	1.10	1.09
	1	27.24	27.04	26.04	25.63	25.03	24.83	24.63		1.21	1.20	1.15	1.13	1.11	1.10	1.09
	2	26.24	26.04	25.63	25.23	24.83	24.63	24.63		1.16	1.15	1.13	1.12	1.10	1.09	1.09
	3	25.63	25.43	25.23	25.03	24.83	24.63	24.42		1.13	1.13	1.12	1.11	1.10	1.09	1.08
	4	25.23	25.23	25.03	24.83	24.63	24.63	24.42		1.12	1.12	1.11	1.10	1.09	1.09	1.08
	5	25.03	25.03	24.83	24.63	24.63	24.42	24.42		1.11	1.11	1.10	1.09	1.09	1.08	1.08
dx=1	0	30.06	29.06	27.04	26.04	25.23	24.83	24.83		1.33	1.29	1.20	1.15	1.12	1.10	1.10
	0.5	28.25	27.65	26.44	25.63	25.03	24.83	24.63		1.25	1.22	1.17	1.13	1.11	1.10	1.09
	1	27.24	26.84	26.04	25.43	25.03	24.83	24.63		1.21	1.19	1.15	1.13	1.11	1.10	1.09
	2	26.04	26.04	25.43	25.23	24.83	24.63	24.63		1.15	1.15	1.13	1.12	1.10	1.09	1.09
	3	25.63	25.43	25.23	25.03	24.83	24.63	24.42		1.13	1.13	1.12	1.11	1.10	1.09	1.08
	4	25.23	25.23	25.03	24.83	24.63	24.42	24.42		1.12	1.12	1.11	1.10	1.09	1.08	1.08
	5	25.03	25.03	24.83	24.63	24.63	24.42	24.42		1.11	1.11	1.10	1.09	1.09	1.08	1.08

Parabolic soil – bonded contact																
L/H=5	dw	0	1	5	10	20	30	40		0	1	5	10	20	30	40
	dθ	ω_{11}	ω_{11}	ω_{11}	ω_{11}	ω_{11}	ω_{11}	ω_{11}		ω_{11}/ω_1	ω_{11}/ω_1	ω_{11}/ω_1	ω_{11}/ω_1	ω_{11}/ω_1	ω_{11}/ω_1	ω_{11}/ω_1
dx=0	0	25.03	24.83	24.22	23.82	23.42	23.22	23.22		1.11	1.10	1.07	1.05	1.04	1.03	1.03
	0.5	24.63	24.42	24.02	23.62	23.42	23.22	23.02		1.09	1.08	1.06	1.05	1.04	1.03	1.02
	1	24.22	24.22	23.82	23.62	23.22	23.22	23.02		1.07	1.07	1.05	1.05	1.03	1.03	1.02
	2	23.82	23.82	23.62	23.42	23.22	23.02	23.02		1.05	1.05	1.05	1.04	1.03	1.02	1.02
	3	23.42	23.42	23.42	23.22	23.02	23.02	23.02		1.04	1.04	1.04	1.03	1.02	1.02	1.02
	4	23.42	23.22	23.22	23.22	23.02	23.02	22.81		1.04	1.03	1.03	1.03	1.02	1.02	1.01
	5	23.22	23.22	23.02	23.02	23.02	23.02	22.81		1.03	1.03	1.02	1.02	1.02	1.02	1.01
dx=0.1	0	24.83	24.63	24.02	23.82	23.42	23.22	23.02		1.10	1.09	1.06	1.05	1.04	1.03	1.02
	0.5	24.42	24.22	23.82	23.62	23.22	23.22	23.02		1.08	1.07	1.05	1.05	1.03	1.03	1.02
	1	24.02	24.02	23.62	23.42	23.22	23.02	23.02		1.06	1.06	1.05	1.04	1.03	1.02	1.02
	2	23.62	23.62	23.42	23.22	23.22	23.02	23.02		1.05	1.05	1.04	1.03	1.03	1.02	1.02
	3	23.42	23.42	23.22	23.22	23.02	23.02	23.02		1.04	1.04	1.03	1.03	1.02	1.02	1.02
	4	23.22	23.22	23.22	23.02	23.02	23.02	22.81		1.03	1.03	1.03	1.02	1.02	1.02	1.01
	5	23.22	23.22	23.02	23.02	23.02	22.81	22.81		1.03	1.03	1.02	1.02	1.02	1.01	1.01
dx=0.5	0	24.63	24.42	24.02	23.62	23.22	23.02	23.02		1.09	1.08	1.06	1.05	1.03	1.02	1.02
	0.5	24.22	24.22	23.82	23.42	23.22	23.02	23.02		1.07	1.07	1.05	1.04	1.03	1.02	1.02
	1	24.02	23.82	23.62	23.42	23.22	23.02	23.02		1.06	1.05	1.05	1.04	1.03	1.02	1.02
	2	23.62	23.62	23.42	23.22	23.02	23.02	22.81		1.05	1.05	1.04	1.03	1.02	1.02	1.01
	3	23.42	23.42	23.22	23.22	23.02	23.02	22.81		1.04	1.04	1.03	1.03	1.02	1.02	1.01
	4	23.22	23.22	23.22	23.02	23.02	22.81	22.81		1.03	1.03	1.03	1.02	1.02	1.01	1.01
	5	23.22	23.22	23.02	23.02	22.81	22.81	22.81		1.03	1.03	1.02	1.02	1.01	1.01	1.01
dx=1	0	24.42	24.42	23.82	23.62	23.22	23.02	23.02		1.08	1.08	1.05	1.05	1.03	1.02	1.02
	0.5	24.22	24.02	23.82	23.42	23.22	23.02	23.02		1.07	1.06	1.05	1.04	1.03	1.02	1.02
	1	24.02	23.82	23.62	23.42	23.22	23.02	23.02		1.06	1.05	1.05	1.04	1.03	1.02	1.02
	2	23.62	23.62	23.42	23.22	23.02	23.02	22.81		1.05	1.05	1.04	1.03	1.02	1.02	1.01
	3	23.42	23.42	23.22	23.22	23.02	23.02	22.81		1.04	1.04	1.03	1.03	1.02	1.02	1.01
	4	23.22	23.22	23.22	23.02	23.02	22.81	22.81		1.03	1.03	1.03	1.02	1.02	1.01	1.01
	5	23.22	23.22	23.02	23.02	22.81	22.81	22.81		1.03	1.03	1.02	1.02	1.01	1.01	1.01

Parabolic soil – bonded contact																
L/H=10	dw	0	1	5	10	20	30	40		0	1	5	10	20	30	40
	dθ	ω_{11}	ω_{11}	ω_{11}	ω_{11}	ω_{11}	ω_{11}	ω_{11}		ω_{11}/ω_1	ω_{11}/ω_1	ω_{11}/ω_1	ω_{11}/ω_1	ω_{11}/ω_1	ω_{11}/ω_1	ω_{11}/ω_1
dx=0	0	24.02	23.82	23.82	23.82	23.82	23.82	23.62		1.06	1.05	1.05	1.05	1.05	1.05	1.05
	0.5	23.82	23.82	23.82	23.82	23.82	23.82	23.62		1.05	1.05	1.05	1.05	1.05	1.05	1.05
	1	23.82	23.82	23.82	23.82	23.82	23.62	23.62		1.05	1.05	1.05	1.05	1.05	1.05	1.05
	2	23.82	23.82	23.82	23.82	23.62	23.62	23.62		1.05	1.05	1.05	1.05	1.05	1.05	1.05
	3	23.82	23.82	23.82	23.82	23.62	23.62	23.62		1.05	1.05	1.05	1.05	1.05	1.05	1.05
	4	23.82	23.82	23.82	23.62	23.62	23.62	23.62		1.05	1.05	1.05	1.05	1.05	1.05	1.05
	5	23.62	23.62	23.62	23.62	23.62	23.62	23.62		1.05	1.05	1.05	1.05	1.05	1.05	1.05
dx=0.1	0	23.82	23.82	23.82	23.82	23.82	23.82	23.62		1.05	1.05	1.05	1.05	1.05	1.05	1.05
	0.5	23.82	23.82	23.82	23.82	23.82	23.62	23.62		1.05	1.05	1.05	1.05	1.05	1.05	1.05
	1	23.82	23.82	23.82	23.82	23.82	23.62	23.62		1.05	1.05	1.05	1.05	1.05	1.05	1.05
	2	23.82	23.82	23.82	23.82	23.62	23.62	23.62		1.05	1.05	1.05	1.05	1.05	1.05	1.05
	3	23.82	23.82	23.82	23.62	23.62	23.62	23.62		1.05	1.05	1.05	1.05	1.05	1.05	1.05
	4	23.82	23.82	23.62	23.62	23.62	23.62	23.62		1.05	1.05	1.05	1.05	1.05	1.05	1.05
	5	23.62	23.62	23.62	23.62	23.62	23.62	23.62		1.05	1.05	1.05	1.05	1.05	1.05	1.05
dx=0.5	0	23.82	23.82	23.82	23.82	23.82	23.62	23.62		1.05	1.05	1.05	1.05	1.05	1.05	1.05
	0.5	23.82	23.82	23.82	23.82	23.82	23.62	23.62		1.05	1.05	1.05	1.05	1.05	1.05	1.05
	1	23.82	23.82	23.82	23.82	23.62	23.62	23.62		1.05	1.05	1.05	1.05	1.05	1.05	1.05
	2	23.82	23.82	23.82	23.82	23.62	23.62	23.62		1.05	1.05	1.05	1.05	1.05	1.05	1.05
	3	23.82	23.82	23.82	23.62	23.62	23.62	23.62		1.05	1.05	1.05	1.05	1.05	1.05	1.05
	4	23.82	23.62	23.62	23.62	23.62	23.62	23.62		1.05	1.05	1.05	1.05	1.05	1.05	1.05
	5	23.62	23.62	23.62	23.62	23.62	23.62	23.62		1.05	1.05	1.05	1.05	1.05	1.05	1.05
dx=1	0	23.82	23.82	23.82	23.82	23.82	23.62	23.62		1.05	1.05	1.05	1.05	1.05	1.05	1.05
	0.5	23.82	23.82	23.82	23.82	23.82	23.62	23.62		1.05	1.05	1.05	1.05	1.05	1.05	1.05
	1	23.82	23.82	23.82	23.82	23.62	23.62	23.62		1.05	1.05	1.05	1.05	1.05	1.05	1.05
	2	23.82	23.82	23.82	23.82	23.62	23.62	23.62		1.05	1.05	1.05	1.05	1.05	1.05	1.05
	3	23.82	23.82	23.82	23.62	23.62	23.62	23.62		1.05	1.05	1.05	1.05	1.05	1.05	1.05
	4	23.82	23.62	23.62	23.62	23.62	23.62	23.62		1.05	1.05	1.05	1.05	1.05	1.05	1.05
	5	23.62	23.62	23.62	23.62	22.61	22.61	22.61		1.05	1.05	1.05	1.05	1.00	1.00	1.00

Linear soil – bonded contact																
L/H=1	dw	0	1	5	10	20	30	40		0	1	5	10	20	30	40
	dθ	ω_{11}	ω_{11}	ω_{11}	ω_{11}	ω_{11}	ω_{11}	ω_{11}		ω_{11}/ω_1	ω_{11}/ω_1	ω_{11}/ω_1	ω_{11}/ω_1	ω_{11}/ω_1	ω_{11}/ω_1	ω_{11}/ω_1
dx=0	0	43.75	40.93	33.49	29.66	26.84	25.63	25.03		1.96	1.83	1.50	1.33	1.20	1.15	1.12
	0.5	38.12	36.10	31.27	28.65	26.44	25.43	24.83		1.70	1.61	1.40	1.28	1.18	1.14	1.11
	1	34.29	32.88	29.86	27.85	26.04	25.23	24.63		1.53	1.47	1.33	1.24	1.16	1.13	1.10
	2	30.26	29.66	28.05	26.84	25.63	24.83	24.42		1.35	1.33	1.25	1.20	1.15	1.11	1.09
	3	28.25	28.05	26.84	26.04	25.23	24.63	24.22		1.26	1.25	1.20	1.16	1.13	1.10	1.08
	4	27.24	26.84	26.24	25.63	24.83	24.42	24.22		1.22	1.20	1.17	1.15	1.11	1.09	1.08
	5	26.44	26.24	25.63	25.23	24.63	24.22	24.02		1.18	1.17	1.15	1.13	1.10	1.08	1.07
dx=0.1	0	42.14	39.32	32.48	29.06	26.44	25.43	24.83		1.88	1.76	1.45	1.30	1.18	1.14	1.88
	0.5	36.71	34.89	30.67	28.25	26.24	25.23	24.63		1.64	1.56	1.37	1.26	1.17	1.13	1.64
	1	33.49	32.28	29.46	27.65	25.83	25.03	24.63		1.50	1.44	1.32	1.24	1.15	1.12	1.50
	2	30.06	29.46	27.85	26.64	25.43	24.83	24.42		1.34	1.32	1.24	1.19	1.14	1.11	1.34
	3	28.25	27.85	26.84	26.04	25.03	24.63	24.22		1.26	1.24	1.20	1.16	1.12	1.10	1.26
	4	27.04	26.84	26.04	25.43	24.83	24.42	24.02		1.21	1.20	1.16	1.14	1.11	1.09	1.21
	5	26.24	26.24	25.63	25.23	24.63	24.22	24.02		1.17	1.17	1.15	1.13	1.10	1.08	1.17
dx=0.5	0	40.73	37.71	31.67	28.65	26.24	25.23	24.63		1.82	1.69	1.42	1.28	1.17	1.13	1.10
	0.5	35.90	34.09	30.26	27.85	26.04	25.03	24.63		1.60	1.52	1.35	1.24	1.16	1.12	1.10
	1	32.88	31.87	29.06	27.24	25.63	24.83	24.42		1.47	1.42	1.30	1.22	1.15	1.11	1.09
	2	29.66	29.26	27.65	26.44	25.23	24.63	24.22		1.33	1.31	1.24	1.18	1.13	1.10	1.08
	3	28.05	27.65	26.64	25.83	25.03	24.42	24.02		1.25	1.24	1.19	1.15	1.12	1.09	1.07
	4	27.04	26.84	26.04	25.43	24.63	24.22	24.02		1.21	1.20	1.16	1.14	1.10	1.08	1.07
	5	26.24	26.04	25.63	25.03	24.42	24.22	23.82		1.17	1.16	1.15	1.12	1.09	1.08	1.06
dx=1	0	40.33	37.51	31.47	28.65	26.24	25.23	24.63		1.80	1.68	1.41	1.28	1.17	1.13	1.10
	0.5	35.70	33.89	30.06	27.85	25.83	25.03	24.42		1.60	1.51	1.34	1.24	1.15	1.12	1.09
	1	32.88	31.67	29.06	27.24	25.63	24.83	24.42		1.47	1.42	1.30	1.22	1.15	1.11	1.09
	2	29.66	29.06	27.65	26.44	25.23	24.63	24.22		1.33	1.30	1.24	1.18	1.13	1.10	1.08
	3	28.05	27.65	26.64	25.83	24.83	24.42	24.02		1.25	1.24	1.19	1.15	1.11	1.09	1.07
	4	27.04	26.84	26.04	25.43	24.63	24.22	24.02		1.21	1.20	1.16	1.14	1.10	1.08	1.07
	5	26.24	26.04	25.63	25.03	24.42	24.02	23.82		1.17	1.16	1.15	1.12	1.09	1.07	1.06

Linear soil – bonded contact																
L/H=2	dw	0	1	5	10	20	30	40		0	1	5	10	20	30	40
	dθ	ω_{11}	ω_{11}	ω_{11}	ω_{11}	ω_{11}	ω_{11}	ω_{11}		ω_{11}/ω_1	ω_{11}/ω_1	ω_{11}/ω_1	ω_{11}/ω_1	ω_{11}/ω_1	ω_{11}/ω_1	ω_{11}/ω_1
dx=0	0	31.87	30.67	27.65	26.04	24.63	24.02	23.82		1.42	1.37	1.24	1.16	1.10	1.07	1.06
	0.5	29.46	28.65	26.64	25.63	24.42	24.02	23.62		1.32	1.28	1.19	1.15	1.09	1.07	1.06
	1	27.85	27.24	26.04	25.23	24.22	23.82	23.62		1.24	1.22	1.16	1.13	1.08	1.06	1.06
	2	26.04	25.83	25.23	24.63	24.02	23.62	23.42		1.16	1.15	1.13	1.10	1.07	1.06	1.05
	3	25.23	25.03	24.63	24.22	23.82	23.62	23.42		1.13	1.12	1.10	1.08	1.06	1.06	1.05
	4	24.63	24.63	24.22	24.02	23.62	23.42	23.22		1.10	1.10	1.08	1.07	1.06	1.05	1.04
	5	24.22	24.22	24.02	23.82	23.62	23.42	23.22		1.08	1.08	1.07	1.06	1.06	1.05	1.04
dx=0.1	0	30.87	29.66	27.24	25.63	24.42	24.02	23.62		1.38	1.33	1.22	1.15	1.09	1.07	1.06
	0.5	28.65	28.05	26.44	25.23	24.22	23.82	23.62		1.28	1.25	1.18	1.13	1.08	1.06	1.06
	1	27.44	27.04	25.83	25.03	24.22	23.82	23.62		1.23	1.21	1.15	1.12	1.08	1.06	1.06
	2	26.04	25.63	25.03	24.42	24.02	23.62	23.42		1.16	1.15	1.12	1.09	1.07	1.06	1.05
	3	25.23	25.03	24.63	24.22	23.82	23.42	23.42		1.13	1.12	1.10	1.08	1.06	1.05	1.05
	4	24.63	24.63	24.22	24.02	23.62	23.42	23.22		1.10	1.10	1.08	1.07	1.06	1.05	1.04
	5	24.22	24.22	24.02	23.82	23.42	23.42	23.22		1.08	1.08	1.07	1.06	1.05	1.05	1.04
dx=0.5	0	29.86	29.06	26.84	25.43	24.42	23.82	23.62		1.33	1.30	1.20	1.14	1.09	1.06	1.06
	0.5	28.25	27.65	26.04	25.03	24.22	23.82	23.62		1.26	1.24	1.16	1.12	1.08	1.06	1.06
	1	27.24	26.64	25.63	24.83	24.02	23.62	23.42		1.22	1.19	1.15	1.11	1.07	1.06	1.05
	2	25.83	25.63	25.03	24.42	23.82	23.62	23.42		1.15	1.15	1.12	1.09	1.06	1.06	1.05
	3	25.03	25.03	24.42	24.22	23.62	23.42	23.22		1.12	1.12	1.09	1.08	1.06	1.05	1.04
	4	24.63	24.42	24.22	23.82	23.62	23.42	23.22		1.10	1.09	1.08	1.06	1.06	1.05	1.04
	5	24.22	24.22	24.02	23.82	23.42	23.22	23.22		1.08	1.08	1.07	1.06	1.05	1.04	1.04
dx=1	0	29.86	28.85	26.64	25.43	24.42	23.82	23.62		1.33	1.29	1.19	1.14	1.09	1.06	1.06
	0.5	28.25	27.65	26.04	25.03	24.22	23.82	23.42		1.26	1.24	1.16	1.12	1.08	1.06	1.05
	1	27.04	26.64	25.63	24.83	24.02	23.62	23.42		1.21	1.19	1.15	1.11	1.07	1.06	1.05
	2	25.83	25.63	24.83	24.42	23.82	23.62	23.42		1.15	1.15	1.11	1.09	1.06	1.06	1.05
	3	25.03	24.83	24.42	24.02	23.62	23.42	23.22		1.12	1.11	1.09	1.07	1.06	1.05	1.04
	4	24.63	24.42	24.22	23.82	23.62	23.42	23.22		1.10	1.09	1.08	1.06	1.06	1.05	1.04
	5	24.22	24.22	24.02	23.62	23.42	23.22	23.22		1.08	1.08	1.07	1.06	1.05	1.04	1.04

Linear soil – bonded contact																
L/H=3	dw	0	1	5	10	20	30	40		0	1	5	10	20	30	40
	dθ	ω_{11}	ω_{11}	ω_{11}	ω_{11}	ω_{11}	ω_{11}	ω_{11}		ω_{11}/ω_1	ω_{11}/ω_1	ω_{11}/ω_1	ω_{11}/ω_1	ω_{11}/ω_1	ω_{11}/ω_1	ω_{11}/ω_1
dx=0	0	27.44	27.04	25.63	24.63	23.82	23.62	23.42		1.18	1.16	1.12	1.09	1.06	1.05	1.04
	0.5	26.44	26.04	25.03	24.42	23.82	23.42	23.22		1.15	1.14	1.10	1.08	1.06	1.05	1.04
	1	25.63	25.43	24.63	24.22	23.62	23.42	23.22		1.10	1.10	1.08	1.06	1.05	1.04	1.04
	2	24.63	24.63	24.22	23.82	23.42	23.22	23.22		1.08	1.07	1.06	1.06	1.05	1.04	1.03
	3	24.22	24.02	23.82	23.62	23.42	23.22	23.02		1.06	1.06	1.06	1.05	1.04	1.03	1.03
	4	23.82	23.82	23.62	23.42	23.22	23.02	23.02		1.06	1.06	1.05	1.05	1.04	1.03	1.03
	5	23.62	23.62	23.42	23.42	23.22	23.02	23.02		1.18	1.16	1.12	1.08	1.06	1.05	1.04
dx=0.1	0	26.44	26.04	25.03	24.22	23.62	23.42	23.22		1.15	1.14	1.10	1.07	1.06	1.04	1.04
	0.5	25.63	25.43	24.63	24.02	23.62	23.22	23.22		1.13	1.12	1.09	1.07	1.05	1.04	1.03
	1	25.23	25.03	24.42	24.02	23.42	23.22	23.02		1.09	1.09	1.07	1.06	1.05	1.04	1.03
	2	24.42	24.42	24.02	23.62	23.42	23.22	23.02		1.07	1.07	1.06	1.05	1.04	1.03	1.03
	3	24.02	24.02	23.82	23.42	23.22	23.02	23.02		1.06	1.06	1.06	1.05	1.04	1.03	1.03
	4	23.82	23.82	23.62	23.42	23.22	23.02	23.02		1.06	1.06	1.05	1.04	1.03	1.03	1.03
	5	23.62	23.62	23.42	23.22	23.02	23.02	23.02		1.19	1.16	1.12	1.09	1.06	1.05	1.04
dx=0.5	0	26.64	26.04	25.03	24.42	23.62	23.42	23.22		1.15	1.14	1.10	1.08	1.06	1.05	1.04
	0.5	25.83	25.43	24.63	24.22	23.62	23.42	23.22		1.13	1.12	1.09	1.07	1.05	1.04	1.04
	1	25.23	25.03	24.42	24.02	23.42	23.22	23.22		1.09	1.09	1.07	1.06	1.05	1.04	1.03
	2	24.42	24.42	24.02	23.62	23.42	23.22	23.02		1.07	1.07	1.06	1.06	1.04	1.03	1.03
	3	24.02	24.02	23.82	23.62	23.22	23.02	23.02		1.06	1.06	1.06	1.05	1.04	1.03	1.03
	4	23.82	23.82	23.62	23.42	23.22	23.02	23.02		1.06	1.06	1.05	1.04	1.03	1.03	1.03
	5	23.62	23.62	23.42	23.22	23.02	23.02	23.02		1.18	1.16	1.12	1.08	1.06	1.05	1.04
dx=1	0	26.44	26.04	25.03	24.22	23.62	23.42	23.22		1.15	1.14	1.10	1.07	1.06	1.04	1.04
	0.5	25.63	25.43	24.63	24.02	23.62	23.22	23.22		1.13	1.12	1.09	1.07	1.05	1.04	1.03
	1	25.23	25.03	24.42	24.02	23.42	23.22	23.02		1.09	1.09	1.07	1.06	1.05	1.04	1.03
	2	24.42	24.42	24.02	23.62	23.42	23.22	23.02		1.07	1.07	1.06	1.05	1.04	1.03	1.03
	3	24.02	24.02	23.82	23.42	23.22	23.02	23.02		1.06	1.06	1.06	1.05	1.04	1.03	1.03
	4	23.82	23.82	23.62	23.42	23.22	23.02	23.02		1.06	1.06	1.05	1.04	1.03	1.03	1.03
	5	23.62	23.62	23.42	23.22	23.02	23.02	23.02		1.18	1.16	1.12	1.09	1.06	1.05	1.04

Linear soil – bonded contact																
L/H=5	dw	0	1	5	10	20	30	40		0	1	5	10	20	30	40
	dθ	ω_{11}	ω_{11}	ω_{11}	ω_{11}	ω_{11}	ω_{11}	ω_{11}		ω_{11}/ω_1	ω_{11}/ω_1	ω_{11}/ω_1	ω_{11}/ω_1	ω_{11}/ω_1	ω_{11}/ω_1	ω_{11}/ω_1
dx=0	0	24.42	24.22	23.82	23.62	23.22	23.02	23.02		1.09	1.08	1.06	1.06	1.04	1.03	1.03
	0.5	24.02	24.02	23.62	23.42	23.22	23.02	23.02		1.07	1.07	1.06	1.05	1.04	1.03	1.03
	1	23.82	23.82	23.62	23.42	23.02	23.02	22.81		1.06	1.06	1.06	1.05	1.03	1.03	1.02
	2	23.62	23.42	23.42	23.22	23.02	23.02	22.81		1.06	1.05	1.05	1.04	1.03	1.03	1.02
	3	23.42	23.22	23.22	23.02	23.02	22.81	22.81		1.05	1.04	1.04	1.03	1.03	1.02	1.02
	4	23.22	23.22	23.02	23.02	22.81	22.81	22.81		1.04	1.04	1.03	1.03	1.02	1.02	1.02
	5	23.02	23.02	23.02	23.02	22.81	22.81	22.81		1.03	1.03	1.03	1.03	1.02	1.02	1.02
dx=0.1	0	24.22	24.02	23.82	23.42	23.22	23.02	22.81		1.08	1.07	1.06	1.05	1.04	1.03	1.02
	0.5	24.02	23.82	23.62	23.42	23.02	23.02	22.81		1.07	1.06	1.06	1.05	1.03	1.03	1.02
	1	23.82	23.62	23.42	23.22	23.02	23.02	22.81		1.06	1.06	1.05	1.04	1.03	1.03	1.02
	2	23.42	23.42	23.22	23.22	23.02	22.81	22.81		1.05	1.05	1.04	1.04	1.03	1.02	1.02
	3	23.22	23.22	23.22	23.02	23.02	22.81	22.81		1.04	1.04	1.04	1.03	1.03	1.02	1.02
	4	23.22	23.22	23.02	23.02	22.81	22.81	22.81		1.04	1.04	1.03	1.03	1.02	1.02	1.02
	5	23.02	23.02	23.02	22.81	22.81	22.81	22.81		1.03	1.03	1.03	1.02	1.02	1.02	1.02
dx=0.5	0	24.02	24.02	23.62	23.42	23.22	23.02	22.81		1.07	1.07	1.06	1.05	1.04	1.03	1.02
	0.5	23.82	23.82	23.62	23.42	23.02	23.02	22.81		1.06	1.06	1.06	1.05	1.03	1.03	1.02
	1	23.82	23.62	23.42	23.22	23.02	22.81	22.81		1.06	1.06	1.05	1.04	1.03	1.02	1.02
	2	23.42	23.42	23.22	23.22	23.02	22.81	22.81		1.05	1.05	1.04	1.04	1.03	1.02	1.02
	3	23.22	23.22	23.22	23.02	22.81	22.81	22.81		1.04	1.04	1.04	1.03	1.02	1.02	1.02
	4	23.22	23.22	23.02	23.02	22.81	22.81	22.81		1.04	1.04	1.03	1.03	1.02	1.02	1.02
	5	23.02	23.02	23.02	22.81	22.81	22.81	22.81		1.03	1.03	1.03	1.02	1.02	1.02	1.02
dx=1	0	24.02	24.02	23.62	23.42	23.02	23.02	22.81		1.07	1.07	1.06	1.05	1.03	1.03	1.02
	0.5	23.82	23.82	23.62	23.22	23.02	23.02	22.81		1.06	1.06	1.06	1.04	1.03	1.03	1.02
	1	23.62	23.62	23.42	23.22	23.02	22.81	22.81		1.06	1.06	1.05	1.04	1.03	1.02	1.02
	2	23.42	23.42	23.22	23.22	23.02	22.81	22.81		1.05	1.05	1.04	1.04	1.03	1.02	1.02
	3	23.22	23.22	23.22	23.02	22.81	22.81	22.81		1.04	1.04	1.04	1.03	1.02	1.02	1.02
	4	23.22	23.22	23.02	23.02	22.81	22.81	22.81		1.04	1.04	1.03	1.03	1.02	1.02	1.02
	5	23.02	23.02	23.02	22.81	22.81	22.81	22.81		1.03	1.03	1.03	1.02	1.02	1.02	1.02

Linear soil – bonded contact																
L/H=10	dw	0	1	5	10	20	30	40		0	1	5	10	20	30	40
	dθ	ω_{11}	ω_{11}	ω_{11}	ω_{11}	ω_{11}	ω_{11}	ω_{11}		ω_{11}/ω_1	ω_{11}/ω_1	ω_{11}/ω_1	ω_{11}/ω_1	ω_{11}/ω_1	ω_{11}/ω_1	ω_{11}/ω_1
dx=0	0	22.81	22.81	22.81	22.81	22.61	22.61	22.61		1.02	1.02	1.02	1.02	1.01	1.01	1.01
	0.5	22.81	22.81	22.81	22.81	22.61	22.61	22.61		1.02	1.02	1.02	1.02	1.01	1.01	1.01
	1	22.81	22.81	22.81	22.61	22.61	22.61	22.61		1.02	1.02	1.02	1.01	1.01	1.01	1.01
	2	22.81	22.81	22.61	22.61	22.61	22.61	22.61		1.02	1.02	1.01	1.01	1.01	1.01	1.01
	3	22.61	22.61	22.61	22.61	22.61	22.61	22.61		1.01	1.01	1.01	1.01	1.01	1.01	1.01
	4	22.61	22.61	22.61	22.61	22.61	22.61	22.61		1.01	1.01	1.01	1.01	1.01	1.01	1.01
	5	22.61	22.61	22.61	22.61	22.61	22.61	22.61		1.01	1.01	1.01	1.01	1.01	1.01	1.01
dx=0.1	0	22.61	22.81	22.81	22.81	22.61	22.61	22.61		1.01	1.02	1.02	1.02	1.01	1.01	1.01
	0.5	22.81	22.81	22.81	22.61	22.61	22.61	22.61		1.02	1.02	1.02	1.01	1.01	1.01	1.01
	1	22.81	22.81	22.81	22.61	22.61	22.61	22.61		1.02	1.02	1.02	1.01	1.01	1.01	1.01
	2	22.81	22.61	22.61	22.61	22.61	22.61	22.61		1.02	1.01	1.01	1.01	1.01	1.01	1.01
	3	22.61	22.61	22.61	22.61	22.61	22.61	22.61		1.01	1.01	1.01	1.01	1.01	1.01	1.01
	4	22.61	22.61	22.61	22.61	22.61	22.61	22.61		1.01	1.01	1.01	1.01	1.01	1.01	1.01
	5	22.61	22.61	22.61	22.61	22.61	22.61	22.61		1.01	1.01	1.01	1.01	1.01	1.01	1.01
dx=0.5	0	22.81	22.81	22.81	22.61	22.61	22.61	22.61		1.02	1.02	1.02	1.01	1.01	1.01	1.01
	0.5	22.81	22.81	22.81	22.61	22.61	22.61	22.61		1.02	1.02	1.02	1.01	1.01	1.01	1.01
	1	22.81	22.81	22.81	22.61	22.61	22.61	22.61		1.02	1.02	1.02	1.01	1.01	1.01	1.01
	2	22.61	22.61	22.61	22.61	22.61	22.61	22.61		1.01	1.01	1.01	1.01	1.01	1.01	1.01
	3	22.61	22.61	22.61	22.61	22.61	22.61	22.61		1.01	1.01	1.01	1.01	1.01	1.01	1.01
	4	22.61	22.61	22.61	22.61	22.61	22.61	22.61		1.01	1.01	1.01	1.01	1.01	1.01	1.01
	5	22.61	22.61	22.61	22.61	22.61	22.61	22.61		1.01	1.01	1.01	1.01	1.01	1.01	1.01
dx=1	0	22.81	22.81	22.81	22.81	22.61	22.61	22.61		1.02	1.02	1.02	1.02	1.01	1.01	1.01
	0.5	22.81	22.81	22.81	22.61	22.61	22.61	22.61		1.02	1.02	1.02	1.01	1.01	1.01	1.01
	1	22.81	22.81	22.61	22.61	22.61	22.61	22.61		1.02	1.02	1.01	1.01	1.01	1.01	1.01
	2	22.61	22.61	22.61	22.61	22.61	22.61	22.61		1.01	1.01	1.01	1.01	1.01	1.01	1.01
	3	22.61	22.61	22.61	22.61	22.61	22.61	22.61		1.01	1.01	1.01	1.01	1.01	1.01	1.01
	4	22.61	22.61	22.61	22.61	22.61	22.61	22.61		1.01	1.01	1.01	1.01	1.01	1.01	1.01
	5	22.61	22.61	22.61	22.61	22.61	22.61	22.61		1.01	1.01	1.01	1.01	1.01	1.01	1.01

Annex G

Amplification factors (AF) as result of the steady state dynamic analysis for different wall and base flexibilities and for the three soil profiles; homogeneous, parabolic and linear.

Homogeneous soil – Amplification Factors															
L/H=1	dw	0	1	5	10	20	30	40		0	1	5	10	20	30	40
	dθ	Bonded contact								Smooth contact						
dx=0	0	5.77	14.02	14.99	12.43	10.71	10.13	9.80		6.25	9.34	10.21	10.09	9.07	8.76	8.65
	0.5	19.58	18.96	14.46	12.17	10.66	10.08	9.82		11.03	14.15	12.62	10.85	9.36	8.94	8.75
	1	18.89	17.42	13.64	11.88	10.54	10.06	9.78		19.52	18.72	13.64	11.23	9.51	9.04	8.79
	2	14.99	14.27	12.30	11.22	10.29	9.86	9.68		16.51	18.25	13.61	11.27	9.68	9.13	8.93
	3	12.64	12.43	11.34	10.66	10.02	9.71	9.52		12.47	16.07	12.88	10.95	9.67	9.21	8.96
	4	11.35	11.23	10.65	10.24	9.79	9.56	9.44		9.03	14.41	11.98	10.57	9.55	9.18	9.01
	5	10.45	10.44	10.10	9.82	9.57	9.40	9.35		7.53	12.84	11.22	10.21	9.44	9.07	8.96
dx=0.1	0	13.58	18.19	16.17	13.21	11.23	10.55	10.22		12.23	8.62	12.42	11.12	9.48	9.05	8.89
	0.5	19.93	19.32	15.10	12.74	11.07	10.45	10.14		15.12	18.32	13.87	11.62	9.68	9.11	8.99
	1	18.56	17.43	14.04	12.23	10.89	10.32	10.06		16.53	20.16	14.46	11.73	9.71	9.26	9.09
	2	14.93	14.34	12.48	11.43	10.52	10.10	9.84		12.12	18.06	13.78	11.41	9.81	9.34	9.13
	3	12.76	12.46	11.45	10.87	10.21	9.94	9.74		8.84	16.09	12.81	10.99	9.83	9.36	9.19
	4	11.48	11.37	10.75	10.34	9.91	9.71	9.58		7.11	14.32	12.09	10.72	9.67	9.28	9.09
	5	10.61	10.47	10.19	10.01	9.72	9.58	9.43		5.55	13.12	11.41	10.31	9.49	9.23	9.07
dx=0.5	0	12.23	14.66	14.52	12.74	11.10	10.43	10.15		12.29	13.86	13.34	11.60	9.71	9.06	8.78
	0.5	15.64	15.86	13.92	12.26	10.88	10.37	10.04		14.51	17.70	14.21	11.82	9.71	9.07	8.77
	1	15.56	15.10	13.22	11.88	10.79	10.30	10.01		14.85	18.93	14.40	11.73	9.65	9.03	8.74
	2	13.81	13.38	12.15	11.19	10.42	10.07	9.83		11.77	18.05	13.86	11.35	9.46	8.90	8.75
	3	12.27	12.02	11.27	10.75	10.16	9.84	9.70		8.59	16.42	12.78	10.64	9.16	8.83	8.67
	4	11.17	11.16	10.60	10.25	9.84	9.68	9.53		6.67	14.88	11.62	9.97	8.94	8.71	8.65
	5	10.37	10.42	10.14	9.88	9.68	9.51	9.43		5.86	13.14	10.63	9.40	8.77	8.66	8.57
dx=1	0	11.21	13.21	13.78	12.23	10.98	10.37	10.00		10.75	13.11	13.02	11.55	9.65	9.01	8.69
	0.5	14.25	14.59	13.36	11.96	10.70	10.26	9.97		13.71	16.27	13.83	11.68	9.66	8.97	8.67
	1	14.52	14.37	12.80	11.66	10.66	10.15	9.91		14.08	17.62	14.11	11.76	9.63	8.91	8.61
	2	13.29	12.95	11.86	11.05	10.27	9.95	9.76		11.64	17.49	13.72	11.35	9.43	8.80	8.55
	3	12.01	11.83	11.10	10.59	10.00	9.76	9.62		9.11	16.26	12.96	10.77	9.13	8.67	8.45
	4	11.04	10.92	10.51	10.16	9.81	9.61	9.49		7.54	15.00	11.95	10.11	8.86	8.46	8.34
	5	10.32	10.30	9.99	9.83	9.58	9.44	9.37		6.20	13.78	11.12	9.56	8.61	8.28	8.25

Homogeneous soil– Amplification Factors																
L/H=2	dw	0	1	5	10	20	30	40		0	1	5	10	20	30	40
	dθ	Bonded contact								Smooth contact						
dx=0	0	6.74	9.09	9.93	9.43	8.95	8.75	8.67		6.14	9.17	11.09	10.15	9.40	9.19	9.10
	0.5	12.49	11.92	10.38	9.57	9.01	8.80	8.66		14.33	13.83	11.74	10.35	9.49	9.25	9.15
	1	12.82	12.03	10.38	9.59	9.03	8.80	8.71		15.48	14.33	11.74	10.37	9.54	9.29	9.19
	2	11.43	11.08	10.03	9.48	8.99	8.82	8.70		13.9	13.14	11.12	10.16	9.53	9.30	9.21
	3	10.38	10.20	9.64	9.21	8.91	8.78	8.68		12.19	11.69	10.51	9.89	9.45	9.29	9.21
	4	9.66	9.61	9.28	9.04	8.81	8.69	8.63		10.92	10.65	9.98	9.63	9.34	9.24	9.19
	5	9.13	9.17	8.98	8.80	8.68	8.61	8.57		10.06	9.94	9.59	9.40	9.24	9.18	9.14
dx=0.1	0	10.04	11.12	10.66	9.93	9.29	9.02	8.91		10.62	12.52	12.01	10.72	9.69	9.36	9.22
	0.5	13.20	12.52	10.78	9.95	9.27	9.03	8.90		15.49	14.57	12.25	10.69	9.68	9.37	9.24
	1	12.96	12.20	10.64	9.86	9.24	9.01	8.90		15.58	14.45	12.00	10.56	9.63	9.35	9.27
	2	11.53	11.19	10.19	9.59	9.14	8.99	8.86		13.76	13.1	11.18	10.18	9.51	9.33	9.31
	3	10.52	10.31	9.75	9.37	9.03	8.89	8.78		12.13	11.64	10.43	9.81	9.42	9.34	9.33
	4	9.78	9.69	9.39	9.14	8.88	8.79	8.69		10.79	10.51	9.88	9.56	9.38	9.32	9.30
	5	9.28	9.21	9.02	8.94	8.75	8.69	8.66		10.01	9.87	9.56	9.41	9.31	9.27	9.24
dx=0.5	0	10.48	11.02	10.59	9.93	9.34	9.15	9.00		11.68	12.36	11.84	10.71	9.64	9.31	9.18
	0.5	12.22	11.82	10.56	9.94	9.31	9.07	8.97		14.10	13.62	11.93	10.6	9.61	9.29	9.16
	1	12.17	11.68	10.48	9.84	9.29	9.09	8.90		14.31	13.59	11.73	10.43	9.54	9.26	9.13
	2	11.21	10.89	10.07	9.63	9.16	8.98	8.87		13.23	12.67	11.00	10.03	9.37	9.17	9.06
	3	10.27	10.21	9.69	9.37	9.07	8.91	8.82		12.01	11.56	10.31	9.66	9.20	9.07	9.00
	4	9.71	9.64	9.35	9.13	8.88	8.82	8.79		10.87	10.53	9.73	9.33	9.06	8.96	8.95
	5	9.25	9.18	9.06	8.92	8.77	8.75	8.63		9.91	9.71	9.27	9.04	8.92	8.88	8.91
dx=1	0	10.15	10.62	10.43	9.83	9.37	9.09	9.00		11.12	11.77	11.52	10.58	9.59	9.25	9.10
	0.5	11.69	11.41	10.42	9.83	9.29	9.07	8.93		13.28	12.95	11.64	10.49	9.54	9.23	9.07
	1	11.77	11.39	10.34	9.75	9.25	9.06	8.95		13.67	13.08	11.52	10.34	9.48	9.19	9.04
	2	10.96	10.76	9.96	9.56	9.14	8.96	8.88		12.91	12.42	10.94	9.99	9.33	9.1	8.99
	3	10.27	10.08	9.61	9.34	9.02	8.89	8.82		11.96	11.53	10.34	9.65	9.16	8.99	8.91
	4	9.63	9.61	9.29	9.09	8.91	8.81	8.71		10.95	10.61	9.79	9.34	9.00	8.89	8.84
	5	9.18	9.21	9.02	8.89	8.76	8.72	8.67		10.08	9.86	9.33	9.06	8.87	8.79	8.77

Homogeneous soil– Amplification Factors															
L/H=3	dw	0	1	5	10	20	30	40	0	1	5	10	20	30	40
	dθ	Bonded contact							Smooth contact						
dx=0	0	6.63	8.07	9.04	8.86	8.57	8.48	8.43	6.27	5.64	5.09	5.26	5.73	6.12	6.39
	0.5	9.99	9.94	9.42	8.98	8.62	8.50	8.41	10.82	8.42	6.38	6.03	6.19	6.44	6.65
	1	10.59	10.24	9.47	9.01	8.65	8.51	8.45	12.12	9.85	7.30	6.64	6.57	6.73	6.87
	2	10.00	9.85	9.27	8.90	8.60	8.50	8.43	11.79	10.54	8.20	7.45	7.15	7.17	7.24
	3	9.43	9.34	9.00	8.77	8.54	8.41	8.41	10.81	10.10	8.49	7.87	7.54	7.50	7.52
	4	8.93	8.92	8.71	8.58	8.44	8.42	8.35	9.93	9.53	8.56	8.08	7.79	7.74	7.72
	5	8.59	8.61	8.51	8.45	8.38	8.30	8.33	9.32	9.11	8.50	8.18	7.96	7.90	7.88
dx=0.1	0	8.50	9.44	9.63	9.26	8.88	8.71	8.65	8.99	7.15	5.82	5.81	6.12	6.44	6.68
	0.5	10.66	10.49	9.78	9.29	8.86	8.71	8.60	12.02	9.34	6.95	6.46	6.50	6.70	6.89
	1	10.78	10.51	9.69	9.22	8.84	8.68	8.58	12.50	10.32	7.72	6.96	6.81	6.93	7.08
	2	10.14	9.93	9.43	9.04	8.75	8.61	8.58	11.72	10.63	8.39	7.58	7.28	7.31	7.40
	3	9.52	9.41	9.12	8.90	8.69	8.58	8.49	10.76	10.11	8.54	7.89	7.59	7.61	7.66
	4	9.07	9.06	8.85	8.70	8.59	8.48	8.46	9.79	9.47	8.47	8.06	7.84	7.84	7.85
	5	8.70	8.68	8.61	8.51	8.47	8.43	8.38	9.24	9.06	8.47	8.18	8.01	7.99	8.00
dx=0.5	0	6.63	8.07	9.04	8.86	8.57	8.47	8.44	10.37	8.21	6.57	6.40	6.55	6.77	6.96
	0.5	9.98	9.94	9.42	8.98	8.63	8.51	8.42	12.13	9.82	7.48	6.92	6.84	6.97	7.11
	1	10.59	10.25	9.47	9.00	8.64	8.52	8.46	12.36	10.53	8.11	7.31	7.07	7.14	7.24
	2	10.00	9.86	9.27	8.90	8.60	8.49	8.42	11.75	10.69	8.63	7.78	7.41	7.40	7.46
	3	9.42	9.33	9.00	8.77	8.55	8.40	8.40	10.95	10.29	8.66	7.98	7.61	7.58	7.60
	4	8.94	8.91	8.71	8.59	8.44	8.41	8.36	10.07	9.63	8.54	8.01	7.74	7.70	7.74
	5	8.58	8.61	8.49	8.44	8.36	8.28	8.34	9.29	9.05	8.35	8.00	7.82	7.80	7.82
dx=1	0	9.28	9.72	9.72	9.38	9.02	8.81	8.75	10.32	8.32	6.74	6.57	6.68	6.87	7.04
	0.5	10.37	10.30	9.77	9.31	8.98	8.79	8.67	11.83	9.78	7.60	7.06	6.94	7.06	7.17
	1	10.43	10.28	9.67	9.23	8.96	8.79	8.71	12.14	10.43	8.19	7.42	7.15	7.20	7.30
	2	10.01	9.89	9.35	9.13	8.84	8.72	8.67	11.66	10.65	8.72	7.86	7.48	7.43	7.47
	3	9.46	9.40	9.15	8.94	8.73	8.66	8.57	11.00	10.33	8.76	8.04	7.67	7.60	7.62
	4	9.11	9.04	8.84	8.71	8.60	8.56	8.51	10.22	9.78	8.65	8.09	7.77	7.71	7.71
	5	8.80	8.75	8.67	8.58	8.53	8.45	8.47	9.52	9.24	8.46	8.07	7.82	7.77	7.79

Homogeneous soil – Amplification Factors															
L/H=5	dw	0	1	5	10	20	30	40	0	1	5	10	20	30	40
	dθ	Bonded contact							Smooth contact						
dx=0	0	5.79	6.71	7.87	8.06	8.09	8.09	8.07	5.85	6.85	8.66	8.88	8.83	8.86	8.88
	0.5	7.66	7.97	8.22	8.20	8.15	8.13	8.11	8.17	8.64	9.25	9.04	8.89	8.89	8.93
	1	8.37	8.45	8.33	8.27	8.15	8.12	8.14	9.38	9.52	9.43	9.08	8.92	8.92	8.95
	2	8.54	8.55	8.36	8.26	8.17	8.14	8.11	9.96	9.91	9.37	9.08	8.95	8.94	8.97
	3	8.35	8.39	8.26	8.20	8.14	8.10	8.09	9.69	9.55	9.15	8.96	8.92	8.94	8.96
	4	8.12	8.19	8.15	8.10	8.07	8.04	8.07	9.17	9.14	8.93	8.85	8.87	8.90	8.97
	5	7.91	7.99	8.02	8.03	7.98	8.04	8.05	8.81	8.80	8.74	8.74	8.80	8.89	8.93
dx=0.1	0	6.83	7.54	8.31	8.40	8.34	8.30	8.28	7.23	8.09	9.34	9.28	9.01	8.95	8.93
	0.5	8.23	8.41	8.54	8.45	8.32	8.26	8.28	9.12	9.44	9.68	9.28	9.01	8.95	8.96
	1	8.70	8.74	8.61	8.44	8.35	8.29	8.25	9.94	9.96	9.75	9.28	8.99	8.94	8.98
	2	8.75	8.72	8.55	8.42	8.28	8.25	8.24	10.13	10.08	9.49	9.12	8.93	8.95	9.03
	3	8.51	8.54	8.43	8.30	8.26	8.19	8.23	9.75	9.62	9.15	8.93	8.89	8.97	9.07
	4	8.26	8.33	8.22	8.23	8.17	8.16	8.18	9.17	9.08	8.84	8.80	8.87	9.00	9.06
	5	8.06	8.14	8.11	8.12	8.10	8.12	8.14	8.77	8.80	8.72	8.77	8.89	8.97	9.04
dx=0.5	0	7.60	8.16	8.64	8.62	8.51	8.47	8.45	8.50	9.15	9.89	9.61	9.16	9.02	8.99
	0.5	8.56	8.73	8.74	8.64	8.51	8.47	8.39	9.93	10.10	10.06	9.55	9.11	9.00	8.97
	1	8.88	8.94	8.79	8.63	8.47	8.43	8.41	10.47	10.44	10.05	9.47	9.06	8.96	8.93
	2	8.82	8.82	8.66	8.52	8.42	8.38	8.38	10.48	10.38	9.69	9.23	8.94	8.91	8.89
	3	8.53	8.61	8.52	8.43	8.31	8.30	8.30	10.11	9.97	9.31	8.97	8.82	8.82	8.84
	4	8.35	8.37	8.35	8.32	8.29	8.29	8.22	9.50	9.35	8.93	8.75	8.70	8.75	8.78
	5	8.15	8.22	8.18	8.18	8.23	8.22	8.21	8.89	8.86	8.61	8.56	8.62	8.66	8.78
dx=1	0	7.73	8.23	8.72	8.66	8.53	8.47	8.46	8.76	9.35	9.97	9.71	9.20	9.05	8.99
	0.5	8.63	8.78	8.77	8.68	8.50	8.49	8.46	10.03	10.18	10.11	9.64	9.17	9.02	8.95
	1	8.90	8.95	8.75	8.64	8.51	8.43	8.43	10.51	10.49	10.10	9.55	9.11	8.99	8.94
	2	8.87	8.85	8.65	8.54	8.42	8.40	8.36	10.59	10.47	9.82	9.31	8.98	8.90	8.88
	3	8.58	8.61	8.48	8.46	8.32	8.33	8.34	10.26	10.10	9.45	9.08	8.87	8.83	8.80
	4	8.34	8.39	8.41	8.33	8.33	8.28	8.25	9.70	9.58	9.10	8.86	8.75	8.72	8.75
	5	8.14	8.23	8.19	8.20	8.21	8.20	8.19	9.17	9.09	8.77	8.64	8.63	8.65	8.69

Homogeneous soil – Amplification Factors															
L/H=10	dw	0	1	5	10	20	30	40	0	1	5	10	20	30	40
	dθ	Bonded contact							Smooth contact						
dx=0	0	4.07	4.62	5.80	6.40	6.83	7.09	7.22	4.12	4.74	6.48	7.42	8.23	8.65	8.91
	0.5	5.19	5.54	6.24	6.62	6.96	7.16	7.26	5.47	5.97	7.22	7.79	8.42	8.79	8.99
	1	5.89	6.09	6.54	6.78	7.05	7.21	7.29	6.56	6.95	7.74	8.11	8.56	8.88	9.08
	2	6.59	6.69	6.84	7.00	7.18	7.28	7.34	7.98	8.12	8.29	8.48	8.81	9.03	9.20
	3	6.80	6.89	7.02	7.12	7.22	7.30	7.38	8.48	8.50	8.53	8.66	8.91	9.15	9.27
	4	6.90	7.00	7.08	7.14	7.24	7.34	7.40	8.53	8.54	8.63	8.75	9.01	9.20	9.33
	5	6.89	7.00	7.08	7.13	7.24	7.35	7.42	8.50	8.55	8.64	8.81	9.04	9.24	9.38
dx=0.1	0	4.64	5.16	6.22	6.75	7.14	7.31	7.45	4.89	5.50	7.17	7.96	8.57	8.89	9.08
	0.5	5.65	5.95	6.60	6.91	7.22	7.36	7.45	6.21	6.65	7.82	8.27	8.68	8.93	9.16
	1	6.23	6.43	6.83	7.03	7.27	7.39	7.45	7.21	7.51	8.23	8.45	8.79	9.01	9.23
	2	6.81	6.91	7.07	7.20	7.35	7.42	7.50	8.32	8.52	8.60	8.65	8.88	9.13	9.33
	3	7.00	7.10	7.19	7.28	7.38	7.44	7.53	8.73	8.76	8.67	8.75	9.00	9.24	9.47
	4	7.07	7.15	7.21	7.29	7.37	7.48	7.52	8.63	8.67	8.65	8.81	9.11	9.38	9.54
	5	7.05	7.13	7.21	7.27	7.39	7.49	7.53	8.56	8.61	8.74	8.91	9.20	9.41	9.56
dx=0.5	0	5.27	5.78	6.72	7.11	7.41	7.57	7.65	5.90	6.55	8.12	8.71	9.03	9.19	9.31
	0.5	6.12	6.39	6.98	7.24	7.49	7.58	7.63	7.19	7.64	8.63	8.90	9.07	9.23	9.34
	1	6.60	6.81	7.11	7.35	7.50	7.59	7.66	8.10	8.40	8.94	8.98	9.09	9.24	9.36
	2	7.08	7.14	7.33	7.45	7.55	7.60	7.67	9.09	9.17	9.10	9.05	9.11	9.26	9.37
	3	7.19	7.28	7.38	7.48	7.55	7.62	7.71	9.37	9.33	9.09	9.01	9.09	9.25	9.34
	4	7.20	7.32	7.40	7.41	7.52	7.60	7.69	9.20	9.14	8.94	8.94	9.08	9.23	9.34
	5	7.20	7.28	7.35	7.43	7.51	7.59	7.65	8.86	8.85	8.78	8.83	9.05	9.18	9.37
dx=1	0	5.27	5.78	6.72	7.11	7.41	7.57	7.65	6.25	6.88	8.40	8.99	9.18	9.32	9.42
	0.5	6.12	6.39	6.98	7.24	7.49	7.58	7.63	7.51	7.92	8.89	9.13	9.23	9.33	9.41
	1	6.60	6.81	7.11	7.35	7.50	7.59	7.66	8.39	8.66	9.16	9.20	9.26	9.34	9.42
	2	7.08	7.14	7.33	7.45	7.55	7.60	7.67	9.31	9.41	9.36	9.21	9.23	9.34	9.41
	3	7.19	7.28	7.38	7.48	7.55	7.62	7.71	9.60	9.60	9.30	9.18	9.21	9.31	9.35
	4	7.20	7.32	7.40	7.41	7.52	7.60	7.69	9.47	9.39	9.16	9.06	9.19	9.25	9.34
	5	7.20	7.28	7.35	7.43	7.51	7.59	7.65	9.18	9.15	8.96	8.99	9.11	9.20	9.33

Parabolic soil– Amplification Factors																
L/H=1	dw	0	1	5	10	20	30	40		0	1	5	10	20	30	40
	dθ	Bonded contact								Smooth contact						
dx=0	0	8.16	10.00	11.64	10.88	9.62	9.01	8.70		3.53	4.35	9.13	9.67	9.25	8.95	3.53
	0.5	11.02	11.62	11.49	10.58	9.45	8.89	8.61		9.32	10.53	11.01	10.36	9.51	9.07	9.32
	1	11.72	11.80	11.08	10.26	9.25	8.78	8.48		13.97	13.42	11.84	10.71	9.64	9.15	13.97
	2	11.06	10.94	10.26	9.63	8.91	8.55	8.31		14.91	14.10	12.09	10.84	9.68	9.21	14.91
	3	9.97	9.96	9.47	9.04	8.55	8.33	8.16		13.61	13.22	11.61	10.63	9.66	9.18	13.61
	4	9.02	9.05	8.79	8.55	8.27	8.10	8.01		12.66	12.23	11.06	10.28	9.53	9.13	12.66
	5	8.20	8.30	8.19	8.10	7.98	7.87	7.83		12.13	11.71	10.56	10.00	9.38	9.03	12.13
dx=0.1	0	6.42	9.42	12.28	11.22	9.71	9.03	8.67		4.94	7.68	10.55	10.43	9.59	9.10	4.94
	0.5	11.67	12.46	12.05	10.85	9.53	8.89	8.59		12.46	12.61	11.73	10.78	9.67	9.14	12.46
	1	12.71	12.72	11.57	10.49	9.29	8.77	8.54		15.51	14.39	12.14	10.89	9.65	9.13	15.51
	2	11.69	11.50	10.55	9.79	8.98	8.55	8.37		15.33	13.94	11.94	10.67	9.63	9.30	15.33
	3	10.35	10.20	9.69	9.21	8.61	8.36	8.16		13.94	13.03	11.22	10.45	9.72	9.37	13.94
	4	9.27	9.24	8.97	8.62	8.33	8.14	8.05		13.46	12.71	10.99	10.38	9.69	9.33	13.46
	5	8.45	8.52	8.36	8.25	8.06	7.95	7.90		13.51	12.96	10.72	10.12	9.51	9.22	13.51
dx=0.5	0	8.05	10.07	11.95	11.04	9.74	9.10	8.74		8.43	9.91	11.18	10.51	9.50	8.98	8.43
	0.5	11.28	11.94	11.65	10.70	9.54	8.95	8.61		13.32	13.21	11.96	10.69	9.45	8.97	13.32
	1	12.03	12.01	11.24	10.38	9.36	8.83	8.54		15.54	14.62	12.26	10.73	9.39	8.92	15.54
	2	11.27	11.14	10.32	9.69	8.93	8.57	8.36		16.06	15.03	12.00	10.37	9.27	8.82	16.06
	3	10.14	10.02	9.52	9.12	8.61	8.33	8.20		15.41	14.46	11.39	9.88	9.06	8.77	15.41
	4	9.13	9.09	8.85	8.60	8.29	8.11	8.03		14.68	13.90	10.87	9.35	8.92	8.72	14.68
	5	8.32	8.38	8.28	8.17	8.02	7.91	7.83		14.21	13.50	10.51	9.05	8.77	8.56	14.21
dx=1	0	8.16	10.00	11.64	10.88	9.62	9.01	8.70		8.79	10.02	11.12	10.52	9.46	8.89	8.79
	0.5	11.02	11.62	11.49	10.58	9.45	8.89	8.61		13.03	12.97	11.92	10.69	9.42	8.85	13.03
	1	11.72	11.80	11.08	10.26	9.25	8.78	8.48		15.04	14.35	12.22	10.70	9.35	8.80	15.04
	2	11.06	10.94	10.26	9.63	8.91	8.55	8.31		15.71	14.82	12.07	10.42	9.18	8.68	15.71
	3	9.97	9.96	9.47	9.04	8.55	8.33	8.16		15.18	14.38	11.63	9.95	8.88	8.54	15.18
	4	9.02	9.05	8.79	8.55	8.27	8.10	8.01		14.65	13.86	11.27	9.44	8.60	8.37	14.65
	5	8.20	8.30	8.19	8.10	7.98	7.87	7.83		14.18	13.51	11.01	8.98	8.33	8.29	14.18

Parabolic soil– Amplification Factors																
L/H=2	dw	0	1	5	10	20	30	40		0	1	5	10	20	30	40
	dθ	Bonded contact								Smooth contact						
dx=0	0	6.45	9.23	9.48	9.85	9.06	8.72	8.54		6.01	7.50	9.66	9.64	9.12	8.82	8.64
	0.5	11.23	11.44	10.52	9.66	8.92	8.66	8.52		10.11	10.52	10.42	9.85	9.19	8.84	8.66
	1	11.47	11.21	10.18	9.45	8.85	8.61	8.44		11.69	11.53	10.63	9.94	9.19	8.84	8.67
	2	10.28	10.11	9.49	9.04	8.64	8.45	8.33		11.62	11.37	10.42	9.79	9.12	8.83	8.65
	3	9.30	9.26	8.89	8.67	8.44	8.30	8.24		10.83	10.69	10.03	9.52	9.01	8.78	8.60
	4	8.61	8.63	8.47	8.37	8.23	8.17	8.17		10.11	10.10	9.62	9.29	8.91	8.71	8.54
	5	8.08	8.14	8.12	8.10	8.07	8.08	8.02		9.48	9.58	9.28	9.04	8.76	8.62	8.47
dx=0.1	0	8.92	10.85	11.16	10.14	9.29	8.91	8.70		8.23	9.34	10.37	10.00	9.22	8.83	8.61
	0.5	11.65	11.72	10.73	9.87	9.12	8.80	8.64		11.18	11.26	10.70	10.02	9.20	8.80	8.59
	1	11.50	11.28	10.28	9.63	8.99	8.71	8.58		11.92	11.70	10.71	9.92	9.12	8.76	8.60
	2	10.32	10.14	9.58	9.17	8.75	8.54	8.47		11.37	11.18	10.25	9.63	8.97	8.72	8.66
	3	9.35	9.32	9.01	8.79	8.52	8.43	8.35		10.47	10.40	9.74	9.28	8.90	8.79	8.73
	4	8.64	8.70	8.53	8.44	8.31	8.26	8.21		9.68	9.74	9.40	9.16	8.93	8.82	8.72
	5	8.12	8.18	8.16	8.16	8.13	8.08	8.10		9.38	9.53	9.29	9.11	8.88	8.74	8.64
dx=0.5	0	9.30	10.57	10.83	9.99	9.25	8.95	8.71		9.18	9.86	10.41	9.91	9.08	8.69	8.46
	0.5	11.09	11.15	10.48	9.80	9.12	8.81	8.71		11.18	11.25	10.65	9.87	8.97	8.64	8.39
	1	11.09	10.92	10.10	9.51	8.96	8.74	8.56		11.90	11.63	10.58	9.74	8.88	8.57	8.37
	2	10.07	10.05	9.44	9.10	8.78	8.59	8.51		11.66	11.29	10.08	9.35	8.68	8.42	8.33
	3	9.23	9.25	8.98	8.78	8.52	8.42	8.33		10.91	10.57	9.51	8.93	8.50	8.33	8.27
	4	8.64	8.61	8.53	8.43	8.29	8.26	8.23		10.18	9.89	8.95	8.53	8.32	8.26	8.25
	5	8.08	8.16	8.15	8.16	8.13	8.13	8.10		9.54	9.24	8.40	8.24	8.20	8.18	8.15
dx=1	0	9.16	10.36	10.63	10.00	9.26	8.88	8.72		9.15	9.81	10.35	9.85	9.05	8.62	8.37
	0.5	10.84	10.95	9.71	9.75	9.15	8.82	8.64		11.02	11.04	10.55	9.79	8.97	8.54	8.35
	1	10.87	10.71	10.11	9.54	8.96	8.71	8.59		11.71	11.47	10.51	9.68	8.87	8.45	8.30
	2	10.03	9.92	9.44	9.07	8.74	8.55	8.45		11.64	11.27	10.13	9.33	8.65	8.30	8.22
	3	9.21	9.18	8.91	8.74	8.51	8.42	8.34		11.05	10.72	9.62	8.96	8.38	8.16	8.10
	4	8.59	8.59	8.50	8.40	8.29	8.23	8.22		10.43	10.11	9.07	8.55	8.13	8.04	7.96
	5	8.03	8.11	8.12	8.12	8.10	8.11	8.08		9.88	9.62	8.55	8.18	7.90	7.91	7.92

Parabolic soil– Amplification Factors																
L/H=3	dw	0	1	5	10	20	30	40		0	1	5	10	20	30	40
	dθ	Bonded contact								Smooth contact						
dx=0	0	5.80	6.77	8.35	8.51	8.23	8.01	7.89		5.83	6.86	8.93	9.26	9.02	8.78	8.60
	0.5	7.76	8.18	8.61	8.49	8.15	7.94	7.83		7.00	9.05	9.62	9.50	9.06	8.76	8.64
	1	8.52	8.61	8.61	8.40	8.07	7.91	7.81		7.11	10.04	9.86	9.54	9.05	8.80	8.61
	2	8.58	8.57	8.38	8.15	7.90	7.75	7.69		5.86	10.35	9.78	9.40	8.94	8.76	8.60
	3	8.19	8.17	8.03	7.91	7.74	7.67	7.58		4.77	9.99	9.47	9.19	8.83	8.69	8.53
	4	7.79	7.80	7.72	7.64	7.57	7.49	7.51		3.80	9.63	9.12	9.01	8.70	8.57	8.48
	5	7.38	7.39	7.41	7.38	7.38	7.41	7.38		3.35	9.23	8.87	8.80	8.61	8.53	8.40
dx=0.1	0	6.83	7.65	8.77	8.76	8.40	8.13	8.01		6.92	8.24	9.54	9.57	9.06	8.75	8.52
	0.5	8.23	8.54	8.87	8.66	8.29	8.10	7.96		7.77	9.76	9.87	9.59	9.02	8.66	8.53
	1	8.72	8.85	8.80	8.52	8.20	7.99	7.90		7.38	10.33	9.92	9.53	8.95	8.66	8.53
	2	8.67	8.64	8.45	8.26	8.03	7.88	7.77		5.93	10.29	9.65	9.23	8.76	8.62	8.59
	3	8.22	8.26	8.11	7.94	7.82	7.69	7.67		4.61	9.76	9.17	8.96	8.73	8.65	8.65
	4	7.82	7.84	7.74	7.71	7.60	7.61	7.56		3.61	9.33	8.95	8.88	8.77	8.72	8.62
	5	7.41	7.46	7.45	7.46	7.43	7.46	7.44		2.92	9.28	8.86	8.87	8.70	8.61	8.57
dx=0.5	0	6.83	7.65	8.77	8.76	8.40	8.13	8.01		9.22	8.95	9.79	9.59	8.93	8.62	8.35
	0.5	8.23	8.54	8.87	8.66	8.29	8.10	7.96		9.65	10.05	10.02	9.53	8.80	8.51	8.33
	1	8.72	8.85	8.80	8.52	8.20	7.99	7.90		8.66	10.55	9.96	9.40	8.73	8.46	8.30
	2	8.67	8.64	8.45	8.26	8.03	7.88	7.77		6.50	10.46	9.59	9.03	8.53	8.28	8.21
	3	8.22	8.26	8.11	7.94	7.82	7.69	7.67		4.98	10.04	9.07	8.69	8.32	8.21	8.20
	4	7.82	7.84	7.74	7.71	7.60	7.61	7.56		4.05	9.54	8.54	8.29	8.13	8.11	8.13
	5	7.41	7.46	7.45	7.46	7.43	7.46	7.44		3.43	8.96	8.02	8.03	8.03	8.06	8.01
dx=1	0	8.35	9.23	9.61	9.32	8.84	8.56	8.46		10.28	9.02	9.82	9.57	8.94	8.51	8.33
	0.5	9.55	9.75	9.49	9.13	8.68	8.55	8.43		9.72	10.08	10.01	9.52	8.84	8.46	8.29
	1	9.66	9.68	9.30	8.99	8.64	8.47	8.38		8.91	10.47	9.98	9.41	8.70	8.37	8.22
	2	9.19	9.17	8.88	8.67	8.44	8.32	8.26		7.32	10.54	9.64	9.08	8.46	8.20	8.12
	3	8.63	8.66	8.50	8.38	8.23	8.19	8.15		5.66	10.16	9.17	8.71	8.22	8.04	7.99
	4	8.12	8.19	8.15	8.13	8.09	8.04	8.05		4.34	9.79	8.71	8.32	7.96	7.90	7.85
	5	7.72	7.79	7.86	7.88	7.90	7.93	7.96		3.61	9.40	8.21	7.95	7.73	7.78	7.80

Parabolic soil – smooth contact– Amplification Factors															
L/H=5	dw	0	1	5	10	20	30	40	0	1	5	10	20	30	40
	dθ	Bonded contact							Smooth contact						
dx=0	0	4.88	5.46	6.74	7.25	7.47	7.49	7.45	1.15	1.14	1.11	1.08	1.04	1.02	1.01
	0.5	6.12	6.46	7.15	7.39	7.48	7.47	7.46	1.09	1.12	1.09	1.06	1.03	1.02	1.01
	1	6.80	6.98	7.34	7.44	7.45	7.43	7.45	1.11	1.10	1.07	1.05	1.02	1.02	1.01
	2	7.25	7.31	7.40	7.42	7.40	7.40	7.39	1.07	1.07	1.05	1.04	1.02	1.01	1.00
	3	7.22	7.28	7.31	7.32	7.30	7.31	7.30	1.05	1.01	1.03	1.02	1.01	1.00	0.99
	4	7.06	7.10	7.14	7.16	7.20	7.23	7.22	1.04	1.03	1.02	1.02	1.00	1.00	0.99
	5	6.82	6.89	6.93	6.99	7.08	7.10	7.17	1.02	1.02	1.02	1.01	1.00	0.99	0.99
dx=0.1	0	5.53	6.07	7.15	7.53	7.65	7.65	7.62	1.14	1.14	1.05	1.06	1.03	1.02	1.00
	0.5	6.54	6.85	7.42	7.60	7.65	7.58	7.58	1.12	1.11	1.08	1.05	0.99	1.01	1.00
	1	7.05	7.27	7.51	7.61	7.63	7.56	7.54	1.10	1.09	1.06	1.05	1.02	1.01	1.00
	2	7.38	7.48	7.55	7.51	7.51	7.49	7.48	1.07	1.06	1.05	1.03	1.01	1.00	0.99
	3	7.35	7.38	7.40	7.39	7.39	7.40	7.37	1.05	1.04	0.99	1.02	1.01	0.96	0.99
	4	7.12	7.19	7.23	7.21	7.27	7.25	7.31	1.03	1.03	1.02	1.01	1.00	0.99	0.99
	5	6.88	6.92	6.97	7.08	7.12	7.19	7.23	1.02	1.02	1.01	1.01	0.96	0.99	0.99
dx=0.5	0	6.01	6.50	7.39	7.72	7.77	7.71	7.72	1.13	1.07	1.09	1.05	1.02	1.01	1.00
	0.5	6.84	7.11	7.62	7.77	7.79	7.73	7.69	1.11	1.09	1.02	1.00	1.02	1.00	0.99
	1	7.26	7.40	7.69	7.74	7.69	7.65	7.63	1.09	1.08	1.05	1.04	1.02	1.00	0.99
	2	7.49	7.59	7.60	7.60	7.57	7.56	7.52	1.06	1.05	1.04	1.02	1.01	0.99	0.99
	3	7.37	7.42	7.47	7.48	7.45	7.44	7.44	1.04	1.04	1.02	1.02	1.00	0.99	0.99
	4	7.16	7.21	7.23	7.26	7.30	7.32	7.36	1.03	1.02	1.02	1.01	0.99	0.99	0.99
	5	6.88	6.92	7.01	7.08	7.14	7.21	7.26	0.98	1.02	1.01	0.96	0.99	0.99	0.99
dx=1	0	6.10	6.58	7.46	7.76	7.80	7.73	7.73	1.13	1.11	1.08	1.01	1.02	1.00	0.99
	0.5	6.91	7.15	7.62	7.77	7.77	7.73	7.69	1.10	1.09	1.06	1.04	1.02	1.00	0.99
	1	7.29	7.46	7.70	7.77	7.72	7.69	7.63	1.09	1.08	1.05	1.03	1.01	1.00	0.99
	2	7.57	7.59	7.64	7.64	7.59	7.57	7.53	1.05	1.05	1.04	1.02	1.00	0.96	0.99
	3	7.40	7.44	7.47	7.45	7.47	7.43	7.45	1.04	1.04	1.02	1.02	0.96	0.95	0.95
	4	7.16	7.20	7.23	7.27	7.30	7.33	7.35	0.99	0.99	0.98	0.96	0.96	0.95	0.94
	5	6.83	6.90	6.99	7.07	7.13	7.20	7.25	0.98	0.98	0.97	0.96	0.95	0.95	0.94

Parabolic soil– Amplification Factors																
L/H=10	dw	0	1	5	10	20	30	40		0	1	5	10	20	30	40
	dθ	Bonded contact								Smooth contact						
dx=0	0	3.43	3.81	4.87	5.58	6.19	6.47	6.62		3.46	3.90	5.60	7.31	9.13	9.80	10.03
	0.5	4.30	4.58	5.34	5.84	6.30	6.52	6.65		4.60	5.05	6.65	8.06	9.47	9.94	10.11
	1	4.92	5.12	5.65	6.02	6.39	6.56	6.68		5.52	6.13	7.28	8.68	9.57	10.03	10.14
	2	5.59	5.72	6.01	6.21	6.44	6.61	6.73		7.49	7.83	8.75	9.41	9.96	10.16	10.19
	3	5.85	5.93	6.14	6.28	6.50	6.61	6.72		8.55	8.83	9.36	9.73	10.06	10.18	10.25
	4	5.88	5.98	6.12	6.28	6.47	6.60	6.72		9.02	9.28	9.51	9.81	10.05	10.17	10.28
	5	5.82	5.92	6.08	6.24	6.45	6.59	6.68		9.15	9.40	9.61	9.82	10.03	10.14	10.37
dx=0.1	0	3.89	4.27	5.27	5.92	6.45	6.69	6.83		4.10	4.60	6.39	8.02	9.51	9.87	9.99
	0.5	4.65	4.94	5.65	6.12	6.52	6.72	6.84		5.28	5.74	7.31	8.64	9.66	9.93	10.03
	1	5.21	5.41	5.91	6.25	6.58	6.73	6.85		6.33	6.75	8.04	9.03	9.77	9.95	10.09
	2	5.78	5.90	6.19	6.38	6.61	6.77	6.86		7.78	8.16	8.94	9.45	9.82	10.03	10.22
	3	6.00	6.07	6.25	6.40	6.62	6.75	6.85		8.56	8.82	9.25	9.56	9.95	10.20	10.37
	4	5.97	6.08	6.23	6.38	6.57	6.70	6.80		8.82	9.10	9.49	9.77	10.12	10.31	10.52
	5	5.89	5.98	6.17	6.33	6.55	6.67	6.76		9.19	9.48	9.72	9.94	10.21	10.32	10.60
dx=0.5	0	4.26	4.65	5.62	6.20	6.69	6.88	6.99		4.79	5.33	7.27	8.73	9.71	9.93	9.92
	0.5	4.96	5.25	5.91	6.37	6.69	6.89	7.01		6.02	6.52	8.09	9.12	9.73	9.86	9.89
	1	5.44	5.64	6.11	6.45	6.73	6.87	7.00		7.13	7.55	8.70	9.38	9.72	9.83	9.86
	2	5.93	6.04	6.32	6.50	6.71	6.88	6.96		8.71	8.91	9.28	9.50	9.66	9.74	9.86
	3	6.06	6.17	6.34	6.50	6.70	6.84	6.92		9.46	9.49	9.32	9.34	9.53	9.67	9.83
	4	6.02	6.11	6.29	6.43	6.65	6.78	6.87		9.74	9.58	9.04	9.09	9.44	9.63	9.80
	5	5.90	6.00	6.18	6.35	6.57	6.71	6.81		9.81	9.54	8.72	9.00	9.32	9.55	9.75
dx=1	0	4.37	4.72	5.69	6.28	6.72	6.91	7.04		4.98	5.55	7.51	8.92	9.82	9.86	9.90
	0.5	5.05	5.32	5.98	6.42	6.77	6.93	7.03		6.22	6.74	8.29	9.27	9.81	9.81	9.84
	1	5.50	5.72	6.18	6.49	6.77	6.93	7.03		7.14	7.55	8.68	9.37	9.79	9.76	9.80
	2	5.97	6.09	6.35	6.55	6.77	6.90	6.98		8.94	9.13	9.48	9.61	9.62	9.62	9.68
	3	6.06	6.19	6.37	6.53	6.72	6.86	6.95		9.74	9.64	9.54	9.44	9.39	9.47	9.54
	4	6.01	6.12	6.28	6.45	6.67	6.79	6.89		10.11	9.97	9.27	9.14	9.16	9.34	9.48
	5	5.86	5.99	6.17	6.34	6.57	6.72	6.82		10.27	10.04	8.92	8.78	8.98	9.25	9.50

Linear soil– Amplification Factors																
L/H=1	dw	0	1	5	10	20	30	40		0	1	5	10	20	30	40
	dθ	Bonded contact								Smooth contact						
dx=0	0	3.33	5.28	9.53	9.63	8.65	8.08	7.76		2.94	4.67	8.57	9.25	9.01	8.69	8.49
	0.5	7.77	8.91	10.07	9.50	8.49	7.97	7.66		8.17	9.24	10.16	9.87	9.18	8.78	8.54
	1	9.84	10.08	9.91	9.20	8.28	7.85	7.56		11.71	11.70	10.96	10.15	9.24	8.82	8.56
	2	9.78	9.63	9.05	8.55	7.88	7.56	7.38		13.33	12.53	11.12	10.15	9.26	8.80	8.57
	3	8.56	8.44	8.15	7.89	7.50	7.30	7.16		13.76	12.89	10.69	9.92	9.14	8.76	8.57
	4	7.41	7.40	7.33	7.25	7.11	7.02	6.99		14.16	13.47	10.93	9.68	8.95	8.66	8.59
	5	6.44	6.45	6.60	6.68	6.74	6.79	6.77		14.74	14.08	11.47	9.91	8.92	8.72	8.65
dx=0.1	0	4.98	7.04	10.31	10.05	8.90	8.27	7.92		5.17	7.01	9.64	9.73	9.16	8.69	8.42
	0.5	8.81	9.71	10.38	9.74	8.67	8.14	7.79		10.61	10.98	10.63	10.10	9.14	8.67	8.43
	1	10.22	10.36	10.08	9.34	8.46	7.96	7.70		13.52	12.93	10.99	10.11	9.13	8.64	8.46
	2	9.75	9.63	9.13	8.61	7.97	7.65	7.45		15.28	14.00	10.79	9.81	8.93	8.69	8.59
	3	8.56	8.49	8.17	7.87	7.58	7.34	7.22		15.68	14.55	10.72	9.48	8.99	8.81	8.70
	4	7.39	7.37	7.34	7.25	7.14	7.02	7.03		16.51	15.58	11.86	9.86	9.06	8.80	8.69
	5	6.27	6.33	6.54	6.63	6.73	6.78	6.79		17.89	17.16	13.14	10.66	9.11	8.82	8.76
dx=0.5	0	6.02	7.71	10.32	10.00	8.86	8.29	7.94		7.02	8.46	10.18	9.89	8.98	8.55	8.27
	0.5	9.01	9.72	10.25	9.62	8.64	8.13	7.81		11.35	11.56	11.09	10.05	8.94	8.48	8.27
	1	10.06	10.19	9.95	9.27	8.42	7.94	7.71		13.64	13.22	11.48	10.03	8.84	8.39	8.24
	2	9.53	9.45	9.02	8.55	7.96	7.65	7.46		15.17	14.35	11.64	9.69	8.62	8.30	8.22
	3	8.36	8.29	8.07	7.82	7.49	7.34	7.19		15.42	14.58	11.90	9.31	8.38	8.23	8.09
	4	7.12	7.14	7.17	7.13	7.05	7.01	6.99		15.49	14.86	12.22	9.59	8.21	8.07	7.95
	5	5.93	6.03	6.28	6.47	6.61	6.68	6.73		15.68	15.20	12.90	10.23	8.26	7.94	7.89
dx=1	0	6.21	7.83	10.10	9.88	8.87	8.29	7.94		7.30	8.60	10.22	9.89	8.99	8.43	8.21
	0.5	8.97	9.64	10.20	9.61	8.63	8.12	7.79		11.30	11.52	11.09	10.05	8.93	8.39	8.16
	1	9.94	10.08	9.87	9.26	8.41	7.94	7.69		13.43	13.05	11.47	10.03	8.84	8.34	8.09
	2	9.55	9.36	8.94	8.51	7.95	7.64	7.45		14.94	14.12	11.80	9.78	8.54	8.16	7.95
	3	8.33	8.21	8.00	7.77	7.44	7.31	7.18		15.08	14.36	11.96	9.70	8.20	7.95	7.89
	4	7.05	7.00	7.06	7.07	7.01	6.98	6.93		15.07	14.51	12.26	9.92	7.88	7.86	7.83
	5	5.76	5.85	6.15	6.36	6.56	6.64	6.67		15.15	14.72	12.61	10.31	7.98	7.81	7.82

Linear soil – Amplification Factors																
L/H=2	dw	0	1	5	10	20	30	40		0	1	5	10	20	30	40
	dθ	Bonded contact								Smooth contact						
dx=0	0	5.33	6.42	8.35	8.47	7.96	7.60	7.42		5.53	6.19	8.05	8.88	9.04	8.79	8.60
	0.5	7.65	8.09	8.58	8.34	7.84	7.53	7.34		7.42	7.91	9.01	9.32	9.15	8.86	8.61
	1	8.49	8.53	8.51	8.16	7.68	7.41	7.26		8.81	9.14	9.57	9.59	9.21	8.88	8.62
	2	8.36	8.28	7.99	7.74	7.38	7.20	7.09		10.23	10.26	10.06	9.75	9.21	8.85	8.61
	3	7.66	7.60	7.42	7.28	7.09	6.96	6.93		10.45	10.41	10.05	9.67	9.11	8.78	8.56
	4	6.90	6.87	6.84	6.83	6.78	6.75	6.74		10.22	10.24	9.83	9.48	8.99	8.69	8.49
	5	6.12	6.17	6.29	6.37	6.45	6.53	6.58		9.88	9.93	9.55	9.27	8.84	8.58	8.42
dx=0.1	0	6.39	7.32	8.74	8.67	8.08	7.73	7.52		6.73	7.33	8.81	9.33	9.20	8.85	8.59
	0.5	8.10	8.45	8.79	8.48	7.92	7.64	7.46		8.39	8.75	9.50	9.61	9.22	8.82	8.59
	1	8.65	8.68	8.62	8.26	7.80	7.48	7.32		9.52	9.69	9.87	9.73	9.17	8.79	8.60
	2	8.36	8.31	8.05	7.80	7.41	7.27	7.15		10.42	10.38	10.05	9.67	9.08	8.81	8.69
	3	7.67	7.59	7.44	7.29	7.10	6.98	6.93		10.34	10.33	9.91	9.49	9.03	8.83	8.68
	4	6.86	6.83	6.85	6.84	6.79	6.79	6.75		9.97	9.98	9.58	9.34	9.05	8.80	8.61
	5	6.09	6.14	6.28	6.38	6.46	6.50	6.62		9.87	9.76	9.51	9.31	8.91	8.69	8.51
dx=0.5	0	6.80	7.63	8.79	8.70	8.09	7.73	7.56		7.77	8.23	9.32	9.55	9.16	8.77	8.46
	0.5	8.16	8.48	8.75	8.43	7.96	7.65	7.43		9.10	9.37	9.84	9.72	9.16	8.72	8.41
	1	8.53	8.62	8.55	8.26	7.79	7.50	7.36		10.02	10.12	10.13	9.76	9.08	8.67	8.36
	2	8.31	8.25	8.00	7.78	7.45	7.26	7.16		10.90	10.77	10.28	9.66	8.91	8.48	8.25
	3	7.55	7.51	7.38	7.24	7.09	7.00	6.94		11.08	10.93	10.17	9.42	8.66	8.32	8.18
	4	6.74	6.72	6.74	6.73	6.74	6.74	6.73		11.10	10.88	9.97	9.15	8.42	8.20	8.09
	5	5.85	5.92	6.08	6.22	6.38	6.46	6.53		11.07	10.83	9.79	8.79	8.19	8.08	7.98
dx=1	0	6.87	7.65	8.71	8.65	8.10	7.78	7.55		7.93	8.36	9.38	9.56	9.18	8.74	8.41
	0.5	8.11	8.46	8.73	8.47	7.96	7.64	7.43		9.17	9.38	9.85	9.74	9.16	8.68	8.35
	1	8.54	8.62	8.54	8.26	7.79	7.51	7.36		10.02	10.08	10.12	9.78	9.11	8.59	8.26
	2	8.27	8.20	7.98	7.76	7.43	7.25	7.15		10.87	10.77	10.30	9.76	8.93	8.45	8.12
	3	7.51	7.45	7.35	7.21	7.06	6.99	6.91		11.14	10.95	10.27	9.57	8.72	8.26	8.00
	4	6.69	6.64	6.67	6.68	6.69	6.70	6.73		11.16	10.98	10.15	9.33	8.42	8.04	7.87
	5	5.75	5.80	5.98	6.13	6.32	6.43	6.49		11.13	10.95	10.01	9.11	8.16	7.83	7.72

Linear soil – Amplification Factors															
L/H=3	dw	0	1	5	10	20	30	40	0	1	5	10	20	30	40
	dθ	Bonded contact							Smooth contact						
dx=0	0	5.34	6.05	7.46	7.74	7.54	7.34	7.21	5.39	6.23	8.14	8.71	8.58	8.35	8.24
	0.5	6.83	7.17	7.72	7.73	7.45	7.27	7.16	5.69	8.00	8.77	8.91	8.59	8.40	8.22
	1	7.48	7.58	7.74	7.61	7.35	7.20	7.09	5.42	8.91	9.09	8.92	8.56	8.34	8.21
	2	7.57	7.55	7.45	7.33	7.11	7.01	6.92	4.53	9.40	9.02	8.84	8.48	8.23	8.15
	3	7.14	7.12	7.07	6.97	6.85	6.80	6.77	3.36	9.19	8.77	8.57	8.30	8.18	8.25
	4	6.60	6.60	6.58	6.59	6.60	6.60	6.63	2.57	8.93	8.39	8.32	8.11	8.12	8.30
	5	5.99	6.01	6.11	6.20	6.32	6.40	6.45	2.38	9.45	8.23	8.08	8.01	8.21	8.42
dx=0.1	0	6.48	7.01	7.92	7.99	7.74	7.51	7.37	6.43	7.23	8.69	8.87	8.58	8.25	8.04
	0.5	7.37	7.60	7.96	7.84	7.62	7.39	7.26	6.24	8.63	9.03	8.96	8.48	8.19	7.98
	1	7.70	7.81	7.87	7.73	7.46	7.29	7.15	5.71	9.28	9.07	8.85	8.38	8.09	7.92
	2	7.61	7.60	7.51	7.35	7.18	7.07	6.99	3.91	9.35	8.87	8.56	8.10	8.02	8.04
	3	7.11	7.08	7.00	6.93	6.88	6.81	6.80	3.31	9.24	8.34	8.18	8.05	8.05	8.22
	4	6.40	6.42	6.46	6.50	6.54	6.58	6.60	2.74	9.74	8.12	8.14	8.09	8.16	8.41
	5	5.63	5.71	5.89	6.03	6.21	6.31	6.40	2.72	11.08	8.86	8.26	8.10	8.27	8.56
dx=0.5	0	6.43	6.93	7.88	7.98	7.71	7.50	7.36	7.15	7.80	8.99	8.97	8.41	8.11	7.84
	0.5	7.30	7.57	7.97	7.87	7.59	7.39	7.27	8.24	8.94	9.25	8.96	8.29	7.95	7.82
	1	7.72	7.78	7.91	7.77	7.45	7.29	7.17	6.88	9.59	9.32	8.79	8.16	7.88	7.80
	2	7.58	7.62	7.53	7.36	7.19	7.07	7.00	6.35	10.01	9.07	8.42	7.88	7.68	7.68
	3	7.10	7.12	7.04	6.96	6.88	6.82	6.83	4.16	10.14	8.76	7.88	7.61	7.61	7.70
	4	6.46	6.49	6.51	6.55	6.58	6.60	6.62	3.29	10.43	8.70	7.37	7.41	7.49	7.68
	5	5.74	5.80	5.97	6.09	6.24	6.37	6.42	2.97	10.95	9.07	7.57	7.28	7.44	7.69
dx=1	0	6.50	7.01	7.92	7.99	7.74	7.51	7.37	6.63	7.90	8.99	8.99	8.45	8.00	7.86
	0.5	7.37	7.60	7.96	7.84	7.62	7.39	7.26	6.62	9.00	9.27	8.93	8.29	7.91	7.75
	1	7.70	7.81	7.87	7.73	7.46	7.29	7.15	6.51	9.57	9.32	8.82	8.16	7.83	7.67
	2	7.61	7.60	7.51	7.35	7.18	7.07	6.99	4.86	10.01	9.19	8.46	7.84	7.59	7.48
	3	7.11	7.08	7.00	6.93	6.88	6.81	6.80	4.48	10.20	8.96	8.00	7.47	7.39	7.46
	4	6.40	6.42	6.46	6.50	6.54	6.58	6.60	3.28	10.40	8.98	7.58	7.13	7.24	7.49
	5	5.63	5.71	5.89	6.03	6.21	6.31	6.40	2.76	10.85	9.23	7.68	7.02	7.27	7.59

Linear soil – Amplification Factors																
L/H=5	dw	0	1	5	10	20	30	40		0	1	5	10	20	30	40
	dθ	Bonded contact								Smooth contact						
dx=0	0	4.48	4.91	6.03	6.57	6.84	6.84	6.83		4.52	5.04	6.76	7.94	8.62	8.70	8.66
	0.5	5.46	5.73	6.37	6.69	6.81	6.82	6.79		5.82	6.28	7.59	8.35	8.69	8.71	8.68
	1	6.03	6.19	6.54	6.70	6.76	6.77	6.73		6.86	7.24	8.11	8.56	8.72	8.68	8.65
	2	6.41	6.45	6.56	6.63	6.64	6.62	6.64		7.98	8.24	8.56	8.67	8.66	8.63	8.71
	3	6.33	6.30	6.37	6.41	6.48	6.49	6.53		8.27	8.46	8.57	8.56	8.54	8.56	8.81
	4	5.99	6.01	6.09	6.17	6.26	6.35	6.40		8.40	8.46	8.40	8.38	8.45	8.63	8.96
	5	5.57	5.61	5.74	5.88	6.05	6.17	6.27		9.19	9.20	8.51	8.29	8.45	8.76	9.13
dx=0.1	0	4.99	5.35	6.35	6.80	6.97	6.95	6.92		5.25	5.77	7.36	8.26	8.64	8.55	8.44
	0.5	5.78	6.02	6.61	6.84	6.90	6.91	6.87		6.49	6.91	7.95	8.52	8.59	8.45	8.40
	1	6.24	6.36	6.70	6.84	6.86	6.86	6.82		7.43	7.70	8.24	8.58	8.55	8.41	8.39
	2	6.49	6.56	6.63	6.69	6.73	6.69	6.71		8.30	8.33	8.46	8.45	8.34	8.34	8.55
	3	6.34	6.36	6.44	6.48	6.52	6.53	6.58		8.75	8.50	8.22	8.19	8.32	8.46	8.86
	4	6.02	6.02	6.10	6.18	6.28	6.36	6.42		9.39	9.24	8.19	8.23	8.40	8.71	9.14
	5	5.54	5.57	5.76	5.86	6.08	6.17	6.28		10.95	10.87	9.32	8.59	8.64	8.96	9.40
dx=0.5	0	5.27	5.62	6.51	6.93	7.05	7.02	6.99		5.84	6.35	7.85	8.52	8.54	8.41	8.28
	0.5	5.94	6.20	6.75	6.95	6.97	6.98	6.93		7.04	7.43	8.37	8.64	8.47	8.32	8.22
	1	6.38	6.48	6.80	6.89	6.92	6.87	6.87		7.94	8.18	8.64	8.62	8.34	8.20	8.19
	2	6.55	6.57	6.69	6.73	6.73	6.72	6.73		9.01	9.07	8.74	8.39	8.06	8.06	8.16
	3	6.37	6.38	6.42	6.48	6.50	6.55	6.58		9.62	9.56	8.65	7.95	7.87	7.93	8.23
	4	5.97	5.95	6.06	6.15	6.27	6.36	6.41		10.24	10.11	8.80	7.52	7.69	7.91	8.27
	5	5.40	5.46	5.63	5.78	6.01	6.16	6.24		10.89	10.79	9.40	7.98	7.73	8.03	8.38
dx=1	0	5.34	5.72	6.57	6.95	7.07	7.06	7.00		5.99	6.50	7.97	8.61	8.60	8.33	8.25
	0.5	6.02	6.26	6.72	6.93	7.03	6.97	6.94		7.14	7.53	8.46	8.70	8.49	8.25	8.14
	1	6.42	6.52	6.81	6.94	6.95	6.88	6.88		8.04	8.29	8.72	8.69	8.36	8.12	8.04
	2	6.59	6.60	6.69	6.73	6.74	6.72	6.72		9.10	9.16	8.88	8.46	8.03	7.91	7.93
	3	6.33	6.35	6.40	6.46	6.49	6.54	6.57		9.70	9.67	8.90	8.07	7.67	7.71	7.97
	4	5.91	5.90	6.01	6.11	6.24	6.33	6.39		10.27	10.14	9.09	7.79	7.38	7.71	8.10
	5	5.31	5.36	5.54	5.71	5.97	6.10	6.21		10.84	10.74	9.51	8.05	7.46	7.84	8.25

Linear soil– Amplification Factors																
L/H=10	dw	0	1	5	10	20	30	40		0	1	5	10	20	30	40
	dθ	Bonded contact								Smooth contact						
dx=0	0	3.24	3.54	4.42	5.06	5.68	5.95	6.11		3.27	3.63	5.04	6.65	8.70	9.51	9.79
	0.5	3.93	4.15	4.80	5.28	5.76	5.99	6.13		4.18	4.55	5.93	7.39	9.06	9.63	9.84
	1	4.42	4.58	5.06	5.42	5.81	6.00	6.11		5.04	5.43	6.73	8.00	9.27	9.69	9.88
	2	4.95	5.04	5.32	5.55	5.82	5.98	6.09		6.52	6.90	7.92	8.74	9.51	9.75	10.06
	3	5.09	5.15	5.35	5.53	5.76	5.92	6.03		7.52	7.88	8.54	9.07	9.54	9.79	10.26
	4	5.03	5.08	5.27	5.44	5.67	5.84	5.95		8.27	8.46	8.79	9.15	9.53	10.00	10.46
	5	4.83	4.88	5.07	5.27	5.53	5.71	5.88		9.58	9.73	9.29	9.30	9.73	10.27	10.73
dx=0.1	0	3.58	3.88	4.72	5.31	5.87	6.13	6.25		3.78	4.17	5.66	7.21	9.00	9.46	9.56
	0.5	4.19	4.40	5.02	5.47	5.91	6.12	6.24		4.70	5.09	6.44	7.85	9.14	9.47	9.57
	1	4.61	4.76	5.23	5.59	5.94	6.13	6.22		5.57	5.92	7.09	8.27	9.23	9.45	9.63
	2	5.07	5.14	5.42	5.65	5.93	6.07	6.17		6.93	7.16	8.03	8.70	9.19	9.48	9.94
	3	5.18	5.23	5.44	5.62	5.85	5.97	6.11		8.04	8.04	8.34	8.75	9.34	9.74	10.34
	4	5.08	5.12	5.29	5.47	5.70	5.86	5.99		9.38	9.39	8.77	9.04	9.55	10.17	10.76
	5	4.79	4.86	5.10	5.28	5.58	5.74	5.91		11.53	11.60	10.34	9.81	10.02	10.57	11.10
dx=0.5	0	3.85	4.11	4.93	5.49	6.02	6.22	6.35		4.24	4.66	6.28	7.80	9.10	9.42	9.45
	0.5	4.38	4.59	5.17	5.64	6.02	6.24	6.34		5.21	5.61	7.04	8.26	9.15	9.35	9.43
	1	4.75	4.91	5.37	5.68	6.03	6.20	6.30		6.12	6.49	7.68	8.54	9.13	9.25	9.40
	2	5.15	5.21	5.50	5.72	5.98	6.13	6.22		7.72	7.98	8.49	8.75	8.99	9.17	9.45
	3	5.20	5.25	5.45	5.63	5.87	6.02	6.11		9.08	9.21	8.92	8.55	8.83	9.11	9.61
	4	5.00	5.09	5.27	5.46	5.71	5.88	6.00		10.28	10.34	9.50	8.33	8.73	9.22	9.73
	5	4.65	4.78	4.99	5.21	5.52	5.72	5.87		11.52	11.56	10.47	9.12	8.98	9.42	9.90
dx=1	0	3.89	4.18	4.97	5.53	6.05	6.26	6.38		4.37	4.79	6.43	7.95	9.19	9.37	9.41
	0.5	4.44	4.62	5.22	5.65	6.07	6.25	6.35		5.33	5.74	7.19	8.39	9.23	9.29	9.31
	1	4.80	4.95	5.38	5.73	6.05	6.22	6.32		6.25	6.63	7.83	8.66	9.18	9.20	9.21
	2	5.15	5.24	5.51	5.74	5.99	6.14	6.22		7.85	8.13	8.67	8.87	8.95	9.00	9.20
	3	5.18	5.25	5.44	5.62	5.86	6.02	6.12		9.20	9.35	9.21	8.73	8.60	8.86	9.33
	4	5.00	5.06	5.25	5.43	5.68	5.86	5.99		10.33	10.43	9.80	8.67	8.41	8.95	9.51
	5	4.63	4.70	4.93	5.17	5.49	5.69	5.86		11.46	11.51	10.59	9.21	8.69	9.21	9.75

Annex H

The derivation of the Equations 6-2 and 6-3 of Chapter 6 is given below (taken by (Kramer 2009):

Considering the free vibration of an infinitely long rod with cross sectional area *A*, Young's Modulus *E*, Poisson's ration v and density ρ as shown in Fig.H-1 and that the od is constrained against radial straining, the free cut of an element of length dx, is described by the relations:

$$\left(\sigma_{xo} + \frac{\partial \sigma_x}{\partial x} dx\right) A - \sigma_{xo} A = \rho A dx \frac{\partial^2 u}{\partial t^2} \qquad \textbf{(H-1)}$$

Where u the displacement in the x-direction. This states that the unbalanced external forces acting on the ends of the element must equal the inertial force induced bz the acceleration of the mass of the element. This yields after some simplifications the one-dimensional equation of motion:

$$\frac{\partial \sigma_x}{\partial x} = \rho \frac{\partial^2 u}{\partial t^2} \qquad \textbf{(H-2)}$$

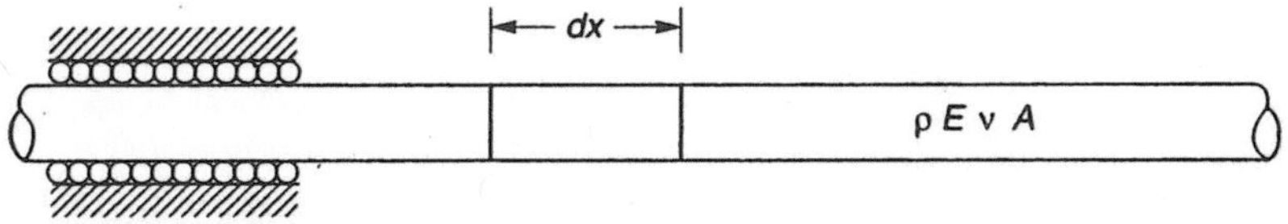

Fig. H-1 Constrained, infinite rod for one-dimensional wave propagation. Constraint against radial straining schematically represented by rollers (Kramer 2009).

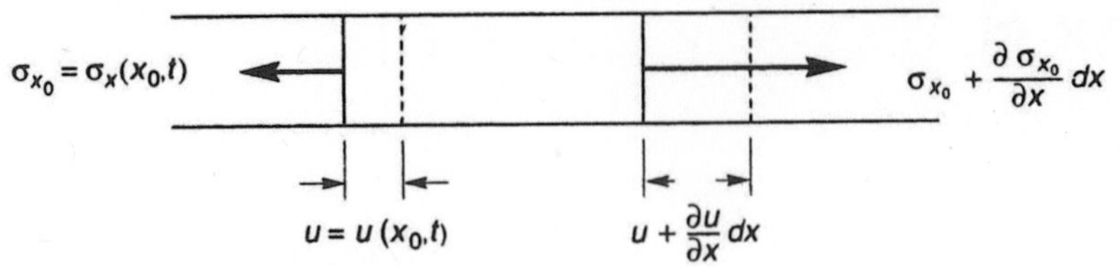

Fig. H-2 Strains and displacements at ends of element of length *dx* and cross-sectional area *A*. (Kramer 2009).

By using the stress-strain relationship:

$$\sigma_x = M\varepsilon_x \quad \textbf{(H-3)}$$

Where the constrained modulus *M* is given by:

$$M = \frac{1-v}{(1+v)(1-2v)}E \quad \textbf{(H-4)}$$

The one-dimensional wave equation can be written in the alternative form:

$$\frac{\partial^2 u}{\partial t^2} = \frac{M}{\rho}\frac{\partial^2 u}{\partial x^2} = \mathrm{v}_p^2\frac{\partial^2 u}{\partial x^2} \quad \textbf{(H-5)}$$

Where v_p is the wave propagation velocity. The wave propagation velocity depends only on the properties of the material (stinfness and density) and is independent of the amplitude of the stress wave. The particle velocity, which is not the same as the wave propagation velocity but is the velocity at which a single point within the rod would move as the wave passes through it, can be shown to be:

$$\dot{u} = \frac{\partial u}{\partial t} = \frac{\varepsilon_x \partial x}{\partial t} = \frac{\sigma_x}{M}\frac{\mathrm{v}_p \partial t}{\partial t} = \frac{\sigma_x}{M}\mathrm{v}_p = \frac{\sigma_x}{\rho \mathrm{v}_p^2}\mathrm{v}_p = \frac{\sigma_x}{\rho \mathrm{v}_p} \quad \textbf{(H-6)}$$

The Equation H-6 shows that the particle velocity is proportional to the axial stress in the rod. The coefficient of proportionality, $\rho \mathrm{v}_{\mathrm{p}}$, is called the specific impedance of the material.

Solution of the one-dimensional equation of motion

The one-dimensional wave equation is a partial differential equation of the form

$$\frac{\partial^2 u}{\partial t^2} = \mathrm{v}^2 \frac{\partial^2 u}{\partial x^2} \quad \textbf{(H-7)}$$

where v is the wave propagation velocity corresponding to the type of the stress wave of interest. The solution of such equation has the form:

$$u(x,t) = f(\mathrm{v}t - x) + g(\mathrm{v}t + x) \quad \textbf{(H-8)}$$

Where *f* and *g* can be arbitrary functions of *(vt-x)* and *(vt+x)* that satisfz the equation H-8. The argument of *f* remains constant when *x* increases with time (at velocity v) and the argument *g* remains constant when *x* decreases with time. Therefore, the solution of Equation H-8 describes a displacement wave [*f(vt-x)*] traveling at velocity v in the positive *x*-direction and another [*g(vt+x)*] traveling at the same spead at the negative *x*-direction.

If the rod is subjected to some steady state harmonic stress $\sigma(\mathrm{t})=\sigma_{\mathrm{o}} \cos \omega t$, where σ_{o} is the stress wave amplitude and ω is the circular frequency of the applied loading, the solution can be expressed using the wave number, $k= \omega/\mathrm{v}$, in the form:

$$u(x,t) = A\cos(\omega t - kx) + B\cos(\omega t + kx) \quad \textbf{(H-9)}$$

and its equivalent form using complex notation is:

$$u(x,t) = Ce^{i(\omega t - kx)} + De^{i(\omega t + kx)} \quad \textbf{(H-10)}$$

Material boundary in an infinite rod

Considering a harmonic stress wave traveling along a constrained rod in the +x direction and approaching an interface between two different materials as shown in Figure H-3. Since the wave is traveling toward the interface, it will be referred to as the incident wave, Since it is traveling in material 1, its wavelength will be $\lambda_1=2\pi/k_1$ and it can be therefore described by

$$\sigma_I(x,t) = \sigma_i e^{i(\omega t - k_1 x)} \quad \textbf{(H-11)}$$

When the incident wave reaches the interface, part of its energy will be transmitted through the interface to continue travelling in the positive x-direction through the material 2. This transmitted wave will have a wavelength $\lambda_2=2\pi/k_2$. The remainder will be reflected at the interface and will travel back through the material 1 in the negative x-direction as a reflected wave. The transmitted and reflected waves can be described by

$$\sigma_T(x,t) = \sigma_t e^{i(\omega t - k_2 x)} \quad \textbf{(H-12a)}$$

$$\sigma_R(x,t) = \sigma_r e^{i(\omega t + k_1 x)} \quad \textbf{(H-12b)}$$

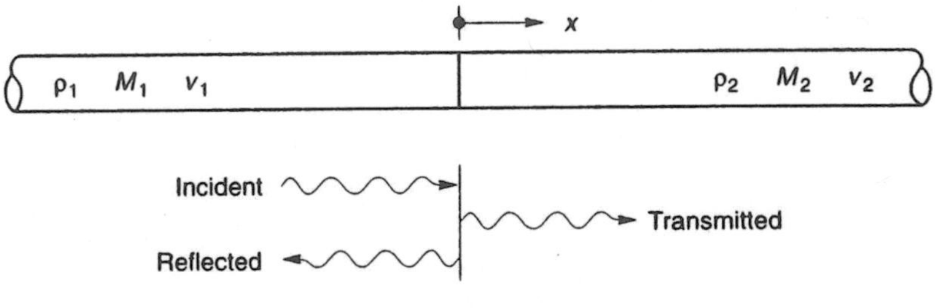

Fig. H-3 One dimensional wave propagation at material interface. Incident and reflected waves travel in opposite directions in material 1. The transmitted wave travels through material 2 in the same direction as the incident wave (Kramer 2009).

Assuming that the displacements associated with each of these waves are of the same harmonic form as the stresses that cause them, that is

$$u_I(x,t) = A_i e^{i(\omega t - k_1 x)} \quad \textbf{(H-13a)}$$

$$u_R(x,t) = A_r e^{i(\omega t + k_1 x)} \quad \textbf{(H-13b)}$$

$$u_T(x,t) = A_t e^{i(\omega t - k_2 x)} \quad \textbf{(H-13c)}$$

Stress-strain and strain-displacement relationships can be used to relate the stress amplitudes to the displacement amplitudes:

$$\sigma_I(x,t) = M_1 \frac{\partial u_I(x,t)}{\partial x} = -ik_1 M_1 A_i e^{i(\omega t - k_1 x)} \quad \textbf{(H-14a)}$$

$$\sigma_R(x,t) = M_1 \frac{\partial u_R(x,t)}{\partial x} = +ik_1 M_1 A_r e^{i(\omega t + k_1 x)} \quad \textbf{(H-14b)}$$

$$\sigma_T(x,t) = M_2 \frac{\partial u_T(x,t)}{\partial x} = -ik_2 M_2 A_T e^{i(\omega t - k_2 x)} \quad \textbf{(H-14c)}$$

From these, the stress amplitudes are related to the displacement amplitudes by

$$\sigma_i = -ik_1 M_1 A_i \quad \textbf{(H-15a)}$$

$$\sigma_r = +ik_1 M_1 A_r \quad \textbf{(H-15b)}$$

$$\sigma_t = -ik_2M_2A_t \tag{H-15c}$$

At the interface, both of compatibility of displacements and continuity of stresses must be satisfied. The former requires that

$$u_I(0,t) + u_R(0,t) = u_T(0,t) \tag{H-16}$$

and the latter that

$$\sigma_I(0,t) + \sigma_R(0,t) = \sigma_T(0,t) \tag{H-17}$$

Substituting equations H-13 and H-12 into equations H-16 and H-17, respectively, indicates that

$$A_i + A_r = A_t \tag{H-18a}$$

$$\sigma_i + \sigma_r = \sigma_t \tag{H-18b}$$

at the interface. Substituting Equations H-15 into Equation H-18b and using the relationship $kM=\omega\rho v$, gives

$$-\rho_1v_1A_i + \rho_1v_1A_r = -\rho_2v_2A_t = -\rho_2v_2(A_i + A_r) \tag{H-19}$$

Equation H-19 can be rearranged to relate the displacement amplitude of the reflected wave to that of the inclined wave:

$$A_r = \frac{\rho_1v_1 - \rho_2v_2}{\rho_1v_1 + \rho_2v_2}A_i = \frac{1 - \rho_2v_2/\rho_1v_1}{1 + \rho_2v_2/\rho_1v_1}A_i \tag{H-20}$$

and knowing A_i and A_r the Equation H-18a can be used to determine A_t as

$$A_t = \frac{2\rho_1 v_1}{\rho_1 v_1 + \rho_2 v_2} A_i = \frac{2}{1 + \rho_2 v_2 / \rho_1 v_1} A_i \tag{H-21}$$

Remember that the product of the density and the wave propagation velocity is the specific impedance of the material. Equations H-20 and H-21 indicate that the partitioning of energy at the interface depends only on the ratio of the specific impedances of the materials on either side of the interface. Defining the impedance ratio as

$$\alpha_z = \rho_2 v_2 / \rho_1 v_1 \tag{H-22}$$

the displacement amplitudes of the reflected and transmitted waves are

$$A_r = \frac{1 - \alpha_z}{1 + \alpha_z} A_i \tag{H-23a}$$

$$A_t = \frac{2}{1 + \alpha_z} A_i \tag{H-23b}$$

After evaluating the effect of the interface on the displacement amplitudes of the reflected and transmitted waves, its effect on stress amplitudes can be investigated. From Equations H-15

$$A_i = -\frac{\sigma_i}{i k_1 M_1} \tag{H-24a}$$

$$A_r = \frac{\sigma_r}{i k_1 M_1} \tag{H-24b}$$

$$A_t = -\frac{\sigma_t}{ik_2 M_2} \tag{H-24c}$$

Substituting Equations H-24 into Equations H-24 and rearranging gives

$$\sigma_r = \frac{\alpha_z - 1}{1 + \alpha_z}\sigma_i \tag{H-25a}$$

$$\sigma_t = \frac{2\alpha_z}{1 + \alpha_z}\sigma_i \tag{H-25b}$$

The importance of the impedance ratio in determining the nature of reflection and transmission at interfaces can clearly be seen. When the impedance ratio is smaller than 1, the incident wave is approaching a "softer" material. If the impedance ratio is bigger than 1, then the incident wave is approaching a "stiffer" material. The relative stress and displacement amplitudes of reflected and transmitted waves at boundaries with several different impedance ratios are given in Table 47.

Table 47 Influence of impedance ratio on displacement and stress amplitudes of reflected and transmitted waves (Kramer 2009).

Impedance Ratio, α_z	Displacement Amplitudes			Stress Amplitudes		
	Incident	Reflected	Transmitted	Incident	Reflected	Transmitted
0	A_i	A_i	$2A_i$	σ_i	$-\sigma_i$	0
$\frac{1}{4}$	A_i	$3A_i/5$	$8A_i/5$	σ_i	$-3\sigma_i/5$	$2\sigma_i/5$
$\frac{1}{2}$	A_i	$A_i/3$	$4A_i/3$	σ_i	$-\sigma_i/3$	$2\sigma_i/3$
1	A_i	0	A_i	σ_i	0	σ_i
2	A_i	$-A_i/3$	$2A_i/3$	σ_i	$\sigma_i/3$	$4\sigma_i/3$
4	A_i	$-3A_i/5$	$2A_i/5$	σ_i	$3\sigma_i/5$	$8\sigma_i/5$
∞	A_i	$-A_i$	0	σ_i	σ_i	$2\sigma_i$

The cases of α_z=0 and α_z=∞ are of particular interest. An impedance ratio of zero implies that the incident wave is approaching a "free end" across which no stress can be transmitted. To satisfy this zero stress boundary condition, the displacement of the boundary (transmitted displacement) must be twice the displacement amplitude of the incident wave (A_t=$2A_i$). The reflected wave has the same amplitude as the incident wave but is of the opposite polarity (σ_r=-σ_i). An infinite impedance ratio implies that the incident wave is approaching a "fixed end" at which no displacement can occur (u_t=0). In that case the stress at the boundary is twice that of the incident wave (σ_t=2σ_i)and the reflected wave has the same amplitude and polarity as the incident wave (A_r=-A_i).

The case of α_z=1, in which the impedances on each side of the boundary are equal is also of interest. Equations H-23 and H-25 indicate that no reflected wave is produced and that the transmitted wave has, as expected, the same amplitude and polarity as the incident wave. Another way of looking at a boundary with an impedance ratio of unity is as a boundary between two identical, semi-infinite rods. A harmonic wave travelling in the positive x-direction (Figure H-4) would impose an axial force on the boundary:

$$F = \sigma_x A = \rho v_m A \dot{u} \tag{H-26}$$

This axial force is identical to that which would exist if the semi-infinite rod on the right side of the boundary were replaced by a dashpot (Figure H-4) of coefficient c=$\rho v_m A$. In other words, the dashpot would absorb all the elastic energy of the incident wave, so the response of the rod o the left would be identical for both cases illustrated in Figure H-4. Considering now that the harmonic stress wave is applied as input motion direct at the dashpot, a factor of 2 has to be considered in order to have wave propagation on the left of the rod as the half of the energy will simultaneously be damped by the dashpot.

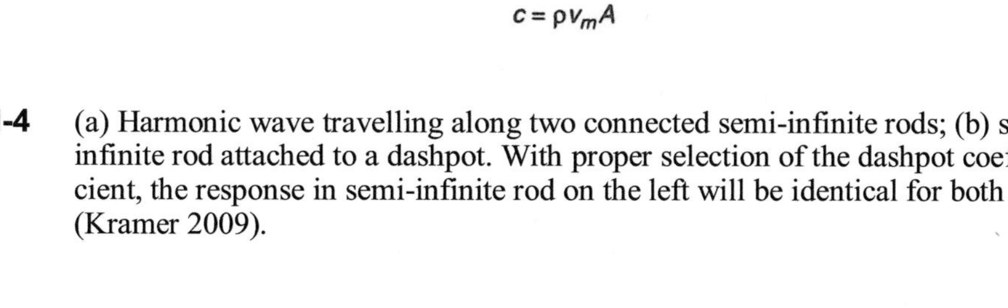

Fig. H-4 (a) Harmonic wave travelling along two connected semi-infinite rods; (b) semi-infinite rod attached to a dashpot. With proper selection of the dashpot coefficient, the response in semi-infinite rod on the left will be identical for both cases (Kramer 2009).

Annex I

Screenshots of the results of the two practical examples. The pictures show the undamaged concrete, the plastic strains and the equivalent plastic strains of the soil and the pore pressure of the reservoir.

I.1 Navigation lock Iffezheim

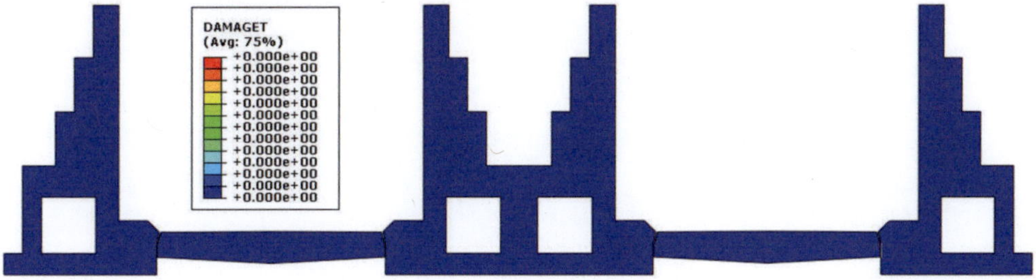

Fig. I-1 The undamaged concrete section of the lock Iffezheim.

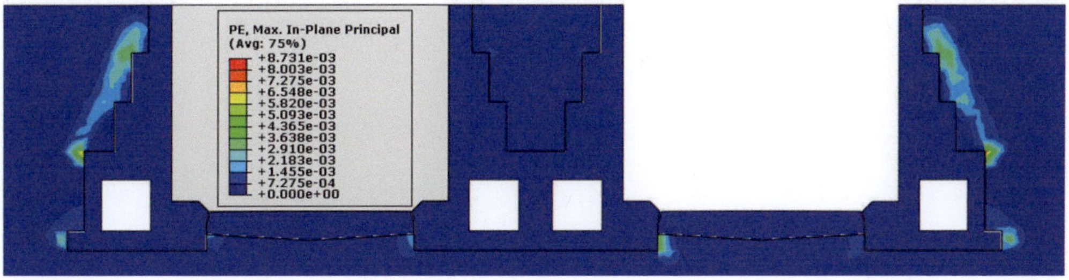

Fig. I-2 Plastic strains in the soil of the lock Iffezheim as result of the static loading.

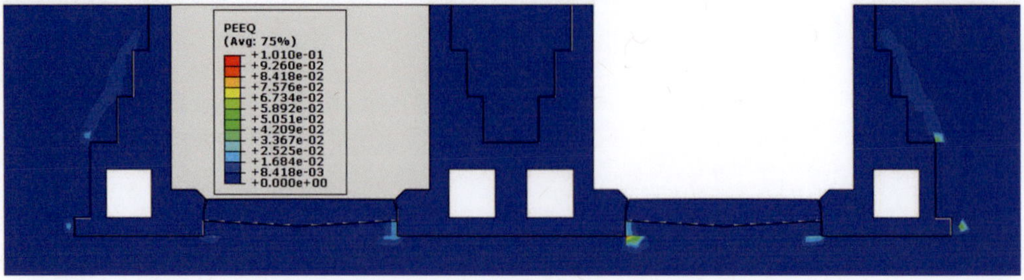

Fig. I-3 Equivalent plastic strains of the soil due to the static loading.

Fig. I-4 Snapshot of the hydrodynamic pressures at the time of the maximum total hydrodynamic pressure.

I.2 Navigation lock Fankel

Fig. I-5 The undamaged concrete section of the lock Fankel.

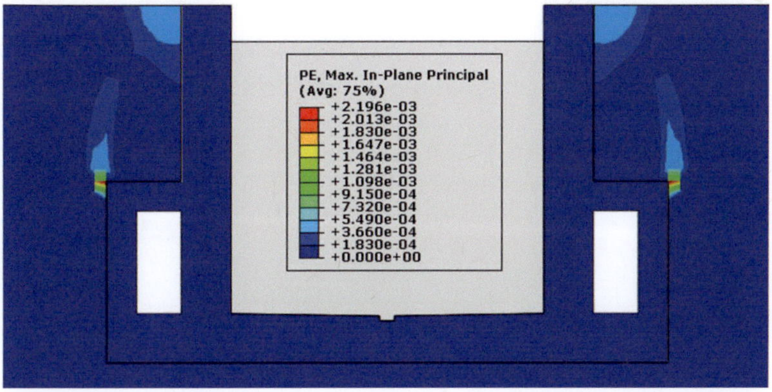

Fig. I-6 Plastic strains at the soil as result of the static loading.

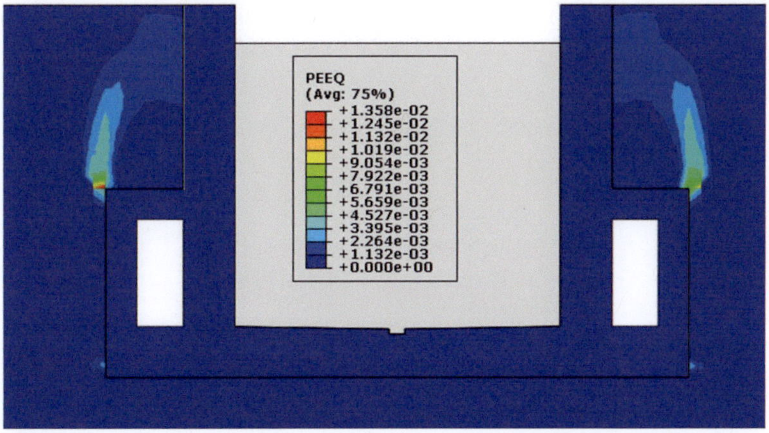

Fig. I-7 Equivalent plastic strains at the soil as result of the static loading.

Fig. I-8 Snapshot of the hydrodynamic pressures at the time of the maximum total hydrodynamic pressure.

Schriftenreihe des
Instituts für Massivbau und Baustofftechnologie

Herausgeber Univ.-Prof. Dr.-Ing. Harald S. Müller
Univ.-Prof. Dr.-Ing. Lothar Stempniewski

Institut für Massivbau und Baustofftechnologie
Universität Karlsruhe (TH)
ISSN 0933-0461

Heft 1 **Manfred Curbach**
Festigkeitssteigerung von Beton bei hohen
Belastungsgeschwindigkeiten. 1987

Heft 2 **Franz-Hermann Schlüter**
Dicke Stahlbetonplatten unter stoßartiger Belastung –
Flugzeugabsturz. 1987

Heft 3 **Marlies Schieferstein**
Der Zugflansch von Stahlbetonplattenbalken unter Längsschub
und Querbiegung bei kritischer Druckbeanspruchung von Beton. 1988

Heft 4 **Thomas Bier**
Karbonatisierung und Realkalisierung von Zementstein und Beton. 1988

Heft 5 **Wolfgang Brameshuber**
Bruchmechanische Eigenschaften von jungem Beton. 1988

Heft 6 **Bericht DFG-Forschungsschwerpunkt**
Durability of Non-Metallic Inanorganic Building Materials. 1988

Heft 7 **Manfred Feyerabend**
Der harte Querstoß auf Stützen aus Stahl und Stahlbeton.1988

Heft 8 **Klaus F. Schönlin**
Permeabilität als Kennwert der Dauerhaftigkeit von Beton. 1989

Heft 9 **Lothar Stempniewski**
Flüssigkeitsgefüllte Stahlbetonbehälter unter Erdbebeneinwirkung.1990

Heft 10 **Jörg Weidner**
Vergleich von Stoffgesetzen granularer Schüttgüter
zur Silodruckermittlung. 1990

Heft 11 **Pingli Yi**
Explosionseinwirkungen auf Stahlbetonplatten. 1991

Schriftenreihe des

Instituts für Massivbau und Baustofftechnologie

Heft 12 **Rainer Kunterding**
Beanspruchung der Oberfläche von Stahlbetonsilos durch Schüttgüter. 1991

Heft 13 **Peter Haardt**
Zementgebundene und kunststoffvergütete Beschichtungen auf Beton. 1991

Heft 14 **Günter Rombach**
Schüttguteinwirkungen auf Silozellen – Exzentrische Entleerung. 1991

Heft 15 **Harald Garrecht**
Porenstrukturmodelle für den Feuchtehaushalt von Baustoffen mit und ohne Salzbefrachtung und rechnerische Anwendung auf Mauerwerk. 1992

Heft 16 **Violandi Vratsanou**
Das nichtlineare Verhalten unbewehrter Mauerwerksscheiben unter Erdbebenbeanspruchung – Hilfsmittel zur Bestimmung der q-Faktoren. 1992

Heft 17 **Carlos Rebelo**
Stochastische Modellierung menschenerzeugter Schwingungen. 1992

Heft 18 **Seminar 29./30. März 1993**
Erdbebenauslegung von Massivbauten unter Berücksichtigung des Eurocode 8. 1993

Heft 19 **Hubert Bachmann**
Die Massenträgheit in einem Pseudo-Stoffgesetz für Beton bei schneller Zugbeanspruchung. 1993

Heft 20 **DBV/AiF-Forschungsbericht H. Emrich**
Zum Tragverhalten von Stahlbetonbauteilen unter Querkraft- und Längszugbeanspruchung. 1993

Heft 21 **Robert Stolze**
Zum Tragverhalten von Stahlbetonplatten mit von den Bruchlinien abweichender Bewehrungsrichtung – Bruchlinien-Rotationskapazität. 1993

Heft 22 **Jie Huang**
Extern vorgespannte Segmentbrücken unter kombinierter Beanspruchung aus Biegung, Querkraft und Torsion. 1994

Schriftenreihe des

Instituts für Massivbau und Baustofftechnologie

Heft 23 **Rolf Wörner**
Verstärkung von Stahlbetonbauteilen mit Spritzbeton. 1994

Heft 24 **Ioannis Retzepis**
Schiefe Betonplatten im gerissenen Zustand. 1995

Heft 25 **Frank Dahlhaus**
Stochastische Untersuchungen von Silobeanspruchungen. 1995

Heft 26 **Cornelius Ruckenbrod**
Statische und dynamische Phänomene bei der Entleerung von Silozellen. 1995

Heft 27 **Shishan Zheng**
Beton bei variierender Dehngeschwindigkeit, untersucht mit einer neuen modifizierten Split-Hopkinson-Bar-Technik. 1996

Heft 28 **Yong-zhi Lin**
Tragverhalten von Stahlfaserbeton. 1996

Heft 29 **DFG**
Korrosion nichtmetallischer anorganischer Werkstoffe im Bauwesen. 1996

Heft 30 **Jürgen Ockert**
Ein Stoffgesetz für die Schockwellenausbreitung in Beton. 1997

Heft 31 **Andreas Braun**
Schüttgutbeanspruchungen von Silozellen unter Erdbebeneinwirkung. 1997

Heft 32 **Martin Günter**
Beanspruchung und Beanspruchbarkeit des Verbundes zwischen Polymerbeschichtungen und Beton. 1997

Heft 33 **Gerhard Lohrmann**
Faserbeton unter hoher Dehngeschwindigkeit. 1998

Heft 34 **Klaus Idda**
Verbundverhalten von Betonrippenstäben bei Querzug. 1999

Heft 35 **Stephan Kranz**
Lokale Schwind- und Temperaturgradienten in bewehrten, oberflächennahen Zonen von Betonstrukturen. 1999

Schriftenreihe des

Instituts für Massivbau und Baustofftechnologie

Heft 36	**Gunther Herold** Korrosion zementgebundener Werkstoffe in mineralsauren Wässern. 1999
Heft 37	**Mostafa Mehrafza** Entleerungsdrücke in Massefluss-Silos – Einflüsse der Geometrie und Randbedingungen. 2000
Heft 38	**Tarek Nasr** Druckentlastung bei Staubexplosionen in Siloanlagen. 2000
Heft 39	**Jan Akkermann** Rotationsverhalten von Stahlbeton-Rahmenecken. 2000
Heft 40	**Viktor Mechtcherine** Bruchmechanische und fraktologische Untersuchungen zur Rißausbreitung in Beton. 2001
Heft 41	**Ulrich Häußler-Combe** Elementfreie Galerkin-Verfahren – Grundlagen und Einsatzmöglichkeiten zur Berechnung von Stahlbetontragwerken. 2001
Heft 42	**Björn Schmidt-Hurtienne** Ein dreiaxiales Schädigungsmodell für Beton unter Einschluß des Dehnrateneffekts bei Hochgeschwindigkeitsbelastung. 2001
Heft 43	**Nazir Abdou** Ein stochastisches nichtlineares Berechnungsverfahren für Stahlbeton mit finiten Elementen. 2002
Heft 44	**Andreas Plokitza** Ein Verfahren zur numerischen Simulation von Betonstrukturen beim Abbruch durch Sprengen. 2002
Heft 45	**Timon Rabczuk** Numerische Untersuchungen zum Fragmentierungsverhalten von Beton mit Hilfe der SPH-Methode. 2002
Heft 46	**Norbert J. Krutzik** Zu Anwendungsgrenzen von FE-Modellen bei der Simulation von Erschütterungen in Kernkraftbauwerken bei Stoßbelastungen. 2002
Heft 47	**Thorsten Timm** Beschuß von flüssigkeitsgefüllten Stahlbehältern. 2002

Schriftenreihe des

Instituts für Massivbau und Baustofftechnologie

Heft 48 **Slobodan Kasic**
Tragverhalten von Segmentbauteilen mit interner und externer Vorspannung ohne Verbund. 2002

Heft 49 **Christoph Kessler-Kramer**
Zugtragverhalten von Beton unter Ermüdungsbeanspruchung. 2002

Heft 50 **Nico Herrmann**
Experimentelle Verifizierung von Prognosen zur Sprengtechnik. 2002

Heft 51 **Michael Baur**
Elastomerlager und nichtlineare Standorteffekte bei Erdbebeneinwirkung. 2003

Heft 52 **Seminar 02. Juli 2004**
DIN 1045-1; Aus der Praxis für die Praxis. 2004

Heft 53 **Abdelkhalek Saber Omar Mohamed**
Behaviour of Retrofitted Masonry Shear Walls Subjected to Cyclic Loading. 2004

Heft 54 **Werner Hörenbaum**
Verwitterungsmechanismen und Dauerhaftigkeit von Sandsteinsichtmauenwerk. 2005

Heft 55 **Seminar Februar 2006**
DIN 4149 – Aus der Praxis für die Praxis. 2006

Heft 56 **Sam Foos**
Unbewehrte Betonfahrbahnplatten unter witterungsbedingten Beanspruchungen. 2006

Heft 57 **Ramzi Maliha**
Untersuchungen zur Rissbildung in Fahrbahndecken aus Beton. 2006

Heft 58 **Andreas Fäcke**
Numerische Simulation des Schädigungsverhaltens von Brückenpfeilern aus Stahlbeton unter Erdbebenlasten. 2006

Heft 59 **Juliane Möller**
Rotationsverhalten von verbundlos vorgespannten Segmenttragwerken. 2006

Schriftenreihe des
Instituts für Massivbau und Baustofftechnologie

Heft 60 **Martin Larcher**
Numerische Simulation des Betonverhaltens unter Stoßwellen mit Hilfe des Elementfreien Galerkin-Verfahrens. 2007

Heft 61 **Christoph Niklasch**
Numerische Untersuchungen zum Leckageverhalten von gerissenen Stahlbetonwänden. 2007

Heft 62 **Halim Khbeis**
Experimentelle und numerische Untersuchungen von Topflagern. 2007

Heft 63 **Sascha Schnepf**
Vereinfachte numerische Simulation des Tragverhaltens ebener mauerwerksausgefachter Stahlbetonrahmen unter zyklischer Belastung. 2007

Heft 64 **Christian Wallner**
Erdbebengerechtes Verstärken von Mauerwerk durch Faserverbundwerkstoffe – experimentelle und numerische Untersuchungen. 2008

Heft 65 **Niklas Puttendörfer**
Ein Beitrag zum Gleitverhalten und zur Sattelausbildung externer Spannglieder. 2008

»»»»»»»»» **Bezug der Hefte 1 – 65 und 67**
Institut für Massivbau und Baustofftechnologie
Karlsruher Institut für Technologie (KIT)
Gotthard-Franz-Str. 3, 76131 Karlsruhe
www.betoninstitut.de

Bezug ab Heft 66
KIT Scientific Publishing
Straße am Forum 2, 76131 Karlsruhe
www.ksp.kit.edu

»»»»»»»»» **Fortführung der Reihe ab Heft 66 unter neuem Namen**

KARLSRUHER REIHE
Massivbau
Baustofftechnologie
Materialprüfung

erschienen bei KIT Scientific Publishing (ISSN 1869-912X)

KARLSRUHER REIHE
Massivbau – Baustofftechnologie – Materialprüfung

Herausgeber Univ.-Prof. Dr.-Ing. Harald S. Müller
Univ.-Prof. Dr.-Ing. Lothar Stempniewski

Institut für Massivbau und Baustofftechnologie
Materialprüfungs- und Forschungsanstalt, MPA Karlsruhe

Karlsruher Institut für Technologie (KIT)
KIT Scientific Publishing
ISSN 1869-912X

Heft 66 **Michael Haist**
Zur Rheologie und den physikalischen Wechselwirkungen
bei Zementsuspensionen. 2009
ISBN 978-3-86644-475-1

Heft 67 **Stephan Steiner**
Beton unter Kontaktdetonation – neue experimentelle Methoden. 2009
(noch erschienen in der Schriftenreihe des Instituts für Massivbau
und Baustofftechnologie, ISSN 0933-0461)

Heft 68 **Christian Münich**
Hybride Multidirektionaltextilien zur Erdbebenverstärkung
von Mauerwerk – Experimente und numerische Untersuchungen
mittels eines erweiterten Makromodells. 2011
ISBN 978-3-86644-734-9

Heft 69 **Viktória Malárics**
Ermittlung der Betonzugfestigkeit aus dem Spaltzugversuch
an zylindrischen Betonproben. 2011
ISBN 978-3-86644-735-6

Heft 70 **Daniela Ruch**
Bestimmung der Last-Zeit-Funktion beim Aufprall
flüssigkeitsgefüllter Stoßkörper. 2011
ISBN 978-3-86644-736-3

Heft 71 **Marc Beitzel**
Frischbetondruck unter Berücksichtigung der
rheologischen Eigenschaften. 2012
ISBN 978-3-86644-783-7

KARLSRUHER REIHE

Massivbau – Baustofftechnologie – Materialprüfung

Die Bände sind unter www.ksp.kit.edu als PDF frei verfügbar oder als Druckausgabe bestellbar.

KARLSRUHER REIHE

Massivbau – Baustofftechnologie – Materialprüfung

Heft 81 **Michael Vogel**
Schädigungsmodell für die Hydroabrasionsbeanspruchung zur probabilistischen Lebensdauerprognose von Betonoberflächen im Wasserbau. 2017
ISBN 978-3-7315-0522-8

Heft 82 **Georgios Maltidis**
Seismic soil structure interaction of navigation locks. 2017
ISBN 978-3-7315-0718-5
